AQA BIOLOGY 2

A-Level Year 2/AL **Student Workbook**

AQA BIOLOGY 2

A-Level Year 2/AL Student Workbook

Meet the Writing Team

Tracey Greenwood
I have been writing resources for students since 1993. I have a Ph.D in biology, specialising in lake ecology and I have taught both graduate and undergraduate biology.

Lissa Bainbridge-Smith
I worked in industry in a research and development capacity for 8 years before joining BIOZONE in 2006. I have an M.Sc from Waikato University.

Kent Pryor
I have a BSc from Massey University majoring in zoology and ecology and taught secondary school biology and chemistry for 9 years before joining BIOZONE as an author in 2009.

Richard Allan
I have had 11 years experience teaching senior secondary school biology. I have a Masters degree in biology and founded BIOZONE in the 1980s after developing resources for my own students.

Thanks to:
The staff at BIOZONE, particularly Nell Travaglia for design and graphics support, Paolo Curray for IT support, Debbie Antoniadis and Tim Lind for office handling and logistics, and the BIOZONE sales team.

First edition 2016
Second printing with corrections

ISBN 978-1-927309-20-9

Published by BIOZONE International Ltd

Printed by REPLIKA PRESS PVT LTD using paper produced from renewable and waste materials

Purchases of this workbook may be made direct from the publisher:

BIOZONE Learning Media (UK) Ltd.

Telephone local: 01283 530 366
Telephone international: +44 1283 530 366
Fax local: 01283 831 900
Fax international: +44 1283 831 900
Email: sales@biozone.co.uk

www.BIOZONE.co.uk

Cover Photograph

The white rhinoceros (*Ceratotherium simum*) comprises two subspecies. The southern white rhinoceros (*C. s. simum*) is found in southern Africa including South Africa whereas the historical range of the northern white rhino (*C. s. cottoni*) included central African countries, including Uganda. Of the two, the southern subspecies is the by far the most common numbering over 20 000. The northern subspecies is on the brink of extinction with just four individuals left.

PHOTO: www.dollarphotoclub.com

Contents

Activity is marked: to be done; when completed

Contents

The control of gene expression

Activity is marked: to be done; when completed

Using This Workbook

This first edition of AQA Biology 2 has been specifically written to meet the content and skills requirements of the second year of AQA A Level Biology. Learning outcomes in the introduction to each chapter provide you with a concise guide to the knowledge and skills requirements for each section of work. Each learning outcome is matched to the activity or activities addressing it. The six required practicals at this level are identified in the chapter introductions by a code (*PR-#*) and supported by activities designed to provide background and familiarity with apparatus, techniques, experimental design, and interpretation of results. A range of activities will help you to build on what you already know, explore new topics, work collaboratively, and practise your skills in data handling and interpretation. We hope that you find the workbook valuable and that you make full use of its features.

▶ The outline of the chapter structure below will help you to navigate through the material in each chapter.

Introduction
- A check list of the knowledge and skills requirements for the chapter.
- A list of key terms.

Activities
- The KEY IDEA provides your focus for the activity.
- Annotated diagrams help you understand the content.
- Questions review the content of the page.

Review
- Create your own summary for review.
- Hints help you to focus on what is important.
- Your summary will consolidate your understanding of the content in the chapter.

Literacy
- Activities are based on, but not restricted to, the introductory key terms list.
- Several types of activities test your understanding of the concepts and biological terms in the chapter.

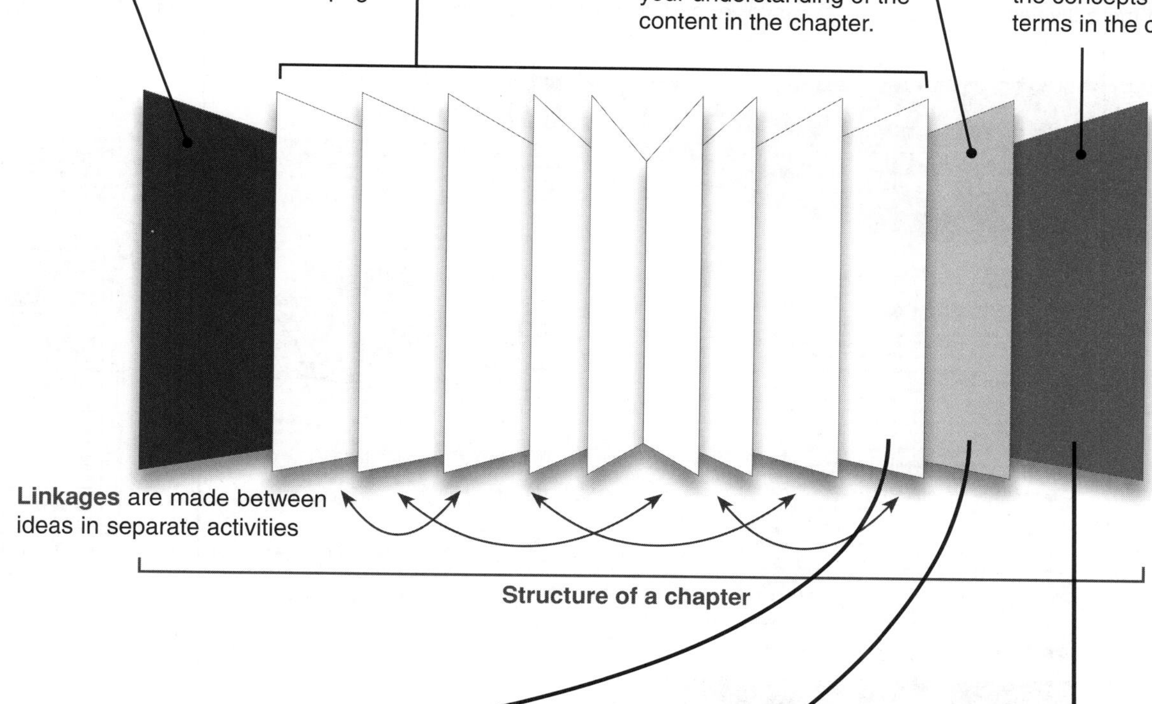

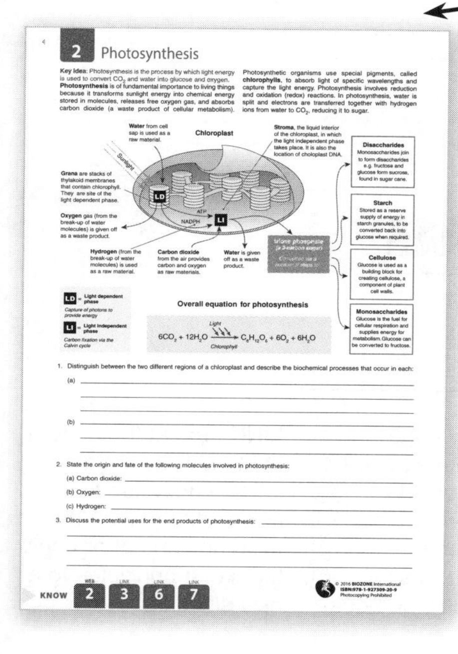

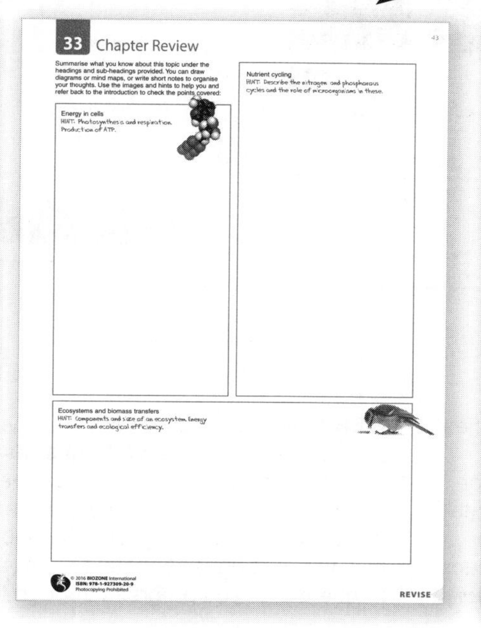

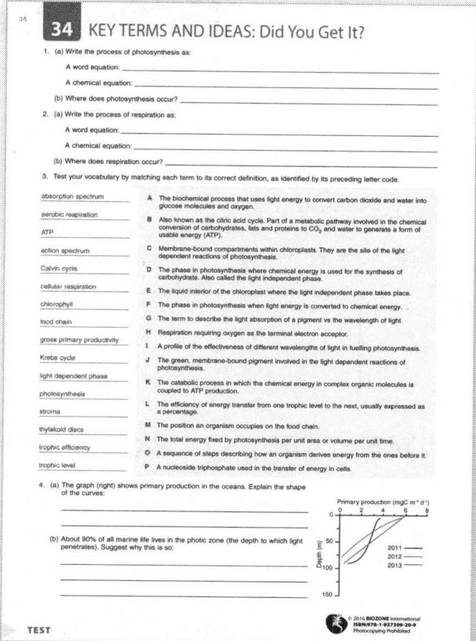

▶ Understanding the activity coding system and making use of the online material identified will enable you to get the most out of this resource. The chapter content is structured to build knowledge and skills but this structure does not necessarily represent a strict order of treatment. Be guided by your teacher, who will assign activities as part of a wider programme of independent and group-based work.

Look out for these features and know how to use them:

The **chapter introduction** provides you with a summary of the knowledge and skills requirements for the topic, phrased as a set of learning outcomes. Use the check boxes to identify and mark off the points as you complete them. The chapter introduction also provides you with a list of key terms for the chapter, from which you can construct your own glossary as you work through the activities.

The **activities** form most of this workbook. They are numbered sequentially and each has a task code identifying the skill emphasised. Each activity has a short introduction with a key idea identifying the main message of the page. Most of the information is associated with pictures and diagrams, and your understanding of the content is reviewed through the questions. Some of the activities involve modelling and group work.

Free response questions allow you to use the information provided to answer questions about the content of the activity, either directly or by applying the same principles to a new situation. In some cases, an activity will assume understanding of prior content.

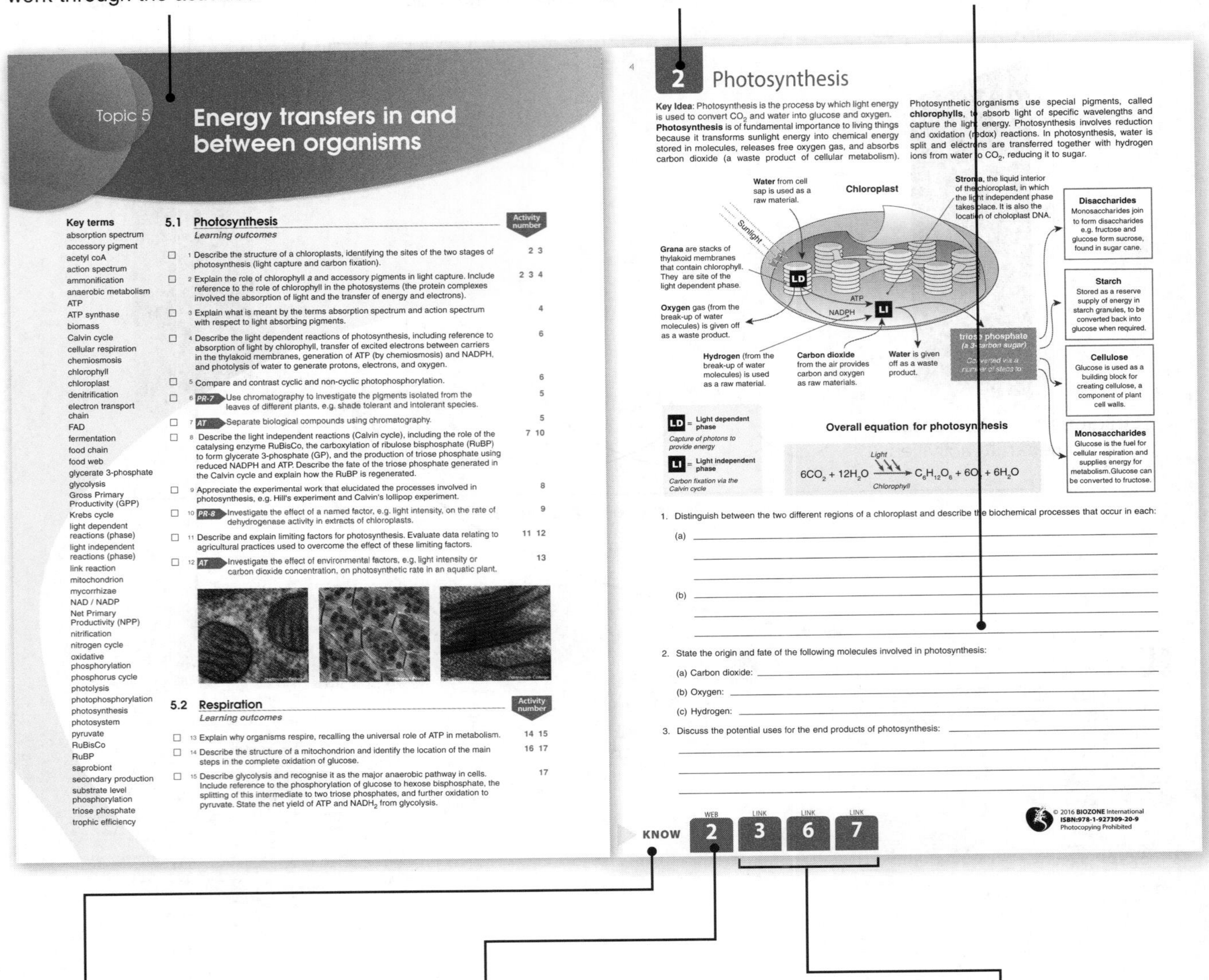

Topic 5

Energy transfers in and between organisms

Key terms
absorption spectrum
accessory pigment
acetyl coA
action spectrum
ammonification
anaerobic metabolism
ATP
ATP synthase
biomass
Calvin cycle
cellular respiration
chemiosmosis
chlorophyll
chloroplast
denitrification
electron transport chain
FAD
fermentation
food chain
food web
glycerate 3-phosphate
glycolysis
Gross Primary Productivity (GPP)
Krebs cycle
light dependent reactions (phase)
light independent reactions (phase)
link reaction
mitochondrion
mycorrhizae
NAD / NADP
Net Primary Productivity (NPP)
nitrification
nitrogen cycle
oxidative phosphorylation
phosphorus cycle
photolysis
photophosphorylation
photosynthesis
photosystem
pyruvate
RuBisCo
RuBP
saprobiont
secondary production
substrate level phosphorylation
triose phosphate
trophic efficiency

5.1 Photosynthesis
Learning outcomes

		Activity number
☐	1 Describe the structure of a chloroplast, identifying the sites of the two stages of photosynthesis (light capture and carbon fixation).	2 3
☐	2 Explain the role of chlorophyll a and accessory pigments in light capture. Include reference to the role of chlorophyll in the photosystems (the protein complexes involved the absorption of light and the transfer of energy and electrons).	2 3 4
☐	3 Explain what is meant by the terms absorption spectrum and action spectrum with respect to light absorbing pigments.	4
☐	4 Describe the light dependent reactions of photosynthesis, including reference to absorption of light by chlorophyll, transfer of excited electrons between carriers in the thylakoid membranes, generation of ATP (by chemiosmosis) and NADPH, and photolysis of water to generate protons, electrons, and oxygen.	6
☐	5 Compare and contrast cyclic and non-cyclic photophosphorylation.	6
☐	6 PR-7 Use chromatography to investigate the pigments isolated from the leaves of different plants, e.g. shade tolerant and intolerant species.	5
☐	7 AT Separate biological compounds using chromatography.	5
☐	8 Describe the light independent reactions (Calvin cycle), including the role of the catalysing enzyme RuBisCo, the carboxylation of ribulose bisphosphate (RuBP) to form glycerate 3-phosphate (GP), and the production of triose phosphate using reduced NADPH and ATP. Describe the fate of the triose phosphate generated in the Calvin cycle and explain how the RuBP is regenerated.	7 10
☐	9 Appreciate the experimental work that elucidated the processes involved in photosynthesis, e.g. Hill's experiment and Calvin's lollipop experiment.	8
☐	10 PR-8 Investigate the effect of a named factor, e.g. light intensity, on the rate of dehydrogenase activity in extracts of chloroplasts.	9
☐	11 Describe and explain limiting factors for photosynthesis. Evaluate data relating to agricultural practices used to overcome the effect of these limiting factors.	11 12
☐	12 AT Investigate the effect of environmental factors, e.g. light intensity or carbon dioxide concentration, on photosynthetic rate in an aquatic plant.	13

5.2 Respiration
Learning outcomes

		Activity number
☐	13 Explain why organisms respire, recalling the universal role of ATP in metabolism.	14 15
☐	14 Describe the structure of a mitochondrion and identify the location of the main steps in the complete oxidation of glucose.	16 17
☐	15 Describe glycolysis and recognise it as the major anaerobic pathway in cells. Include reference to the phosphorylation of glucose to hexose bisphosphate, the splitting of this intermediate to two triose phosphates, and further oxidation to pyruvate. State the net yield of ATP and $NADH_2$ from glycolysis.	17

4

2 Photosynthesis

Key Idea: Photosynthesis is the process by which light energy is used to convert CO_2 and water into glucose and oxygen.

Photosynthesis is of fundamental importance to living things because it transforms sunlight energy into chemical energy stored in molecules, releases free oxygen gas, and absorbs carbon dioxide (a waste product of cellular metabolism). Photosynthetic organisms use special pigments, called **chlorophylls**, to absorb light of specific wavelengths and capture the light energy. Photosynthesis involves reduction and oxidation (redox) reactions. In photosynthesis, water is split and electrons are transferred together with hydrogen ions from water to CO_2, reducing it to sugar.

Overall equation for photosynthesis

$$6CO_2 + 12H_2O \xrightarrow[\text{Chlorophyll}]{\text{Light}} C_6H_{12}O_6 + 6O_2 + 6H_2O$$

1. Distinguish between the two different regions of a chloroplast and describe the biochemical processes that occur in each:

(a) ____________

(b) ____________

2. State the origin and fate of the following molecules involved in photosynthesis:

(a) Carbon dioxide: ____________

(b) Oxygen: ____________

(c) Hydrogen: ____________

3. Discuss the potential uses for the end products of photosynthesis: ____________

KNOW | WEB 2 | LINK 3 | LINK 6 | LINK 7

© 2016 BIOZONE International
ISBN:978-1-927309-20-9
Photocopying Prohibited

A **TASK CODE** on the page tab identifies the type of activity. For example, is it primarily information-based (KNOW), or does it involve modelling (PRAC) or data handling (DATA)? A full list of codes is given on the following page but the codes themselves are relatively self explanatory.

WEB tabs at the bottom of the activity page alert the reader to the **Weblinks** resource, which provides external, online support material for the activity, usually in the form of an animation, video clip, photo library, or quiz. Bookmark the Weblinks page (see next page) and visit it frequently as you progress through the workbook.

LINK tabs at the bottom of the activity page identify activities that are related in that they build on content or apply the same principles to a new situation.

Using the Tab System

The tab system is a useful system for quickly identifying related content and online support. Links generally refer to activities that build on the information in the activity in depth or extent. In the example below, the weblink for the activity (2) provides an animation explaining photosynthesis. Activity 3 describes the structure of the chloroplast and activities 6 and 7 cover the light dependent and light independent reactions of photosynthesis in detail. Sometimes, a link will reflect on material that has been covered earlier as a reminder for important terms that have already been defined or for a formula that may be required to answer a question. The weblinks code is always the same as the activity number on which it is cited. On visiting the weblink page (below), find the number and it will correspond to one or more external websites providing a video or animation of some aspect of the activity's content. Occasionally, the weblink may access a reference paper of provide a bank of photographs where images are provided in colour, e.g. for plant and animal histology.

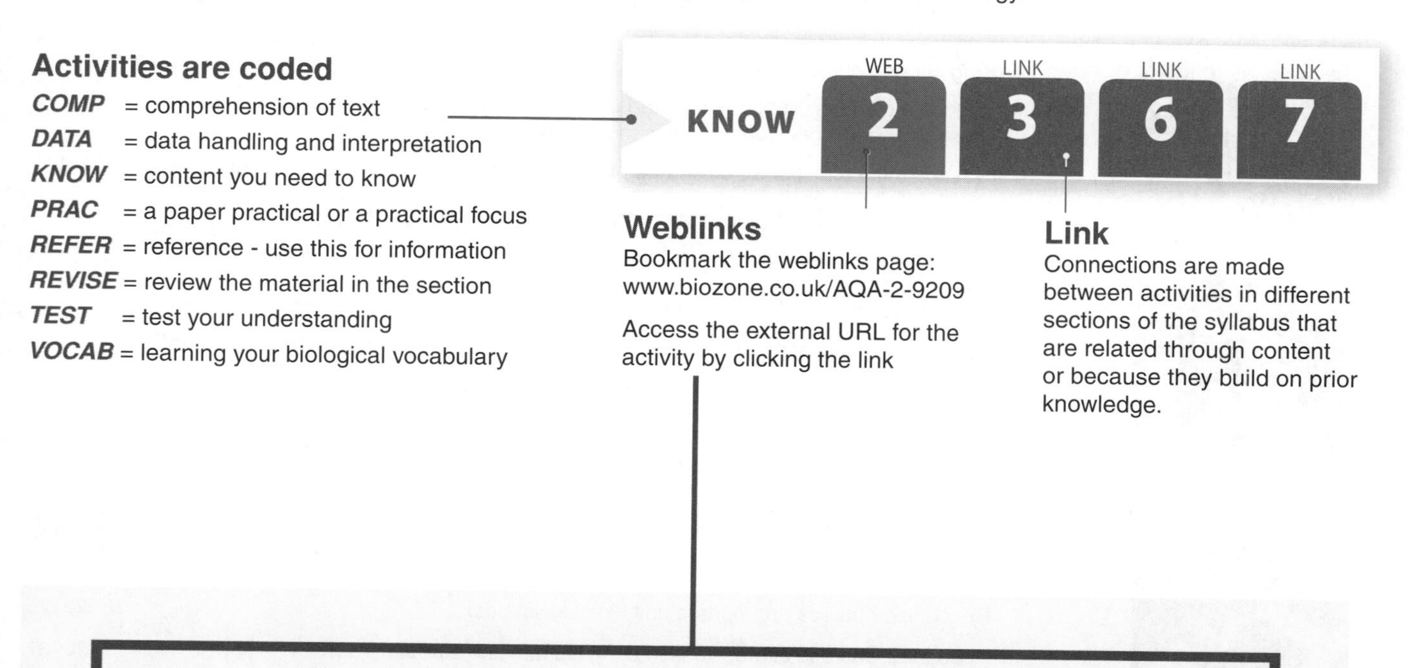

Activities are coded

COMP = comprehension of text
DATA = data handling and interpretation
KNOW = content you need to know
PRAC = a paper practical or a practical focus
REFER = reference - use this for information
REVISE = review the material in the section
TEST = test your understanding
VOCAB = learning your biological vocabulary

Weblinks

Bookmark the weblinks page:
www.biozone.co.uk/AQA-2-9209

Access the external URL for the activity by clicking the link

Link

Connections are made between activities in different sections of the syllabus that are related through content or because they build on prior knowledge.

www.biozone.co.uk/weblink/AQA-2-9209

This WEBLINKS page provides links to **external web sites** with supporting information for the activities. These sites are separate to those provided in the BIOLINKS area of BIOZONE's web site. Almost exclusively, they are narrowly focused animations and video clips directly relevant to the activity on which they are cited. They provide great support to aid student understanding of basic concepts, especially for visual learners.

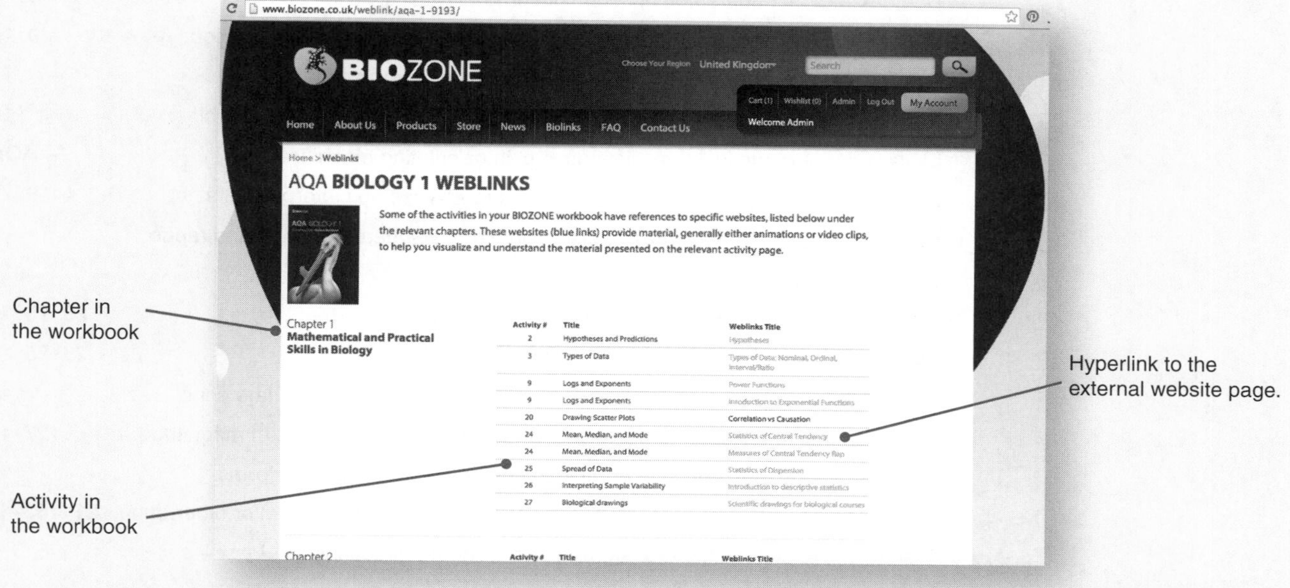

Bookmark weblinks by typing in the address: it is not accessible directly from BIOZONE's website

Corrections and clarifications to current editions are always posted on the weblinks page

Summary of Practical Skills for AQA AL

▸ The practical and mathematical skills for AQA A Level are outlined below and supported in the activities indicated. See AQA Biology 1 for the practical skills to be assessed in written papers and use of apparatus and techniques.

PS A-Level required practical activities

Required practicals supported as indicated

			Activity number
☐	7	Use chromatography to investigate the pigments from leaves of different plants.	5
☐	8	Investigate the effect of a named factor on the rate of dehydrogenase activity in extracts of chloroplasts.	9
☐	9	Investigate the effect of a named variable on the respiration rate of yeast cultures.	20 21
☐	10	Investigate the effect of an environmental variable on the movement of an animal using either a choice chamber or a maze.	44
☐	11	Produce a glucose dilution series and use colorimetry to produce a calibration curve to identify the glucose concentration in an unknown "urine" sample.	76
☐	12	Investigate the effect of an environmental factor of the distribution of a species.	140 143

1.2 Practical skills for assessment by endorsement

Learning outcomes supported via the required practical activities above

- ☐ 1 Follow instructions and carry out experimental techniques and procedures.
- ☐ 2 Apply investigative approaches and methods when using instruments and equipment.
- ☐ 3 Use a range of practical equipment and materials safely, identifying risks.
- ☐ 4 Make and record observations accurately in order to obtain sufficient reliable data.
- ☐ 5 Use appropriate tools to process data, carry out research, and report findings.

MS Mathematical skills

Supported as noted but also throughout AQA Biology 1 & 2

				Activity number
☐	0	1	Recognise and use appropriate units in calculations.	13 20 25
☐		2	Recognise and use expressions in both decimal and standard form.	25 27
☐		3	Carry out calculations involving fractions, percentages, and ratios.	96 105 138
☐		4	Estimate results to assess if calculated values are appropriate.	25
☐		5	Use calculator to find and use power, exponential, and logarithmic functions.	127 128
☐	1	1	Use an appropriate number of significant figures in reporting calculations.	25
☐		2	Find arithmetic means for a range of data.	21 142
☐		3	Represent and interpret frequency data in the form of bar graphs and histograms.	99 109 142
☐		4	Demonstrate an understanding of simple probability, e.g. as in genetic inheritance.	84 87 96
☐		5	Understand sampling and analyse data collected by an appropriate method.	138 140 142
☐		6	Calculate or compare mean, mode, and median for sample data.	142
☐		7	Plot and interpret scatter graphs to identify correlation between two variables.	109
☐		8	Make order of magnitude calculations, e.g. in calculating magnification.	AQA1
☐		9	Select and apply appropriate statistical tests to analyse and interpret data.	43 44 94 95
☐		10	Understand and use measures of dispersion, e.g. standard deviation and range.	99
☐		11	Identify and determine uncertainties in measurements.	9 13
☐	2	1	Demonstrate understanding of the symbols $=, <, \ll, \gg, >, \propto, \sim$	23 25 105
☐		2	Manipulate equations to change the subject.	23 25 105
☐		3	Substitute numerical values into algebraic equations using appropriate units.	125
☐		4	Use logarithms in relation to quantities ranging over several orders of magnitude.	127 128
☐	3	1	Translate information between graphical, numerical, and algebraic forms.	126
☐		2	Select an appropriate format to plot two variables from experimental or other data.	21 143 167
☐		3	Predict or sketch the shape of a graph with a linear relationship ($y = mx + c$).	21
☐		4	Determine the intercept of a graph.	AQA1
☐		5	Calculate rate of change from a graph showing a linear relationship.	13 21
☐		6	Draw and use the slope of a tangent to a curve as a measure of rate of change.	AQA1
☐	4	1	Calculate the circumferences, surface areas, and volumes of regular shapes.	AQA1

Topic 5

Energy transfers in and between organisms

Key terms

absorption spectrum
accessory pigment
acetyl coA
action spectrum
ammonification
anaerobic metabolism
ATP
ATP synthase
biomass
Calvin cycle
cellular respiration
chemiosmosis
chlorophyll
chloroplast
denitrification
electron transport chain
FAD
fermentation
food chain
food web
glycerate 3-phosphate
glycolysis
Gross Primary Productivity (GPP)
Krebs cycle
light dependent reactions (phase)
light independent reactions (phase)
link reaction
mitochondrion
mycorrhizae
NAD / NADP
Net Primary Productivity (NPP)
nitrification
nitrogen cycle
oxidative phosphorylation
phosphorus cycle
photolysis
photophosphorylation
photosynthesis
photosystem
pyruvate
RuBisCo
RuBP
saprobiont
secondary production
substrate level phosphorylation
triose phosphate
trophic efficiency

5.1 Photosynthesis

Learning outcomes — **Activity number**

☐ 1 Describe the structure of a chloroplast, identifying the sites of the two stages of photosynthesis (light capture and carbon fixation). — 2 3

☐ 2 Explain the role of chlorophyll *a* and accessory pigments in light capture. Include reference to the role of chlorophyll in the photosystems (the protein complexes involved the absorption of light and the transfer of energy and electrons). — 2 3 4

☐ 3 Explain what is meant by the terms absorption spectrum and action spectrum with respect to light absorbing pigments. — 4

☐ 4 Describe the light dependent reactions of photosynthesis, including reference to absorption of light by chlorophyll, transfer of excited electrons between carriers in the thylakoid membranes, generation of ATP (by chemiosmosis) and NADPH, and photolysis of water to generate protons, electrons, and oxygen. — 6

☐ 5 Compare and contrast cyclic and non-cyclic photophosphorylation. — 6

☐ 6 **PR-7** Use chromatography to investigate the pigments isolated from the leaves of different plants, e.g. shade tolerant and intolerant species. — 5

☐ 7 **AT** Separate biological compounds using chromatography. — 5

☐ 8 Describe the light independent reactions (Calvin cycle), including the role of the catalysing enzyme RuBisCo, the carboxylation of ribulose bisphosphate (RuBP) to form glycerate 3-phosphate (GP), and the production of triose phosphate using reduced NADPH and ATP. Describe the fate of the triose phosphate generated in the Calvin cycle and explain how the RuBP is regenerated. — 7 10

☐ 9 Appreciate the experimental work that elucidated the processes involved in photosynthesis, e.g. Hill's experiment and Calvin's lollipop experiment. — 8

☐ 10 **PR-8** Investigate the effect of a named factor, e.g. light intensity, on the rate of dehydrogenase activity in extracts of chloroplasts. — 9

☐ 11 Describe and explain limiting factors for photosynthesis. Evaluate data relating to agricultural practices used to overcome the effect of these limiting factors. — 11 12

☐ 12 **AT** Investigate the effect of environmental factors, e.g. light intensity or carbon dioxide concentration, on photosynthetic rate in an aquatic plant. — 13

Dartmouth College

Kristian Peters

Dartmouth College

5.2 Respiration

Learning outcomes — **Activity number**

☐ 13 Explain why organisms respire, recalling the universal role of ATP in metabolism. — 14 15

☐ 14 Describe the structure of a mitochondrion and identify the location of the main steps in the complete oxidation of glucose. — 16 17

☐ 15 Describe glycolysis and recognise it as the major anaerobic pathway in cells. Include reference to the phosphorylation of glucose to hexose bisphosphate, the splitting of this intermediate to two triose phosphates, and further oxidation to pyruvate. State the net yield of ATP and $NADH_2$ from glycolysis. — 17

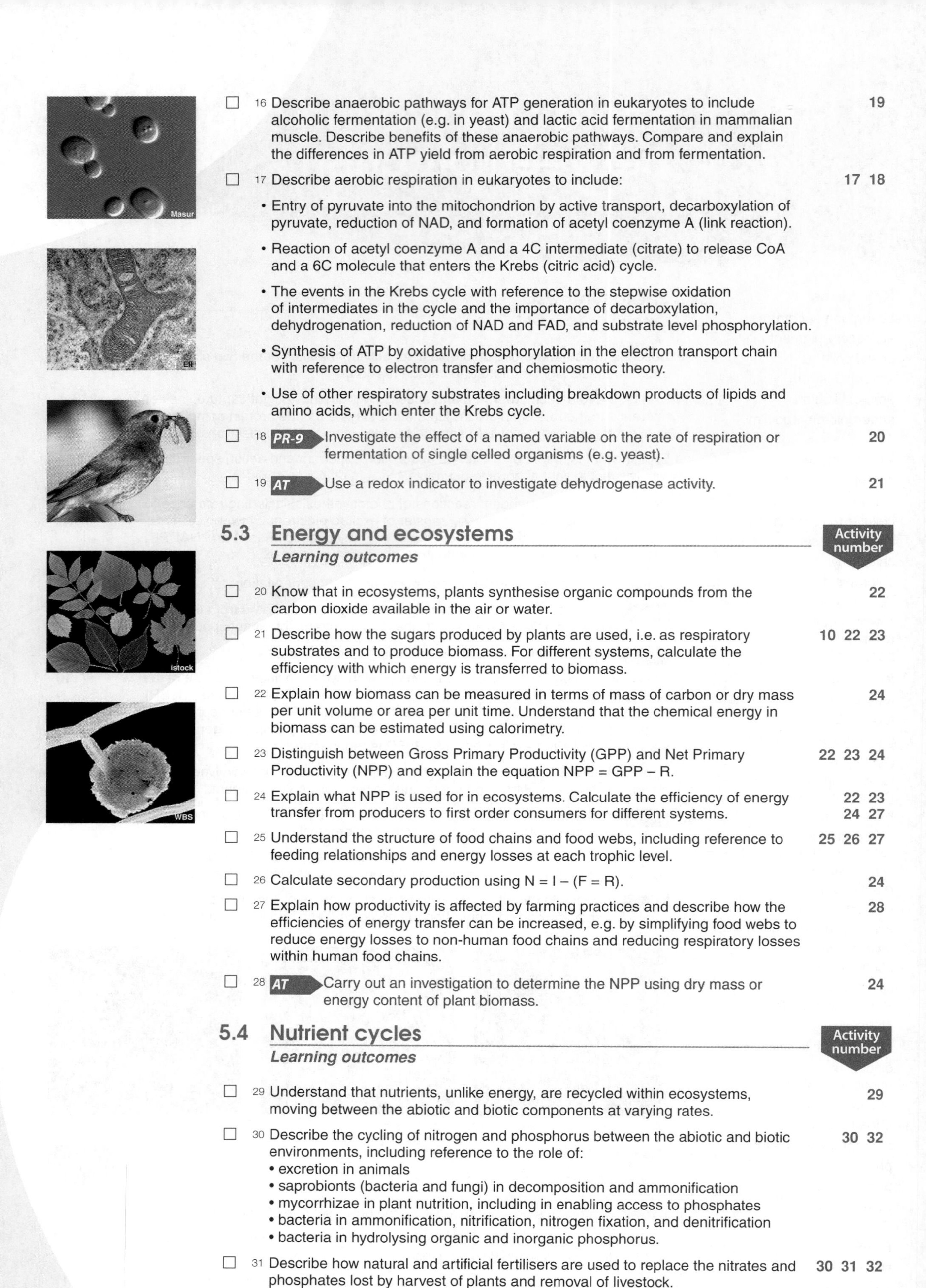

☐ 16 Describe anaerobic pathways for ATP generation in eukaryotes to include alcoholic fermentation (e.g. in yeast) and lactic acid fermentation in mammalian muscle. Describe benefits of these anaerobic pathways. Compare and explain the differences in ATP yield from aerobic respiration and from fermentation. — 19

☐ 17 Describe aerobic respiration in eukaryotes to include: — 17 18

- Entry of pyruvate into the mitochondrion by active transport, decarboxylation of pyruvate, reduction of NAD, and formation of acetyl coenzyme A (link reaction).
- Reaction of acetyl coenzyme A and a 4C intermediate (citrate) to release CoA and a 6C molecule that enters the Krebs (citric acid) cycle.
- The events in the Krebs cycle with reference to the stepwise oxidation of intermediates in the cycle and the importance of decarboxylation, dehydrogenation, reduction of NAD and FAD, and substrate level phosphorylation.
- Synthesis of ATP by oxidative phosphorylation in the electron transport chain with reference to electron transfer and chemiosmotic theory.
- Use of other respiratory substrates including breakdown products of lipids and amino acids, which enter the Krebs cycle.

☐ 18 **PR-9** Investigate the effect of a named variable on the rate of respiration or fermentation of single celled organisms (e.g. yeast). — 20

☐ 19 **AT** Use a redox indicator to investigate dehydrogenase activity. — 21

5.3 Energy and ecosystems

Learning outcomes — Activity number

☐ 20 Know that in ecosystems, plants synthesise organic compounds from the carbon dioxide available in the air or water. — 22

☐ 21 Describe how the sugars produced by plants are used, i.e. as respiratory substrates and to produce biomass. For different systems, calculate the efficiency with which energy is transferred to biomass. — 10 22 23

☐ 22 Explain how biomass can be measured in terms of mass of carbon or dry mass per unit volume or area per unit time. Understand that the chemical energy in biomass can be estimated using calorimetry. — 24

☐ 23 Distinguish between Gross Primary Productivity (GPP) and Net Primary Productivity (NPP) and explain the equation NPP = GPP – R. — 22 23 24

☐ 24 Explain what NPP is used for in ecosystems. Calculate the efficiency of energy transfer from producers to first order consumers for different systems. — 22 23 24 27

☐ 25 Understand the structure of food chains and food webs, including reference to feeding relationships and energy losses at each trophic level. — 25 26 27

☐ 26 Calculate secondary production using N = I – (F = R). — 24

☐ 27 Explain how productivity is affected by farming practices and describe how the efficiencies of energy transfer can be increased, e.g. by simplifying food webs to reduce energy losses to non-human food chains and reducing respiratory losses within human food chains. — 28

☐ 28 **AT** Carry out an investigation to determine the NPP using dry mass or energy content of plant biomass. — 24

5.4 Nutrient cycles

Learning outcomes — Activity number

☐ 29 Understand that nutrients, unlike energy, are recycled within ecosystems, moving between the abiotic and biotic components at varying rates. — 29

☐ 30 Describe the cycling of nitrogen and phosphorus between the abiotic and biotic environments, including reference to the role of: — 30 32
- excretion in animals
- saprobionts (bacteria and fungi) in decomposition and ammonification
- mycorrhizae in plant nutrition, including in enabling access to phosphates
- bacteria in ammonification, nitrification, nitrogen fixation, and denitrification
- bacteria in hydrolysing organic and inorganic phosphorus.

☐ 31 Describe how natural and artificial fertilisers are used to replace the nitrates and phosphates lost by harvest of plants and removal of livestock. — 30 31 32

☐ 32 Describe and explain the environmental issues associated with fertiliser use, including leaching and eutrophication. — 31

1 Energy in Cells

Key Idea: Photosynthesis uses energy from the sun to produce glucose. Glucose breakdown produces ATP, which is used by all cells to provide the energy for metabolism.

A summary of the flow of energy within a plant cell is illustrated below. Heterotrophic cells (animals and fungi) have a similar flow except the glucose is supplied by ingestion or absorption of food molecules rather than by photosynthesis. The energy not immediately stored in chemical bonds is lost as heat. Note that ATP provides the energy for most metabolic reactions, including photosynthesis.

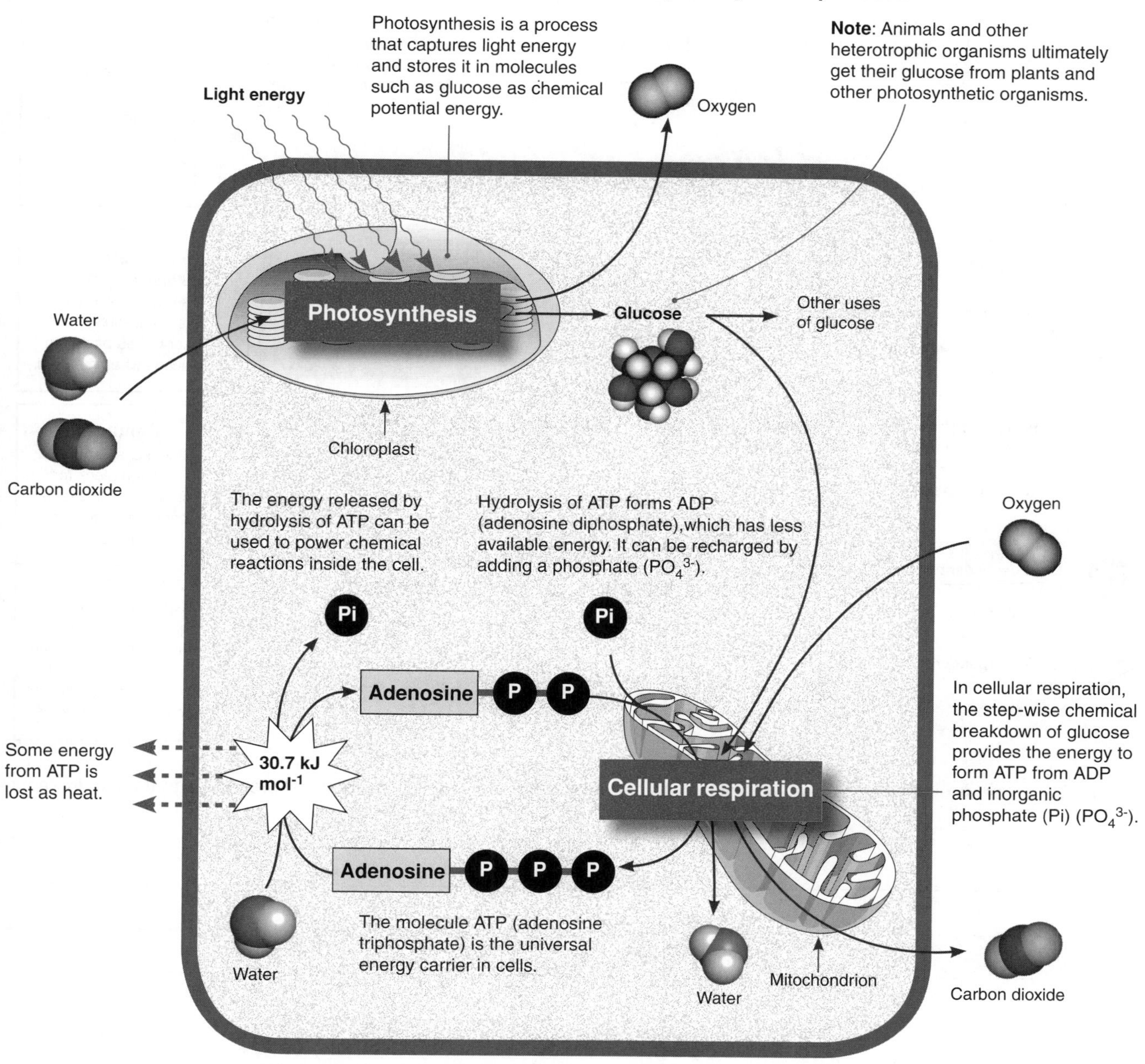

1. (a) What are the raw materials for photosynthesis? ______________________

   (b) What are the raw materials for respiration? ______________________

2. What is the immediate source of energy for reforming ATP from ADP? ______________________

3. What is the ultimate source of energy for plants? ______________________

4. What is the ultimate source of energy for animals? ______________________

ISBN: 978-1-927309-20-9

2 Photosynthesis

Key Idea: Photosynthesis is the process by which light energy is used to convert CO_2 and water into glucose and oxygen.

Photosynthesis is of fundamental importance to living things because it transforms sunlight energy into chemical energy stored in molecules, releases free oxygen gas, and absorbs carbon dioxide (a waste product of cellular metabolism). Photosynthetic organisms use special pigments, called **chlorophylls**, to absorb light of specific wavelengths and capture the light energy. Photosynthesis involves reduction and oxidation (redox) reactions. In photosynthesis, water is split and electrons are transferred together with hydrogen ions from water to CO_2, reducing it to sugar.

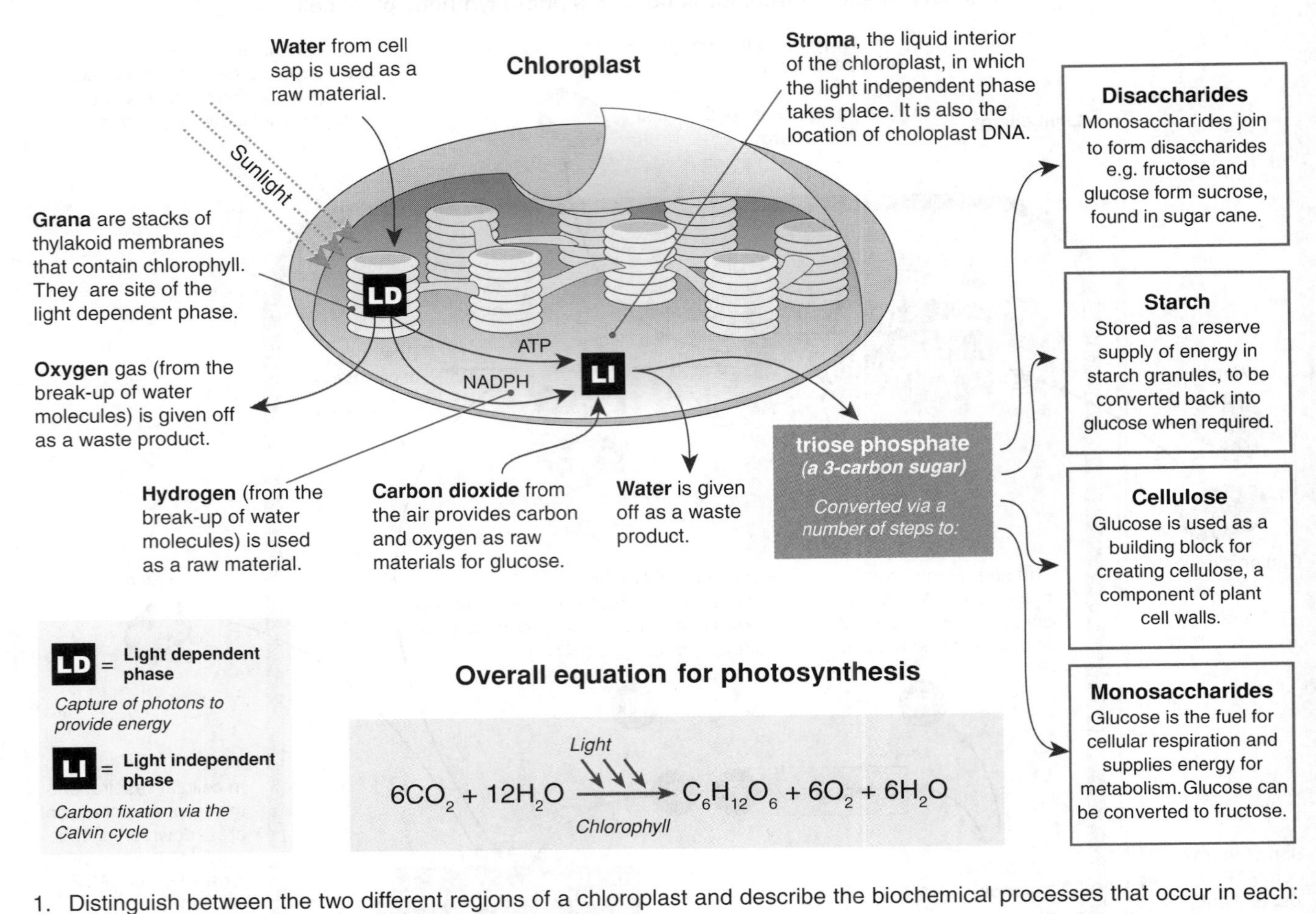

LD = **Light dependent phase**

Capture of photons to provide energy

LI = **Light independent phase**

Carbon fixation via the Calvin cycle

Overall equation for photosynthesis

$$6CO_2 + 12H_2O \xrightarrow[\text{Chlorophyll}]{\text{Light}} C_6H_{12}O_6 + 6O_2 + 6H_2O$$

1. Distinguish between the two different regions of a chloroplast and describe the biochemical processes that occur in each:

   (a) ___

   (b) ___

2. State the origin and fate of the following molecules involved in photosynthesis:

   (a) Carbon dioxide: ___

   (b) Oxygen: ___

   (c) Hydrogen: ___

3. Discuss the potential uses for the end products of photosynthesis: ___

ISBN:978-1-927309-20-9

3 Chloroplasts

Key Idea: Chloroplasts have a complicated internal membrane structure that provides the sites for the light dependent reactions of photosynthesis.

Chloroplasts are the specialised plastids in which photosynthesis occurs. A mesophyll leaf cell contains between 50-100 chloroplasts. The chloroplasts are generally aligned so that their broad surface runs parallel to the cell wall to maximise the surface area available for light absorption. Chloroplasts have an internal structure characterised by a system of membranous structures called **thylakoids** arranged into stacks called **grana**. Special pigments, called **chlorophylls** and **carotenoids**, are bound to the membranes as part of light-capturing photosystems. They absorb light of specific wavelengths and thereby capture the light energy.

The structure of a chloroplast

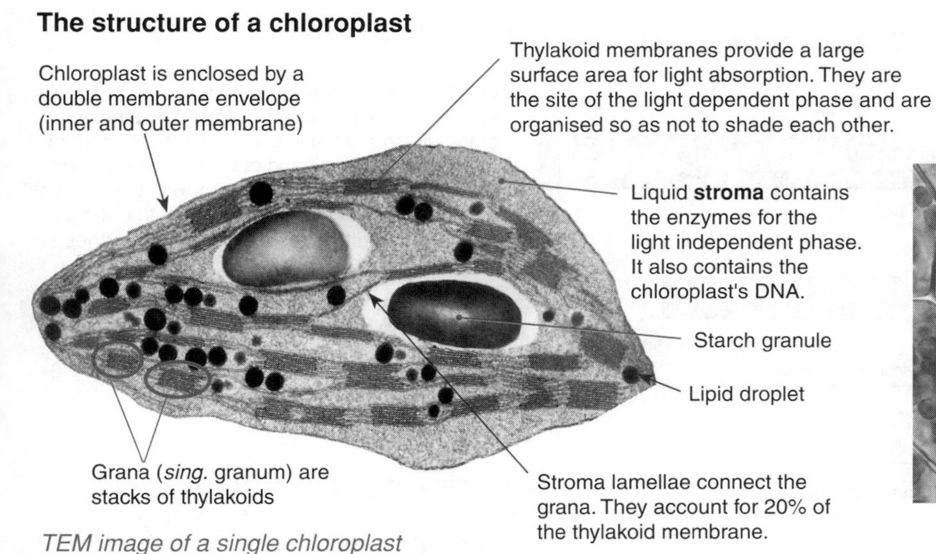

TEM image of a single chloroplast

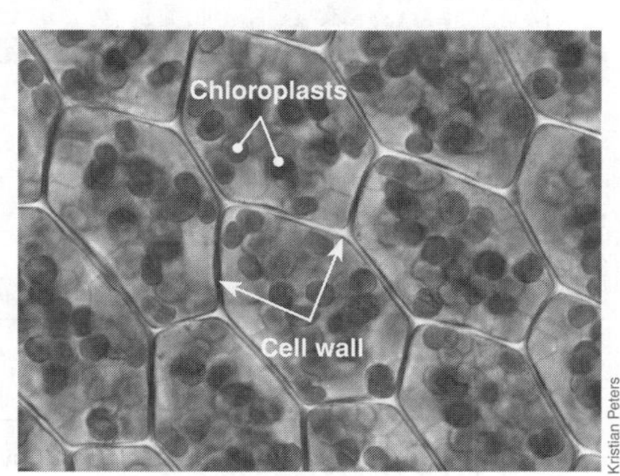

Chloroplasts visible in plant cells

1. Label the transmission electron microscope image of a chloroplast below:

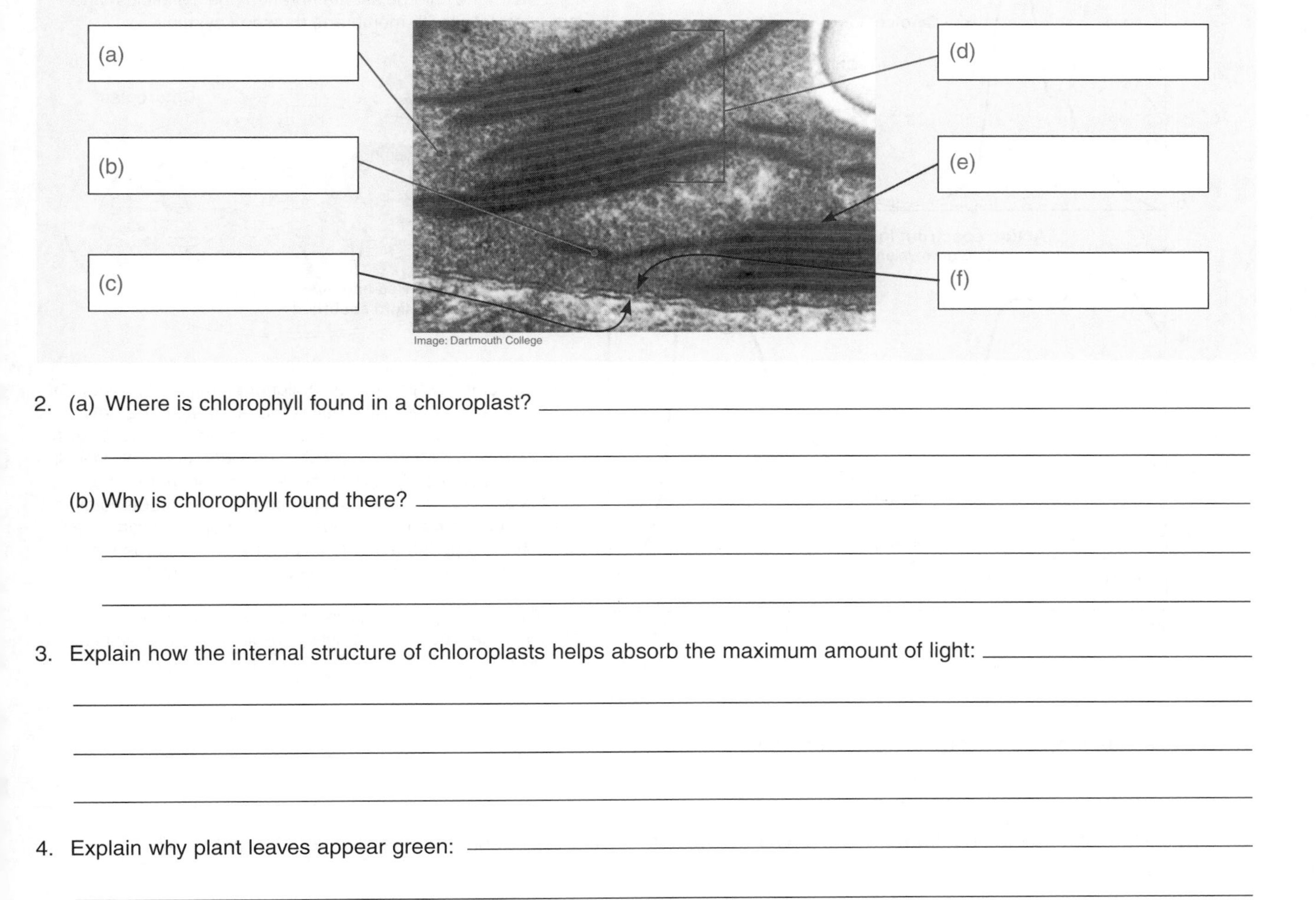

2. (a) Where is chlorophyll found in a chloroplast? ______________________

   (b) Why is chlorophyll found there? ______________________

3. Explain how the internal structure of chloroplasts helps absorb the maximum amount of light: ______________________

4. Explain why plant leaves appear green: ______________________

ISBN: 978-1-927309-20-9

4 Pigments and Light Absorption

Key Idea: Chlorophyll pigments absorb light of specific wavelengths and capture light energy for photosynthesis.

Substances that absorb visible light are called **pigments**, and different pigments absorb light of different wavelengths. The ability of a pigment to absorb particular wavelengths of light can be measured with a spectrophotometer. The light absorption vs the wavelength is called the **absorption spectrum** of that pigment. The absorption spectrum of different photosynthetic pigments provides clues to their role in photosynthesis, since light can only perform work if it is absorbed. An **action spectrum** profiles the effectiveness of different wavelengths of light in fuelling photosynthesis. It is obtained by plotting wavelength against a measure of photosynthetic rate (e.g. O_2 production).

The electromagnetic spectrum

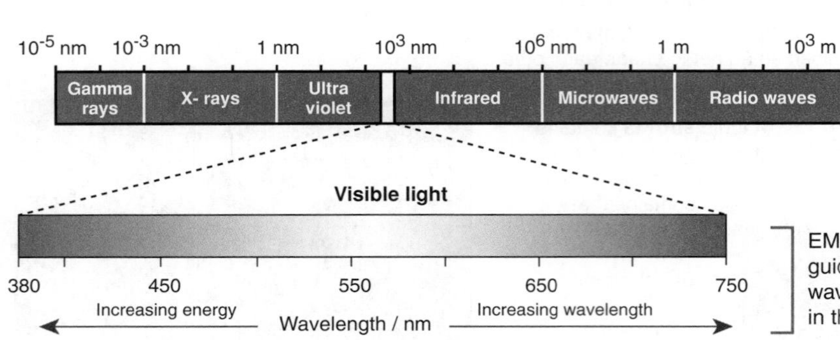

Light is a form of energy known as electromagnetic radiation (EMR). The segment of the electromagnetic spectrum most important to life is the narrow band between about 380 nm and 750 nm. This radiation is known as visible light because it is detected as colours by the human eye. It is visible light that drives photosynthesis.

EMR travels in waves, where wavelength provides a guide to the energy of the photons. The greater the wavelength of EMR, the lower the energy of the photons in that radiation.

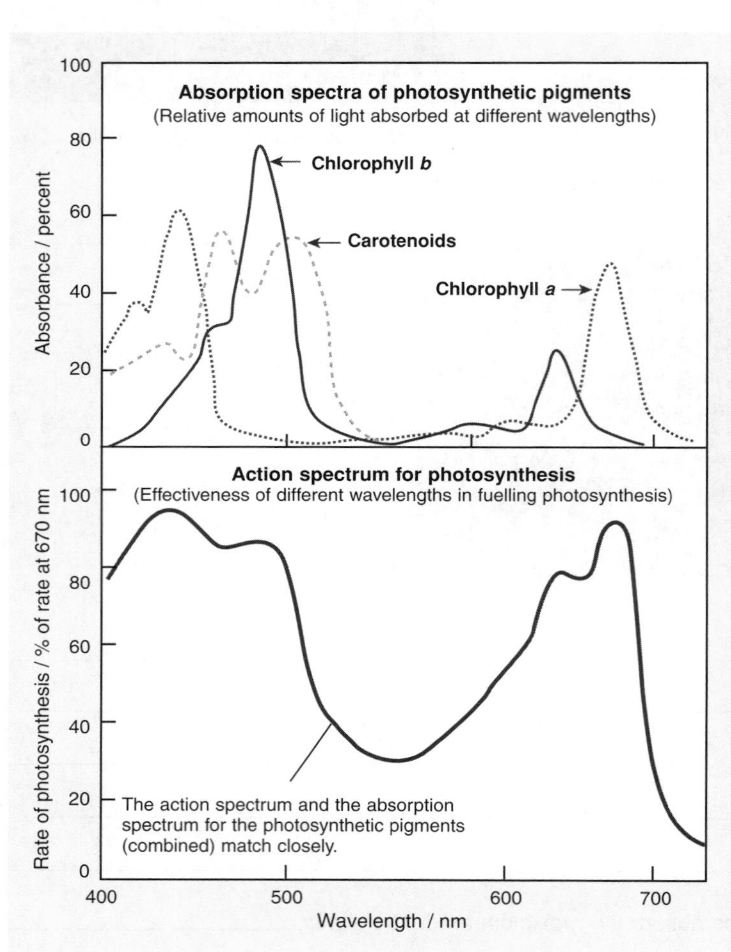

The photosynthetic pigments of plants

The photosynthetic pigments of plants fall into two categories: **chlorophylls** (which absorb red and blue-violet light) and **carotenoids** (which absorb strongly in the blue-violet and appear orange, yellow, or red). The pigments are located on the chloroplast membranes (the thylakoids) and are associated with membrane transport systems.

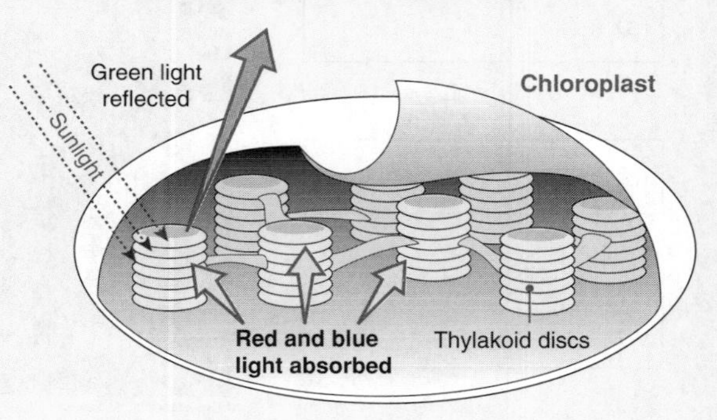

The pigments of chloroplasts in higher plants (above) absorb blue and red light, and the leaves therefore appear green (which is reflected). Each photosynthetic pigment has its own characteristic absorption spectrum (top left). Only chlorophyll *a* participates directly in the light reactions of photosynthesis, but the accessory pigments (chlorophyll *b* and carotenoids) can absorb wavelengths of light that chlorophyll *a* cannot and pass the energy (photons) to chlorophyll *a*, thus broadening the spectrum that can effectively drive photosynthesis.

Left: Graphs comparing absorption spectra of photosynthetic pigments compared with the action spectrum for photosynthesis.

1. What is meant by the absorption spectrum of a pigment? ______________________

____

2. Why doesn't the action spectrum for photosynthesis exactly match the absorption spectrum of chlorophyll *a*?

____

ISBN:978-1-927309-20-9

5 Separation of Pigments by Chromatography

Key Idea: Photosynthetic pigments can be separated from a mixture using chromatography.

Chromatography involves passing a mixture dissolved in a mobile phase (a solvent) through a stationary phase, which separates the molecules according to their specific characteristics (e.g. size or charge). In thin layer chromatography, the stationary phase is a thin layer of adsorbent material (e.g. silica gel or cellulose) attached to a solid plate. A sample is placed near the bottom of the plate which is placed in an appropriate solvent (the mobile phase).

Separation of photosynthetic pigments

The four primary pigments of green plants can be easily separated and identified using thin layer chromatography. The pigments from the leaves are first extracted by crushing leaves, together with acetone, using a mortar and pestle. The extract is dotted on to the chromatography plate. Acetone is used as the mobile phase (solvent). During thin layer chromatography, the pigments separate out according to differences in their relative solubilities. Two major classes of pigments are detected: the two greenish chlorophyll pigments and two yellowish carotenoid pigments.

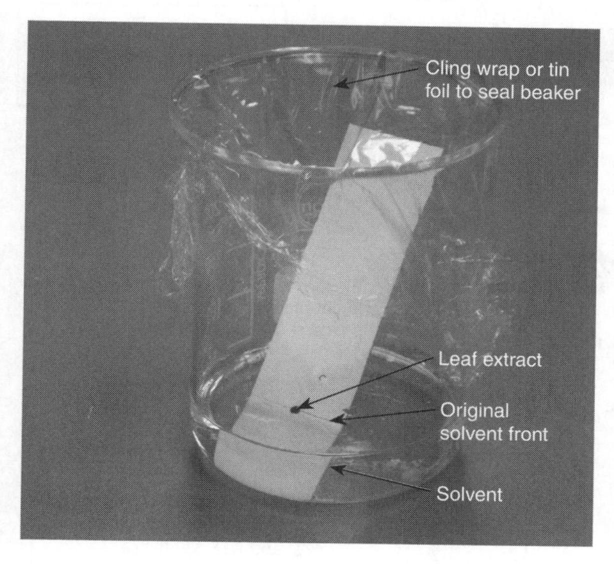

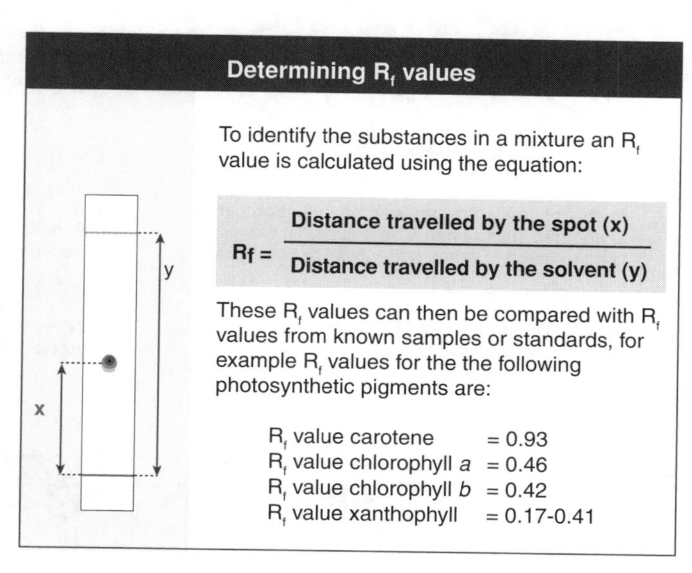

Determining R_f values

To identify the substances in a mixture an R_f value is calculated using the equation:

$$R_f = \frac{\text{Distance travelled by the spot (x)}}{\text{Distance travelled by the solvent (y)}}$$

These R_f values can then be compared with R_f values from known samples or standards, for example R_f values for the the following photosynthetic pigments are:

R_f value carotene = 0.93
R_f value chlorophyll *a* = 0.46
R_f value chlorophyll *b* = 0.42
R_f value xanthophyll = 0.17-0.41

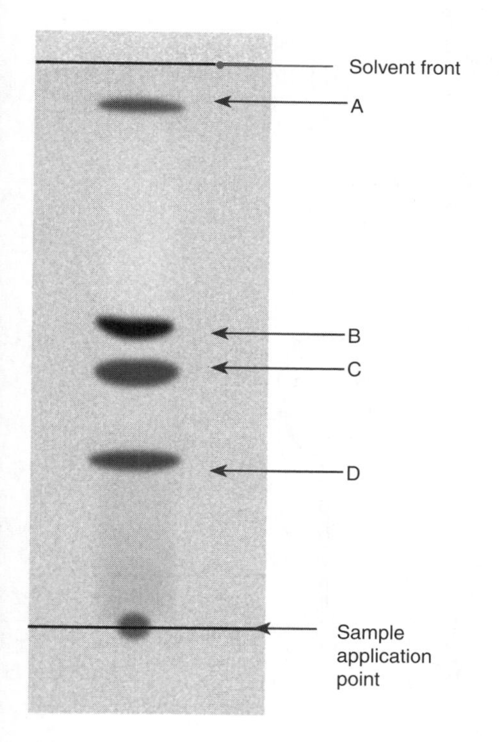

1. (a) Calculate the R_f values for the pigments A-D on the chromatography plate shown left.

   (b) Use the R_f values to identify the pigments:

   A: R_f value: ____________________

   Pigment: ____________________

   B: R_f value: ____________________

   Pigment: ____________________

   C: R_f value: ____________________

   Pigment: ____________________

   D: R_f value: ____________________

   Pigment: ____________________

2. A student carried out a chromatography experiment in class. The instructions said to leave the plate in the solvent for 30 minutes, but the student instead removed the plate after 20 minutes. How would this affect the R_f values and pigment separations obtained?

____

____

____

____

ISBN: 978-1-927309-20-9

6 Light Dependent Reactions

Key Idea: In light dependent reactions of photosynthesis, the energy from photons of light is used to drive the reduction of $NADP^+$ and the production of ATP.

Like cellular respiration, photosynthesis is a redox process, but in photosynthesis, water is split, and electrons and hydrogen ions, are transferred from water to CO_2, reducing it to sugar. The electrons increase in potential energy as they move from water to sugar. The energy to do this is provided by light. Photosynthesis has two phases. In the **light dependent reactions**, light energy is converted to chemical energy (ATP and NADPH). In the **light independent reactions**, the chemical energy is used to synthesise carbohydrate. The light dependent reactions most commonly involve **non-cyclic phosphorylation**, which produces ATP and NADPH in roughly equal quantities. The electrons lost are replaced from water. In **cyclic phosphorylation**, the electrons lost from photosystem II are replaced by those from photosystem I. ATP is generated, but not NADPH.

Non-cyclic phosphorylation

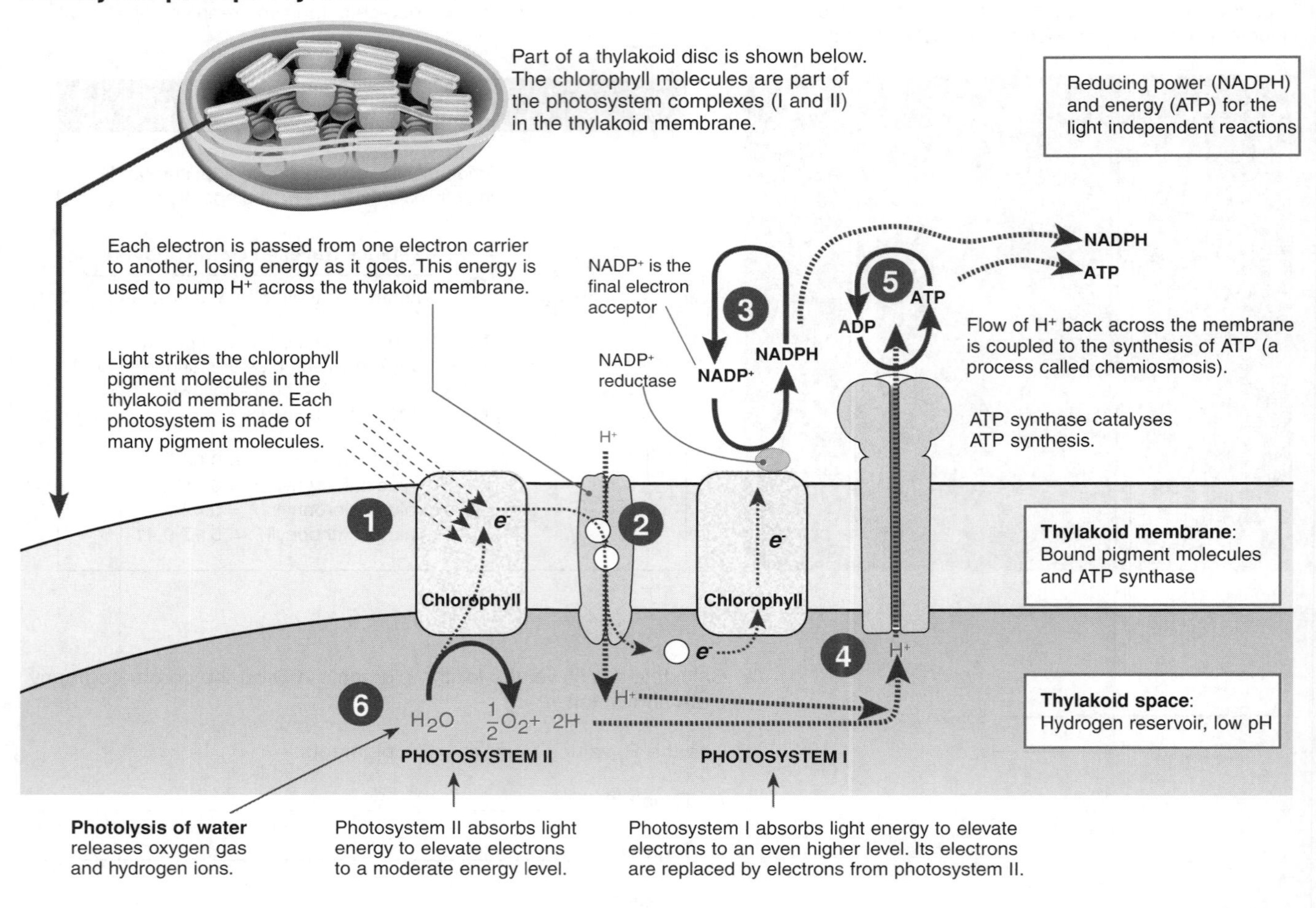

Cyclic phosphorylation

Cyclic phosphorylation involves only photosystem I and NADPH is not generated. Electrons from photosystem I are shunted back to the electron carriers in the membrane. This pathway produces ATP only. The Calvin cycle uses more ATP than NADPH, so cyclic phosphorylation makes up the difference. It is activated when NADPH levels build up, and remains active until enough ATP is made to meet demand.

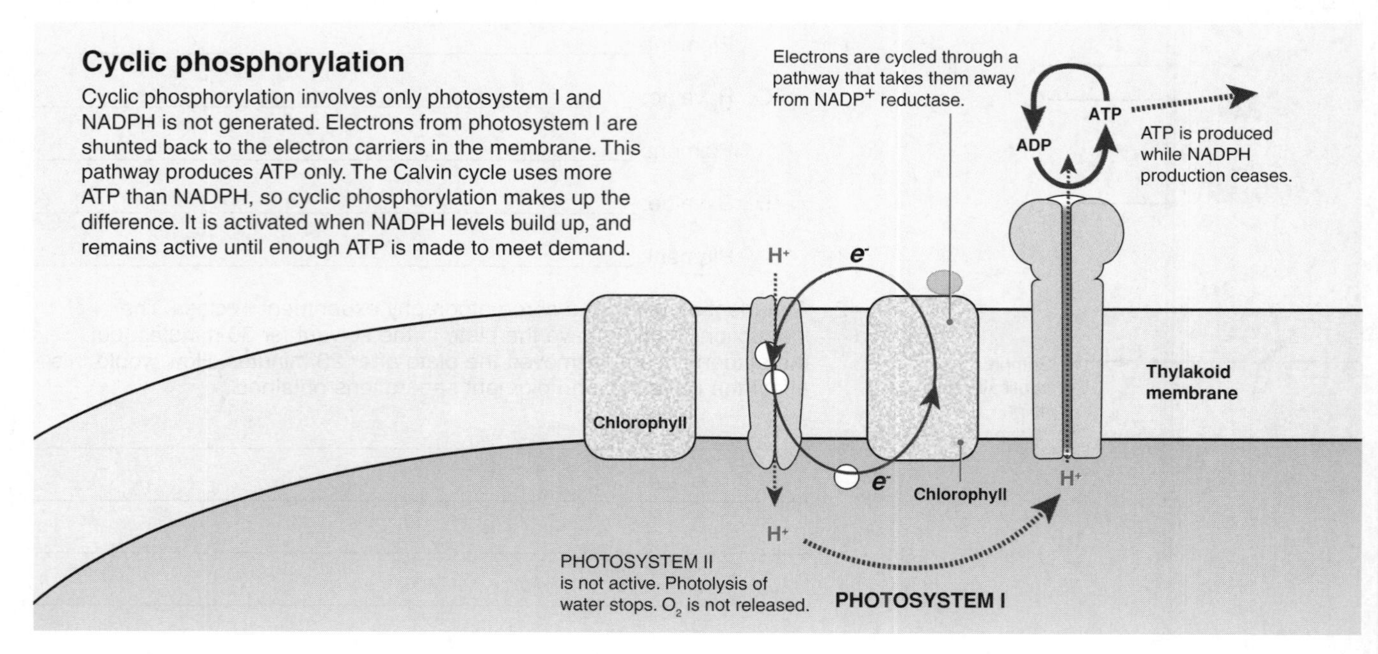

ISBN:978-1-927309-20-9

1. Describe the role of the carrier molecule **NADP** in photosynthesis: ______

2. Explain the role of chlorophyll molecules in photosynthesis: ______

3. Summarise the events of the light dependent reactions and identify where they occur: ______

4. Describe how ATP is produced as a result of light striking chlorophyll molecules during the light dependent phase:

5. (a) Explain what you understand by the term non-cyclic phosphorylation: ______

   (b) Suggest why this process is also known as non-cyclic photophosphorylation: ______

6. (a) Describe how cyclic photophosphorylation differs from non-cyclic photophosphorylation: ______

   (b) Both cyclic and non-cyclic pathways operate to varying degrees during photosynthesis. Since the non-cyclic pathway produces both ATP and NAPH, explain the purpose of the cyclic pathway of electron flow:

7. Complete the summary table of the light dependent reactions of photosynthesis

	Non-cyclic phosphorylation	Cyclic phosphorylation
Photosystem involved		
Energy carrier(s) produced		
Photolysis of water (yes / no)		
Production of oxygen (yes / no)		

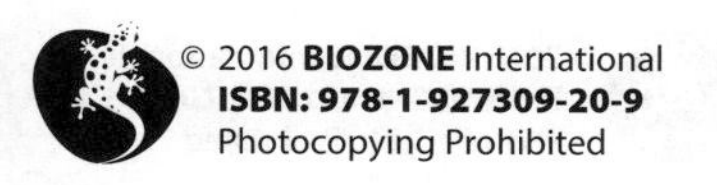

ISBN: 978-1-927309-20-9

7 Light Independent Reactions

Key Idea: The light independent reactions of photosynthesis take place in the stroma of the chloroplast and do not require light to proceed.

In the **light independent reactions** (the **Calvin cycle**) hydrogen (H^+) is added to CO_2 and a 5C intermediate to make carbohydrate. The H^+ and ATP are supplied by the light dependent reactions. The Calvin cycle uses more ATP than NADPH, but the cell uses cyclic phosphorylation (which does not produce NADPH) when it runs low on ATP to make up the difference.

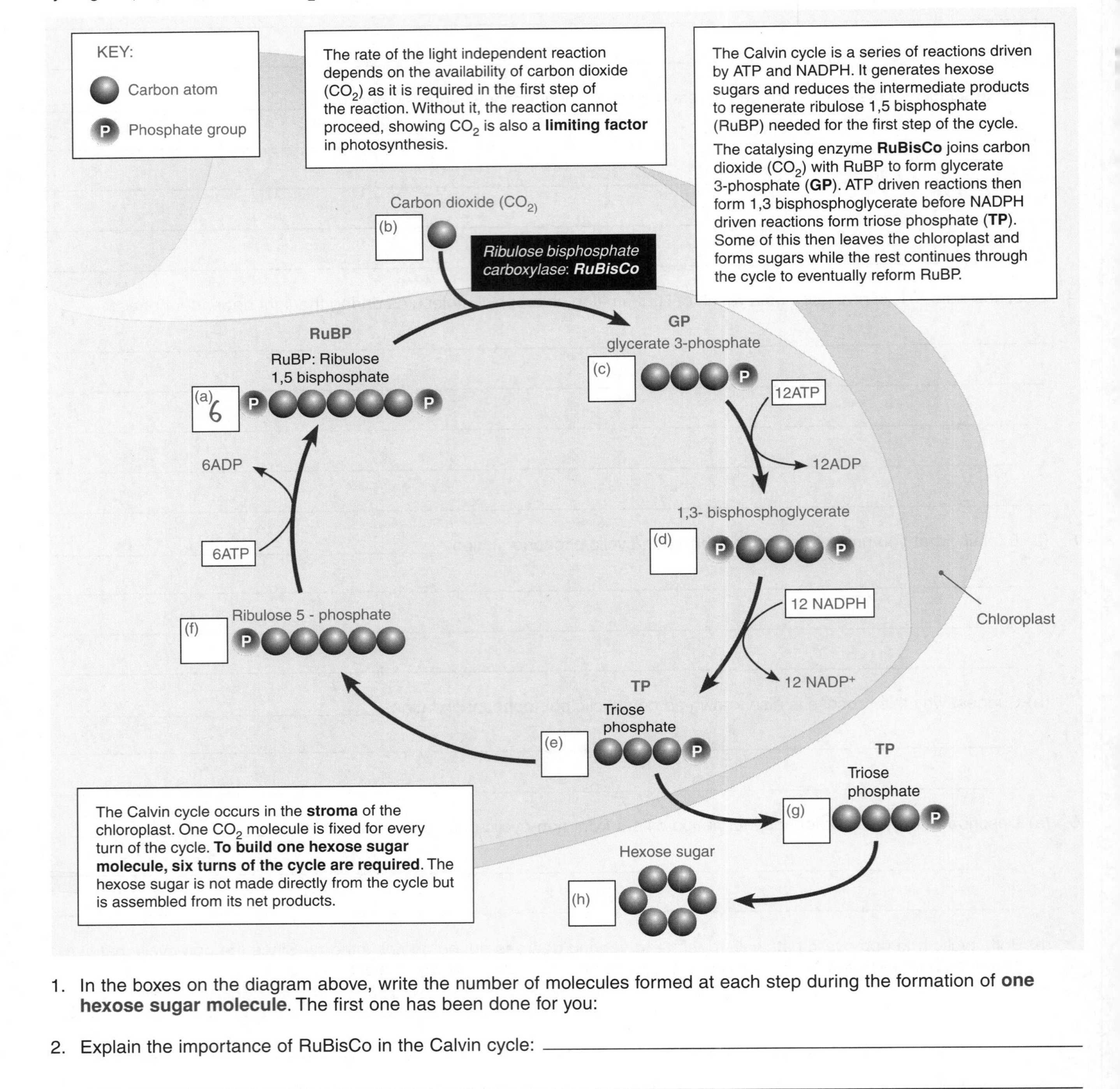

1. In the boxes on the diagram above, write the number of molecules formed at each step during the formation of **one hexose sugar molecule**. The first one has been done for you:

2. Explain the importance of RuBisCo in the Calvin cycle: __________

3. Identify the actual end product on the Calvin cycle: __________

4. Write the equation for the production of one hexose sugar molecule from carbon dioxide: __________

5. Explain why the Calvin cycle is likely to cease in the dark for most plants, even though it is independent of light:

ISBN:978-1-927309-20-9

8 Experimental Evidence for Photosynthesis

Key Idea: Hill's experiment using isolated chloroplasts and Calvin's "lollipop" experiment provided important information on the process of photosynthesis.

In the 1930s Robert Hill devised a way of measuring oxygen evolution and the rate of photosynthesis in isolated chloroplasts. During the 1950s Melvin Calvin led a team using radioisotopes of carbon to work out the steps of the light independent reactions (the Calvin cycle).

Robert Hill's experiment

The dye **DCPIP** (2,6-dichlorophenol-indophenol) is blue. It is reduced by H^+ ions and forms $DCPIPH_2$ (colourless). Hill made use of this dye to show that O_2 is produced during photosynthesis even when CO_2 is not present.

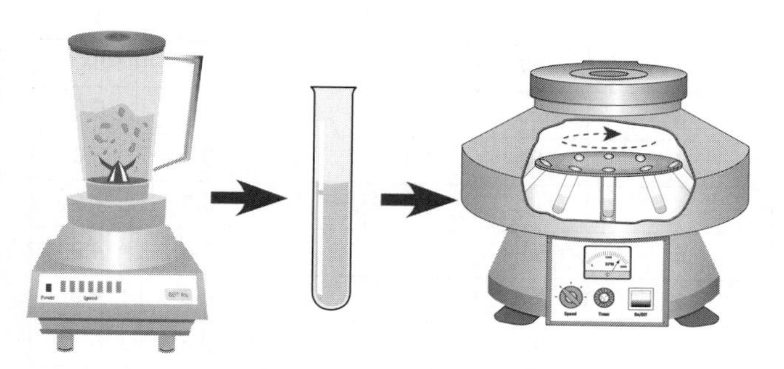

Leaves are homogenised to form a slurry. The slurry is filtered to remove any debris. The filtered extract is then centrifuged at low speed to remove the larger cell debris and then at high speed to separate out the chloroplasts.

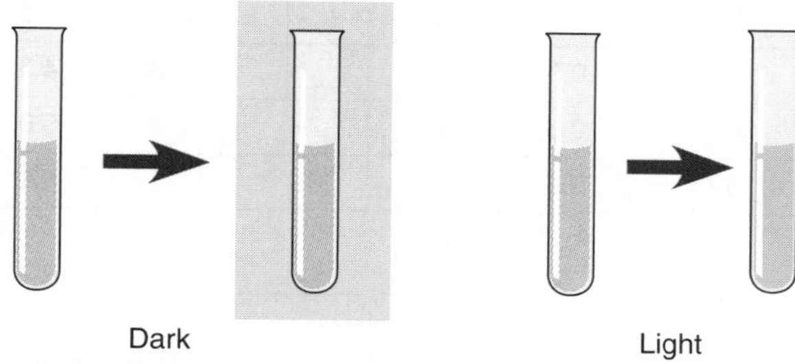

The chloroplasts are resuspended in a buffer. The blue dye **DCPIP** is added to the suspension. In a test tube left in the dark, the dye remains unchanged. In a test tube exposed to the light, the blue dye fades and the test tube turns green again. The rate of colour change can be measured by measuring the light absorbance of the suspension. The rate is proportional to the rate at which oxygen is produced.

Hill's experiment showed that water must be the source of oxygen (and therefore electrons). It is split by light to produce H^+ ions (which reduce DCPIP) and O^{2-} ions. The equation below summarises his findings:

$$H_2O + A \rightarrow AH_2 + \frac{1}{2} O_2$$

where A is the electron acceptor (*in vivo* this is $NADP^+$)

Calvin's lollipop experiment

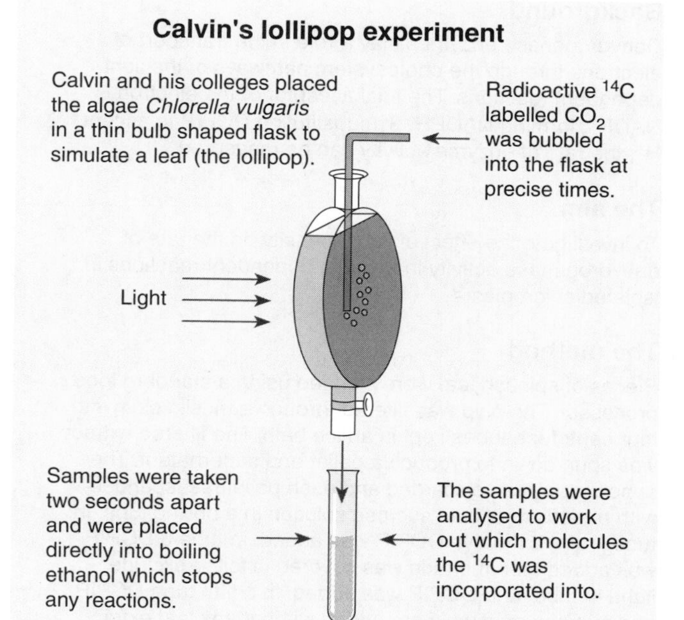

Two-dimensional chromatography was used to separate the molecules in each sample. The sample is run in one direction, then rotated 90 degrees and run again with a different solvent. This separates out molecules that might be close to each other.

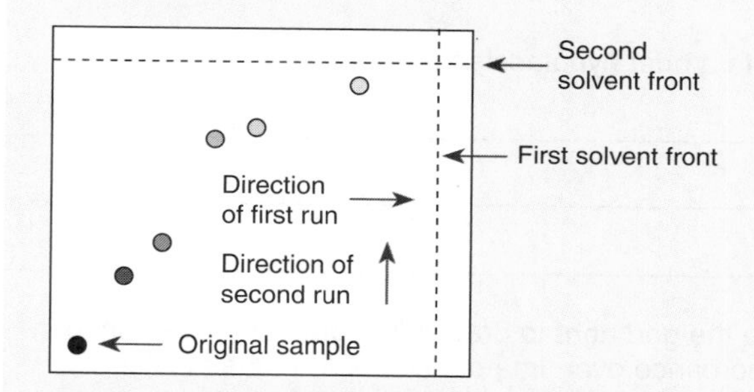

By identifying the order that the molecules incorporating the ^{14}C appeared it was possible to work out the steps of the now called Calvin cycle. This could only be done by taking samples only seconds apart.

1. Write an equation for the formation of $DCPIPH_2$ from DCPIP: ______________________________

____

2. What important finding about photosynthesis did Hill's experiment show? ______________________________

____

3. Why did the samples in Calvin's lollipop experiment need to be taken just seconds apart? ______________________________

____

____

____

ISBN: 978-1-927309-20-9

9 Investigating Enzymes in Photosynthesis

Key Idea: Replacing $NADP^+$ with DCPIP as the electron acceptor allows the effect of light on the rate of the light dependent reactions to be measured indirectly.

$NADP^+$ is the electron acceptor for the light dependent reaction. By substituting the dye DCPIP, which fades from blue to colourless as it accepts electrons, it is possible to indirectly measure the rate of the light dependent reactions and therefore the rate of enzyme activity during the reactions.

Background

Dehydrogenase enzymes play a role in the transport of electrons through the photosystem pathways of the light dependent reactions. The final acceptor of the electron is $NADP^+$, forming NADPH. By substituting DCPIP to accept H^+, the rate of enzyme activity can be measured.

The aim

To investigate the effect of light intensity on the rate of dehydrogenase activity in the light dependent reactions in isolated chloroplasts.

The method

Pieces of spinach leaf were blended using a standard food processor. The pulp was filtered through a muslin cloth into four centrifuge tubes kept in an ice bath. The filtered extract was spun down to produce a pellet and supernatant. The supernatant was discarded and each pellet resuspended with a medium of cold sucrose solution in a boiling tube. In tube 1 and 2 the dye DCPIP was added. In tube 3 DCPIP was added then the tube was covered in foil to exclude light. In tube 4 no DCPIP was added. In a fifth tube DCPIP and sucrose medium were added without any leaf extract. Tubes 1 and 3 were exposed to high intensity light. Tube 2 was exposed to a lower intensity light. The absorbance of all the tubes was measured using a colorimeter at time 0 and every minute for 15 minutes. The absorbance of tube 3 was measured at the beginning and end of the experiment only.

Results

	Tube number / absorbance				
Time / min	**1**	**2**	**3**	**4**	**5**
0	5.0	5.0	5.0	0.3	5.0
1	4.8	5.0	-	0.3	5.0
2	4.7	4.9	-	0.3	5.0
3	4.6	4.8	-	0.3	5.0
4	4.3	4.8	-	0.4	5.0
5	4.0	4.7	-	0.3	4.9
6	3.8	4.6	-	0.4	4.9
7	3.4	4.6	-	0.2	4.9
8	3.0	4.5	-	0.3	5.0
9	2.6	4.4	-	0.4	5.1
10	2.2	4.4	-	0.3	5.0
11	1.9	4.3	-	0.2	4.9
12	1.4	4.1	-	0.2	5.0
13	0.9	4.0	-	0.3	4.8
14	0.6	4.0	-	0.3	5.0
15	0.5	3.8	4.7	0.4	5.0

1. Write a brief hypothesis for this experiment:

2. Use the grid right to draw a line graph of the change in absorbance over time of each of the tubes tested.

3. (a) What was the purpose of tube 4?

   (b) What was the purpose of tube 5?

4. Why was the absorbance of tube 3 only measured at the start and end of the investigation?

5. Why did the absorbance of tubes 4 and 5 vary?

6. Write a conclusion for the investigation:

ISBN:978-1-927309-20-9

10 The Fate of Triose Phosphate

Key Idea: The triose phosphate molecules produced in photosynthesis can be combined and rearranged to form monosaccharides such as glucose. Glucose is an important energy source and a precursor of many other molecules.

The triose phosphate molecules produced in photosynthesis can be combined and rearranged to form the hexose monosaccharide glucose. Glucose has three main fates: immediate use to produce ATP molecules (available energy for work), storage for later ATP production, or for use in building other molecules. Plants use the glucose they make in photosynthesis to build all the molecules they require. Animals obtain their glucose by consuming plants or other animals. Other molecules (e.g. amino acids and fatty acids) are also obtained by animals this way.

The fate of glucose

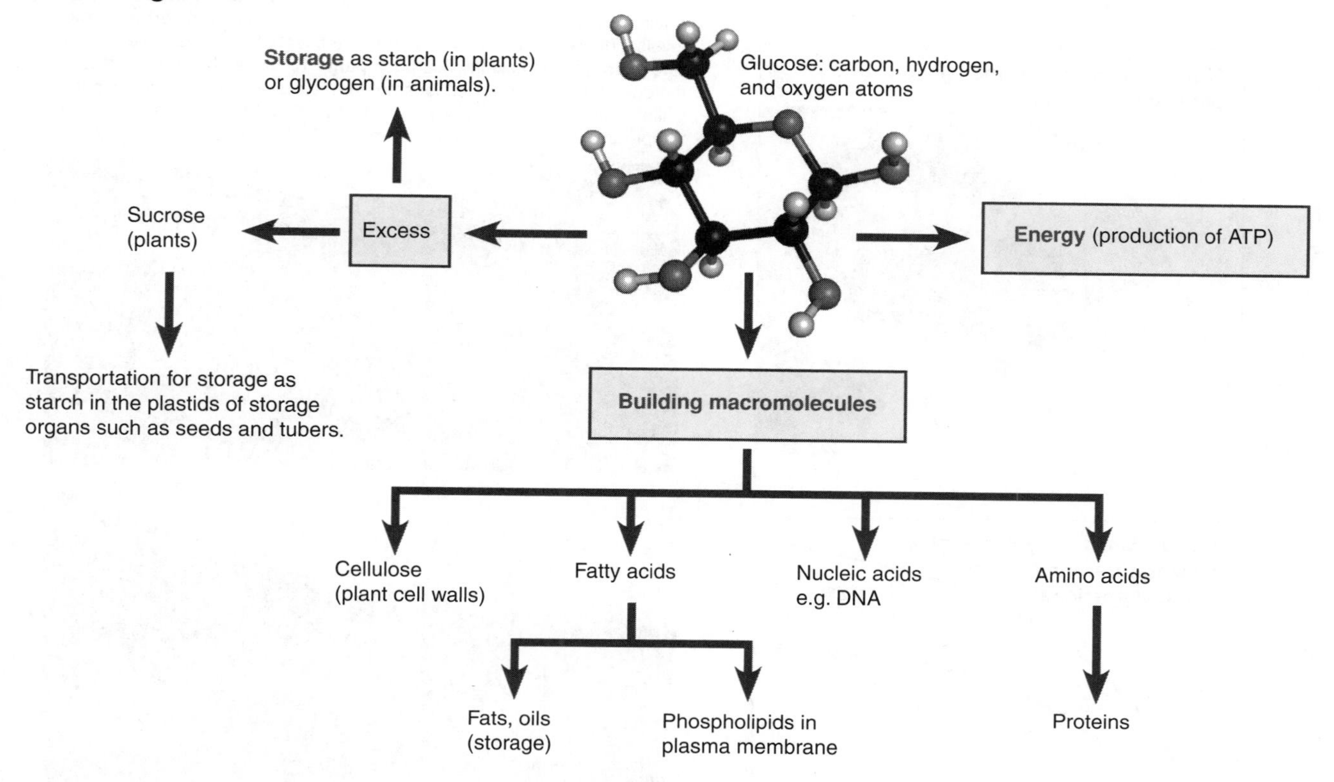

How do we know how glucose is used?

- Labelling the carbon atoms in a glucose molecule with isotopes shows how glucose is incorporated into other molecules.
- An isotope is an element (e.g. carbon) whose atoms have a particular number of neutrons in their nucleus. The different number of neutrons allows the isotopes to be identified by their density (e.g. a carbon atom with 13 neutrons is denser than a carbon atom with 12 neutrons).
- Some isotopes are radioactive. These radioactive isotopes can be traced using X-ray film or devices that detect the disintegration of the isotopes, such as Geiger counters.

The carbon atom

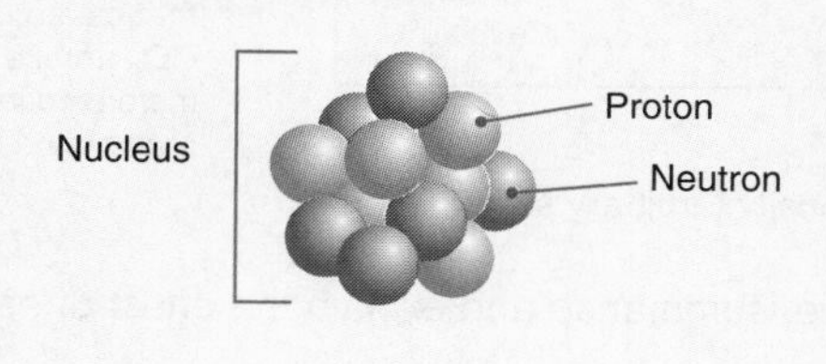

The nucleus of an atom is made up of neutrons and protons. For any element, the number of protons remains the same, but the number of neutrons can vary. Electrons (not shown) are found outside the nucleus.

Naturally occurring C isotopes

^{12}C	^{13}C	^{14}C
6 protons	6 protons	6 protons
6 neutrons	7 neutrons	8 neutrons
Stable. 99.9% of all C isotopes.	Stable	Radioactive

1. How many triose phosphate molecules are used to form a glucose molecule? ______

2. What are the three main fates of glucose? ______

______

3. Identify a use for glucose in a plant that does not occur in animals: ______

4. How can isotopes of carbon be used to find the fate of glucose molecules? ______

______

ISBN: 978-1-927309-20-9

11 Factors Affecting Photosynthesis

Key Idea: Environmental factors, such as light availability and carbon dioxide level, affect photosynthetic rate.

The rate at which plants can photosynthesise is dependent on environmental factors, particularly light and carbon dioxide (CO_2), but also temperature. In the plant's natural environment, fluctuations in these factors (and others) influence photosynthetic rate, so that the rate varies daily and seasonally. The effect of each factor can be tested experimentally by altering one while holding the others constant. Usually, either light or CO_2 level is limiting. Humans can overcome the limitations of low light or CO_2 by growing plants in a controlled environment.

Factors Affecting Photosynthetic Rate

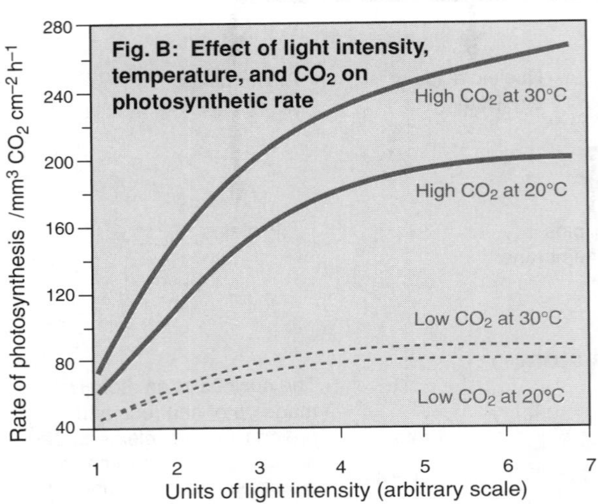

The rate of photosynthesis is affected strongly by abiotic factors in the environment, particularly light intensity, temperature, and carbon dioxide levels. The two graphs (left) illustrate the effect of these variables on the rate of photosynthesis in cucumber plants. Figure A shows the effect of increasing light intensity at constant temperature and CO_2 level. Figure B illustrates how this response is influenced by CO_2 concentration and temperature. 30°C is at the upper range of tolerance for many plants.

Glasshouse environments can artificially boost CO_2 levels

High wind increases water loss from stomata

Plants acquire CO_2 from the atmosphere through their stomata. They also lose water through their stomata when they are open, so conditions that cause them to close their stomata to reduce water loss also reduce CO_2 uptake and lower the photosynthetic rate. Such conditions include increased wind speed and water stress. Glasshouses are used to manipulate the physical environment to maximise photosynthetic rates.

1. Based on the figures above, summarise and explain the effect of each of the following factors on photosynthetic rate:

   (a) CO_2 concentration: ______________________________

   ____

   (b) Light intensity: ______________________________

   ____

   (c) Temperature: ______________________________

   ____

2. Why does photosynthetic rate decline when the CO_2 level is reduced? ______________________________

   ____

   ____

3. Why might hot, windy conditions reduce photosynthetic rates? ______________________________

   ____

ISBN:978-1-927309-20-9

12 Overcoming Limiting Factors in Photosynthesis

Key Idea: The growth of plants in greenhouses can be increased by manipulating abiotic factors.

Manipulating abiotic factors can maximise crop yields for economic benefit. For example, covering the soil with black plastic reduces weed growth, increases soil temperature, and boosts production. A more complete control of the abiotic conditions is achieved by growing crops in a controlled-environment system such as a greenhouse. Temperature, carbon dioxide (CO_2) concentration, and light intensity are optimised to maximise the rate of photosynthesis and therefore production. Greenhouses also allow specific abiotic factors to be manipulated to trigger certain life cycle events such as flowering. CO_2 enrichment dramatically increases the growth of greenhouse crops providing that other important abiotic factors (such as nutrients) are not limiting. The economic viability of such environments is determined by the increased production it provides compared with the capital cost of the equipment and the ongoing operating costs.

The growing environment can be controlled or modified to varying degrees. Black plastic sheeting can be laid over the soil to control weeds and absorb extra solar heat. Tunnel enclosures (such as those above) may be used to reduce light intensity and airflow, prevent frost damage, and reduce damage by pests.

Large, commercial greenhouses have elaborate computer-controlled watering systems linked to sensors that measure soil moisture, air temperature, and humidity. Coupled with a timer, they deliver optimal water conditions for plant growth by operating electric solenoid valves attached to the irrigation system.

Air flow through a greenhouse is essential to providing a homogeneous air temperature. Air flow also ensures an even distribution of carbon dioxide gas throughout the enclosure. A general airflow from one end of the enclosure to the other is maintained by a large number of fans all blowing in the same direction.

Carbon dioxide enrichment

Carbon dioxide (CO_2) is a raw material used in photosynthesis. If the supply of CO_2 is cut off or reduced, plant growth and development are curtailed. The amount of CO_2 in air is normally 0.03% (250-330 ppm). Most plants will stop growing when the CO_2 level falls below 150 ppm. Even at 220 ppm, a slowing of plant growth is noticeable (see graph, right).

Controlled CO_2 atmospheres, which boost the CO_2 concentration to more than 1000 ppm, significantly increase the rate of formation of dry plant matter and total yield (e.g. of flowers or fruit). Extra carbon dioxide can be generated (at a cost) by burning hydrocarbon fuels, using compressed, bottled CO_2 or dry ice, or by fermentation or decomposition of organic matter.

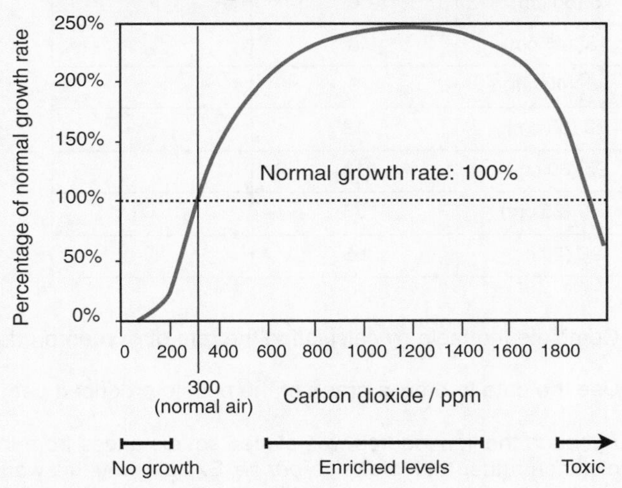

1. Explain why CO_2 enrichment has the capacity to radically increase crop production: ____________________

2. State the two primary considerations influencing the economic viability of controlled environments: ____________________

3. List the abiotic factors that are controlled in a greenhouse environment: ____________________

ISBN: 978-1-927309-20-9

13 Investigating Photosynthetic Rate

Key Idea: Measuring the production of oxygen provides a simple means of measuring the rate of photosynthesis.

The rate of photosynthesis can be investigated by measuring the substances involved in photosynthesis. These include measuring the uptake of carbon dioxide, the production of oxygen, or the change in biomass over time. Measuring the rate of oxygen production provides a good approximation of the photosynthetic rate and is relatively easy to carry out.

The aim

To investigate the effect of light intensity on the rate of photosynthesis in an aquatic plant, *Cabomba aquatica*.

Cabomba aquatica, a common aquarium plant

Piotr Kuczynski CC 3.0

The method

- 0.8-1.0 grams of *Cabomba* stem were weighed on a balance. The stem was cut and inverted to ensure a free flow of oxygen bubbles.
- The stem was placed into a beaker filled with a solution containing 0.2 mol L^{-1} sodium hydrogen carbonate (to supply carbon dioxide). The solution was at approximately 20°C. A funnel was inverted over the *Cabomba* and a test tube filled with the sodium hydrogen carbonate solution was inverted on top to collect any gas produced.
- The beaker was placed at distances (20, 25, 30, 35, 40, 45, 50 cm) from a 60W light source and the light intensity measured with a lux meter at each interval.
- Before recording data, the *Cabomba* stem was left to acclimatise to the new light level for 5 minutes. Because the volumes of oxygen gas produced are very low, bubbles were counted for a period of three minutes at each distance.

The results

Light intensity / lx (distance)	Bubbles counted in three minutes	Bubbles per minute
5 (50 cm)	0	
13 (45 cm)	6	
30 (40 cm)	9	
60 (35 cm)	12	
95 (30 cm)	18	
150 (25 cm)	33	
190 (20 cm)	35	

1. Complete the table by calculating the rate of oxygen production (bubbles of oxygen gas per minute):
2. Use the data to draw a graph of the bubble produced per minute vs light intensity:
3. Although the light source was placed set distances from the *Cabomba* stem, light intensity in lux was recorded at each distance rather than distance *per se*. Explain why this would be more accurate:

   ____

   ____

   ____

4. The sample of gas collected during the experiment was tested with a glowing splint. The splint reignited when placed in the gas. What does this confirm about the gas produced?

   ____

5. What could be a more accurate way of measuring the gas produced in the experiment? ______________________________

   ____

   ____

   ____

ISBN:978-1-927309-20-9

14 The Role of ATP in Cells

Key Idea: ATP transports chemical energy within the cell for use in metabolic processes.

All organisms require energy to be able to perform the metabolic processes required for them to function and reproduce. This energy is obtained by cellular respiration, a set of metabolic reactions which ultimately convert biochemical energy from 'food' into the nucleotide **adenosine triphosphate** (ATP). ATP is considered to be a universal energy carrier, transferring chemical energy within the cell for use in metabolic processes such as biosynthesis, cell division, cell signalling, thermoregulation, cell mobility, and active transport of substances across membranes.

Adenosine triphosphate (ATP)

The ATP molecule consists of three components; a purine base (**adenine**), a pentose sugar (**ribose**), and **three phosphate groups** which attach to the 5' carbon of the pentose sugar. The structure of ATP is described below.

The bonds between the phosphate groups contain electrons in a high energy state which store a large amount of energy. The energy is released during ATP hydrolysis. Typically, hydrolysis is coupled to another cellular reaction to which the energy is transferred. The end products of the reaction are adenosine diphosphate (ADP) and an inorganic phosphate (Pi).

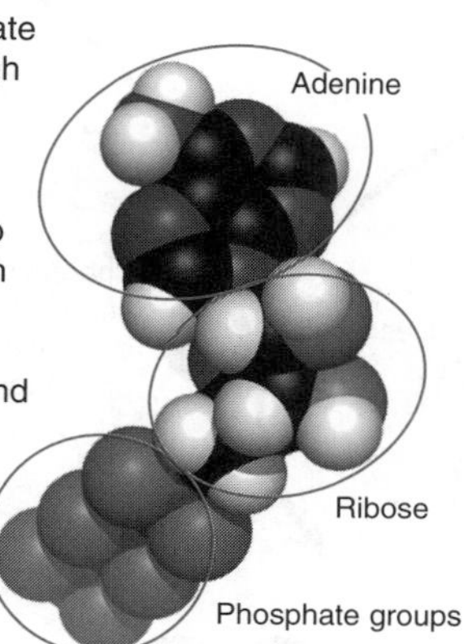

Note that energy is released during the formation of bonds during the hydrolysis reaction, not the breaking of bonds between the phosphates (which requires energy input).

The mitochondrion

Cellular respiration and ATP production occur in mitochondria. A mitochondrion is bounded by a double membrane. The inner and outer membranes are separated by an intermembrane space, compartmentalising the regions where the different reactions of cellular respiration take place.

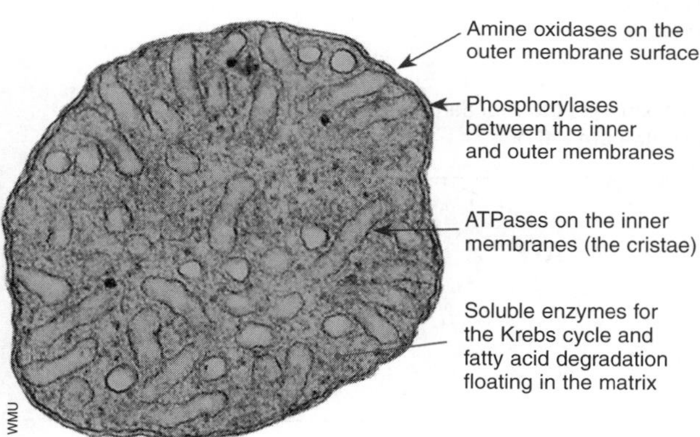

ATP powers metabolism

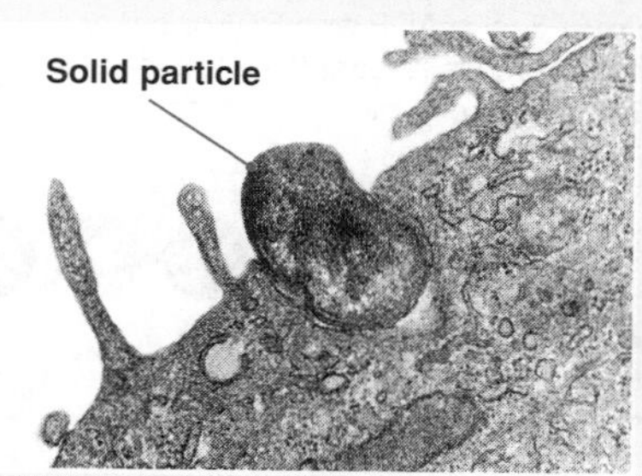

The energy released from the removal of a phosphate group of ATP is used for active transport of molecules and substances across the plasma membrane. **Phagocytosis** (left), which involves the engulfment of solid particles, is an example.

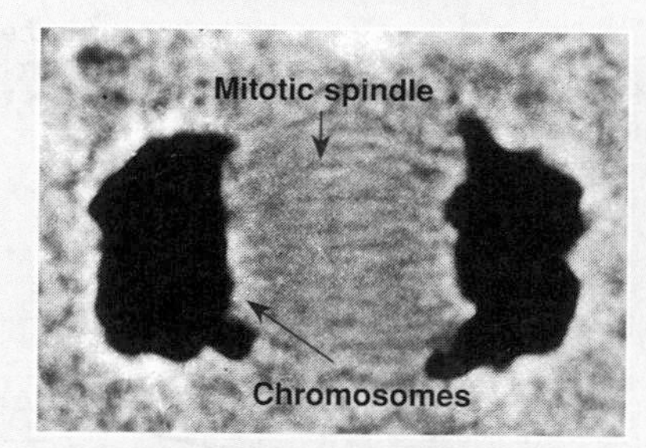

Cell division (**mitosis**), as observed in this onion cell, requires ATP to proceed. Formation of the mitotic spindle and chromosome separation are two aspects of cell division which require energy from ATP hydrolysis to occur.

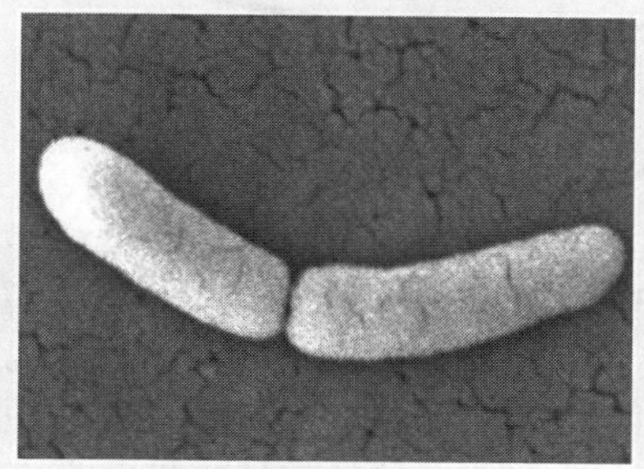

ATP is required when bacteria divide by binary fission (left). For example, ATP is required in DNA replication and to synthesise components of the peptidoglycan cell wall.

Maintaining a stable body temperature requires energy. Muscular activity (shivering, movement) helps to maintain body heat. Cooling requires energy expenditure too. For example, sweating is an energy requiring process involving secretion from glands in the skin.

1. Why do organisms need to respire? ______________________________

____

2. (a) Describe the general role of mitochondria in cell respiration: ______________________________

____

(b) Explain the importance of compartmentalisation in the mitochondrion: ______________________________

____

3. Explain why thermoregulation is associated with energy expenditure: ______________________________

____

____

ISBN: 978-1-927309-20-9

15 ATP and Energy

Key Idea: ATP is the universal energy carrier in cells. Energy is stored in the covalent bonds between phosphate groups.

The molecule ATP (adenosine triphosphate) is the universal energy carrier for the cell. ATP can release its energy quickly by hydrolysis of the terminal phosphate. This reaction is catalysed by the enzyme ATPase. Once ATP has released its energy, it becomes ADP (adenosine diphosphate), a low energy molecule that can be recharged by adding a phosphate. The energy to do this is supplied by the controlled breakdown of respiratory substrates in cellular respiration.

How does ATP provide energy?

ATP releases its energy during hydrolysis. Water is split and added to the terminal phosphate group resulting in ADP and Pi. For every mole of ATP hydrolysed **30.7 kJ** of energy is released. Note that energy is released during the formation of chemical bonds not from the breaking of chemical bonds.

The reaction of A + B is endergonic. It requires energy to proceed and will not occur spontaneously.

A B

Adenosine P P P

Mitochondrion

ATP is reformed during the reactions of **cellular respiration** (i.e. glycolysis, Krebs cycle, and the electron transport chain).

Hydrolysis is the addition of water. ATP hydrolysis gives ADP + Pi (HPO_4^{2-}) + H^+.

ATPase

The enzyme **ATPase** is able to couple the hydrolysis of ATP directly to the formation of a phosphorylated intermediate A-Pi

Adenosine P P + A Pi

A-Pi is more reactive than A. It is now able to react with B.

Inorganic phosphate → Pi + A B

A-Pi reacts with B and Pi is released.

In reality these reactions occur virtually simultaneously.

Note! The phosphate bonds in ATP are often referred to as high energy bonds. This can be misleading. The bonds contain *electrons in a high energy state* (making the bonds themselves relatively weak). A small amount of energy is required to break the bonds, but when the intermediates recombine and form new chemical bonds a large amount of energy is released. The final product is less reactive than the original reactants.

In many textbooks the reaction series above is simplified and the intermediates are left out:

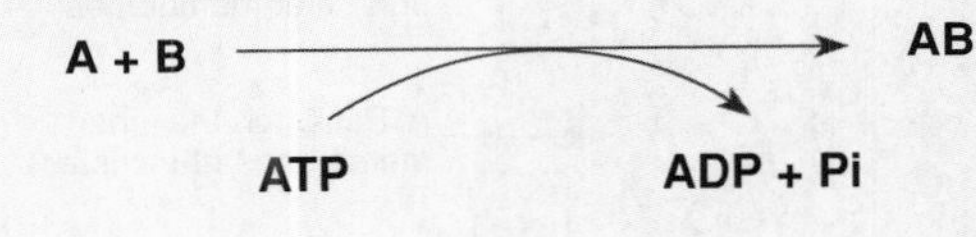

1. (a) How does ATP supply energy to power metabolism? ______

   (b) In what way is the ADP/ATP system like a rechargeable battery? ______

2. What is the immediate source of energy for reforming ATP from ADP? ______

3. Explain the purpose of the folded inner membrane in mitochondria: ______

4. Explain why highly active cells (e.g. sperm cells) have large numbers of mitochondria: ______

ISBN:978-1-927309-20-9

16 ATP Production in Cells

Key Idea: Cellular respiration is the process by which the energy in glucose is transferred to ATP.

Cellular respiration can be **aerobic** (requires oxygen) or **anaerobic** (does not require oxygen). Some plants and animals can generate ATP anaerobically for short periods of time. Other organisms (anaerobic bacteria) use only anaerobic respiration and live in oxygen-free environments. Cellular respiration occurs in the cytoplasm and mitochondria.

An overview of ATP production in cells

Respiration involves three metabolic stages (plus a link reaction) summarised below. The first two stages are the catabolic pathways that decompose glucose and other organic fuels. In the third stage, the electron transport chain accepts electrons from the first two stages and passes these from one electron acceptor to another. The energy released at each stepwise transfer is used to make ATP. The final electron acceptor in this process is molecular oxygen.

1. **Glycolysis**. In the cytoplasm, glucose is broken down into two molecules of pyruvate.
2. **The link reaction**. In the mitochondrial matrix, pyruvate is split and added to coenzyme A.
3. **Krebs cycle**. In the mitochondrial matrix, a derivative of pyruvate is decomposed to CO_2.
4. **Electron transport and oxidative phosphorylation**. This occurs in the inner membranes of the mitochondrion and accounts for almost 90% of the ATP generated by respiration.

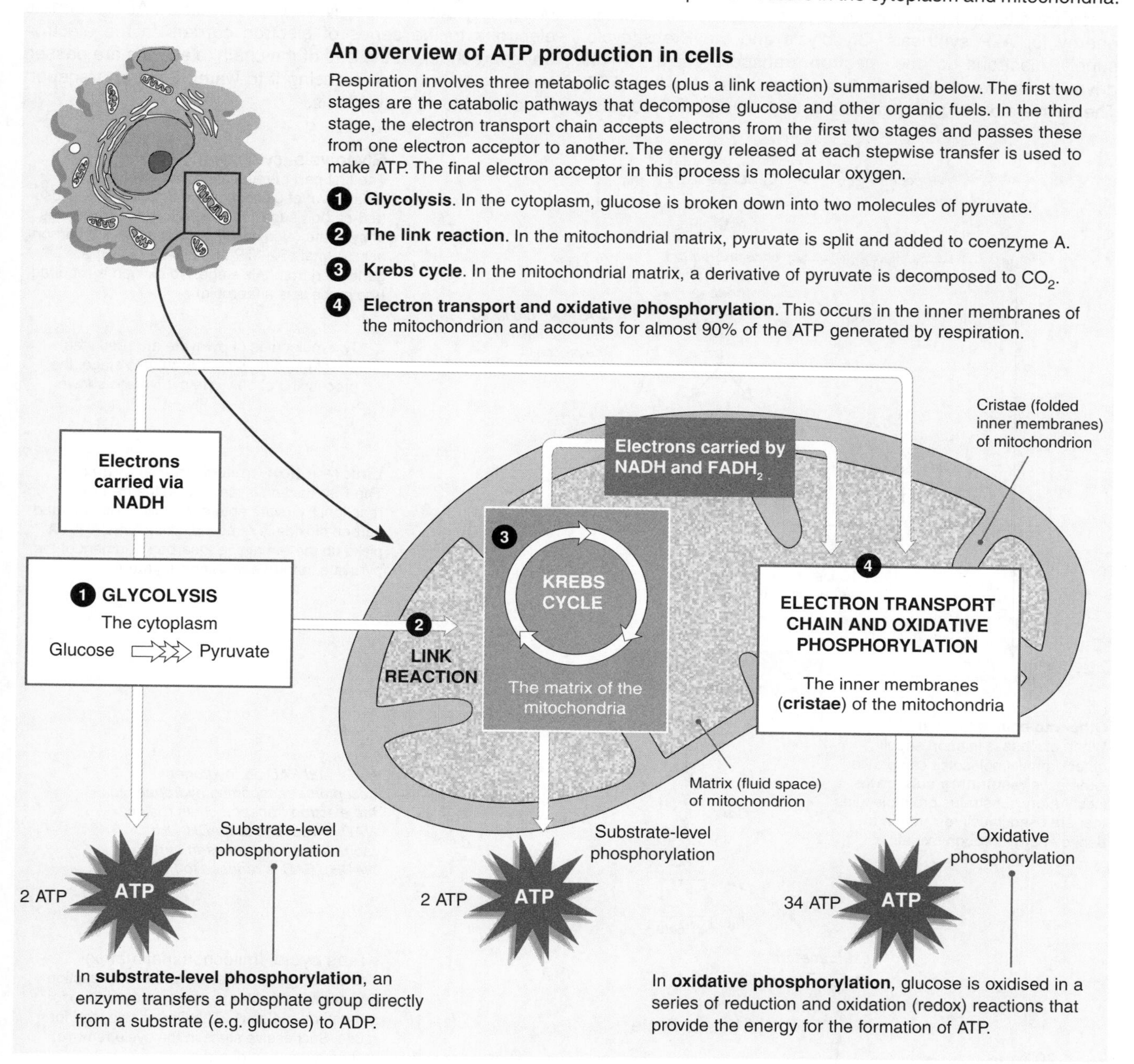

1. Describe precisely in which part of the cell the following take place:

   (a) Glycolysis: ____________________

   (b) The link reaction: ____________________

   (c) Krebs cycle reactions: ____________________

   (d) Electron transport chain: ____________________

2. How does ATP generation in glycolysis and the Krebs cycle differ from ATP generation via the electron transport chain?

____

____

____

ISBN: 978-1-927309-20-9

17 The Biochemistry of Respiration

Key Idea: During cellular respiration, the energy in glucose is transferred to ATP in a series of enzyme controlled steps. The oxidation of glucose is a catabolic, energy yielding pathway. The breakdown of glucose and other organic fuels (such as fats and proteins) to simpler molecules releases energy for ATP synthesis. Glycolysis and the Krebs cycle supply electrons to the electron transport chain, which drives oxidative phosphorylation. Glycolysis nets two ATP. The conversion of pyruvate (the end product of glycolysis) to acetyl CoA links glycolysis to the Krebs cycle. One "turn" of the cycle releases carbon dioxide, forms one ATP, and passes electrons to three NAD^+ and one FAD. Most of the ATP generated in cellular respiration is produced by oxidative phosphorylation when NADH + H^+ and $FADH_2$ donate electrons to the series of electron carriers in the electron transport chain. At the end of the chain, electrons are passed to molecular oxygen, reducing it to water. Electron transport is coupled to ATP synthesis.

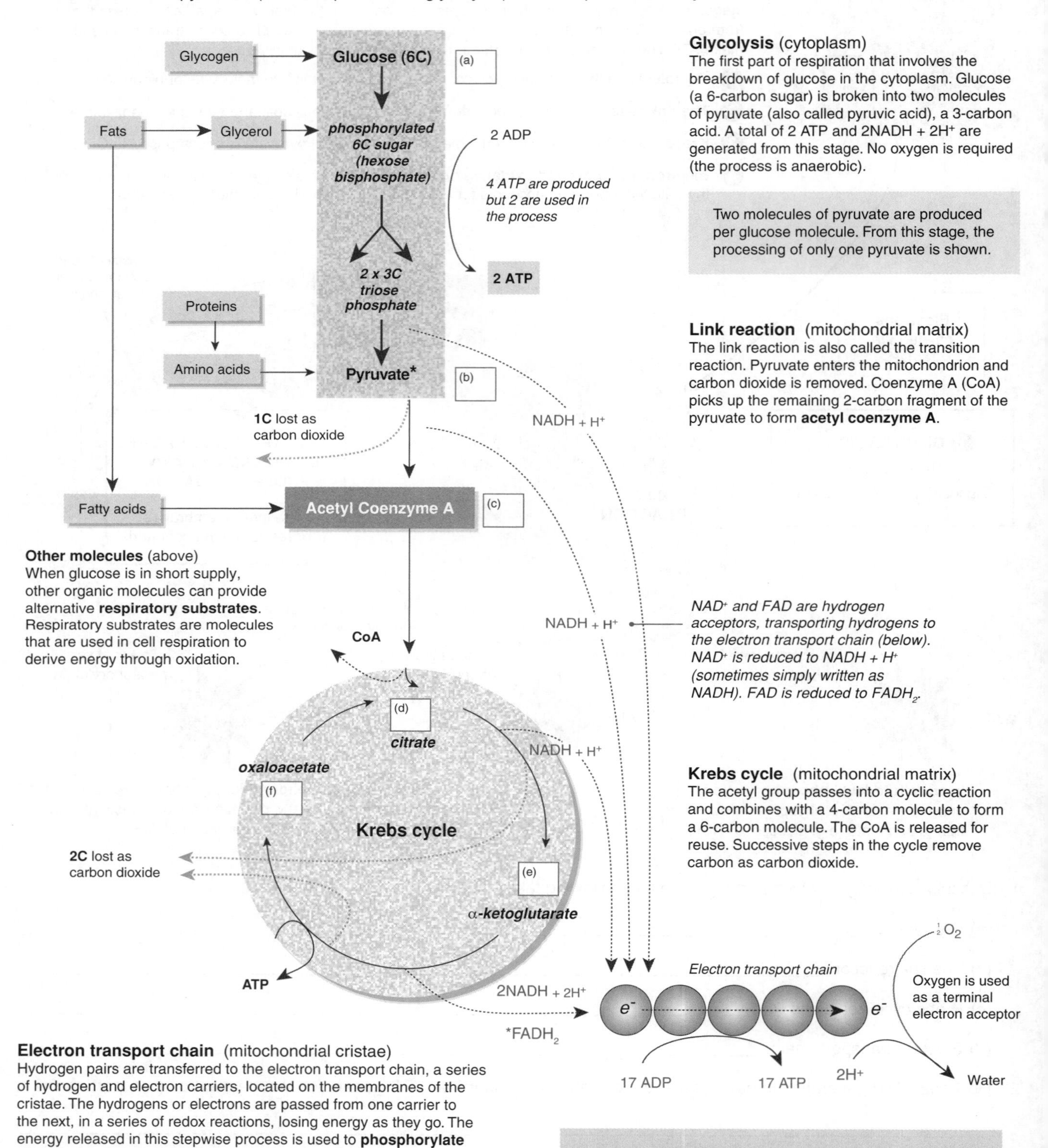

Glycolysis (cytoplasm)
The first part of respiration that involves the breakdown of glucose in the cytoplasm. Glucose (a 6-carbon sugar) is broken into two molecules of pyruvate (also called pyruvic acid), a 3-carbon acid. A total of 2 ATP and 2NADH + $2H^+$ are generated from this stage. No oxygen is required (the process is anaerobic).

Two molecules of pyruvate are produced per glucose molecule. From this stage, the processing of only one pyruvate is shown.

Link reaction (mitochondrial matrix)
The link reaction is also called the transition reaction. Pyruvate enters the mitochondrion and carbon dioxide is removed. Coenzyme A (CoA) picks up the remaining 2-carbon fragment of the pyruvate to form **acetyl coenzyme A**.

Other molecules (above)
When glucose is in short supply, other organic molecules can provide alternative **respiratory substrates**. Respiratory substrates are molecules that are used in cell respiration to derive energy through oxidation.

NAD^+ and FAD are hydrogen acceptors, transporting hydrogens to the electron transport chain (below). NAD^+ is reduced to NADH + H^+ (sometimes simply written as NADH). FAD is reduced to $FADH_2$.

Krebs cycle (mitochondrial matrix)
The acetyl group passes into a cyclic reaction and combines with a 4-carbon molecule to form a 6-carbon molecule. The CoA is released for reuse. Successive steps in the cycle remove carbon as carbon dioxide.

Electron transport chain (mitochondrial cristae)
Hydrogen pairs are transferred to the electron transport chain, a series of hydrogen and electron carriers, located on the membranes of the cristae. The hydrogens or electrons are passed from one carrier to the next, in a series of redox reactions, losing energy as they go. The energy released in this stepwise process is used to **phosphorylate** ADP to form ATP. Oxygen is the final electron acceptor and is reduced to water (hence the term **oxidative phosphorylation**).
Note FAD enters the electron transport chain at a lower energy level than NAD, and only 2ATP are generated per $FADH_2$.

Total ATP yield per glucose
Glycolysis: 2 ATP, *Krebs cycle*: 2 ATP, *Electron transport*: 34 ATP

ISBN:978-1-927309-20-9

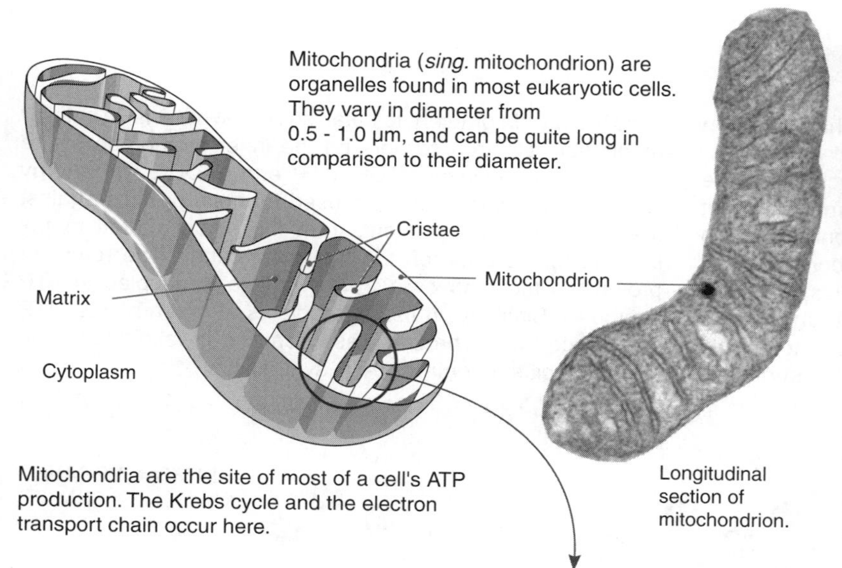

Mitochondria (*sing.* mitochondrion) are organelles found in most eukaryotic cells. They vary in diameter from 0.5 - 1.0 µm, and can be quite long in comparison to their diameter.

Mitochondria are the site of most of a cell's ATP production. The Krebs cycle and the electron transport chain occur here.

Longitudinal section of mitochondrion.

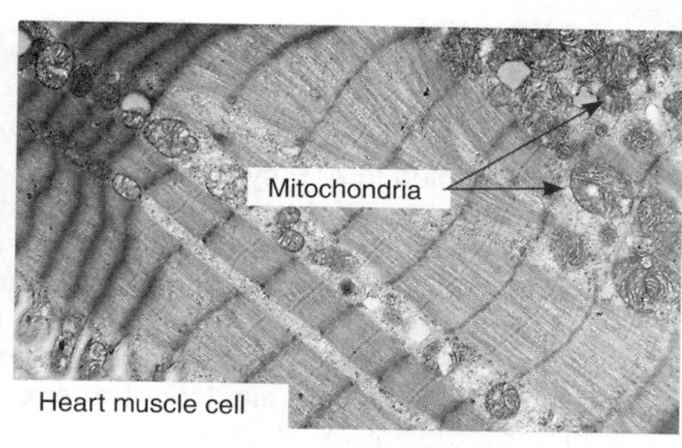

Cells that require a lot of ATP for cellular processes have a lot of mitochondria. Sperm cells contain a large number of mitochondria near the base of the tail. Liver cells have around 2000 mitochondria per cell, taking up 25% of the cytoplasmic space. Heart muscle cells (above) may have 40% of the cytoplasmic space taken up by mitochondria.

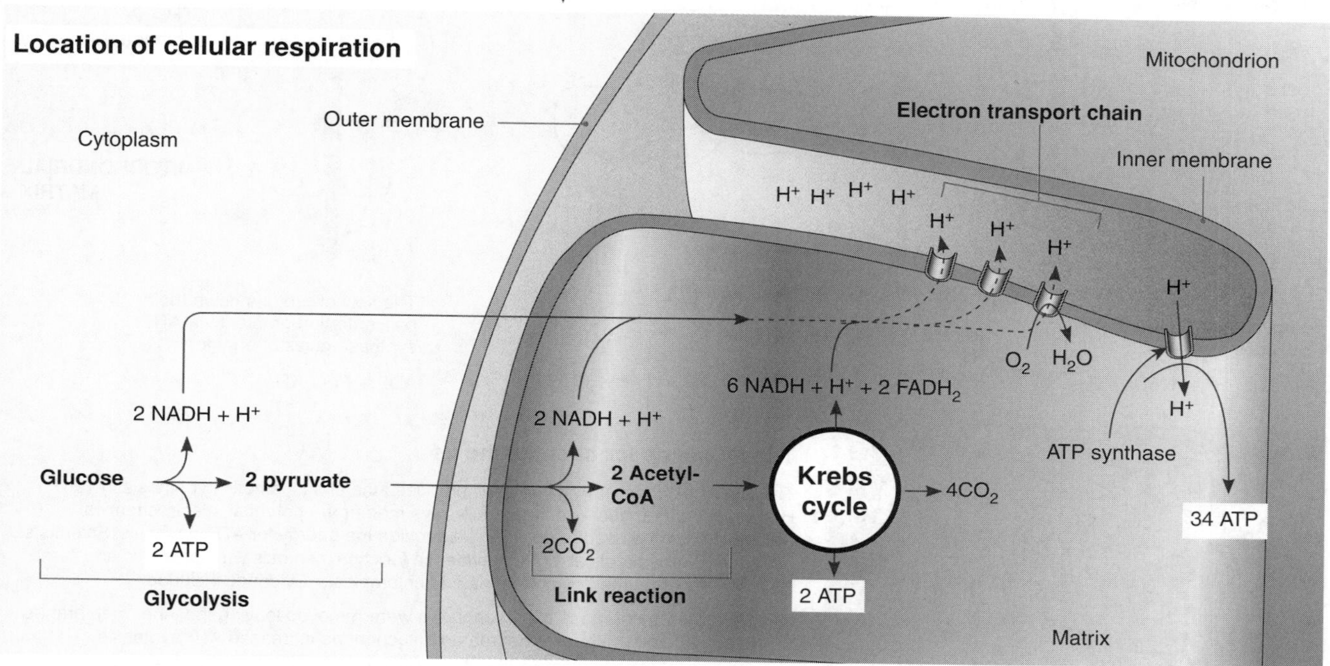

1. In the longitudinal section of a mitochondrion (above), label the matrix and cristae.
2. Explain the purpose of the link reaction: ______________________________
3. On the diagram of cell respiration (previous page), state the number of carbon atoms in each of the molecules (a)-(f):
4. How many ATP molecules **per molecule of glucose** are generated during the following stages of respiration?

   (a) Glycolysis: __________ (b) Krebs cycle: __________ (c) Electron transport chain: __________ (d) Total: __________
5. Explain what happens to the carbon atoms lost during respiration: ______________________________
6. Explain what happens during oxidative phosphorylation: ______________________________

ISBN: 978-1-927309-20-9

18 Chemiosmosis

Key Idea: Chemiosmosis is the process in which electron transport is coupled to ATP synthesis.

Chemiosmosis occurs in the membranes of mitochondria, the chloroplasts of plants, and across the plasma membrane of bacteria. Chemiosmosis involves the establishment of a proton (hydrogen) gradient across a membrane. The concentration gradient is used to drive ATP synthesis. Chemiosmosis has two key components: an **electron transport chain** sets up a proton gradient as electrons pass along it to a final electron acceptor, and an enzyme called **ATP synthase** uses the proton gradient to catalyse ATP synthesis. In cellular respiration, electron carriers on the inner membrane of the mitochondrion oxidise NADH + H^+ and $FADH_2$. The energy released from this process is used to move protons against their concentration gradient, from the mitochondrial matrix into the space between the two membranes. The return of protons to the matrix via ATP synthase is coupled to ATP synthesis. Similarly, in the chloroplasts of green plants, ATP is produced when protons pass from the thylakoid lumen to the chloroplast stroma via ATP synthase.

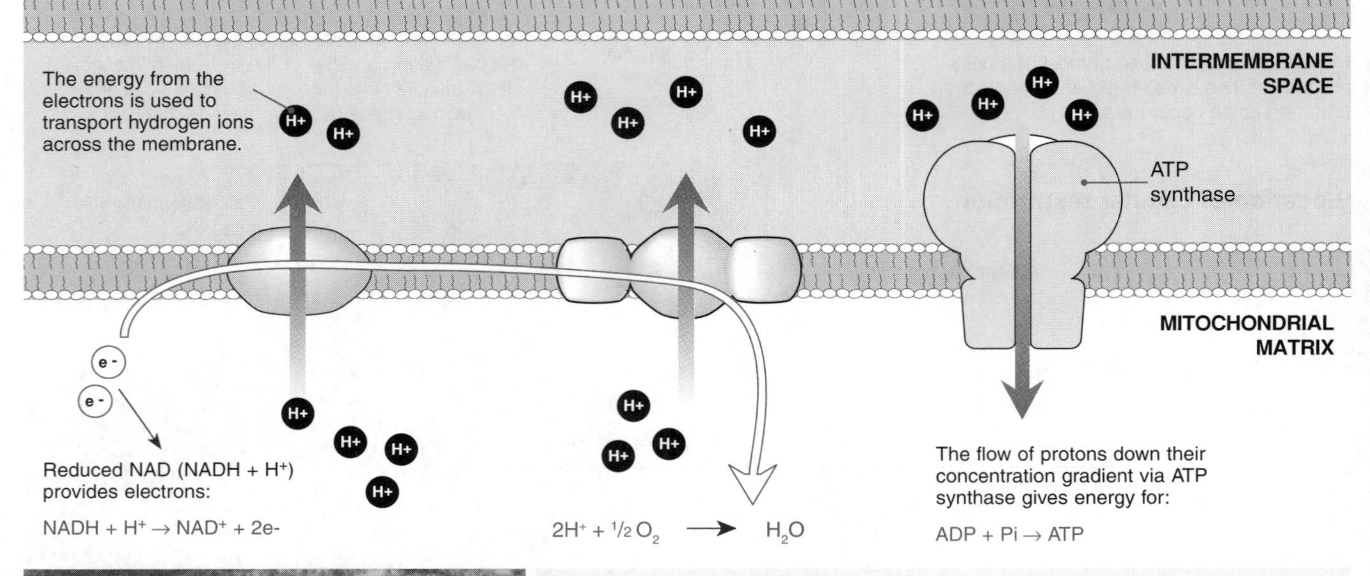

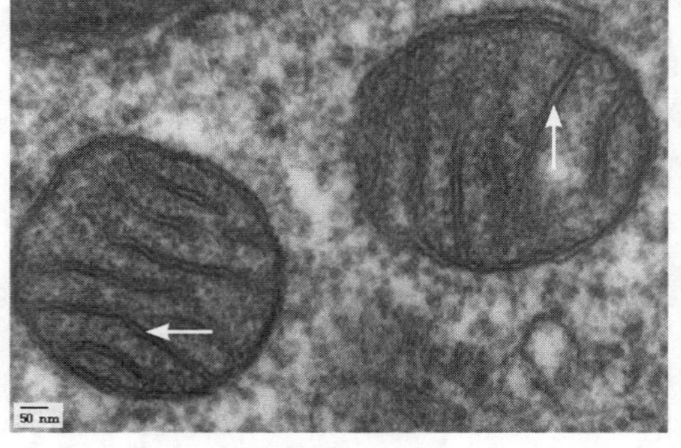

The intermembrane spaces can be seen (arrows) in this transverse section of miotchondria.

The evidence for chemiosmosis

The British biochemist Peter Mitchell proposed the chemiosmotic hypothesis in 1961. He proposed that, because living cells have membrane potential, electrochemical gradients could be used to do work, i.e. provide the energy for ATP synthesis. Scientists at the time were skeptical, but the evidence for chemiosmosis was extensive and came from studies of isolated mitochondria and chloroplasts. Evidence included:

- The outer membranes of mitochondria were removed leaving the inner membranes intact. Adding protons to the treated mitochondria increased ATP synthesis.
- When isolated chloroplasts were illuminated, the medium in which they were suspended became alkaline.
- Isolated chloroplasts were kept in the dark and transferred first to a low pH medium (to acidify the thylakoid interior) and then to an alkaline medium (low protons). They then spontaneously synthesised ATP (no light was needed).

1. Summarise the process of chemiosmosis: ____________________

2. Why did the addition of protons to the treated mitochondria increase ATP synthesis? ____________________

3. Why did the suspension of isolated chloroplasts become alkaline when illuminated? ____________________

4. (a) What was the purpose of transferring the chloroplasts first to an acid then to an alkaline medium? ____________________

   (b) Why did ATP synthesis occur spontaneously in these treated chloroplasts? ____________________

ISBN:978-1-927309-20-9

19 Anaerobic Pathways

Key Idea: Glucose can be metabolised aerobically and anaerobically to produce ATP. The ATP yield from aerobic processes is higher than from anaerobic processes.
Aerobic respiration occurs in the presence of oxygen. Organisms can also generate ATP anaerobically (without oxygen) by using a molecule other than oxygen as the terminal electron acceptor for the pathway. In alcoholic fermentation, the electron acceptor is ethanal. In lactic acid fermentation, which occurs in mammalian muscle even when oxygen is present, the electron acceptor is pyruvate itself.

Alcoholic fermentation

In alcoholic fermentation, the H^+ acceptor is ethanal which is reduced to ethanol with the release of carbon dioxide (CO_2). Yeasts respire aerobically when oxygen is available but can use alcoholic fermentation when it is not. At ethanol levels above 12-15%, the ethanol produced by alcoholic fermentation is toxic and this limits their ability to use this pathway indefinitely. The root cells of plants also use fermentation as a pathway when oxygen is unavailable but the ethanol must be converted back to respiratory intermediates and respired aerobically.

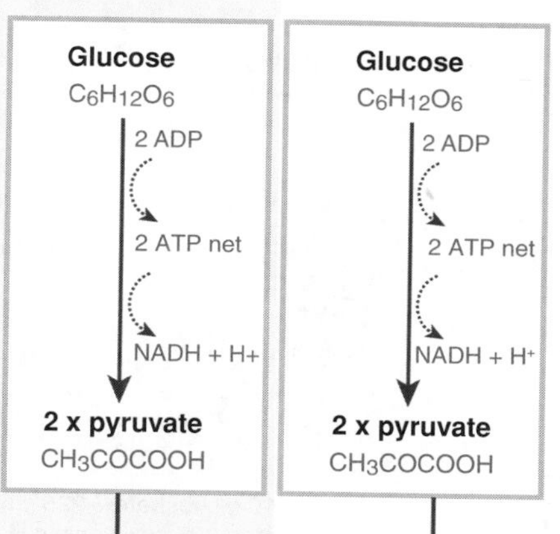

Lactic acid fermentation

Skeletal muscles produce ATP in the absence of oxygen using lactic acid fermentation. In this pathway, pyruvate is reduced to lactic acid, which dissociates to form lactate and H^+. The conversion of pyruvate to lactate is reversible and this pathway operates alongside the aerobic system all the time to enable greater intensity and duration of activity. Lactate can be metabolised in the muscle itself or it can enter the circulation and be taken up by the liver to replenish carbohydrate stores. This 'lactate shuttle' is an important mechanism for balancing the distribution of substrates and waste products.

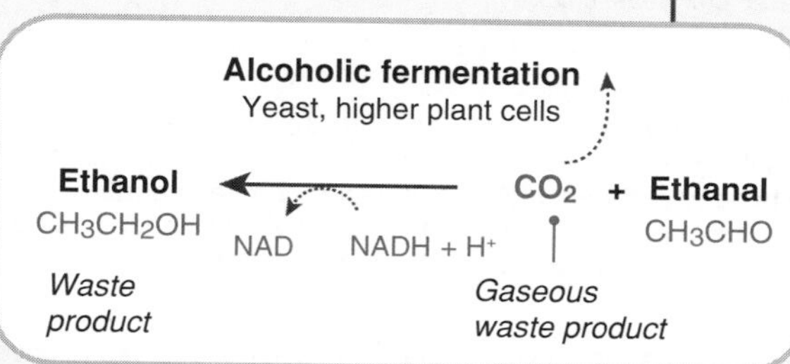

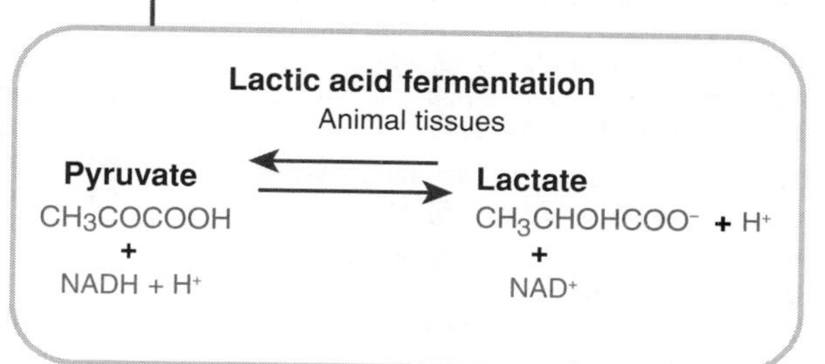

The alcohol and CO_2 produced from alcoholic fermentation form the basis of the brewing and baking industries. In baking, the dough is left to ferment and the yeast metabolises sugars to produce ethanol and CO_2. The CO_2 causes the dough to rise.

Yeasts are used to produce almost all alcoholic beverages (e.g. wine and beers). The yeast used in the process breaks down the sugars into ethanol (alcohol) and CO_2. The alcohol produced is a metabolic by-product of fermentation by the yeast.

Wintec

The lactate shuttle in vertebrate skeletal muscle works alongside the aerobic system to enable maximal muscle activity. Lactate moves from its site of production to regions within and outside the muscle (e.g. liver) where it can be respired aerobically.

1. Describe the key difference between aerobic respiration and fermentation: ______________________

____

2. (a) Refer to page 20 and determine the efficiency of fermentation compared to aerobic respiration: __________ %

   (b) Why is the efficiency of these anaerobic pathways so low? ______________________

   ____

3. Why can't alcoholic fermentation go on indefinitely? ______________________

____

____

ISBN: 978-1-927309-20-9

20 Investigating Fermentation in Yeast

Key Idea: Brewer's yeast preferentially uses alcoholic fermentation when there is excess sugar, releasing CO_2, which can be collected as a measure of fermentation rate.

Brewer's yeast is a facultative anaerobe (meaning it can respire aerobically or use fermentation). It will preferentially use alcoholic fermentation when sugars are in excess. One would expect glucose to be the preferred substrate, as it is the starting molecule in cellular respiration, but brewer's yeast is capable of utilising a variety of sugars, including disaccharides, which can be broken down into single units.

The aim

To investigate the suitability of different mono- and disaccharide sugars as substrates for alcoholic fermentation in yeast.

The hypothesis

If glucose is the preferred substrate for fermentation in yeast, then the rate of fermentation will be highest when the yeast is grown on glucose rather than on other sugars.

Background

The rate at which brewer's or baker's yeast (*Saccharomyces cerevisiae*) metabolises carbohydrate substrates is influenced by factors such as temperature, solution pH, and type of carbohydrate available.

The literature describes yeast metabolism as optimal in warm, acid (pH 4-6) environments.

High levels of sugars suppress aerobic respiration in yeast, so yeast will preferentially use the fermentation pathway in the presence of excess substrate.

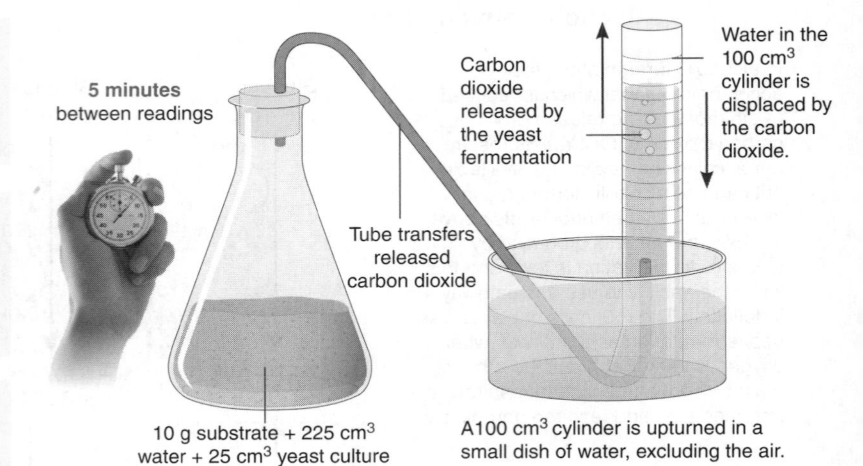

The apparatus

In this experiment, all substrates tested used the same source culture of 30 g active yeast dissolved in 150 cm^3 of room temperature (24°C) tap water. 25 g of each substrate to be tested was added to 225 cm^3 room temperature (24°C) tap water buffered to pH 4.5. Then 25 cm^3 of source culture was added to the test solution. The control contained yeast solution but no substrate.

The substrates

Glucose is a monosaccharide, maltose (glucose-glucose), sucrose (glucose-fructose), and lactose (glucose-galactose) are disaccharides.

Substrate / Time / min	Volume of carbon dioxide collected / cm^3				
	None	Glucose	Maltose	Sucrose	Lactose
0	0	0	0	0	0
5	0	0	0.8	0	0
10	0	0	0.8	0	0
15	0	0	0.8	0.1	0
20	0	0.5	2.0	0.8	0
25	0	1.2	3.0	1.8	0
30	0	2.8	3.6	3.0	0
35	0	4.2	5.4	4.8	0
40	0	4.6	5.6	4.8	0
45	0	7.4	8.0	7.2	0
50	0	10.8	8.9	7.6	0
55	0	13.6	9.6	7.7	0
60	0	16.1	10.4	9.6	0
65	0	22.0	12.1	10.2	0
70	0	23.8	14.4	12.0	0
75	0	26.7	15.2	12.6	0
80	0	32.5	17.3	14.3	0
85	0	37.0	18.7	14.9	0
90	0	39.9	21.6	17.2	0

Experimental design and results adapted from Tom Schuster, Rosalie Van Zyl, & Harold Coller, California State University Northridge 2005

1. Write the equation for the fermentation of glucose by yeast:

 ____

2. The results are presented on the table left. Using the final values, calculate the rate of CO_2 production per minute for each substrate:

 (a) None: ______________________

 (b) Glucose: ______________________

 (c) Maltose: ______________________

 (d) Sucrose: ______________________

 (e) Lactose: ______________________

3. What assumptions are being made in this experimental design and do you think they were reasonable?

 ____

LINK 19
DATA

ISBN:978-1-927309-20-9

4. Use the tabulated data to plot an appropriate graph of the results on the grid provided:

5. (a) Identify the independent variable: ______

   (b) State the range of values for the independent variable: ______

   (c) Name the unit for the independent variable: ______

6. (a) Identify the dependent variable: ______

   (b) Name the unit for the dependent variable: ______

7. (a) Summarise the results of the fermentation experiment: ______

   (b) Why do you think CO_2 production was highest when glucose was the substrate? ______

   (c) Suggest why fermentation rates were lower on maltose and sucrose than on glucose:

   (d) Suggest why there may have been no fermentation on the lactose substrate: ______

8. Predict what would happen to CO_2 production rates if the yeast cells were respiring aerobically: ______

ISBN: 978-1-927309-20-9

21 Investigating Aerobic Respiration in Yeast

Key Idea: Respiration rate can be investigated using the redox indicator triphenyl tetrazolium chloride (TTC).

In simple organisms such as yeast, respiration rate can be affected by different factors, including temperature and availability of respiratory substrates. Yeast is a facultative anaerobe and will respire aerobically when glucose levels are low and oxygen is available. The effect of temperature on yeast respiration rate can be determined using the redox indicator triphenyl tetrazolium chloride (TTC). TTC is a colourless hydrogen acceptor and intercepts the hydrogen ions produced during respiration. It turns red when oxidised and the rate of colour change indicates the rate of respiration.

Background

During respiration, dehydrogenase enzymes remove hydrogens from glucose and pass them to hydrogen acceptors. TTC intercepts these hydrogens and turns red. The rate of colour change indicates the rate of enzyme activity and thus the rate of respiration.

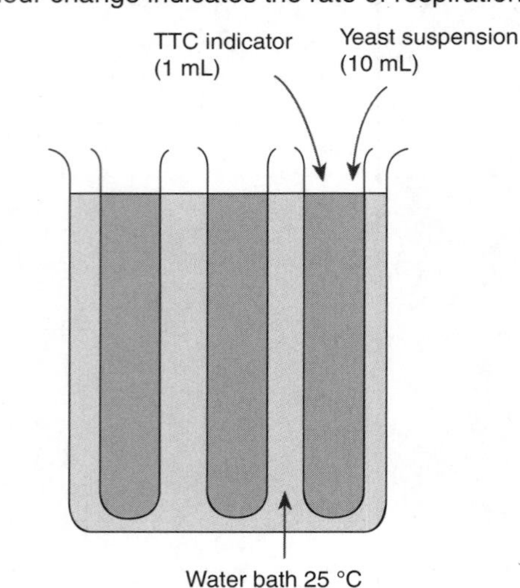

Aim and hypothesis

To investigate the effect of temperature on the rate of aerobic respiration in yeast.
If respiration is occurring, the rate can be indicated by the amount of colour change in the indicator TTC.

The method

A set of three test tubes (tubes 1-3) were set up containing 10 mL of dilute yeast suspension (10 g yeast, 1.5 g glucose per litre) and 1 mL of TTC. At this concentration of glucose and in the presence of oxygen, yeast will respire aerobically.

The test tubes were placed into a water bath at 25 °C. Two more sets of three test tubes were prepared in the same way. The second set was placed into a water bath at 40 °C (tubes 4-6), and the third set was placed into a water bath at 55°C (tubes 7-9). The rate of colour change was measured over 4.5 hours by measuring the absorbance of each tube with a colorimeter (0 being clear or low absorbance and 10 being fully opaque or high absorbance). A control tube containing only yeast and glucose was included in the experiment (tube 10).

Caution needs to be taken with TTC use as it can cause skin irritation. Gloves should be used.

	Tube number/ absorbance units									
	25 °C			40 °C			55°C			Control
Time / hr	1	2	3	4	5	6	7	8	9	10
1.5	1.13	1.02	1.20	2.34	2.33	2.29	4.11	4.05	4.17	0.40
3.0	1.96	1.88	2.04	5.85	5.89	5.80	8.86	8.90	8.82	0.51
4.5	2.76	2.69	2.81	7.84	7.88	7.80	9.77	9. 87	9.74	0.62

1. Calculate the mean absorbance for each of the times and temperatures above and enter the values in the table below:

	Mean absorbance		
Time / hr	25 °C	40 °C	55 °C
1.5			
3.0			
4.5			

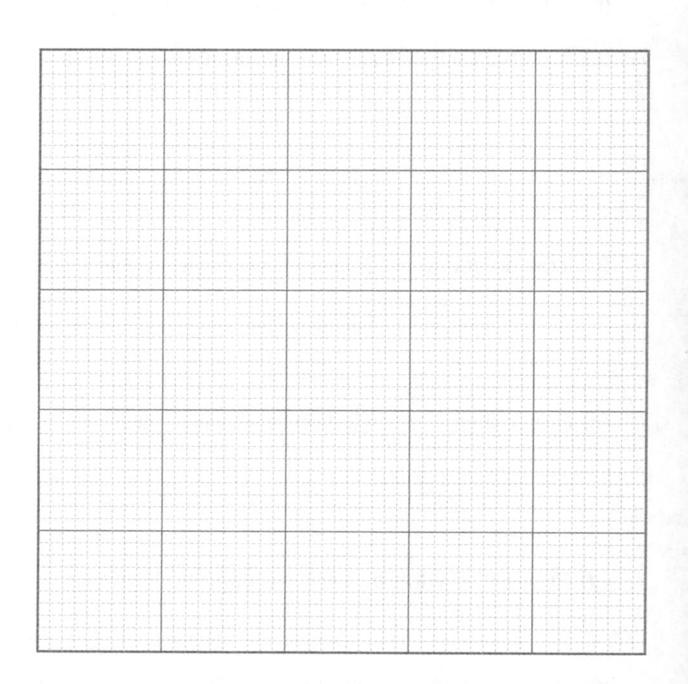

2. Use the table in (1) to plot a graph of absorbance over time for the three temperatures measured:

3. Why did the absorbance of the control tube change?

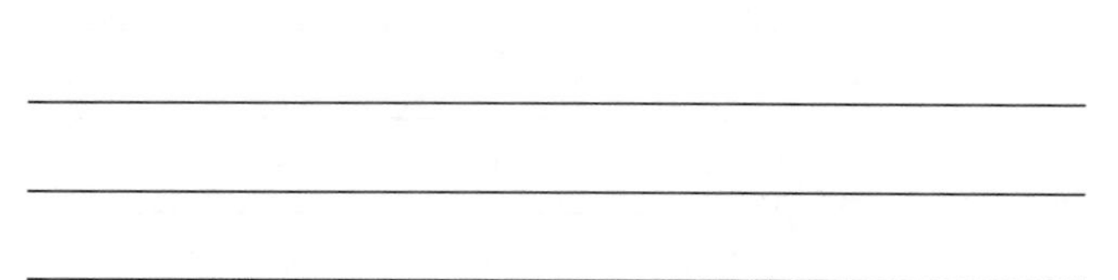

4. How does temperature affect the respiration rate of yeast?

ISBN:978-1-927309-20-9

22 Energy in Ecosystems: Food Chains

Key Idea: A food chain is a model to illustrate the feeding relationships between organisms.

Organisms in ecosystems interact by way of their feeding (trophic) relationships. These interactions can be shown in a **food chain**, which is a simple model to illustrate how energy or biomass, in the form of food, passes from one organism to the next. Each organism in the chain is a source of energy for the next. The levels of a food chain are called **trophic levels**. An organism is assigned to a trophic level based on its position in the food chain. Organisms may occupy different trophic levels in different food chains or during different stages of their life. Arrows link the organisms in a food chain and their direction shows the flow of energy and biomass through the trophic levels. Most food chains begin with a producer, which is eaten by a primary consumer (**herbivore**). Higher level consumers (e.g. **carnivores**) eat other consumers.

Producers (**autotrophs**) e.g. plants, algae, and autotrophic bacteria, make their own food from simple inorganic substances, often by photosynthesis using energy from the sun. Inorganic nutrients are obtained from the abiotic environment, such as the soil and atmosphere.

Consumers (heterotrophs) e.g. animals, get their energy by eating other organisms. Consumers are ranked according to the trophic level they occupy, i.e. 1st order, 2nd order, and classified according to diet (e.g. carnivores eat animal tissue, omnivores eat plant and animal tissue).

Detritivores and **saprotrophs** both gain nutrients from digesting detritus (dead organic matter). Detritivores ingest (eat) and digest detritus inside their bodies. Saprotrophs break it down using enzymes, which are secreted and work externally to their bodies. Nutrients are then absorbed by the organism.

1. (a) Draw arrows on the diagram below to show how the energy flows through the organisms in the food chain. Label each arrow with the process involved in the energy transfer. Draw arrows to show how energy is lost by respiration.

   (b) What is the original energy source for this food chain? ____________________

   (c) How is this energy source converted to biomass? ____________________

2. Energy flows through food chains. In what form is it transferred between trophic levels: ____________________

   ____

3. Describe how the following obtain energy, and give an example of each:

   (a) Producers: ____________________

   (b) Consumers: ____________________

   (c) Detritivores: ____________________

   (d) Saprotrophs: ____________________

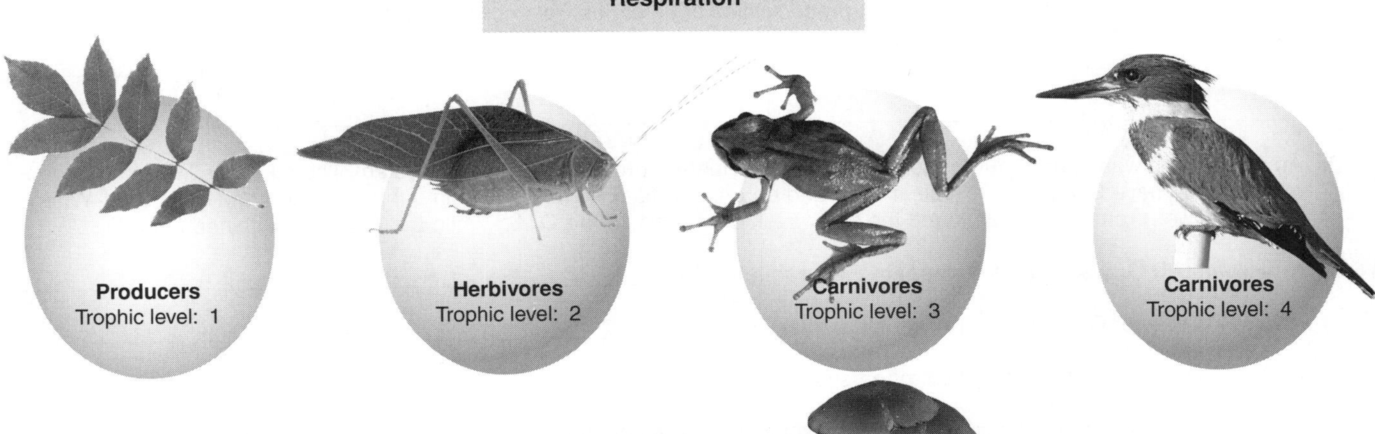

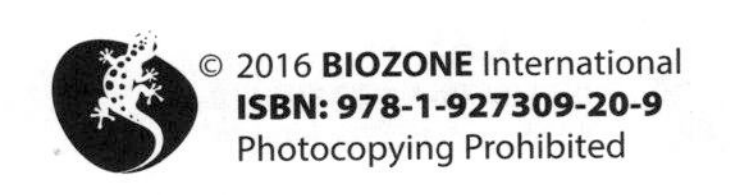

ISBN: 978-1-927309-20-9

23 What is Primary Productivity?

Key Idea: Measuring the amount of photosynthesis and the amount of respiration per unit volume per unit time enables us to determine the gross primary productivity of a system.

The energy entering an ecosystem is determined by the rate at which producers can convert sunlight energy or inorganic compounds into chemical energy. Photosynthesis accounts for most of the energy entering most of Earth's ecosystems.

The total energy fixed by photosynthesis per unit area or volume per unit time is the **gross primary productivity** (GPP) and it is usually expressed as J m^{-2} (or kJ m^{-2}), or as g m^{-2}. However, some of this energy is required for respiration. Subtracting respiration from GPP gives the **net primary productivity** (NPP). This represents the energy or biomass (mass of biological material) available to consumers.

The productivity of ecosystems

The **gross primary productivity** of an ecosystem will depend on the capacity of the producers to capture and fix carbon in organic compounds. In most ecosystems, this is limited by constraints on photosynthesis (availability of light, nutrients, or water for example). The **net primary productivity** (NPP) is then determined by how much of this goes into plant biomass per unit time, after respiratory needs are met. This will be the amount available to the next trophic level.

Production vs productivity: What's the difference?

Strictly speaking, the primary production of an ecosystem is distinct from its productivity, which is a rate. However because values for production (accumulated biomass) are usually given for a certain period of time in order to be meaningful, the two terms are often used interchangeably.

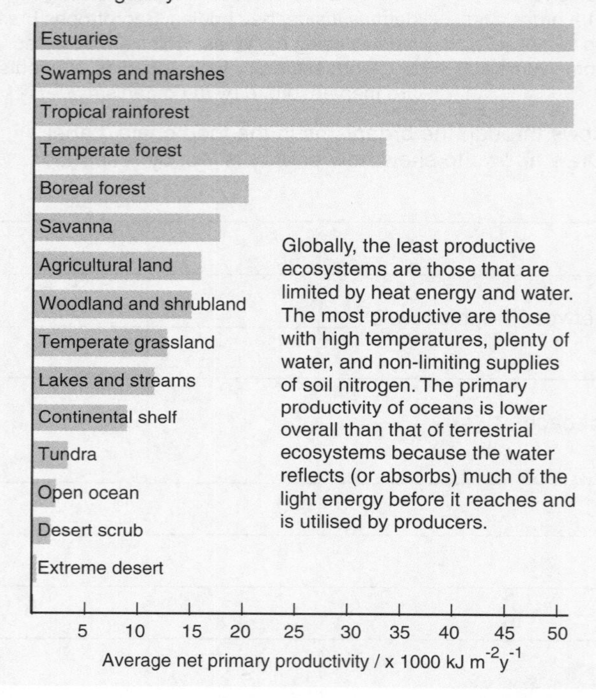

Globally, the least productive ecosystems are those that are limited by heat energy and water. The most productive are those with high temperatures, plenty of water, and non-limiting supplies of soil nitrogen. The primary productivity of oceans is lower overall than that of terrestrial ecosystems because the water reflects (or absorbs) much of the light energy before it reaches and is utilised by producers.

Measuring productivity

Measuring gross primary productivity (GPP) can be difficult due to the effect of on-going respiration, which uses up some of the organic material produced (glucose). One method for measuring GPP is to measure the difference in production between plants kept in the dark and those in the light.

A simple method for measuring GPP in phytoplankton is illustrated below:

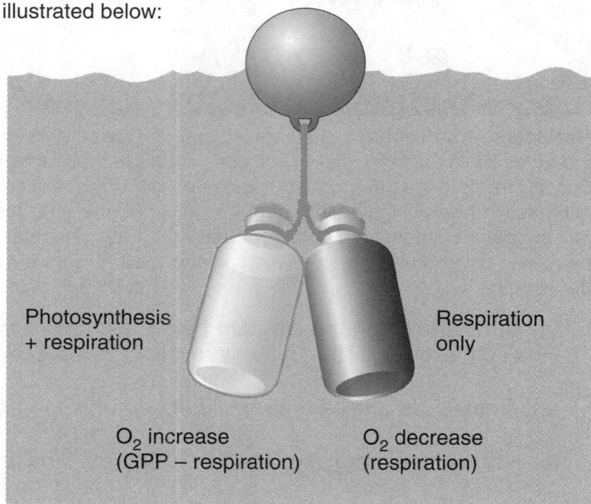

Two bottles are lowered into the ocean or lake to a specified depth, filled with water, and then stoppered. One bottle is transparent, the other is opaque. The O_2 concentration of the water surrounding the bottles is measured and the bottles are left for a specified amount of time. The phytoplankton in the transparent bottle will photosynthesise, increasing the O_2 concentration, and respire, using some of that O_2. The phytoplankton in the opaque bottle will only respire. The final measured difference in O_2 between the bottles gives the amount of O_2 produced by the phytoplankton in the specified time (including that used for respiration).

The amount of O_2 used allows us to determine the amount of glucose produced and therefore the GPP of the phytoplankton.

1. Suggest possible reasons for the high net productivities of estuaries, swamps, and tropical rainforests:

2. An experiment was carried out to measure the gross primary production of a lake system. The lake was initially measured to have 8 mg O_2 L^{-1}. A clear flask and an opaque flask were lowered into the lake filled and stoppered. When the flasks were retrieved it was found the clear flask contained 10 mg O_2 L^{-1} while the opaque contained 5 mg O_2 L^{-1}.

(a) How much O_2 was used (respired) in the opaque flask?

(b) Was is the net O_2 production in the clear flask?

(c) What is the gross O_2 production in the system?

(d) For every 10 g of O_2 formed during photosynthesis, 9.4 grams of glucose is formed. How much glucose formed during the experiment?

KNOW WEB 23 LINK 24 LINK 25

ISBN:978-1-927309-20-9

24 Productivity and Trophic Efficiency

Key Idea: The net primary productivity of an ecosystem determines the amount of biomass available to primary consumers and varies widely between different ecosystems. The energy entering ecosystems is fixed by producers at a rate that depends on limiting factors such as temperature and the availability of light and water. This energy is converted to biomass by anabolic reactions. The rate of biomass production (net primary productivity), is the biomass produced per area per unit time. Trophic (or ecological) efficiency refers to the efficiency of energy transfer from one trophic level to the next. The trophic efficiencies of herbivores vary widely, depending on how much of the producer biomass is consumed and assimilated (incorporated into new biomass). In some natural ecosystems this can be surprisingly high. Humans intervene in natural energy flows by simplifying systems and reducing the number of transfers occurring between trophic levels.

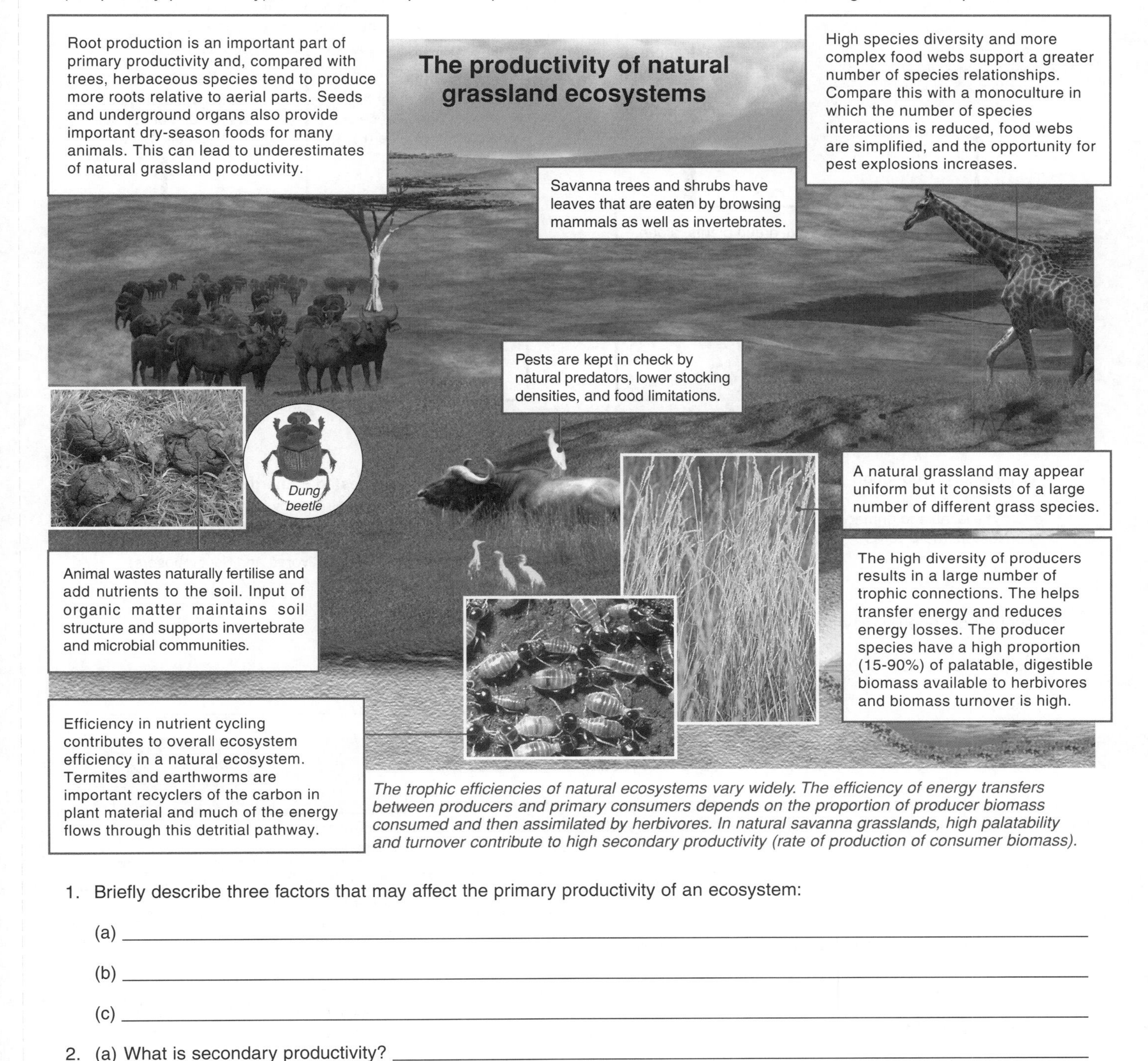

The trophic efficiencies of natural ecosystems vary widely. The efficiency of energy transfers between producers and primary consumers depends on the proportion of producer biomass consumed and then assimilated by herbivores. In natural savanna grasslands, high palatability and turnover contribute to high secondary productivity (rate of production of consumer biomass).

1. Briefly describe three factors that may affect the primary productivity of an ecosystem:

   (a) ______

   (b) ______

   (c) ______

2. (a) What is secondary productivity? ______

   (b) Describe features of the natural tropical grassland (savanna) above that contribute to high secondary productivity:

   ______

   ______

   ______

   ______

ISBN: 978-1-927309-20-9

LINK 28 LINK 25 LINK 23 WEB 24 KNOW

25 Energy Inputs and Outputs

Key Idea: The efficiency of energy transfers in ecosystems can be quantified if we know the amount of energy entering and leaving the different trophic levels.

The GPP of any ecosystem will be determined by the efficiency with which solar energy is captured by photosynthesis. The efficiency of subsequent energy transfers will determine the amount of energy available to consumers. These energy transfers can be quantified using dry mass or calorimetry.

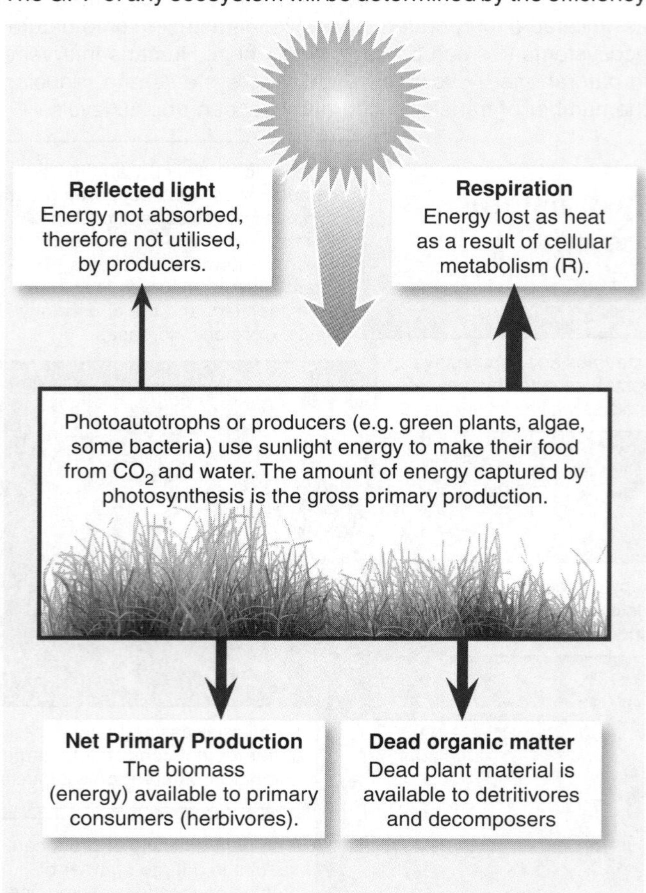

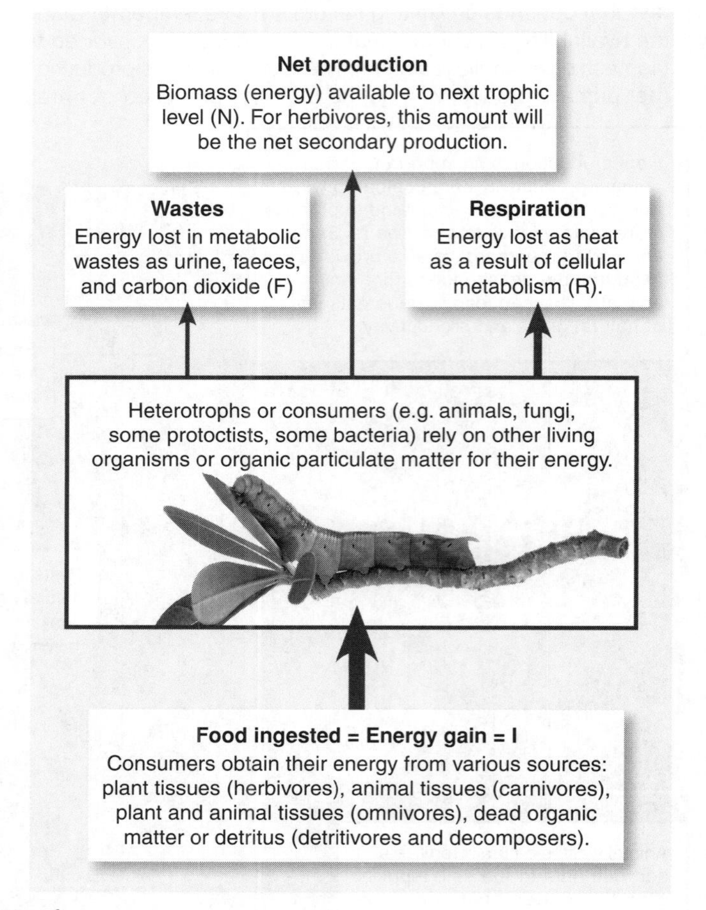

1. Describe how energy may be lost from organisms in the form of:

   (a) Wastes: ______

   (b) Respiration: ______

2. The energy budgets of two agricultural systems (4000 m^2 area) were measured over a growing season of 100 days. The results are tabulated right.

	Corn field	Mature pasture
	kJ x 10^6	kJ x 10^6
Incident solar radiation	8548	1971
Plant utilisation		
Net primary production (NPP)	105.8	20.7
Respiration (R)	32.2	3.7
Gross primary production (GPP)	138.0	24.4

   (a) For each system, calculate the percentage efficiency of energy utilisation (how much incident solar radiation is captured by photosynthesis):

   Corn: ______

   Mature pasture: ______

   (b) For each system, calculate the percentage losses to respiration:

   Corn: ______

   Mature pasture: ______

   (c) For each system, calculate the percentage efficiency of net primary production:

   Corn: ______ Mature pasture: ______

   (d) Which system has the greatest efficiency of energy transfer to biomass? ______

3. Net production in consumers (N), or secondary production, can be expressed as N = I - (F+R). Red meat contains approximately 700 kJ per 100 grams. If N = 20% of the energy gain (I), how much energy is lost as F and R?

______

ISBN:978-1-927309-20-9

Estimating NPP in *Brassica rapa*

Background

Brassica rapa (right) is a fast growing brassica species, which can complete its life cycle in as little as 40 days if growth conditions are favourable. A class of students wished to estimate the gross and net primary productivity of a crop of these plants using wet and dry mass measurements made at three intervals over 21 days.

The method

- Seven groups of three students each grew 60 *B. rapa* plants in plant trays under controlled conditions. On day 7, each group made a random selection of 10 plants and removed them, with roots intact. The 10 plants were washed, blotted dry, and then weighed collectively (giving wet mass).
- The 10 plants were placed in a ceramic drying bowl and placed in a drying oven at 200°C for 24 hours, then weighed (giving dry mass).
- On day 14 and again on day 21, the procedure was repeated with a further 10 plants (randomly selected).
- The full results for group 1 are presented in Table 1. You will complete the calculation columns.

Table 1: Group 1's results for growth of 10 *B. rapa* plants over 21 days

Age in days	Wet mass of 10 plants /g	Dry mass of 10 plants /g	Percent biomass	Energy in 10 plants / kJ	Energy per plant / kJ	NPP / kJ plant^{-1} d^{-1}
7	19.6	4.2				
14	38.4	9.3				
21	55.2	15.5				

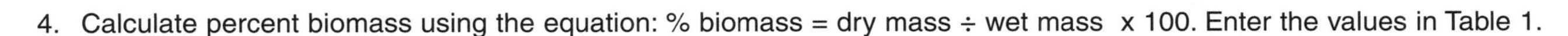

4. Calculate percent biomass using the equation: % biomass = dry mass ÷ wet mass x 100. Enter the values in Table 1.

5. Each gram of dry biomass is equivalent to 18.2 kJ of energy. Calculate the amount of energy per 10 plants and per plant for plants at 7, 14, and 21 days. Enter the values in Table 1.

6. Calculate the Net Primary Productivity per plant, i.e. the amount of energy stored as biomass per day (kJ plant^{-1} d^{-1}). Enter the values in Table 1. We are using per plant in this exercise as we do not have a unit area of harvest.

7. The other 6 groups of students completed the same procedure and, at the end of the 21 days, the groups compared their results for NPP. The results are presented in Table 2, right.

   Transfer group 1's NPP results from Table 1 to complete the table of results and calculate the mean NPP for *B. rapa*.

8. On the grid, plot the class mean NPP vs time.

Table 2: Class results for NPP of *B. rapa* over 21 days

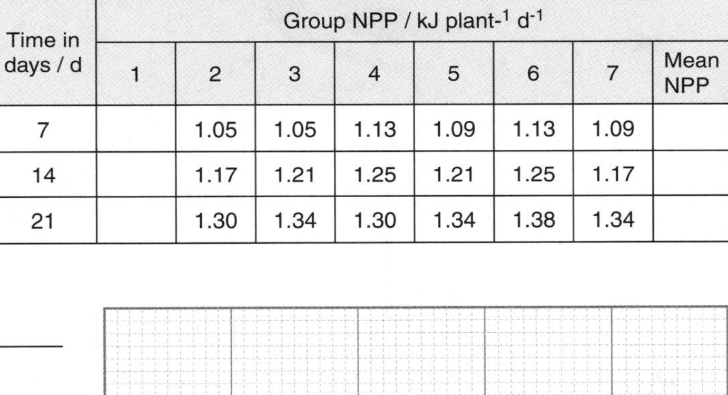

Time in days / d	Group NPP / kJ plant^{-1} d^{-1}							
	1	2	3	4	5	6	7	Mean NPP
7		1.05	1.05	1.13	1.09	1.13	1.09	
14		1.17	1.21	1.25	1.21	1.25	1.17	
21		1.30	1.34	1.30	1.34	1.38	1.34	

9. (a) What is happening to the NPP over time?

   ____

   (b) Explain why this is happening: ______________________________

   ____

   ____

   ____

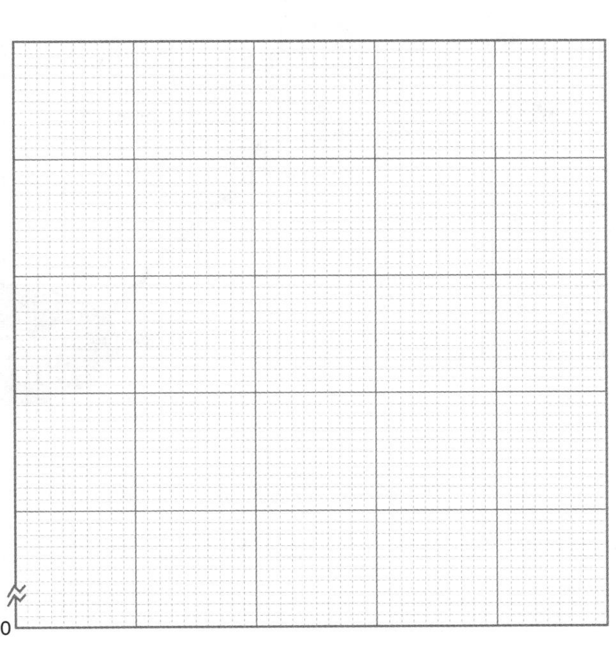

10. What would you need to know to determine the gross primary productivity of *B. rapa*?

    ____

11. As a group, devise a methodology to determine the net secondary production and respiratory losses of 10, 12 day old cabbage white caterpillars feeding on 30 g of brussels sprouts for 3 days. How would you calculate the efficiency of energy transfer from producers to consumers?
    Staple your methodology to this page. You will need to know:

    Energy value of plant material: dry mass x 18.2 kJ.
    Energy value of animal material: dry mass x 23.0 kJ
    Energy value of egested waste (frass): mass x 19.87 kJ

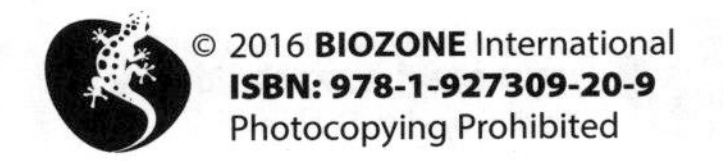

ISBN: 978-1-927309-20-9

26 Food Webs

Key Idea: A food web depicts all the interconnected food chains in an ecosystem. Sunlight is converted to biomass by plants and passed through subsequent trophic levels.

The different food chains in an ecosystem are interconnected to form a complex web of feeding interactions called a **food web.** Sunlight is the initial energy source for almost all ecosystems. Sunlight provides a continuous, but variable, energy supply, which is fixed in carbon compounds by photosynthesis, providing the building blocks and energy for biological materials. Energy stored in this biomass is passed through trophic levels. At each level, some of the energy is lost as heat to the environment so that progressively less is available at each level. This limits the number of links in most food chains to less than six.

- In any community, no species exists independently of others. All organisms, dead or alive, are potential sources of food for other organisms. Within a community, there are hundreds of feeding relationships, and most species participate in several food chains. The different food chains in an ecosystem tend to form food webs, a complex series of interactions showing the feeding relationships between organisms in an ecosystem.
- The complexity of feeding relationships in a community contributes to its structure and specific features. A simple community, like those that establish on bare soil after a landslide, will have a simpler web of feeding relationships than a mature forest. A food web model (below) can be used to show the trophic linkages between different organisms in a community and can be applied to any ecosystem.

Key to food web (below)

- Dashed arrow: Flow of nutrients from the living components to detritus or the nutrient pool.
- Solid arrow: Consumer–resource interactions.
- Hatched arrow: Losses of each food web component from the system and external input of limiting nutrients.

A simple food web

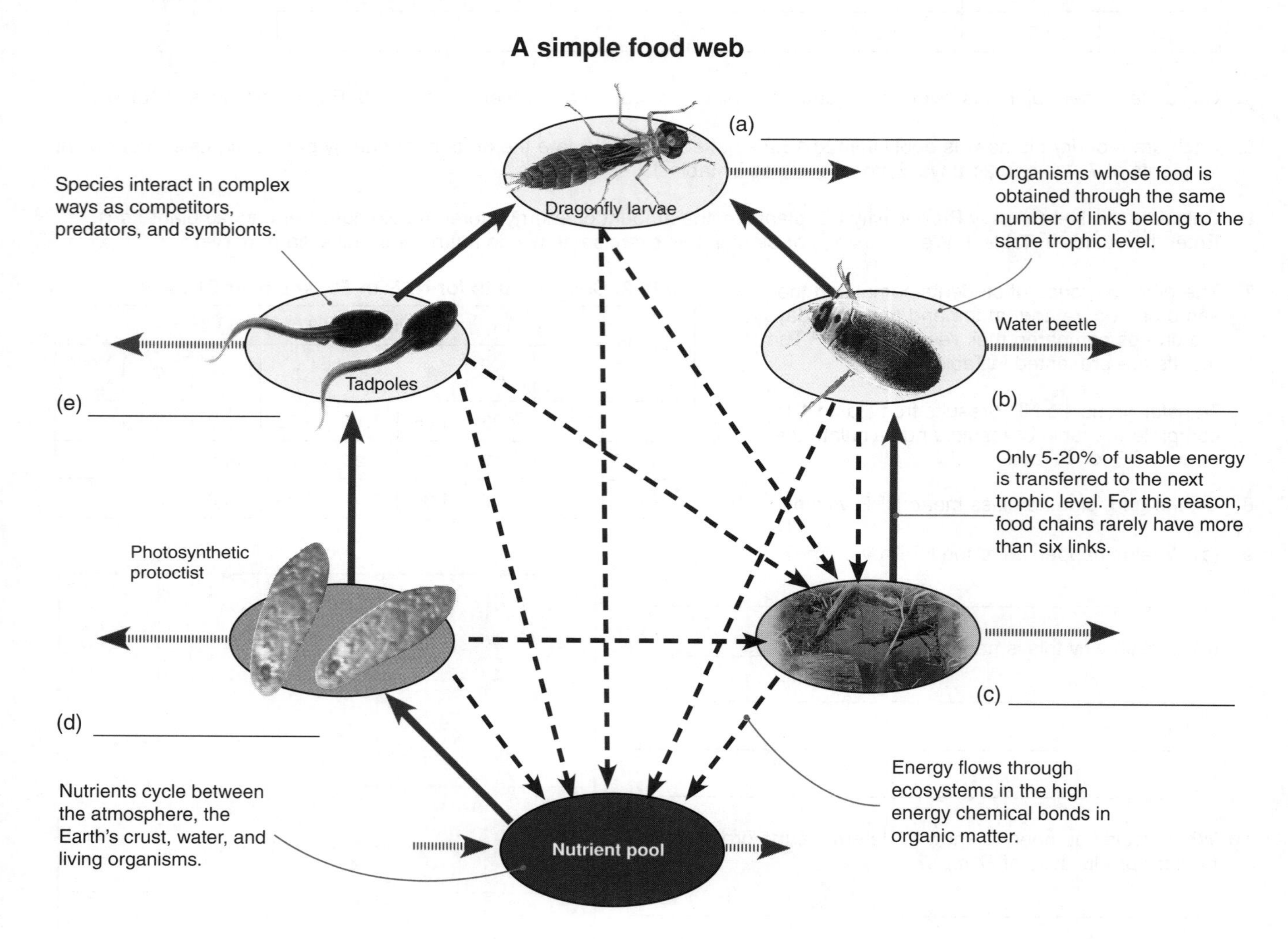

1. (a) - (e) Complete the food web above by adding the following labels: carnivore, herbivore, autotroph, detritus, detritivore:

2. Why do most food chains have fewer than six links? ____________________

____

____

____

ISBN:978-1-927309-20-9

27 Energy Budget in an Ecosystem

Key Idea: Chemical energy in the bonds of molecules flows through an ecosystem between trophic levels. Only 5-20% of energy is transferred from one trophic level to the next.

Energy cannot be created or destroyed, only transformed from one form (e.g. light energy) to another (e.g. chemical energy in the bonds of molecules). This means that the flow of energy through an ecosystem can be measured. Each time energy is transferred from one trophic level to the next (by eating, defaecation, etc.), some energy is given out as heat to the environment, usually during cellular respiration. Living organisms cannot convert heat to other forms of energy, so the amount of energy available to one trophic level is always less than the amount at the previous level. Potentially, we can account for the transfer of energy from its input (as solar radiation) to its release as heat from organisms, because energy is conserved. Recall that the percentage of energy transferred from one trophic level to the next is the **trophic efficiency**. It varies between 5% and 20% and measures the efficiency of energy transfer. An average figure of 10% trophic efficiency is often used. This is called the **ten percent rule**.

Energy flow through an ecosystem

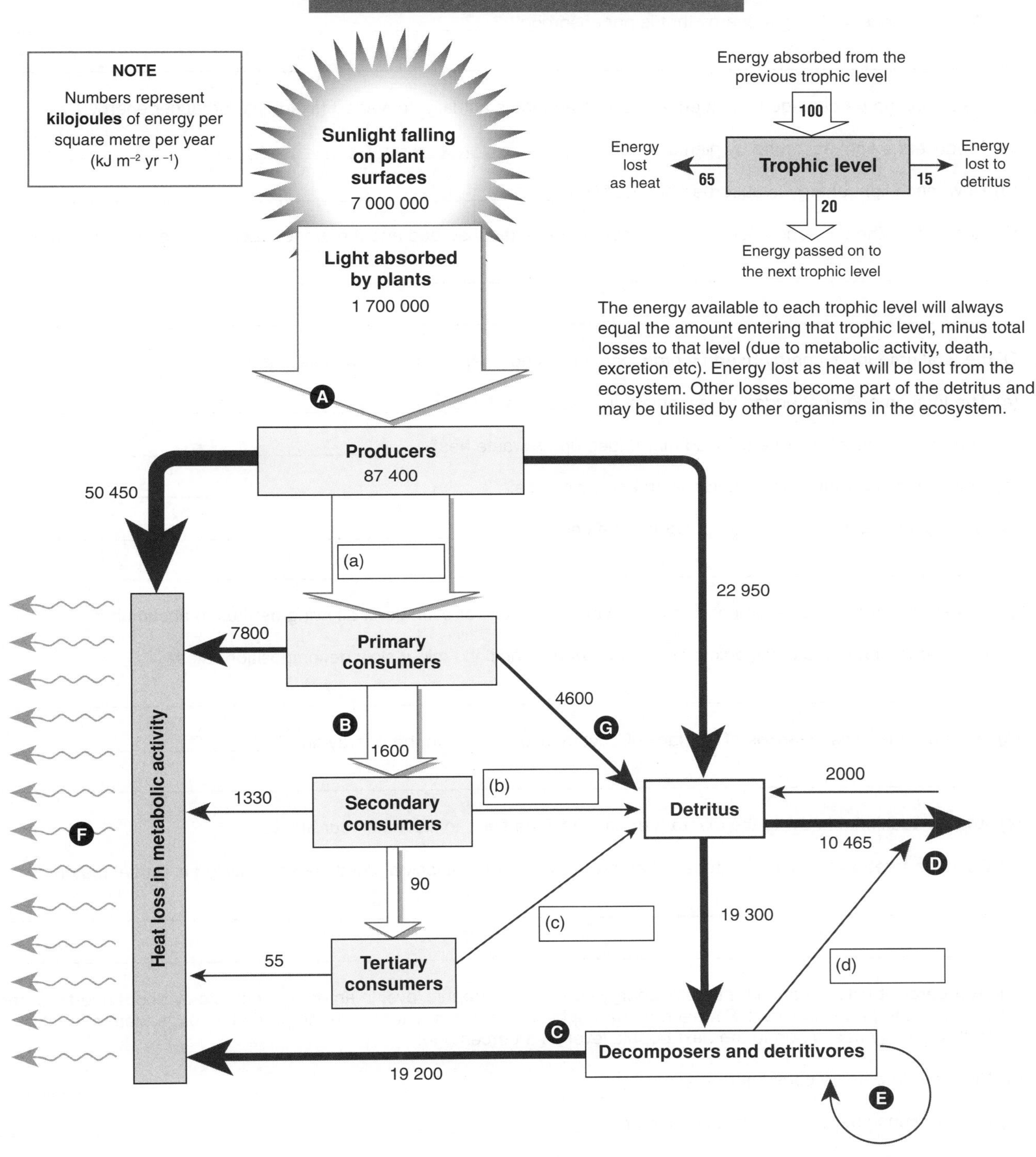

The energy available to each trophic level will always equal the amount entering that trophic level, minus total losses to that level (due to metabolic activity, death, excretion etc). Energy lost as heat will be lost from the ecosystem. Other losses become part of the detritus and may be utilised by other organisms in the ecosystem.

1. Study the diagram above illustrating energy flow through a hypothetical ecosystem. Use the example at the top of the page as a guide to calculate the missing values (a)–(d) in the diagram. Note that the sum of the energy inputs always equals the sum of the energy outputs. Place your answers in the spaces provided on the diagram.

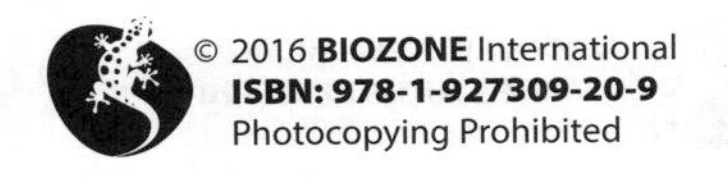

ISBN: 978-1-927309-20-9

2. What is the original source of energy for this ecosystem? ______

3. Identify the processes occurring at the points labelled **A – G** on the diagram:

A. ______ E. ______

B. ______ F. ______

C. ______ G. ______

D. ______

4. (a) Calculate the percentage of light energy falling on the plants that is absorbed at point **A**:

Light absorbed by plants ÷ sunlight falling on plant surfaces x 100 = ______

(b) What happens to the light energy that is not absorbed? ______

5. (a) Calculate the percentage of light energy absorbed that is actually converted (fixed) into producer energy:

Producers ÷ light absorbed by plants x 100 = ______

(b) How much light energy is absorbed but not fixed: ______

(c) Account for the difference between the amount of energy absorbed and the amount actually fixed by producers:

______

6. Of the total amount of energy **fixed** by producers in this ecosystem (at point **A**) calculate:

(a) The total amount that ended up as metabolic waste heat (in kJ): ______

(b) The percentage of the energy fixed that ended up as waste heat: ______

7. (a) State the groups for which detritus is an energy source: ______

(b) How could detritus be removed or added to an ecosystem? ______

8. Under certain conditions, decomposition rates can be very low or even zero, allowing detritus to accumulate:

(a) From your knowledge of biological processes, what conditions might slow decomposition rates?

______

(b) What are the consequences of this lack of decomposer activity to the energy flow? ______

(c) Add an additional arrow to the diagram on the previous page to illustrate your answer.

(d) Describe three examples of materials that have resulted from a lack of decomposer activity on detrital material:

______

9. The **ten percent rule** states that the total energy content of a trophic level in an ecosystem is only about one-tenth (or 10%) that of the preceding level. For each of the trophic levels in the diagram on the preceding page, determine the amount of energy passed on to the next trophic level as a percentage:

(a) Producer to primary consumer: ______

(b) Primary consumer to secondary consumer: ______

(c) Secondary consumer to tertiary consumer: ______

(d) Which of these transfers is the most efficient? ______

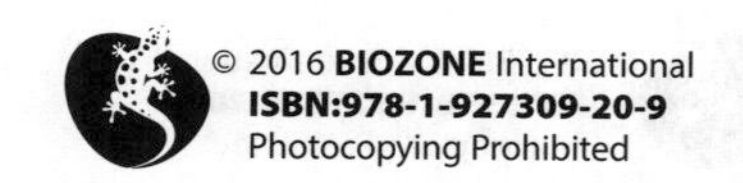

ISBN:978-1-927309-20-9

28 Agriculture and Productivity

Key Idea: Agricultural systems increase productivity by adding nutrients and eliminating competition.

The conversion of much of the Earth's natural grasslands and forested regions to agricultural land did little to increase overall productivity until the introduction of fertilisers in the 1940s. The accelerating demands of the world's population on food resources creates a need to produce more food, more quickly, at minimal cost. For the most part, this has involved intensive, industrialised agricultural systems, where high inputs of energy are used to increase the yield per unit of land farmed. Such systems apply not just to crop plants, but to animals too, which are raised to slaughter weight at high densities in confined areas (a technique called **factory farming**). Producing food from a limited amount of land presents several challenges: to maximise energy flow to the production of harvestable biomass (yield) while minimising energy losses to pests and diseases, to ensure sustainability of the practice, and (in the case of animals) to meet certain standards of welfare and safety. Intensive agriculture makes use of pesticides and fertilisers, as well as antibiotics and hormones to achieve these aims, often with deleterious effects on the environment and on crop and animal health.

Some features of industrialised agriculture

Antibiotics are used in the intensive farming of poultry for egg and meat production. Proponents regard antibiotics as an important tool in managing animals in high density situations. However, widespread, low-dose antibiotic use has contributed to the spread of antibiotic resistance in bacteria.

The application of inorganic fertilisers has been a major factor in the increased yields of intensive agriculture. Removing nutrient limitation can maximise the amount of energy captured and converted to plant biomass. However, excessive application can lead to undesirable enrichment of water bodies and contamination of groundwater.

Fertilisers can be sprayed using aerial topdressing in inaccessible areas.

Crop pests and diseases, which divert energy from production of biomass, are controlled with pesticides and fungicides. However, excessive use of chemicals can lead to pesticide resistance and contamination of land and water.

Clearing land of trees for agriculture can lead to slope instability, soil erosion, and land degradation.

Antibiotics are used to treat diseases such as mastitis in dairy cattle. Milk must be withheld until all antibiotic residues have disappeared.

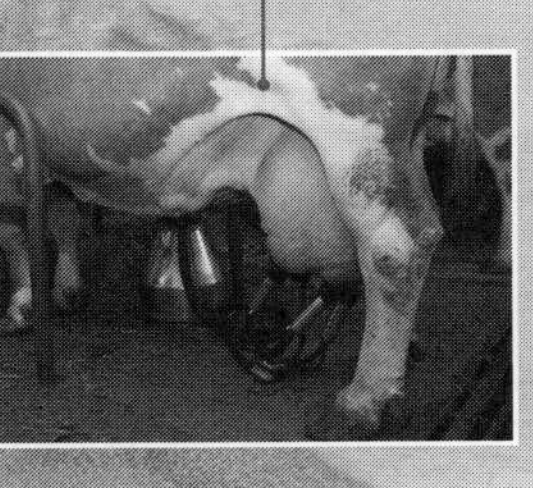

Feedlots are a type of confined animal feeding operation which is used for rapidly feeding livestock, notably cattle, to slaughter weight. Diet for stock in feedlots are very dense in energy to encourage rapid growth and deposition of fat in the meat (marbling). As in many forms of factory farming, antibiotics are used to combat disease in the crowded environment.

1. Identify some of the challenges a farmer faces in maximising crop yields:

2. In terms of energy flows, how do pesticides contribute to higher crop yields?

3. Animals raised for food production in intensive systems are often supplied with antibiotics at low doses in the feed:

   (a) Explain the benefits of this practice to production:

   (b) Discuss the environmental, health, and animal welfare issues associated with this practice:

ISBN: 978-1-927309-20-9

Agriculture and productivity

Increasing net productivity in agriculture (increasing yield) is a matter of manipulating and maximising energy flow through a reduced number of trophic levels. On a farm, the simplest way to increase the net primary productivity is to produce a monoculture. Monocultures reduce competition between the desirable crop and weed species, allowing crops to put more energy into biomass. Other agricultural practices designed to increase productivity in crops include pest (herbivore) control and spraying to reduce disease. Higher productivity in feed-crops also allows greater secondary productivity (e.g. in livestock). Here, similar agricultural practices make sure the energy from feed-crops is efficiently assimilated by livestock.

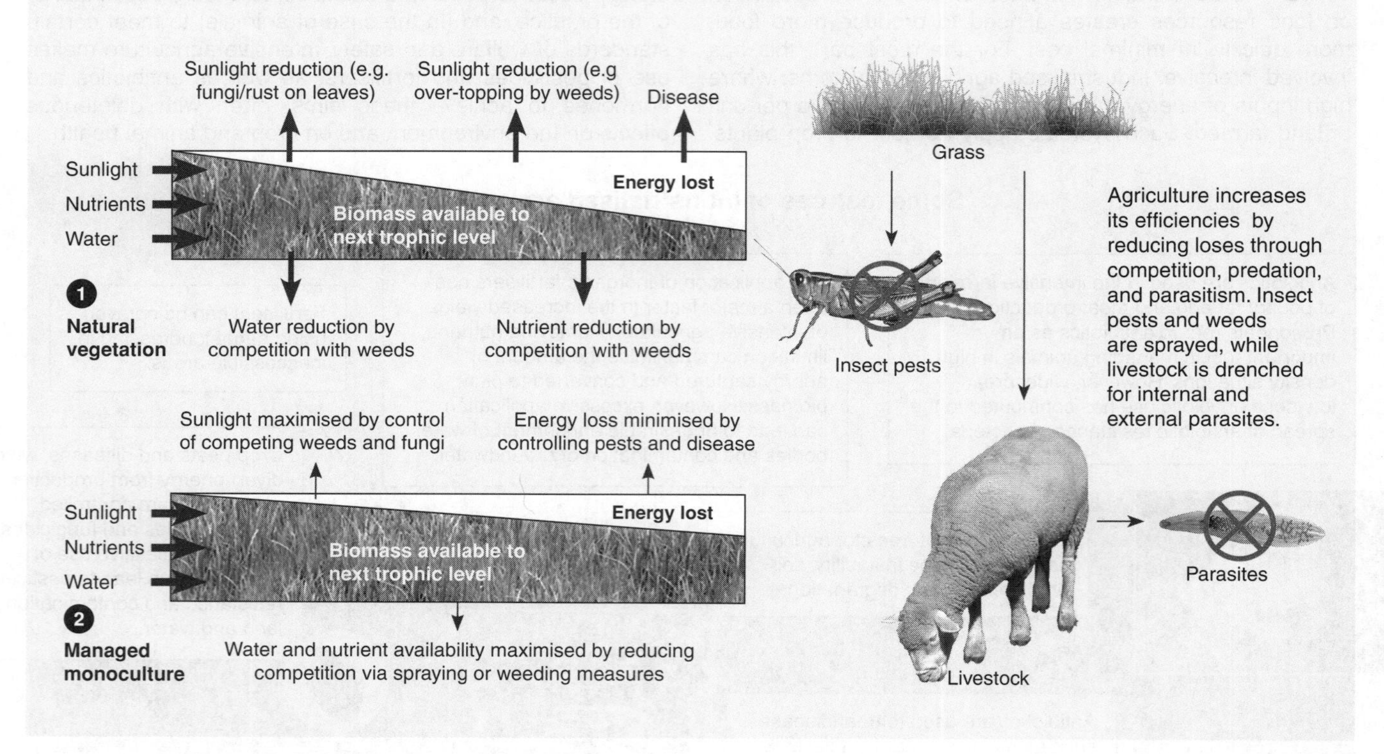

4. Describe the differences in energy input, energy transfers, and interspecific relationships between a natural grassland or woodland ecosystem and an intensive crop farm:

5. In terms of total energy inputs, how might intensive agriculture be considered less efficient than a natural system?

6. (a) How could a farmer maximise the net primary productivity of a particular crop?

(b) How could a farmer maximise livestock productivity?

ISBN:978-1-927309-20-9

29 Nutrient Cycles

Key Idea: Matter cycles through the biotic and abiotic compartments of Earth's ecosystems in nutrient cycles.

Nutrient cycles move and transfer chemical elements (e.g. carbon, hydrogen, nitrogen, and oxygen) through the abiotic and biotic components of an ecosystem. Commonly, nutrients must be in an ionic (rather than elemental) form in order for plants and animals to have access to them. The supply of nutrients in an ecosystem is finite and limited. Macronutrients are required in large amounts by an organism, whereas micronutrients are needed in much smaller quantities.

Essential nutrients

Macronutrient	Common form	Function
Carbon (C)	CO_2	Organic molecules
Oxygen (O)	O_2	Respiration
Hydrogen (H)	H_2O	Cellular hydration
Nitrogen (N)	N_2, NO_3^-, NH_4^+	Proteins, nucleic acids
Potassium (K)	K^+	Principal ion in cells
Phosphorus (P)	$H_2PO_4^-$, HPO_4^{2-}	Nucleic acids, lipids
Calcium (Ca)	Ca^{2+}	Membrane permeability
Magnesium (Mg)	Mg^{2+}	Chlorophyll
Sulfur (S)	SO_4^{2-}	Proteins
Micronutrient	**Common form**	**Function**
Iron (Fe)	Fe^{2+}, Fe^{3+}	Chlorophyll, blood
Manganese (Mn)	Mn^{2+}	Enzyme activation
Molybdenum (Mo)	MoO_4^-	Nitrogen metabolism
Copper (Cu)	Cu^{2+}	Enzyme activation
Sodium (Na)	Na^+	Ion in cells
Silicon (Si)	$Si(OH)_4$	Support tissues

Tropical rainforest

Temperate woodland

The speed of nutrient cycling can vary markedly. Some nutrients are cycled slowly, others quickly. The environment and diversity of an ecosystem can also have a large effect on the speed at which nutrients are recycled.

The role of organisms in nutrient cycling

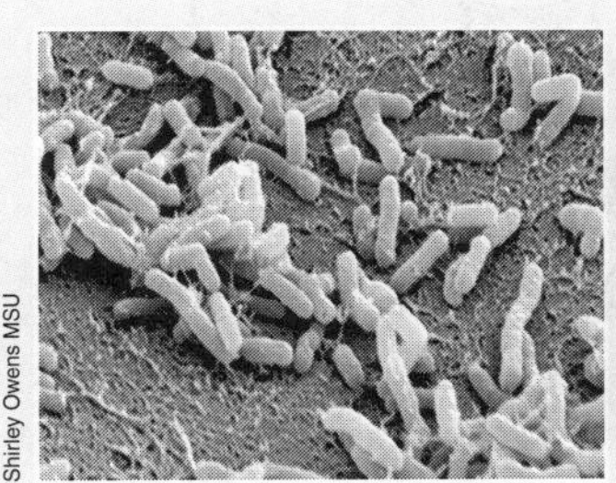

Bacteria
Bacteria play an essential role in nutrient cycles. They act as decomposers, but can also convert nutrients into forms accessible to plants and animals.

Fungi
Fungi are saprophytes and are important decomposers, returning nutrients to the soil or converting them into forms accessible to plants and animals.

Plants
Plants have a role in absorbing nutrients from the soil and making them directly available to browsing animals. They also add their own decaying matter to soils.

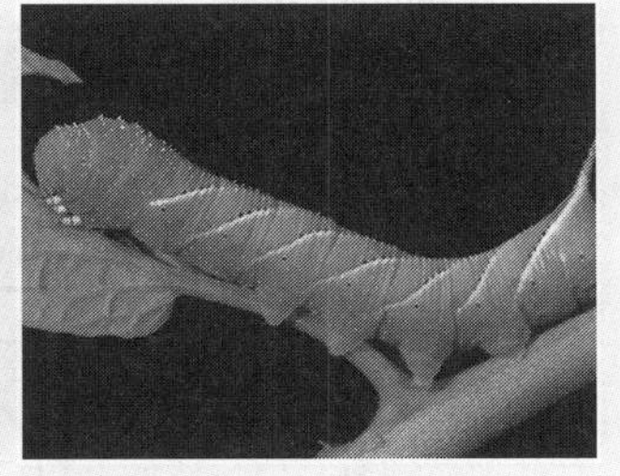

Animals
Animals utilise and break down materials from bacteria, plants and fungi and return the nutrients to soils and water via their wastes and when they die.

1. Describe the role of each of the following in nutrient cycling:

   (a) Bacteria: ____________________

   (b) Fungi: ____________________

   (c) Plants: ____________________

   (d) Animals: ____________________

2. Why are soils in tropical rainforests nutrient deficient relative to soils in temperate woodlands? ____________________

3. Distinguish between macronutrients and micronutrients: ____________________

ISBN: 978-1-927309-20-9

LINK 32 LINK 30 WEB 29 KNOW

30 The Nitrogen Cycle

Key Idea: The nitrogen cycle describes how nitrogen is converted between its various chemical forms. Nitrogen gas is converted to nitrates which are taken up by plants. Heterotrophs obtain their nitrogen by eating other organisms. Nitrogen is an essential component of proteins and nucleic acids and required by all living things. The Earth's atmosphere is about 80% nitrogen gas (N_2), but molecular nitrogen is so stable that it is only rarely available directly to organisms and is often in short supply in biological systems. Bacteria transfer nitrogen between the biotic and abiotic environments. Some can fix atmospheric nitrogen, while others convert ammonia to nitrate, making it available to plants. Lightning discharges also cause the oxidation of nitrogen gas to nitrate. Nitrogen-fixing bacteria are found free in the soil (*Azotobacter*) and in symbioses with some plants in root nodules (*Rhizobium*). Denitrifying bacteria reverse this activity and return fixed nitrogen to the atmosphere. Humans intervene in the nitrogen cycle by applying nitrogen fertilisers to the land.

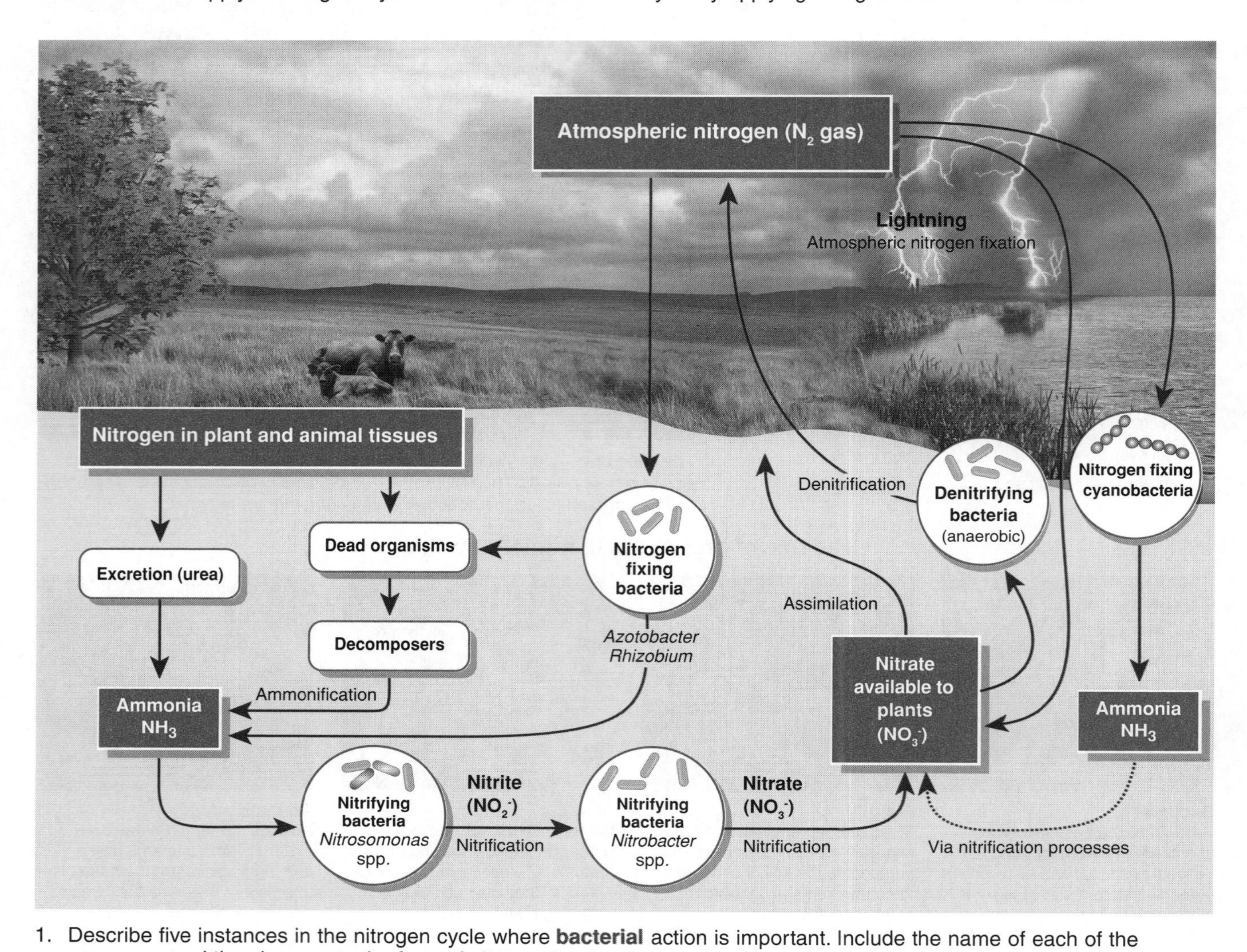

1. Describe five instances in the nitrogen cycle where **bacterial** action is important. Include the name of each of the processes and the changes to the form of nitrogen involved:

(a) ______

(b) ______

(c) ______

(d) ______

(e) ______

ISBN:978-1-927309-20-9

Nitrogen fixation in root nodules

Root nodules are a root **symbiosis** between a higher plant and a bacterium. The bacteria fix atmospheric nitrogen and are extremely important to the nutrition of many plants, including the economically important legume family. Root nodules are extensions of the root tissue caused by entry of a bacterium. In legumes, this bacterium is *Rhizobium*. Other bacterial genera are involved in the root nodule symbioses in non-legumes.

The bacteria in these symbioses live in the nodule where they fix atmospheric nitrogen and provide the plant with most, or all, of its nitrogen requirements. In return, they have access to a rich supply of carbohydrate. The fixation of atmospheric nitrogen to ammonia occurs within the nodule, using the enzyme **nitrogenase**. Nitrogenase is inhibited by oxygen and the nodule provides a low O_2 environment in which fixation can occur.

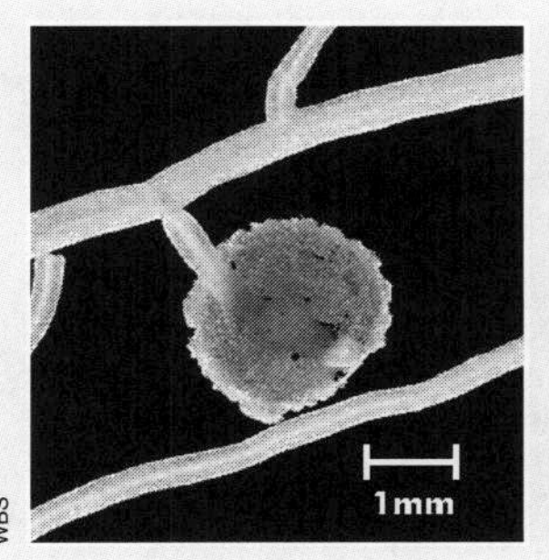

Two examples of legume nodules caused by *Rhizobium.* The images above show the size of a single nodule (left), and the nodules forming clusters around the roots of *Acacia* (right).

Human intervention in the nitrogen cycle

The largest interventions in the nitrogen cycle by humans occur through farming and effluent discharges. Other interventions include burning, which releases nitrogen oxides into the atmosphere, and irrigation and land clearance, which leach nitrate ions from the soil.

Farmers apply organic nitrogen fertilisers to their land in the form of green crops and manures, replacing the nitrogen lost through cropping and harvest. Until the 1950s, atmospheric nitrogen could not be made available to plants except through microbial nitrogen fixation (left). However, during WW II, Fritz Haber developed the Haber process, combining nitrogen and hydrogen gas to form gaseous ammonia. The ammonia is converted into ammonium salts and sold as inorganic fertiliser. This process, although energy expensive, made inorganic nitrogen fertilisers readily available and revolutionised farming practices and crop yields.

Two examples of human intervention in the nitrogen cycle. The photographs above show the aerial application of a commercial fertiliser (left), and the harvesting of an agricultural crop (right).

2. Identify three processes that **fix** atmospheric nitrogen:

   (a) ______________ (b) ______________ (c) ______________

3. What process releases nitrogen gas into the atmosphere? ______________

4. What is the primary reservoir for nitrogen? ______________

5. What form of nitrogen is most readily available to most plants? ______________

6. Name one essential organic compound that plants need nitrogen for: ______________

7. How do animals acquire the nitrogen they need? ______________

   ____

8. Why might farmers plough a crop of legumes into the ground rather than harvest it? ______________

   ____

9. Describe five ways in which humans may intervene in the nitrogen cycle and the effects of these interventions:

   (a) ______________

   ____

   (b) ______________

   ____

   (c) ______________

   ____

   (d) ______________

   ____

   (e) ______________

   ____

ISBN: 978-1-927309-20-9

31 Nitrogen Pollution

Key Idea: Excessive nitrogen in the environment can cause groundwater contamination and eutrophication of waterways. Excess nitrogen in the environment (**nitrogen pollution**) has several effects depending on the compound formed. Nitrogen gas (N_2) makes up almost 80% of the atmosphere but is unreactive at normal pressure and temperature. However, at the high pressures and temperatures of factory processes and combustion engines, N_2 forms nitrogen oxides, which contribute to air pollution. Nitrates in fertilisers are washed into groundwater and slowly make their way to lakes and rivers and eventually out to sea. This process can take time to become noticeable because groundwater may take decades to reach a waterway. Likewise, once nitrate loads are reduced, it may take years for a recovery to occur.

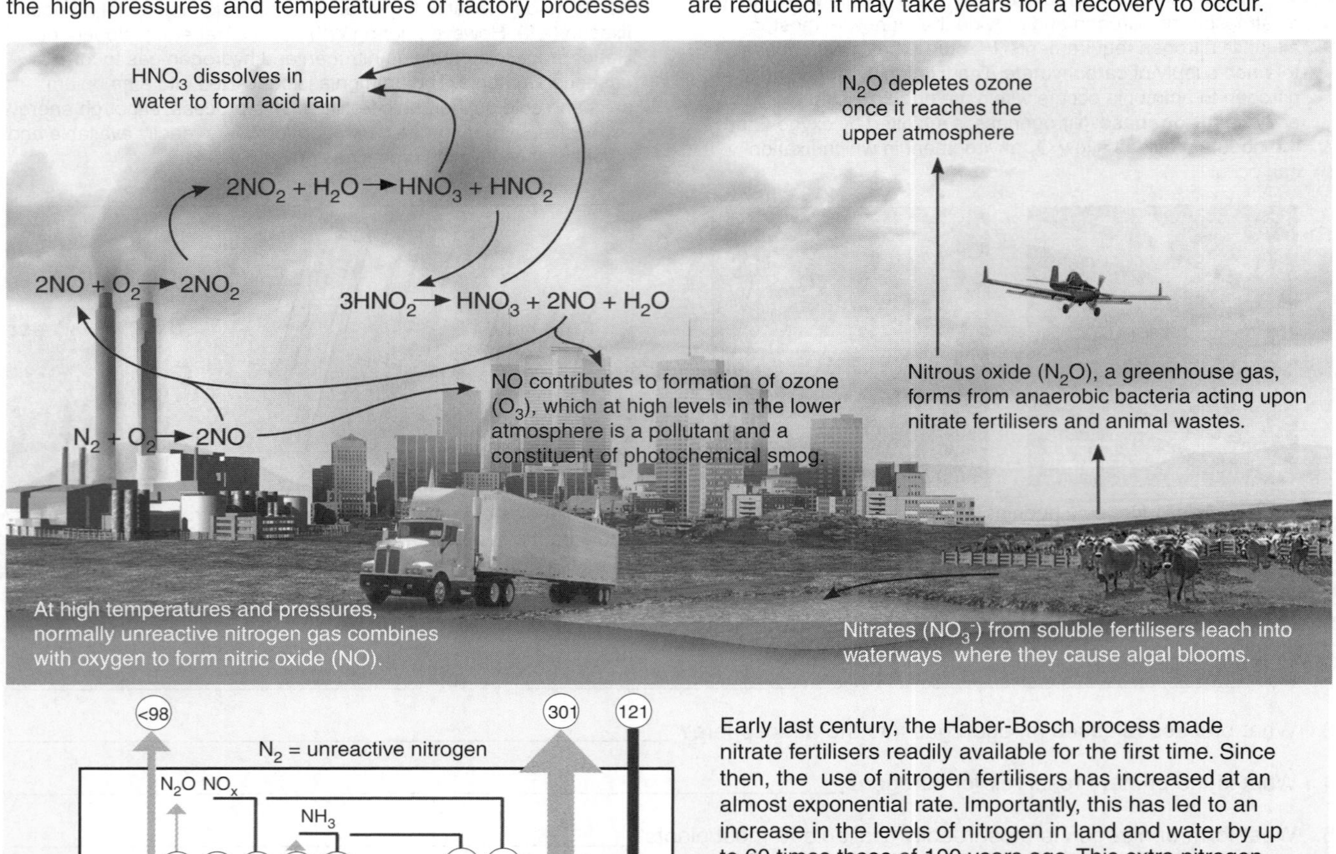

N_2 = unreactive nitrogen

<98 | 301 | 121 | 141 | N_2O NO_x | NH_3 | 8 | 13 | 6.6 | 15 | 11 | 2.3 | 6.2 | Land and fresh water | Oceans | 27 | >0

Nitrates = reactive nitrogen

1860

N_2 = unreactive nitrogen

115 | 322 | 121 | 268 | N_2O NO_x | NH_3 | 11 | 46 | 25 | 53 | 39 | 12 | 21 | Land and fresh water | Oceans | 48 | 60

Nitrates = reactive nitrogen

1995

Changes in nitrogen inputs and outputs between 1860 and 1995 in million Tonne (modified from Galloway *et al* 2004)

Early last century, the Haber-Bosch process made nitrate fertilisers readily available for the first time. Since then, the use of nitrogen fertilisers has increased at an almost exponential rate. Importantly, this has led to an increase in the levels of nitrogen in land and water by up to 60 times those of 100 years ago. This extra nitrogen load is one of the causes of accelerated enrichment of water bodies (**eutrophication**). An increase in algal production also results in higher decomposer activity and, consequently, oxygen depletion, fish deaths, and depletion of aquatic biodiversity. Many aquatic microorganisms also produce toxins, which may accumulate in the water, fish, and shellfish. The diagrams (left) show the increase in nitrates in water sources from 1860 to 1995. The rate at which nitrates are added has increased faster than the rate at which nitrates are returned to the atmosphere as unreactive N_2 gas. This has led to the widespread accumulation of nitrogen.

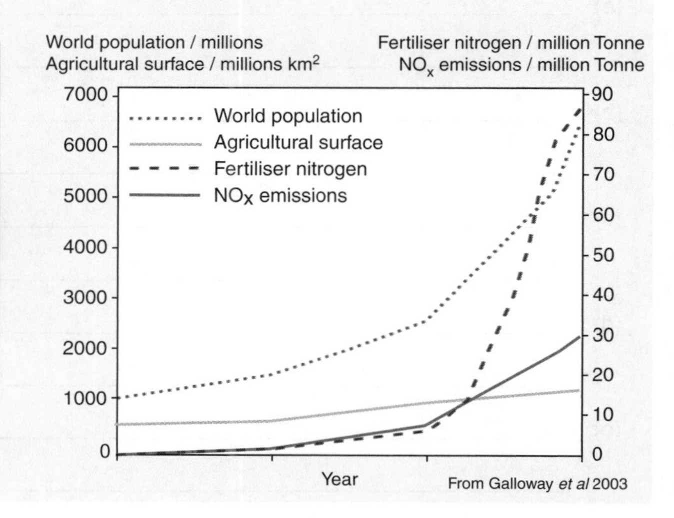

From Galloway *et al* 2003

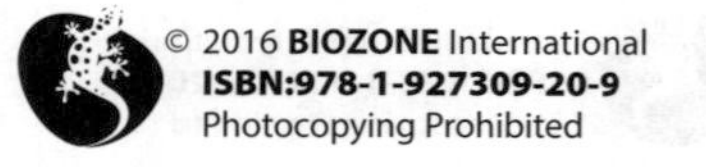

ISBN:978-1-927309-20-9

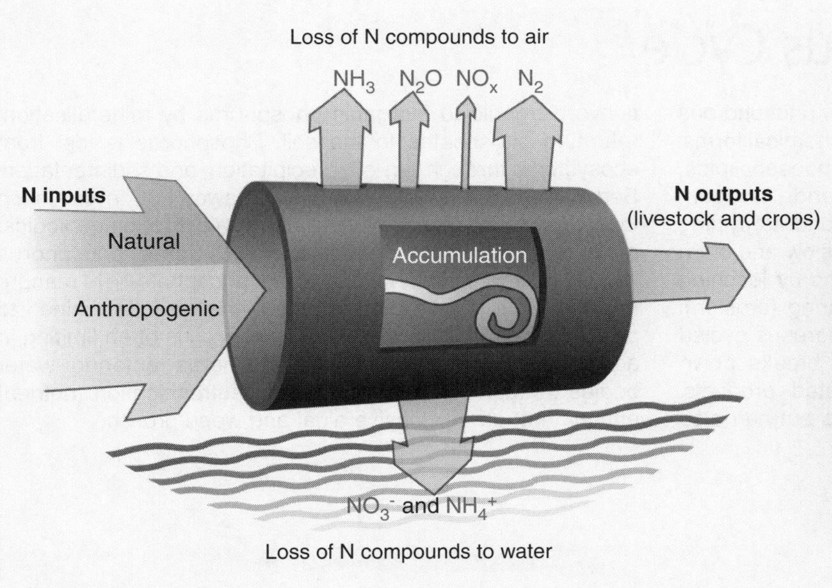

The "hole in the pipe" model (left) demonstrates inefficiencies in nitrogen fertiliser use. Nitrogen that is added to the soil and not immediately taken up by plants is washed into waterways or released into the air by bacterial action. These losses can be minimised to an extent by using slow release fertilisers during periods of wet weather and by careful irrigation practices.

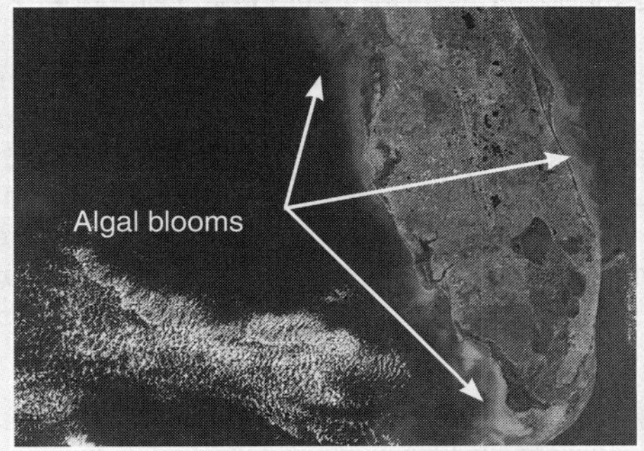

Satellite photo of algal blooms around Florida. Excessive nitrogen contributes to algal blooms in both coastal and inlands waters. *Image: NASA*

1. Describe the effect each of the following nitrogen compounds have on air and water quality:

   (a) NO: ______

   (b) N_2O: ______

   (c) NO_2: ______

   (d) NO_3^-: ______

2. Explain why the formation of NO can cause large scale and long term environmental problems: ______

3. Why would an immediate halt in the use of nitrogen fertilisers not cause an immediate stop in their effects?

4. (a) Calculate the increase in nitrogen deposition in the oceans from 1860 to 1995 and compare this to the increase in release of nitrogen from the oceans.

   (b) What is the effect of this increase on the oceans? ______

5. (a) Why do nitrogen inputs tend to be so much more than outputs in crops and from livestock? ______

   (b) Suggest how the nitrogen losses could be minimised: ______

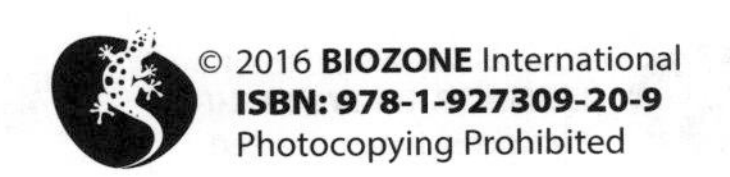

ISBN: 978-1-927309-20-9

32 The Phosphorus Cycle

Key Idea: The phosphorus cycle describes how phosphorous moves through the environment in its various chemical forms. Phosphorus is an essential component of phospholipids, nucleic acids, and ATP. Unlike carbon and nitrogen, phosphorus has no atmospheric component and its cycling through the biotic and abiotic environments is slow and often localised. Small losses from terrestrial systems by leaching are generally balanced by gains from weathering (erosion). In aquatic and terrestrial ecosystems, phosphorus is cycled through food webs. Bacterial decomposition breaks down the remains of dead organisms and excreted products. Bacteria can immobilise inorganic phosphorus but can also convert organic to inorganic phosphorus by mineralisation, returning phosphates to the soil. Phosphorus is lost from ecosystems through run-off, precipitation, and sedimentation. Sedimentation may lock phosphorus away but, in the much longer term, it can become available again through geological processes such as geological uplift. Some phosphorus returns to the land as guano, the phosphate-rich manure (typically of bats and marine birds) which is often mined to produce phosphate fertilisers. Phosphorus is often limiting in aquatic systems, so excess phosphorus entering water bodies through runoff contributes to eutrophication (nutrient enrichment) and excessive algal and weed growth.

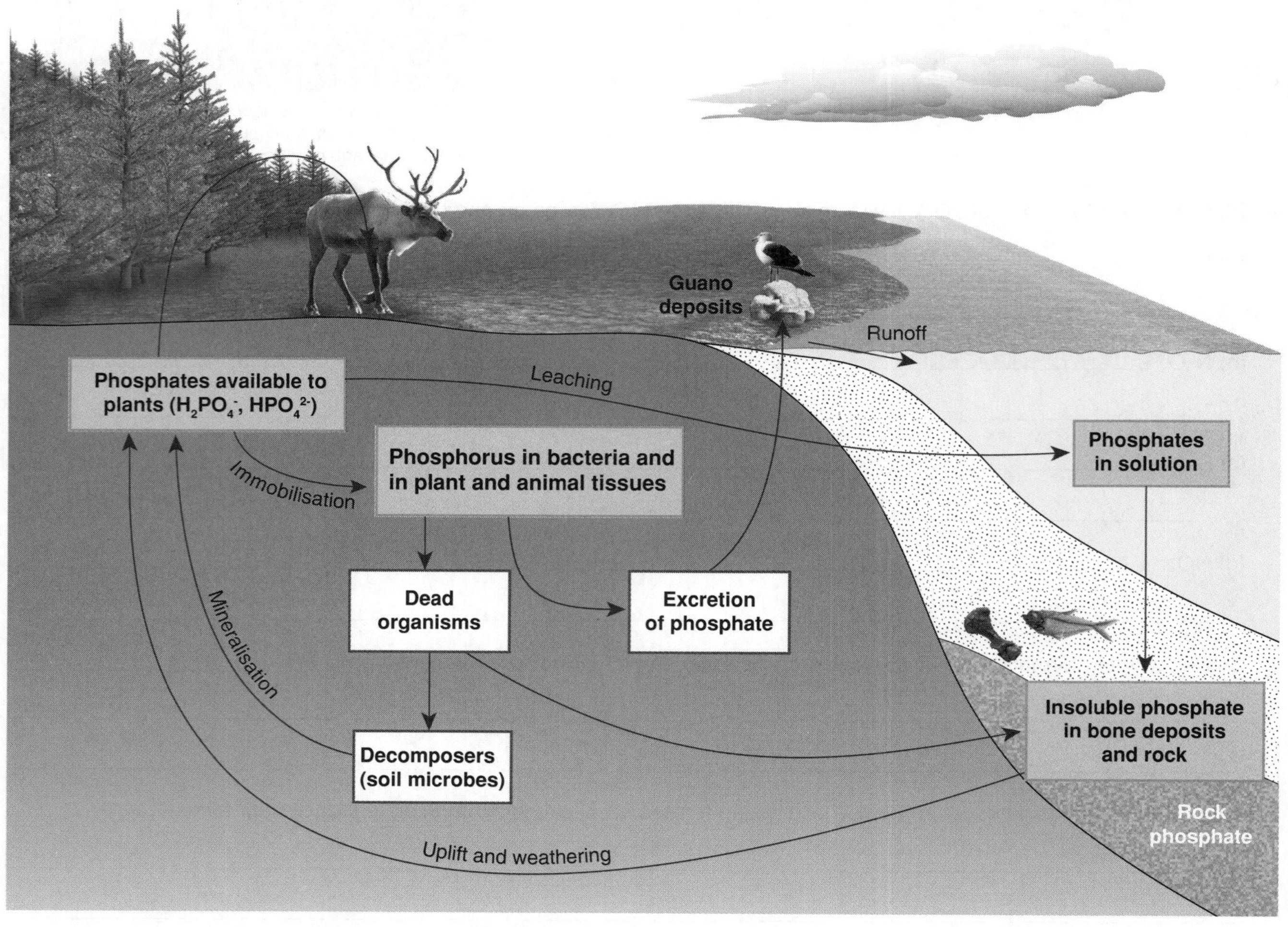

1. On the diagram, add arrows and labels to show where human activity might intervene in the phosphorus cycle.

2. Identify and describe:

   (a) Two instances in the phosphorus cycle where bacterial action is important: ____________________

   ____

   ____

   (b) Two types of molecules found in living organisms which include phosphorus as a part of their structure:

   ____

   (c) The origin of three forms of inorganic phosphate making up the geological reservoir: ____________________

   ____

3. Describe the processes that must occur in order to make rock phosphate available to plants again: ____________________

____

ISBN:978-1-927309-20-9

33 Chapter Review

Summarise what you know about this topic under the headings and sub-headings provided. You can draw diagrams or mind maps, or write short notes to organise your thoughts. Use the images and hints to help you and refer back to the introduction to check the points covered:

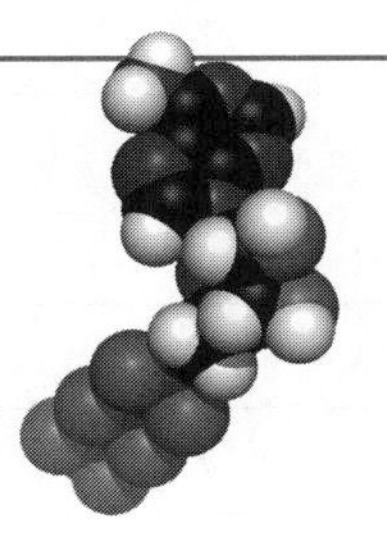

Energy in cells

HINT: Photosynthesis and respiration. Production of ATP.

Nutrient cycling

HINT: Describe the nitrogen and phosphorous cycles and the role of microorganisms in these.

Ecosystems and biomass transfers

HINT: Components and size of an ecosystem. Energy transfers and ecological efficiency.

ISBN: 978-1-927309-20-9

34 KEY TERMS AND IDEAS: Did You Get It?

1. (a) Write the process of photosynthesis as:

   A word equation: ______

   A chemical equation: ______

   (b) Where does photosynthesis occur? ______

2. (a) Write the process of respiration as:

   A word equation: ______

   A chemical equation: ______

   (b) Where does respiration occur? ______

3. Test your vocabulary by matching each term to its correct definition, as identified by its preceding letter code.

absorption spectrum	A	The biochemical process that uses light energy to convert carbon dioxide and water into glucose molecules and oxygen.
aerobic respiration	B	Also known as the citric acid cycle. Part of a metabolic pathway involved in the chemical conversion of carbohydrates, fats and proteins to CO_2 and water to generate a form of usable energy (ATP).
ATP	C	Membrane-bound compartments within chloroplasts. They are the site of the light dependent reactions of photosynthesis.
action spectrum	D	The phase in photosynthesis where chemical energy is used for the synthesis of carbohydrate. Also called the light independent phase.
Calvin cycle	E	The liquid interior of the chloroplast where the light independent phase takes place.
cellular respiration	F	The phase in photosynthesis when light energy is converted to chemical energy.
chlorophyll	G	The term to describe the light absorption of a pigment vs the wavelength of light.
food chain	H	Respiration requiring oxygen as the terminal electron acceptor.
gross primary productivity	I	A profile of the effectiveness of different wavelengths of light in fuelling photosynthesis.
Krebs cycle	J	The green, membrane-bound pigment involved in the light dependent reactions of photosynthesis.
light dependent phase	K	The catabolic process in which the chemical energy in complex organic molecules is coupled to ATP production.
photosynthesis	L	The efficiency of energy transfer from one trophic level to the next, usually expressed as a percentage.
stroma	M	The position an organism occupies on the food chain.
thylakoid discs	N	The total energy fixed by photosynthesis per unit area or volume per unit time.
trophic efficiency	O	A sequence of steps describing how an organism derives energy from the ones before it.
trophic level	P	A nucleoside triphosphate used in the transfer of energy in cells.

4. (a) The graph (right) shows primary production in the oceans. Explain the shape of the curves:

   ______

   ______

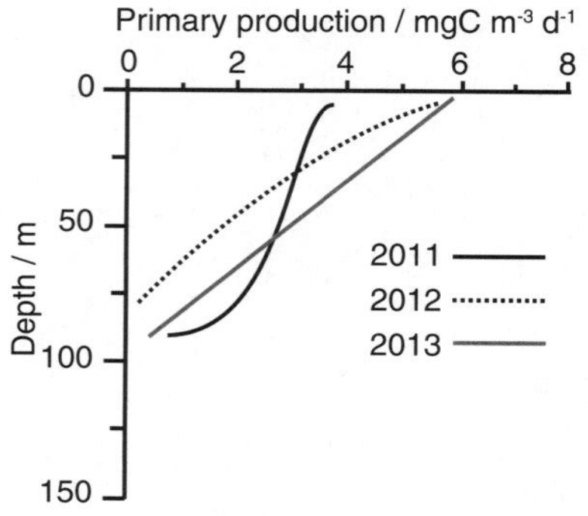

   (b) About 90% of all marine life lives in the photic zone (the depth to which light penetrates). Suggest why this is so:

   ______

   ______

   ______

ISBN:978-1-927309-20-9

Topic 6

Organisms respond to changes in their environment

Key terms

action potential
adrenaline
atrioventricular node
baroreceptor
blood glucose
bundle of His
chemoreceptor
cyclic AMP
diabetes mellitus
generator potential
glucagon
gluconeogenesis
glycogenesis
glycogenolysis
gravitropism
homeostasis
insulin
kinesis (pl. kineses)
motor neurone
myelin
myogenic
negative feedback
nephron
neuromuscular junction
neurone
osmoregulation
Pacinian corpuscle
phototropism
positive feedback
pulse rate
Purkyne tissue
reflex
refractory period
resting potential
saltatory conduction
second messenger
sensory receptor
sinoatrial node
stimulus (pl. stimuli)
synapse
synaptic integration
taxis (pl. taxes)
tropism

6.1 Stimuli and responses

Learning outcomes — Activity number

		Learning outcome	Activity number
☐	1	Using examples, explain why appropriate responses to stimuli enhance survival in both plants and animals.	35 36 37
☐	2	Describe the role of plant growth factors in tropisms in flowering plants. Describe the effect of indoleacetic acid on cell elongation in angiosperm shoots and roots as an explanation for their phototropic and gravitropic responses.	38-41
☐	3	Describe taxes and kineses in animals and explain their adaptive role.	42
☐	4	PR-10 Investigate the effect of an environmental variable (e.g. light or humidity) on the movement of an animal using a choice chamber or maze.	43 44
☐	5	Describe the role of reflexes, as exemplified by a three neurone reflex, such as the pain withdrawal reflex.	45
☐	6	Describe the properties of receptors with reference to the Pacinian corpuscle. Include the basic structure of the Pacinian corpuscle to show how deformation leads to production of generator potentials.	46 47
☐	7	Describe the structure of human retina to show how differences in sensitivity to light, sensitivity to colour, and visual acuity are achieved.	48 49
☐	8	AT Design an experiment to investigate sensory reception in human skin, e.g. resolution of touch receptors (the two point threshold).	50
☐	9	Describe the myogenic stimulation of the heart, including the role of the sinoatrial node (SAN), the atrioventricular node (AVN), and the Purkyne tissue.	51
☐	10	Describe the extrinsic regulation of heart beat, including the role of chemoreceptors, pressure receptors, and the autonomic nervous system.	52
☐	11	AT Investigate the effect of a named variable on human pulse rate.	53

Tangopaso

Fankhauser

6.2 Nervous coordination

Learning outcomes — Activity number

		Learning outcome	Activity number
☐	12	Annotate a diagram to describe the structure of a myelinated motor neurone.	54
☐	13	Describe how the resting potential is established in a motor neurone. Explain how an action potential is generated and how it is propagated in myelinated and non-myelinated axons. Describe the nature and significance of the refractory period in producing discrete impulses and limiting the frequency of impulses.	55
☐	14	Describe factors affecting the speed of impulse conduction, including myelination and saltatory conduction, axon diameter, and temperature.	54 55
☐	15	Describe the structure of a synapse and a neuromuscular junction. Describe the transmission of an impulse across a cholinergic synapse and explain features of synaptic transmission including unidirectionality, summation, and inhibition. Compare impulse transmission across a cholinergic synapse and a neuromuscular junction.	56 57
☐	16	Use information provided to predict and explain the effects of specific drugs on the functioning of a synapse.	58

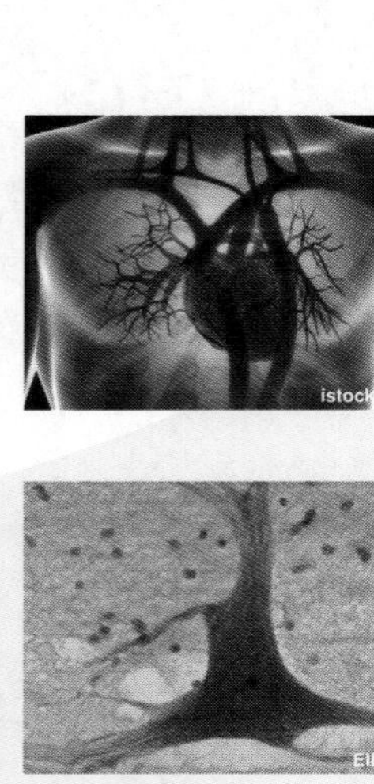

6.3 Muscles as effectors

Learning outcomes

Activity number

☐ 17 Describe the action of antagonistic muscle pairs on an incompressible skeleton. — 59

☐ 18 Describe the gross and microscopic structure of skeletal muscle tissue. Describe the ultrastructure of the myofibrils of a muscle fibre (cell), including reference to the thick and thin (myo)filaments. — 60

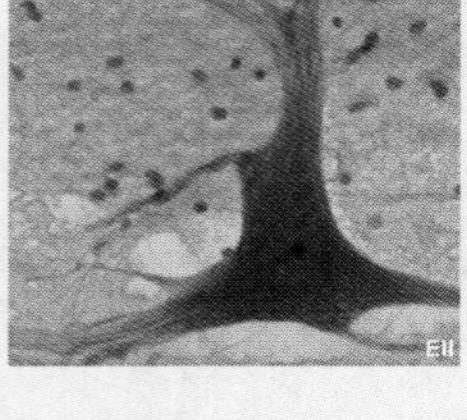

☐ 19 **AT** Compare the ultrastructure of skeletal muscle tissue in diagrams and prepared slides viewed with an optical microscope. — 60

☐ 20 Describe the roles of actin, myosin, calcium ions, and ATP in contraction of a myofibril. Explain actinomyosin cross bridge formation with reference to calcium ions and tropomyosin (blocking complex). — 61

☐ 21 Describe the roles of ATP and phosphocreatine in muscle contraction. — 62

☐ 22 Explain muscle fatigue in terms of fall in pH and decline in ATP supply. — 63

☐ 23 **AT** Investigate the effect of repeated muscular contraction on muscle fatigue. — 64

☐ 24 Compare and contrast the structure, location, and general properties of slow and fast twitch skeletal muscle fibres. — 63

6.4 Homeostasis

Learning outcomes

Activity number

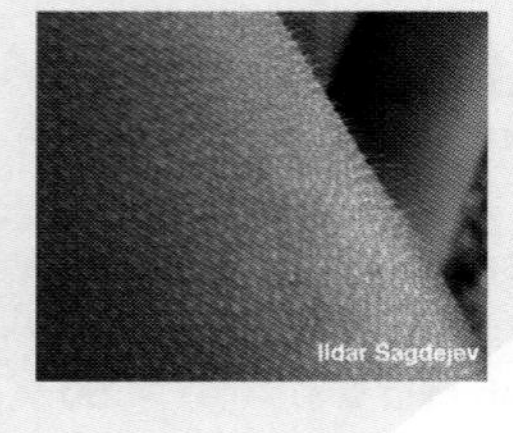

☐ 25 Define homeostasis and explain the importance of maintaining a stable core temperature, stable blood pH, and stable blood glucose concentration. — 65

☐ 26 Describe the role of negative feedback systems in regulating physiological systems, explaining how they operate to maintain a steady state. Giving an example, explain how separate mechanisms involving negative feedback can control departures in different directions from the steady state. — 65 66

☐ 27 Recognise positive feedback as a mechanism that escalates a physiological response. Describe situations involving positive feedback and explain how the escalation in response is eventually ended. — 67

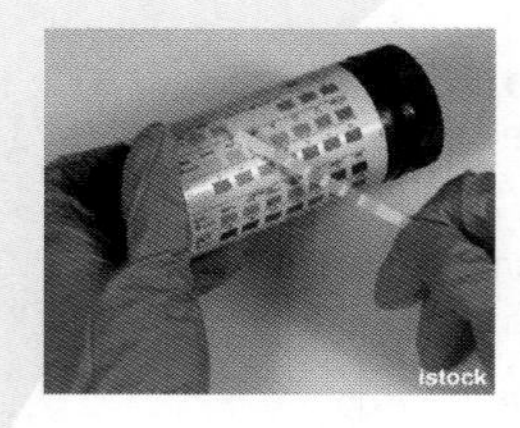

☐ 28 Understand how hormones act as signalling molecules to directly or indirectly bring about a cellular response. — 68 69

☐ 29 Describe the factors that influence blood glucose concentration, including the role of insulin and glucagon. — 69 70

☐ 30 Describe the role of the liver in carbohydrate metabolism: glycogenolysis, glycogenesis, and gluconeogenesis. — 71

☐ 31 Explain the action of insulin and glucagon at the cellular level. — 72

☐ 32 Explain the role of adrenaline as a signal molecule in the signal transduction pathway that results in the conversion of glycogen to glucose. Describe the second messenger model of adrenaline and glucagon action, involving adenyl cyclate, cyclic AMP, and protein kinase. — 68 73

☐ 33 Describe the causes and control of types I and II diabetes. — 74 75

☐ 34 **PR-11** Produce a dilution series of a glucose solution and use colorimetric techniques to produce a calibration curve to determine the concentration of glucose in an unknown urine sample. — 76

☐ 35 Define osmoregulation. Describe the role of the hypothalamus, posterior pituitary, and antidiuretic hormone in osmoregulation. — 77

☐ 36 Describe the structure and function of the kidney nephron, including the role of the glomerulus, the reabsorption of glucose and water from the initial filtrate, the formation of a salt gradient in the medulla by the loop of Henle, and the reabsorption of water to form the final filtrate. — 78

35 Detecting Changing States

Key Idea: Sensory receptors allow the body to respond to a range of stimuli in the internal and external environments.

A **stimulus** is any physical or chemical change in the environment capable of provoking a response in an organism. Organisms respond to stimuli in order to survive. Stimuli may be either external (outside the organism) or internal (within its body). Some of the sensory receptors that animals (including humans) use to detect stimuli are shown below. Sensory receptors respond only to specific stimuli, so the sense organs an animal has determines how it perceives the world.

Hair cells in the vestibule of the inner ear respond to **gravity** by detecting the rate of change and direction of the head and body. Other hair cells in the cochlea of the inner ear detect **sound** waves. The sound is directed and amplified by specialised regions of the outer and middle ear.

Photoreceptor cells in the eyes detect colour, intensity, and movement of **light**.

Olfactory receptors in the nose detect airborne chemicals. The human nose has about 5 million of these receptors, a bloodhound nose has more than 200 million. The taste buds of the tongue detect dissolved chemicals (gustation). Tastes are combinations of five basic sensations: sweet, salt, sour, bitter, and savoury (umami receptor).

Chemoreceptors in certain blood vessels, e.g. carotid arteries, monitor carbon dioxide levels (and therefore pH) of the blood. Breathing and heart rate increase or decrease (as appropriate) to adjust blood composition.

Baroreceptors in the walls of some arteries, e.g. aorta, monitor blood pressure. Heart rate and blood vessel diameter are adjusted accordingly.

Pressure deforms the skin surface and stimulates sensory receptors in the dermis. These receptors are especially abundant on the lips and fingertips.

Proprioreceptors (stretch receptors) in the muscles, tendons, and joints monitor limb position, **stretch**, and **tension**. The muscle spindle is a stretch receptor that monitors the state of muscle contraction and enables muscle to maintain its length.

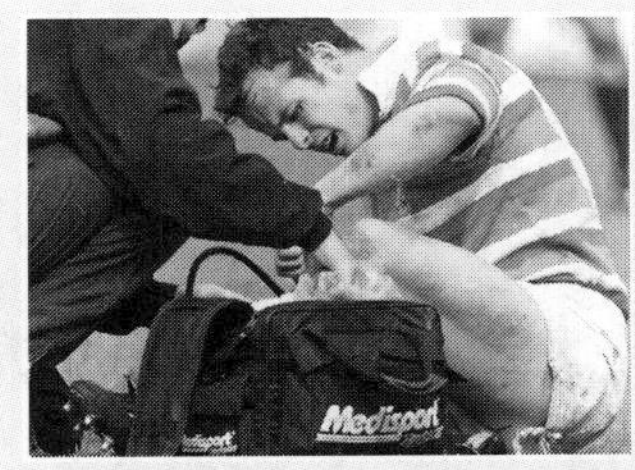

Pain and temperature are detected by nerve endings in the skin. Deep tissue injury is sometimes felt on the skin as referred pain.

Humans rely heavily on hearing when learning to communicate; without it, speech and language development are more difficult.

The vibration receptors in the limbs of arthropods are sensitive to movement: either sound or vibration (from struggling prey).

The chemosensory Jacobson's organ in the roof of the mouth of reptiles (e.g. snakes) enables them to detect chemical stimuli.

Breathing and heart rates are regulated in response to sensory input from chemoreceptors.

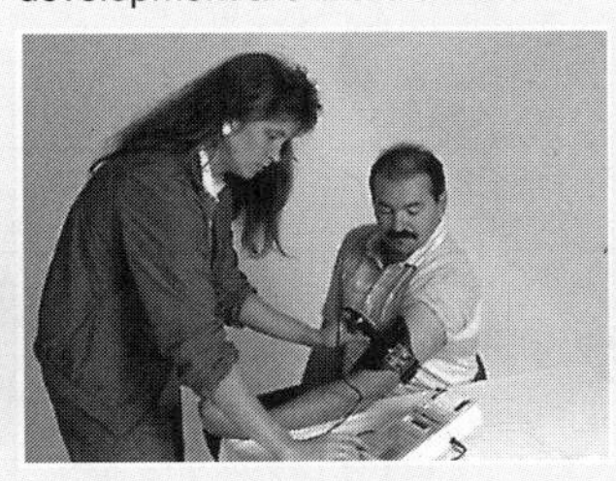
Baroreceptors and osmoreceptors act together to maintain blood pressure and volume.

Many insects, such as these ants, rely on chemical sense for location of food and communication.

Jacobson's organ is also present in mammals and is used to detect sexual receptivity in mates.

1. What is a stimulus and how are stimuli perceived by an organism? ______________________________

____

2. (a) Name one external stimulus and its sensory receptor: ______________________________

(b) Name one internal stimulus and its sensory receptor: ______________________________

ISBN: 978-1-927309-20-9

36 Plant Responses

Key Idea: Plants respond to their environment by either growing to or away from a stimulus or by producing a response that affects some physiological process.

Even though most plants are firmly rooted in the ground, they can still respond to changes in their external environment, mainly through changes in patterns of growth. These responses may involve relatively sudden physiological changes, as in flowering, or a steady growth response, such as a **tropism**. Many of these responses involve annual, seasonal, or circadian (daily) rhythms.

TROPISMS
Tropisms are growth responses made by plants to directional external stimuli, where the direction of the stimulus determines the direction of the growth response. A tropism may be positive (towards the stimulus), or negative (away from the stimulus). Common stimuli for plants include light, gravity, touch, and chemicals.

LIFE CYCLE RESPONSES
Plants use seasonal changes (such as falling temperatures or decreasing daylength) as cues for starting or ending particular life cycle stages. Such changes are mediated by plant growth factors, such as phytochrome and gibberellin and enable the plant to avoid conditions unfavourable to growth or survival. Examples include flowering, dormancy and germination, and leaf fall.

RAPID RESPONSES TO ENVIRONMENTAL STIMULI
Plants are capable of quite rapid responses. Examples include the closing of stomata in response to water loss, opening and closing of flowers in response to temperature, and **nastic responses** (rapid responses that, unlike tropisms, are independent of stimulus direction). These responses may follow a circadian rhythm and are protective in that they reduce the plant's exposure to abiotic or grazing stress.

PLANT COMPETITION AND ALLELOPATHY
Although plants are rooted in the ground, they can still compete with other plants to gain access to resources. Some plants produce chemicals that inhibit the growth of neighbouring plants. Such chemical inhibition is called allelopathy. Plants also compete for light and may grow aggressively to shade out slower growing competitors.

PLANT RESPONSES TO HERBIVORY
Many plant species have responded to grazing or browsing pressure with evolutionary adaptations enabling them to survive constant cropping. Examples include rapid growth to counteract the constant loss of biomass (grasses), sharp spines or thorns to deter browsers (acacias, cacti), or toxins in the leaf tissues (eucalyptus).

1. Identify the stimuli plants typically respond to: ______

2. How do plants benefit by responding appropriately to the environment? ______

3. Describe one adaptive response of plants to each of the following stressors in the environment:

(a) Low soil water: ______

(b) Falling autumn air temperatures: ______

(c) Browsing animals: ______

(d) Low air temperatures at night: ______

ISBN:978-1-927309-20-9

37 Tropisms and Growth Responses

Key Idea: Tropisms are directional growth responses to external stimuli. They may be positive (towards a stimulus) or negative (away from a stimulus).

Tropisms are plant growth responses to external stimuli, in which the stimulus direction determines the direction of the growth response. Tropisms are identified according to the stimulus involved, e.g. photo- (light), geo- (gravity), hydro- (water), and are identified as positive (towards the stimulus) or negative (away from the stimulus). Tropisms act to position the plant in the most favourable available environment.

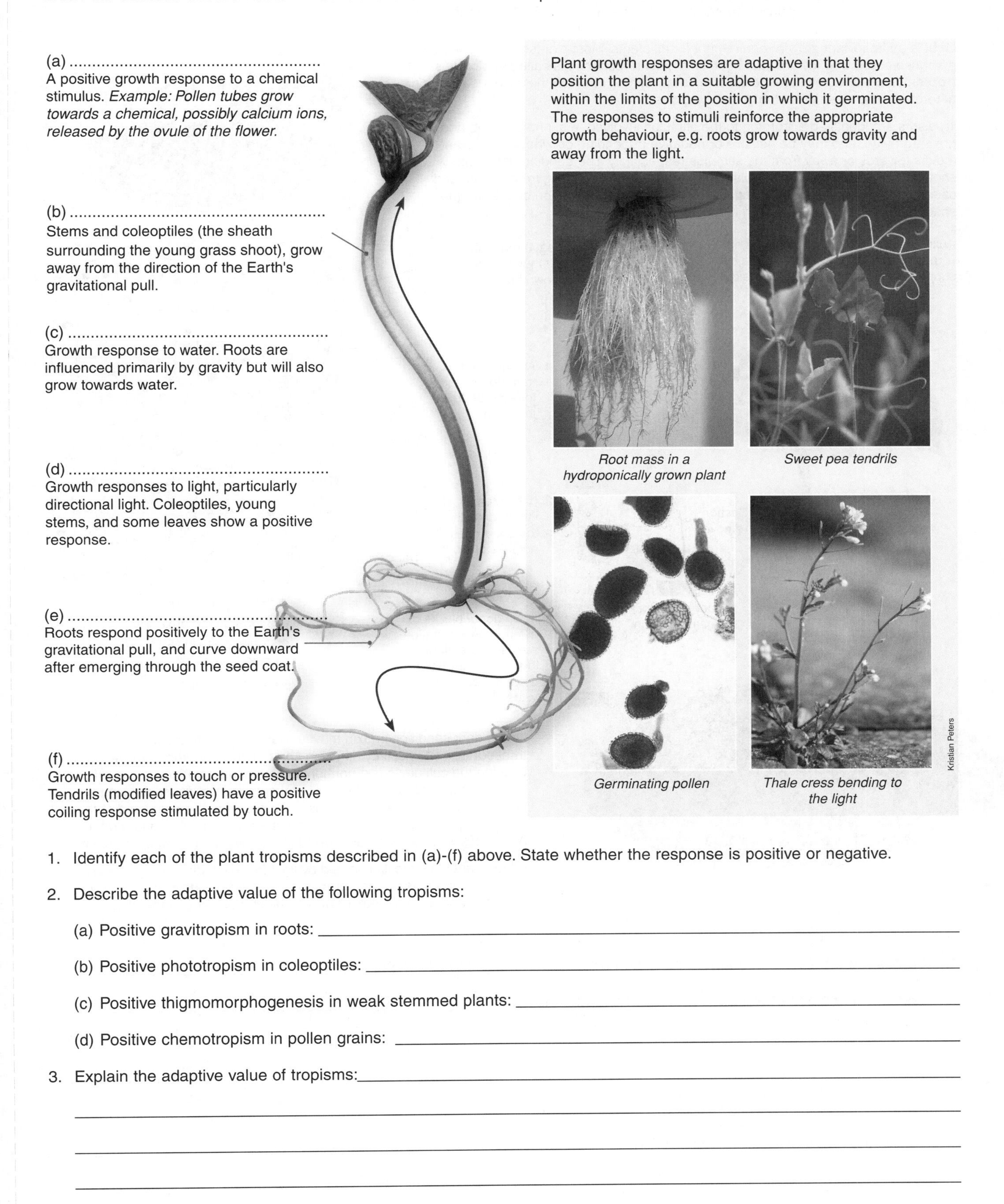

Root mass in a hydroponically grown plant

Sweet pea tendrils

Germinating pollen

Thale cress bending to the light

1. Identify each of the plant tropisms described in (a)-(f) above. State whether the response is positive or negative.

2. Describe the adaptive value of the following tropisms:

   (a) Positive gravitropism in roots: ______________________________

   (b) Positive phototropism in coleoptiles: ______________________________

   (c) Positive thigmomorphogenesis in weak stemmed plants: ______________________________

   (d) Positive chemotropism in pollen grains: ______________________________

3. Explain the adaptive value of tropisms: ______________________________

____

____

____

ISBN: 978-1-927309-20-9

38 Auxins and Shoot Growth

Key Idea: Auxin is a plant hormone involved in the plant's response to the environment and differential growth.

Auxins are **phytohormones** (plant growth substances) that have a central role in a wide range of growth and developmental responses in vascular plants. Indole-3-acetic acid (IAA) is the most potent native auxin in intact plants. The response of any particular plant tissue to IAA depends on the tissue itself, the concentration of the hormone, the timing of its release, and the presence of other phytohormones. Gradients in auxin concentration during growth prompt differential responses in specific tissues and contribute to directional growth.

Light is an important growth requirement for all plants. Most plants show an adaptive response of growing towards the light. This growth response is called phototropism. A **tropism** is a plant growth response to external stimuli in which the stimulus direction determines the direction of the growth response.

Tropisms are identified according to the stimulus involved, e.g. photo- (light) and gravi- (gravity), and may be positive or negative depending on whether the plant moves towards or away from the stimulus respectively. The bending of the plants shown on the right is a phototropism in response to light shining from the left and caused by the plant hormone **auxin**. Auxin causes the elongation of cells on the shaded side of the stem, causing it to bend.

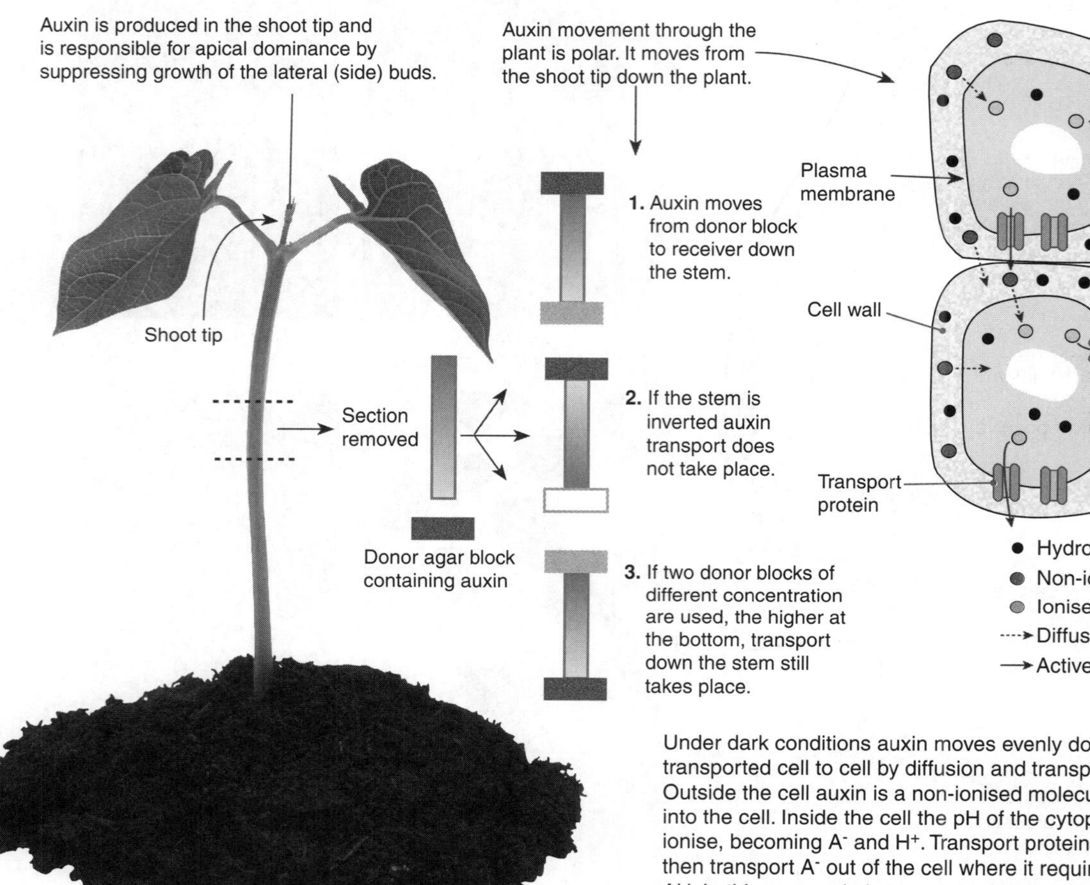

Under dark conditions auxin moves evenly down the stem. It is transported cell to cell by diffusion and transport proteins (above right). Outside the cell auxin is a non-ionised molecule (AH) which can diffuse into the cell. Inside the cell the pH of the cytoplasm causes auxin to ionise, becoming A^- and H^+. Transport proteins at the basal end of the cell then transport A^- out of the cell where it requires an H^+ ion and reforms AH. In this way auxin is transported in one direction through the plant.

When plant cells are illuminated by light from one direction transport proteins in the plasma membrane on the shaded side of the cell are activated and auxin is transported to the shaded side of the plant.

1. What is the term given to the tropism being displayed in the photo (top right)? ______

2. Describe one piece of evidence that demonstrates the transport of auxin is polar: ______

______

______

______

3. What is the effect of auxin on cell growth? ______

______

ISBN:978-1-927309-20-9

39 Investigating Phototropism

Key Idea: Experimental evidence supports the hypothesis that auxin is responsible for tropic responses in stems.

Phototropism in plants was linked to a growth promoting substance in the 1920s. Early experiments using severed coleoptiles gave evidence for the hypothesis that a plant hormone called auxin was responsible for tropic responses in stems. These experiments (below) have been criticised as being too simplistic, although their conclusions have been shown to be valid. Auxins promote cell elongation and are inactivated by light. Thus, when a stem is exposed to directional light, auxin becomes unequally distributed either side of the stem. The stem responds to the unequal auxin concentration by differential growth, i.e. it bends. The mechanisms behind this response are now well understood.

1. **Directional light:** A pot plant is exposed to direct sunlight near a window and as it grows, the shoot tip turns in the direction of the sun. When the plant was rotated, it adjusted by growing towards the sun in the new direction.

   (a) What hormone regulates this growth response?

   ___

   (b) What is the name of this growth response?

   ___

   (c) How do the cells behave to bring about this change in shoot direction at:

   Point **A**? ___

   Point **B**? ___

   (d) Which side (A or B) would have the highest hormone concentration and why?

   ___

   ___

   (e) Draw a diagram of the cells as they appear across the stem from point A to B (in the rectangle on the right).

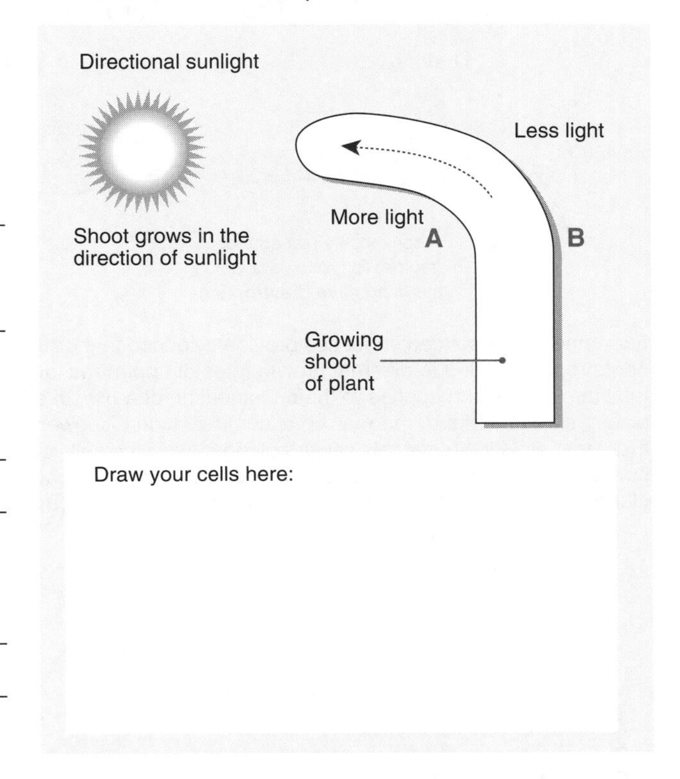

2. **Light excluded from shoot tip:** When a tin-foil cap is placed over the top of the shoot tip, light is prevented from reaching the shoot tip. When growing under these conditions, the direction of growth does not change towards the light source, but grows straight up. State what conclusion can you come to about the source and activity of the hormone that controls the growth response:

   ___

   ___

   ___

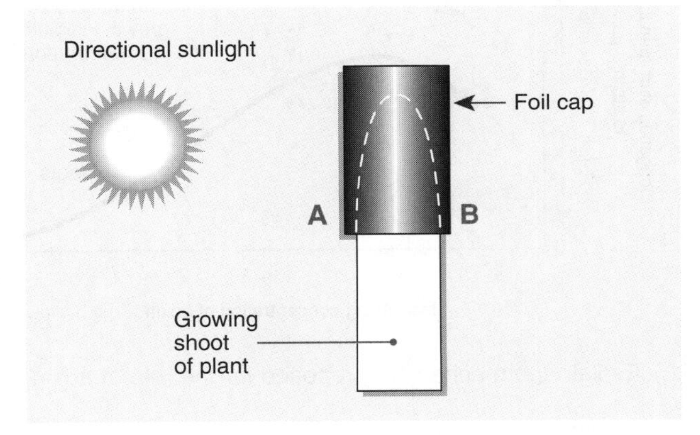

3. **Cutting into the transport system:** Two identical plants were placed side-by-side and subjected to the same directional light source. Razor blades were cut half-way into the stem, thereby interfering with the transport system of the stem. Plant A had the cut on the same side as the light source, while Plant B was cut on the shaded side. Predict the growth responses of:

   Plant **A**: ___

   ___

   ___

   Plant **B**: ___

   ___

   ___

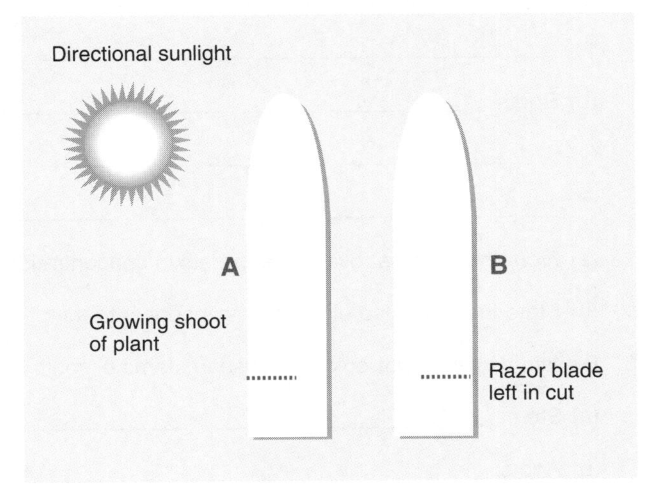

ISBN: 978-1-927309-20-9

40 Investigating Gravitropism

Key Idea: Auxin appears to have a role in the gravitropic responses of roots, but its effect may depend on the presence of other plant growth regulators.

The importance of the plant hormone auxin as a plant growth regulator, as well as its widespread occurrence in plants, led to it being proposed as the primary regulator in the gravitropic (*aka* geotropic) response. The basis of auxin's proposed role in gravitropism is outlined below. The mechanism is appealing in its simplicity but has been widely criticised because of the use of coleoptiles. The coleoptile (the sheath surrounding the young grass shoot) is a specialised, short-lived structure and is probably not representative of plant tissues generally.

The role of auxins in gravitropic responses

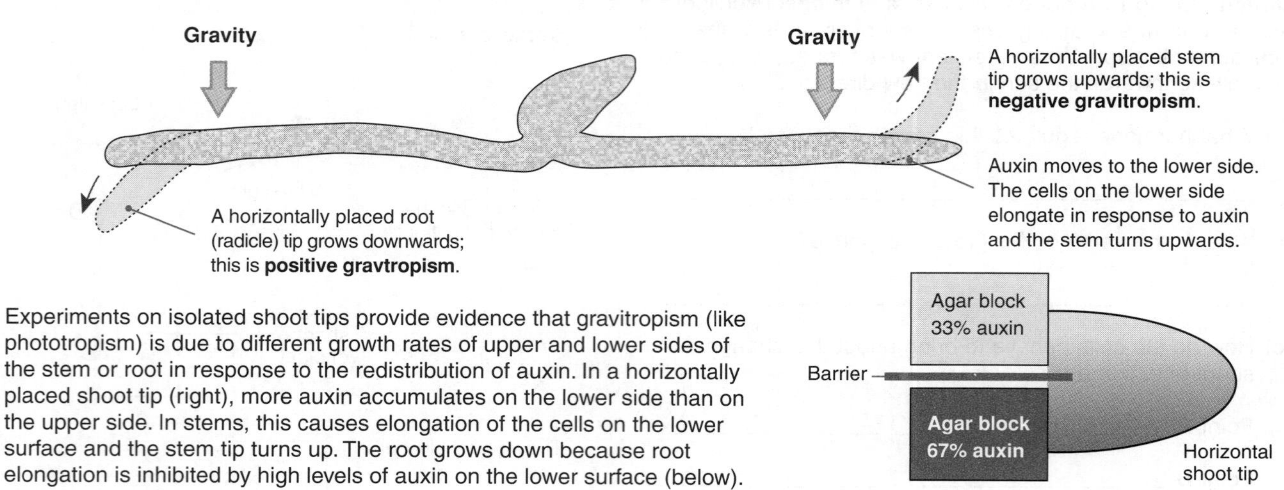

Experiments on isolated shoot tips provide evidence that gravitropism (like phototropism) is due to different growth rates of upper and lower sides of the stem or root in response to the redistribution of auxin. In a horizontally placed shoot tip (right), more auxin accumulates on the lower side than on the upper side. In stems, this causes elongation of the cells on the lower surface and the stem tip turns up. The root grows down because root elongation is inhibited by high levels of auxin on the lower surface (below).

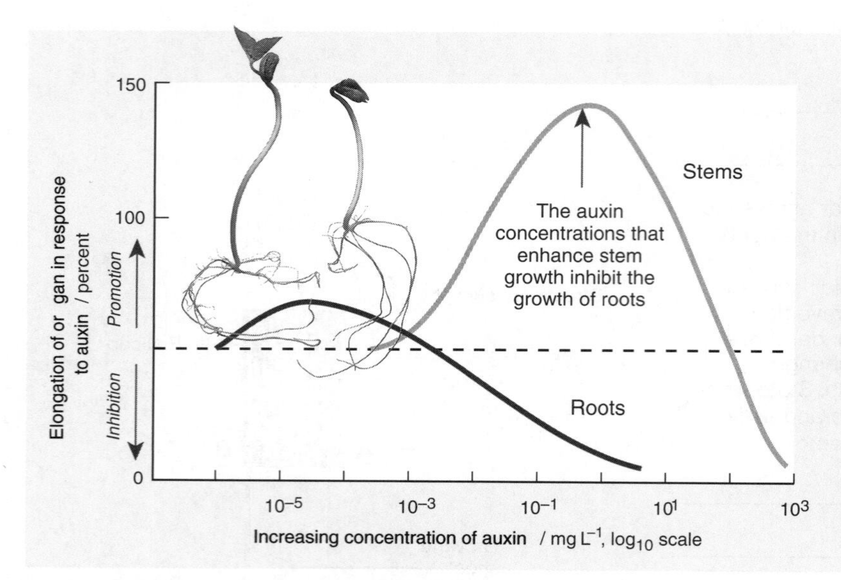

Auxin concentration and root growth

In a horizontally placed seedling, auxin moves to the lower side in stems and roots. The stem tip grows upwards and the root tip grows down. Root elongation is inhibited by the same level of auxin that stimulates stem growth (graph left). The higher auxin levels on the lower surface cause growth inhibition there. The longest cells are then on the upper surface and the root turns down. This simple explanation for gravitropism has been criticised because the concentrations of auxins measured in the upper and lower surfaces of horizontal stems and roots are too small to account for the growth movements observed. Other studies indicate that growth inhibitors may interact with auxin in geotropic responses.

1. Explain the mechanism proposed for the role of auxin in the geotropic response in:

   (a) Shoots (stems): ______________________________

   ____

   (b) Roots: ______________________________

   ____

   ____

2. (a) From the graph above, state the auxin concentration at which root growth becomes inhibited: ______________

   (b) State the response of stem at this concentration: ______________________________

3. Explain why the geotropic response in stems or roots is important to the survival of a seedling:

   (a) Stems: ______________________________

   (b) Roots: ______________________________

ISBN:978-1-927309-20-9

41 Investigation of Gravitropism in Seeds

Key Idea: The effect of gravity on the direction of root growth can be easily studied using sprout seeds. The direction of root growth will change if the seedling's orientation is altered. The experiment described below is a simple but effective way in which to investigate gravitropism in seedlings. Using the information below, analyse results and draw conclusions about the effect of gravity on the directional growth of seedling roots.

The aim

To investigate the effect of gravity on the direction of root growth in seedlings.

Hypothesis

Roots will always grow towards the Earth's gravitational pull, even when the seedling's orientation is changed.

Method

A damp kitchen paper towel was folded and placed inside a clear plastic sandwich bag. Two sprout seeds were soaked in water for five minutes and then placed in the centre of the paper towel. The bag was sealed. The plastic bag was then placed on a piece of cardboard which was slightly larger than the plastic bag. The plastic bag was stretched tightly so the plastic held the seeds in place, and secured with staples to the cardboard.

The cardboard was placed upright against a wall. Once the first root from each seed reached 2 cm long, the cardboard was turned 90° degrees.

Daily observations and photographs were made of the root length and direction throughout the duration of the experiment. Photos of one seedling from days 5 and 11 are shown right.

Results

The students took photographs to record changes in growth during the course of the experiment. One seedling at day 5 and 11 is shown below.

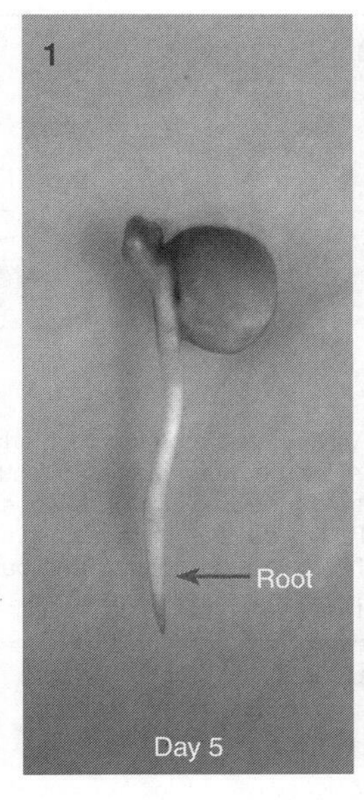

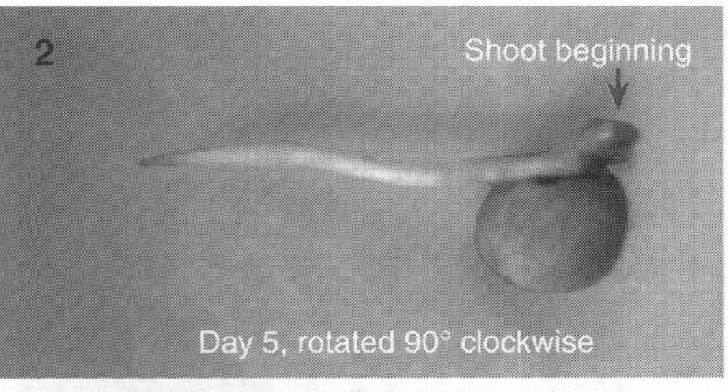

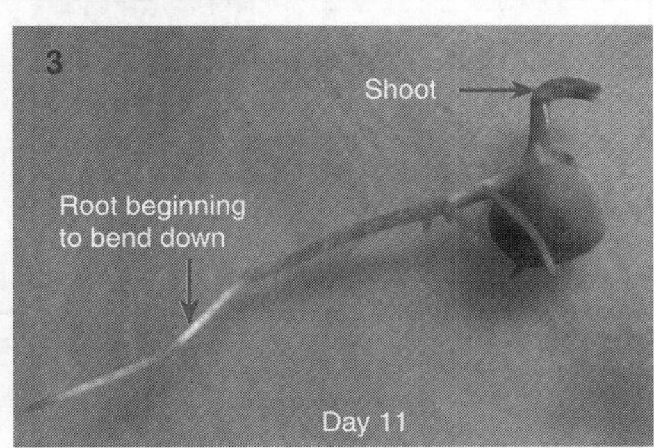

Photo 1: This photo was 5 five days after the seed began to germinate.

Photo 2: After photo 1 was taken, the cardboard was rotated 90°.

Photo 3: This photo was taken 6 days after the seed was rotated 90°.

1. (a) What direction did the root first begin to grow in? ______________________

   (b) Describe what happened to the root when the students rotated the cardboard 90°: ______________________

   ____

   ____

   (c) Explain why this occurred: ______________________

   ____

   ____

   (d) Predict the result after six more days growth if the students rotated the seedling in photo 3 90° clockwise. Draw your answer in the space right:

2. During the course of the experiment a shoot developed.

   (a) In what direction did the shoot grow at first? ______________________

   ____

   (b) In what direction did the shoot grow after rotation 90° (photo 3)?

   ____

   (c) Why did this occur? ______________________

   ____

ISBN: 978-1-927309-20-9

42 Taxes and Kineses

Key Idea: Taxes and kineses are innate locomotory behaviours involving movements in response to external stimuli, e.g. gravity, light, chemicals, water, touch, or temperature. A **kinesis** (*pl.* kineses) is a non-directional response to a stimulus in which the speed of movement or the rate of turning is proportional to the stimulus intensity. Kineses do not involve orientation directly to the stimulus and are typical of many invertebrates and protozoa. **Taxes** (*sing.* taxis) involve orientation and movement in response to a directional stimulus or a gradient in stimulus intensity. Taxes often involve moving the head until the sensory input from both sides is equal (**klinotaxis**). Many taxes involve a simultaneous response to more than one stimulus, e.g. fish orientate dorsal side up in response to both light and gravity. Orientation responses are always classed according to whether they are towards the stimulus (positive) or away from it (negative). Simple orientation responses are **innate** (genetically programmed). More complex orientation responses may involve learning (the behaviour may be modified based on experience).

When confronted with a vertical surface, snails will reorientate themselves so that they climb vertically upwards. The adaptive advantage of this may be to help the snail find food or shelter, or to avoid overly wet surfaces.

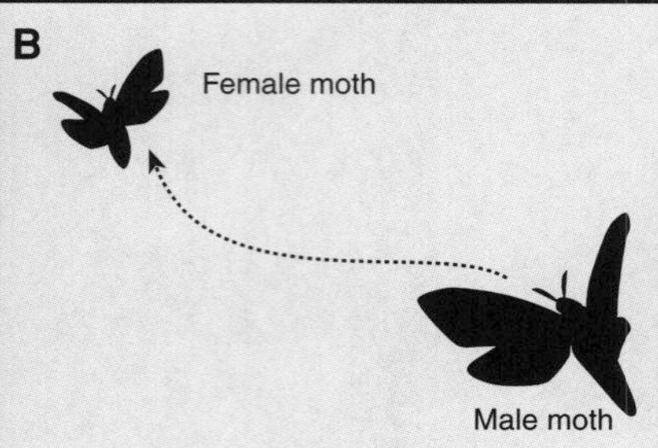

A flying male moth, encountering an odour (pheromone) trail left by a female, will turn and fly upwind until it reaches the female. This behaviour increases the chances of the male moth mating and passing on its genes to the next generation.

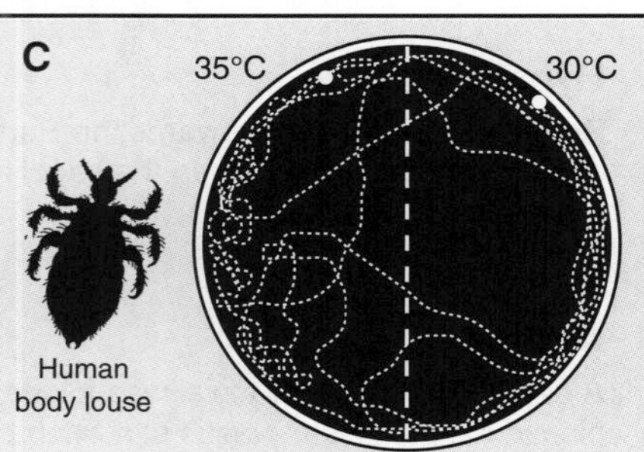

In a circular chamber, lice make relatively few turns at their preferred temperature of 30°C, but many random turns at 35°C. This response enables the lice to increase their chances of finding favourable conditions and remaining in them once found.

Spiny lobsters will back into tight crevices so that their body is touching the crevice sides. The antennae may be extended out. Behaving in such a way gives the lobsters greater protection from predators.

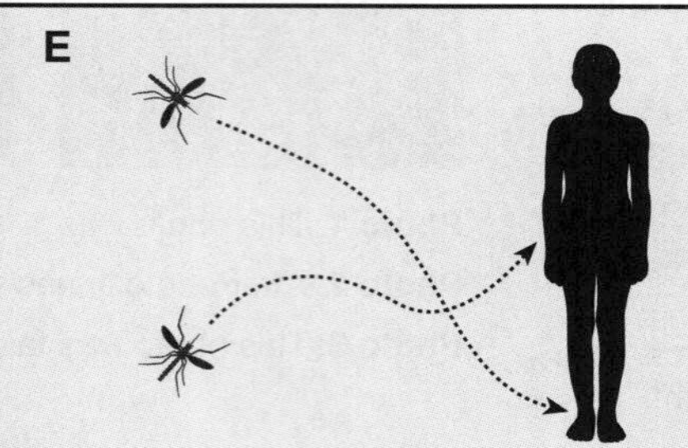

At close range, mosquitoes use the temperature gradient generated by the body heat of a host to locate exposed flesh. This allows the female to find the blood needed for the development of eggs.

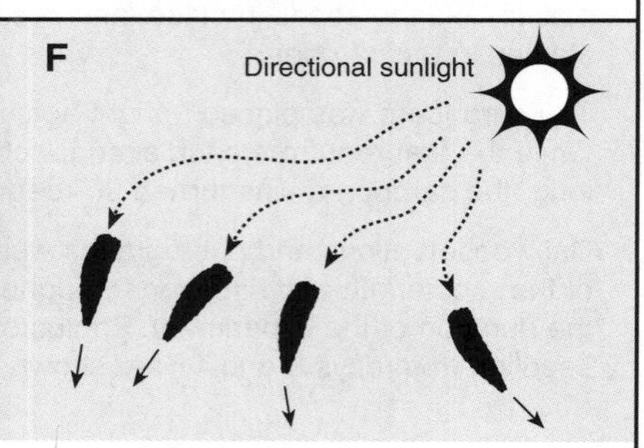

Blowfly maggots will turn and move rapidly away from a directional light source. Light usually indicates hot dry areas and the maggots avoid predators and desiccation (drying out) by avoiding the light.

1. Distinguish between a **kinesis** and a **taxis**, describing examples to illustrate your answer:

2. Describe the adaptive value of simple orientation behaviours such as kineses:

3. Name the physical stimulus for each of the following prefixes used in naming orientation responses:

(a) Gravi- (b) Hydro- (c) Thigmo-

(d) Photo- (e) Chemo- (f) Thermo-

4. For each of the above examples (**A-F**), describe the orientation response and whether it is positive or negative:

(a) **A:** (d) **D:**

(b) **B:** (e) **E:**

(c) **C:** (f) **F:**

ISBN:978-1-927309-20-9

Kinesis in woodlice

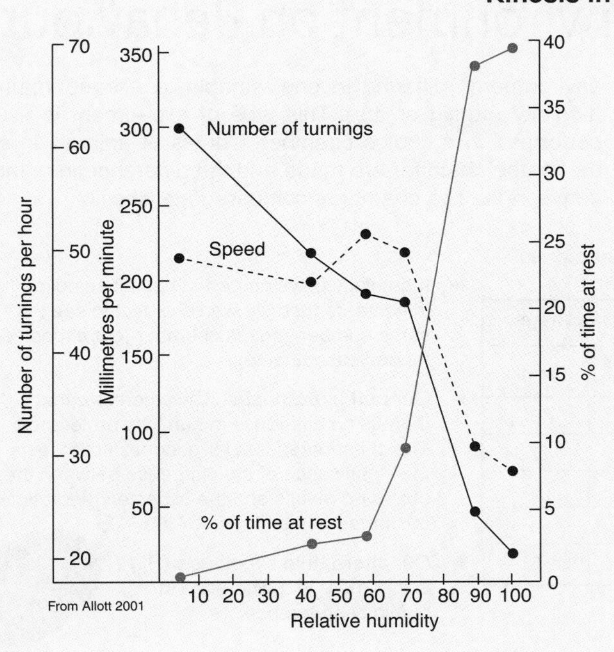

From Allott 2001

Woodlice are commonly found living in damp conditions under logs or bark. Many of the behavioural responses of woodlice are concerned with retaining moisture. Unlike most other terrestrial arthropods, they lack a waterproof cuticle, so water can diffuse through the exoskeleton, making them vulnerable to drying out. When exposed to low humidity, or high temperature or high light levels, woodlice show a kinesis response to return them to their preferred, high humidity environment.

5. Use the data on woodlice above to answer the following questions:

   (a) At which relative humidities do the following occur:

      i. Largest number of turnings per hour: ______

      ii. Highest speed of movement: ______

      iii. Largest percentage of time at rest: ______

   (b) Explain the significance of these movements: ______

   ______

   ______

   ______

   (c) What is the preferred range of relative humidity for the woodlice?

   ______

6. The diagrams on the right show the movement of nematodes on plates where a salt (NH_4Cl) was added (**A**) and on a plate where no NH_4Cl was added (**B**).

   (a) Describe the movements of the nematodes in plates A and B:

   ______

   ______

   ______

   (b) Name the orientation behaviour shown in plate A: ______

   ______

   (c) Describe an advantage of this kind of behaviour to nematodes:

   ______

   ______

   ______

   ______

   ______

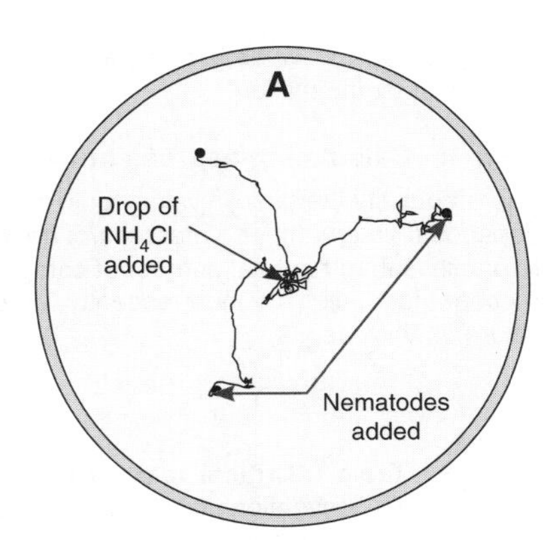

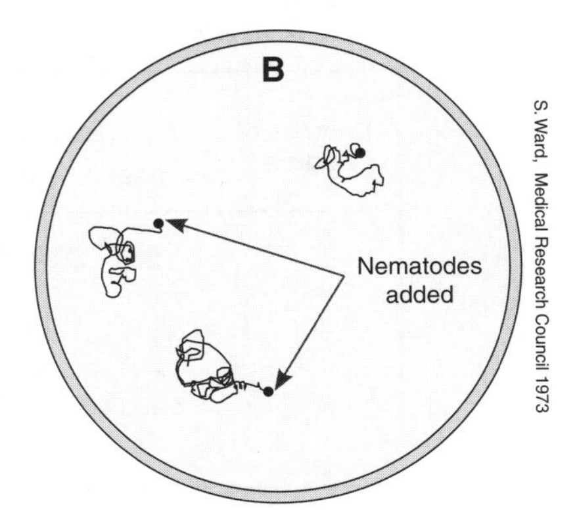

S. Ward, Medical Research Council 1973

43 Testing the Effect of Environment on Behaviour

Key Idea: The chi-squared test for goodness of fit can be used to test differences in habitat preference in animals.
The effect of the environment on simple responses such as kineses can be statistically tested using chi-squared for goodness of fit. An animal is given a choice between two environments differing in one variable, e.g. temperature of humidity (humid or dry). This type of experiment is usually performed in a choice chamber. Counts of animals in each part of the chamber are made and the difference in numbers between the two chambers compared statistically.

Students carried out an investigation into habitat preference in woodlice. In particular, they were wanting to know if the woodlice preferred a humid atmosphere to a dry one, as this may explain their observed distribution in damp areas in natural habitats. They designed a simple investigation to test this idea. The woodlice were randomly placed into a choice chamber for 5 minutes where they could choose between dry and humid conditions (atmosphere). The investigation consisted of five trials with ten woodlice used in each trial. Their results are shown in the table right:

Habitat preference in woodlice

Trial	Atmosphere	
	Dry	Humid
1	2	8
2	3	7
3	4	6
4	1	9
5	5	5

- If humidity plays no part in habitat selection in woodlice then we would expect to see the same number in each chamber (our expected theoretical outcome).
- Our **null hypothesis** (H_0) is therefore that there is no difference in humidity preference. The chi-squared test for goodness of fit tests the significance of the difference between the observed results and the expected theoretical outcome.
- Our **alternative** hypothesis (H_A) is that there is a difference in humidity preference.

Using χ^2, the probability of this result being consistent with the expected result could be tested. Worked example as follows:

Step 1: Calculate the expected value (E)

In this case, this is the sum of the observed values divided by the number of categories.

$$\frac{50}{2} = 25$$

Step 2: Calculate O – E

The difference between the observed and expected values is calculated as a measure of the deviation from a predicted result. Since some deviations are negative, they are all squared to give positive values. This step is usually performed as part of a tabulation (right, darker blue column).

Step 3: Calculate the value of χ^2

$$\chi^2 = \sum \frac{(O - E)^2}{E}$$

Where: O = the observed result
E = the expected result
Σ = sum of

The calculated χ^2 value is given at the bottom right of the last column in the tabulation.

Step 4: Calculating degrees of freedom

The probability that any χ^2 value could be exceeded by chance depends on the number of degrees of freedom (df). Df is equal to one less than the total number of categories (this is the number that could vary independently without affecting the last value). In this case: *2 – 1 = 1.*

Category	O	E	O - E	(O - E)^2	$\frac{(O - E)^2}{E}$
Humid	35	25	10	100	4
Dry	15	25	-10	100	4
	Total = 50			χ^2 →	$\Sigma = 8$

Step 5: Using the χ^2 table

- On the χ^2 table (part reproduced in Table 1 below) with 1 degree of freedom, the calculated value for χ^2 of 8 corresponds to a probability of between 0.01 and 0.001 (see arrow). This means that by chance alone a χ^2 value of 8 could be expected between 1% and 0.1% of the time.
- The probability of between 0.1 and 0.01 is lower than the 0.05 value which is generally regarded as significant.
- The null hypothesis can be rejected and we have reason to believe that the observed results differ significantly from the expected (at $P = 0.05$).
- The alternative hypothesis that there is a difference in humidity preference can be accepted at $P = 0.05$.

Table 1: Critical values of χ^2 at different levels of probability.
By convention, the critical probability for rejecting the null hypothesis (H_0) is 5%. If the test statistic is less than the tabulated critical value for $P = 0.05$ we cannot reject H_0 and the result is not significant.
If the test statistic is greater than the tabulated value for $P = 0.05$ we reject H_0 in favour of the alternative hypothesis.

Degrees of freedom	Level of probability (P)									
	0.98	0.95	0.80	0.50	0.20	0.10	0.05	0.02	0.01	0.001
1	0.001	0.004	0.064	0.455	1.64	2.71	3.84	5.41	6.64	10.83
2	0.040	0.103	0.466	1.386	3.22	4.61	5.99	7.82	9.21	13.82
3	0.185	0.352	1.005	2.366	4.64	6.25	7.82	9.84	11.35	16.27
4	0.429	0.711	1.649	3.357	5.99	7.78	9.49	11.67	13.28	18.47
5	0.752	0.145	2.343	4.351	7.29	9.24	11.07	13.39	15.09	20.52
	← *Do not reject H_0*							*Reject H_0* →		

χ^2 (arrow between 6.64 and 10.83)

REFER

ISBN:978-1-927309-20-9

44 Choice Chamber Investigation

Key Idea: Choice chambers are a simple way to test animal behaviours such as a simple orientation behaviour.

Choice chambers are a simple way to investigate behaviour in animals. A simple choice chamber consists of two distinct areas enclosing opposing environments, e.g. warm and cool, dry and humid, light and dark. Animals are placed in the middle of the chamber and given time to move to their preferred area before numbers in each chamber are counted.

Background

Students carried out two investigations on woodlice. The first was to determine woodlouse preference for light or dark. The second was to test preference for warm or cool environments.

Aim: investigation 1

To investigate if woodlice prefer a light or dark environment.

The method

A choice chamber was set up using two joined petri dishes, one painted black, the other left clear. The chamber was kept at room temperature (21°C). Ten woodlice were placed into the joining segment of the chamber and left for ten minutes to orientate themselves. The numbers of woodlice in each chamber were then recorded. The experiment was carried out a total of four times.

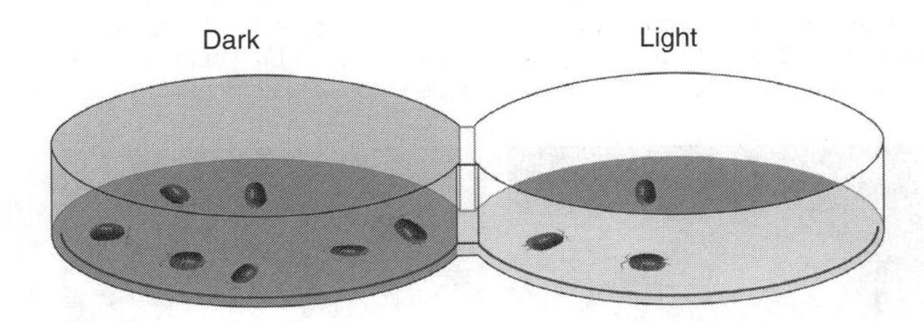

Results

	Number of woodlice in chamber	
Trial	**Dark**	**Light**
1	7	3
2	9	1
3	8	2
4	9	1

Aim: investigation 2

To investigate if woodlice prefer a warm or cool environment.

The method

A choice chamber was set up painted entirely black. One side was heated to 27°C by placing a heat pad underneath. The other side was kept cool at 14°C by placing a towel soaked in cool water around the chamber. Ten woodlice were placed into the joining segment of the chamber and left for ten minutes to orientate. The numbers of woodlice in each chamber were then recorded. The experiment was carried out a total of four times.

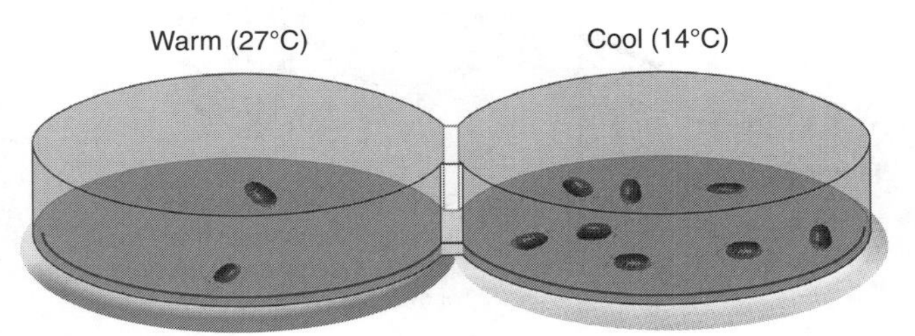

Results

	Number of woodlice in chamber	
Trial	**Warm**	**Cool**
1	2	8
2	3	7
3	2	8
4	1	9

1. State the null hypothesis for test 1: ____________________

2. (a) Calculate a χ^2 value for test 1: ____________ (b) Calculate the degrees of freedom: ____________

3. Use the $\chi 2$ table in the previous activity to decide if the test is significant at P = 0.01: ____________________

4. State the null hypothesis for test 2: ____________________

5. (a) Calculate a χ^2 value for test 1: ____________ (b) Calculate the degrees of freedom: ____________

6. Use the $\chi 2$ table in the previous activity to decide if the test is significant at P = 0.01: ____________________

7. Use the results of the two tests to make a statement about the habitat preference of woodlice: ____________________

ISBN: 978-1-927309-20-9

LINK 40 KNOW

45 Reflexes

Key Idea: A reflex is an automatic response to a stimulus and involves only a few neurones and a central processing point.

A **reflex** is an automatic response to a stimulus. Reflexes require no conscious thought and so act quickly to protect the body from harmful stimuli. Reflexes are controlled by a neural pathway called a reflex arc. A reflex arc involves a small number of neurones (nerve cells) and a central nervous system processing point, which is usually the spinal cord, but sometimes the brain stem. Reflexes are classified according to the number of CNS synapses involved. **Monosynaptic reflexes** involve only one CNS synapse (e.g. knee jerk reflex), whereas **polysynaptic reflexes** involve two or more (e.g. pain withdrawal reflex). Both are spinal reflexes. The pupil reflex and the corneal (blink) reflex are cranial reflexes.

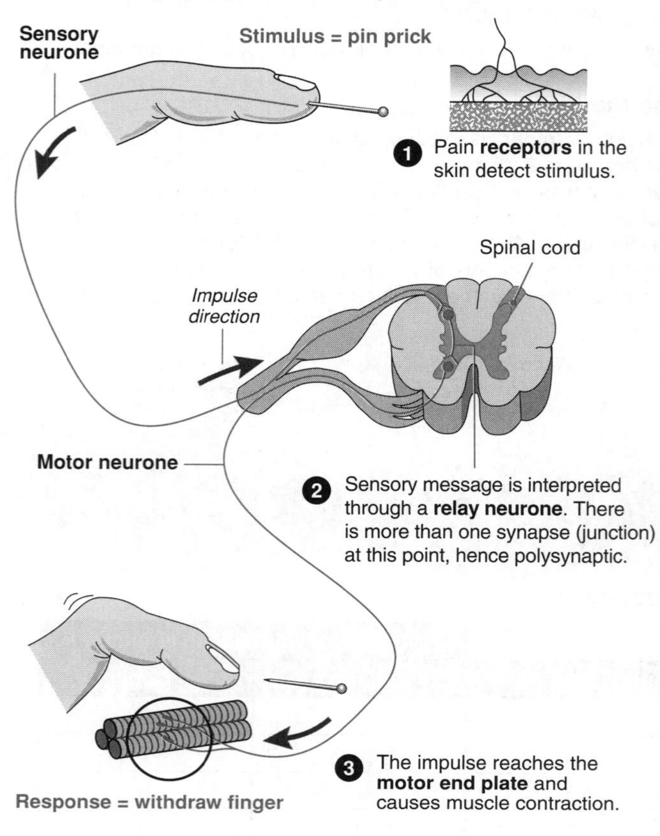

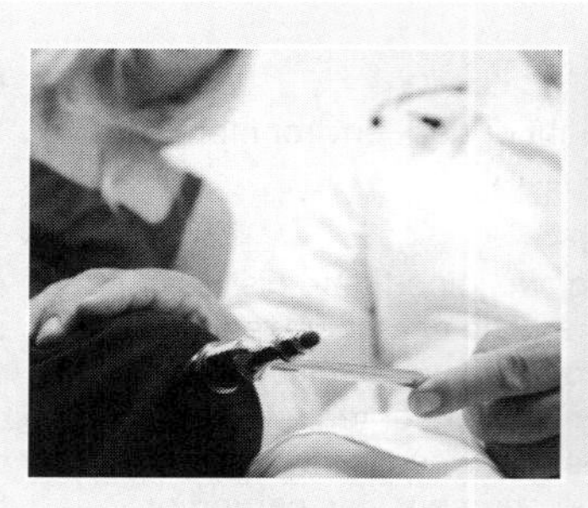

The patellar (knee jerk) reflex is a simple deep tendon reflex that is used to test the function of the femoral nerve and spinal cord segments L2-L4. It helps to maintain posture and balance when walking.

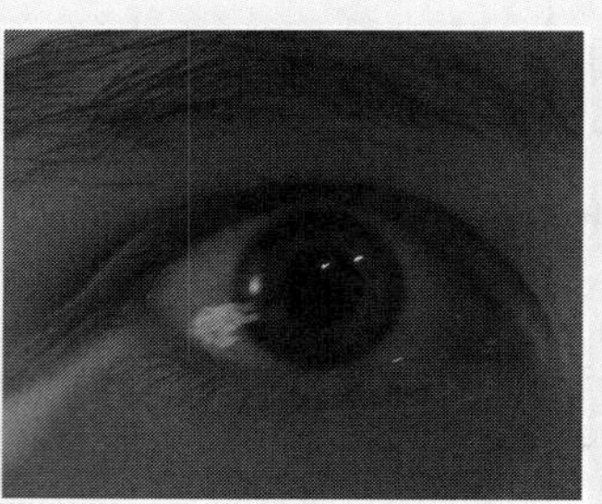

The corneal (blink) reflex is a rapid involuntary blinking of both eyelids occurring when the cornea is stimulated, e.g. by touching. It is mediated by the brainstem and can be used to evaluate coma.

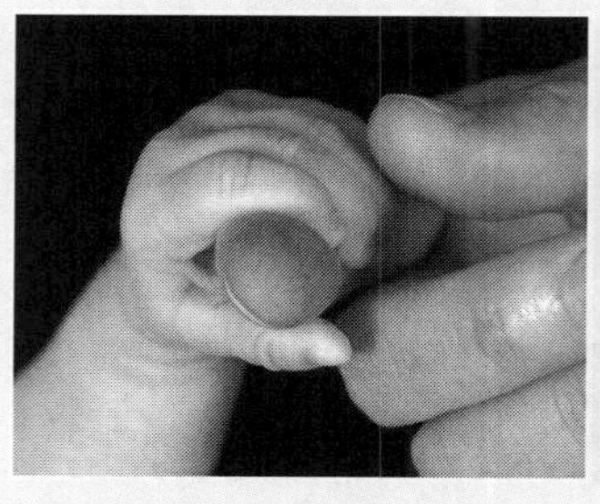

Normal newborns exhibit a number of primitive reflexes in response to particular stimuli. These include the grasp reflex (above) and the startle reflex in which a sudden noise will cause the infant to extend its arms, legs, and head, and cry.

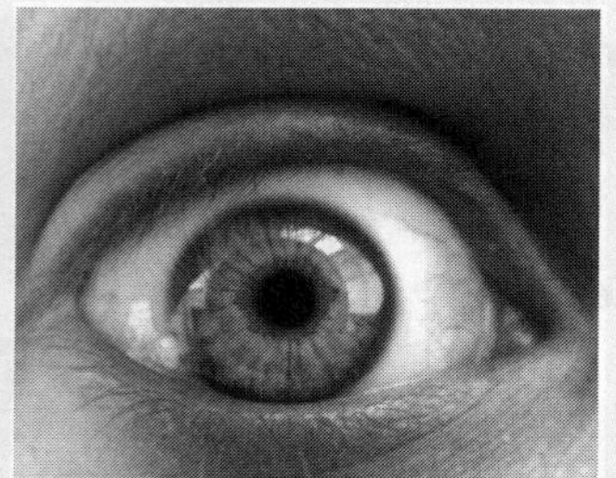

The pupillary light reflex refers to the rapid expansion or contraction of the pupils in response to the intensity of light falling on the retina. It is a polysynaptic cranial reflex and can be used to test for brain death.

1. Explain why higher reasoning or conscious thought are not necessary or desirable features of reflex behaviours:

2. Distinguish between a spinal reflex and a cranial reflex and give an example of each:

3. Describe the survival value of the following reflexes:

(a) Knee-jerk reflex:

(b) Corneal blink reflex:

(c) Grasp reflex:

(d) Pupillary light reflex:

ISBN:978-1-927309-20-9

46 The Basis of Sensory Perception

Key Idea: Sensory receptors covert stimuli into electrical or chemical signals.

Sensory receptors are specialised to detect stimuli and respond by producing an electrical (or chemical) discharge. In this way they act as **biological transducers**, converting the energy from a stimulus into an electrochemical signal. They can do this because the stimulus opens (or closes) ion channels and leads to localised changes in membrane potential called **receptor potentials**. Receptor potentials are graded and not self-propagating, but sense cells can amplify them, generating self-propagating nerve impulses (action potentials). Ultimately the stimulus is transduced into nerve impulses, the frequency of which is dependent on stimulus strength. The simplest sensory receptors are single sensory neurones (e.g. nerve endings). More complex sense cells form synapses with their sensory neurones (e.g. taste buds). Sensory receptors are classified according to the stimuli to which they respond (e.g. photoreceptors respond to light).

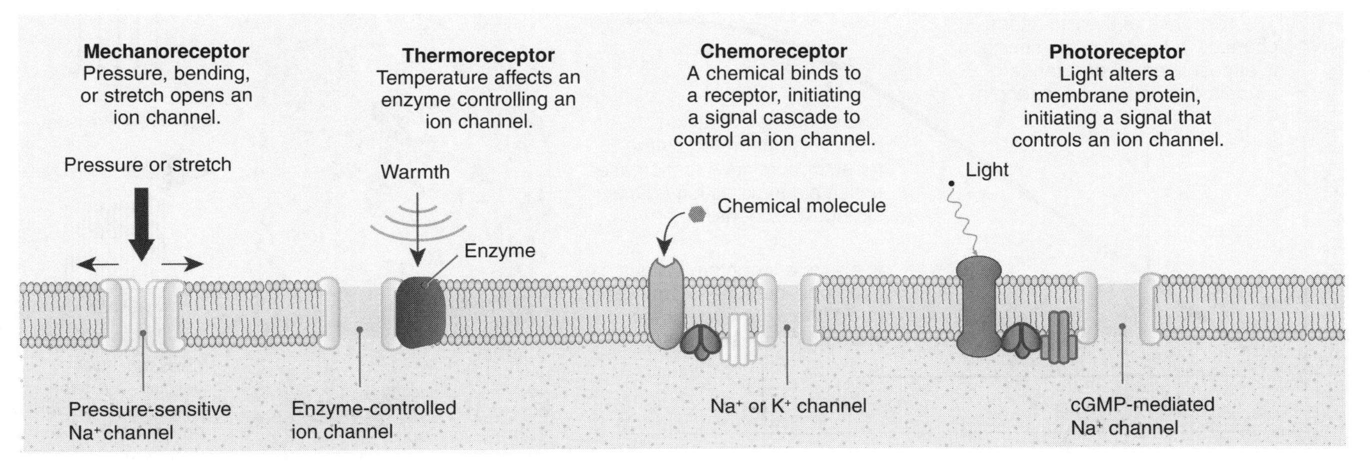

Signal transduction

Sensory cells convert one type of stimulus energy (e.g. pressure) into an electrical signal by altering the flow of ions across the plasma membrane and generating receptor potentials. In many cases (as in the Pacinian corpuscle), this leads directly to the generation of action potentials in the sensory cell. In some receptor cells, the receptor potential leads to the release of signal molecules called **neurotransmitters**, which directly or indirectly lead to action potentials in a post-synaptic cell.

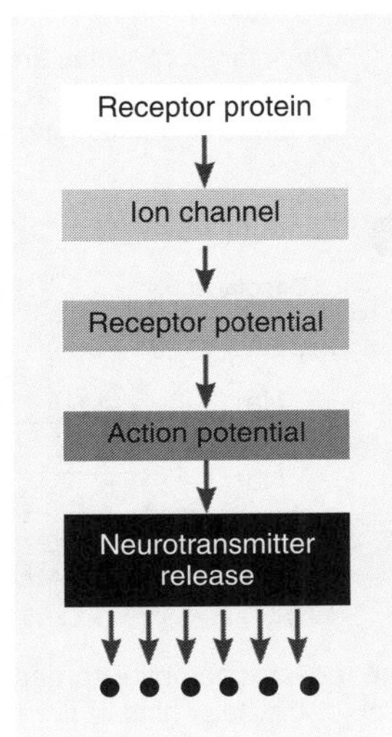

The Pacinian corpuscle

Pacinian corpuscles are pressure receptors in deep subcutaneous tissues of the body. They are relatively large but structurally simple, consisting of a sensory nerve ending (dendrite) surrounded by a capsule of connective tissue layers. Pressure deforms the capsule, stretching the nerve ending and leading to a localised depolarisation called a **receptor potential.** Receptor potentials are graded and do not spread far, although they may sum together and increase in amplitude.
The sense cell converts the receptor potentials to action potentials in the spike generating zone at the start of the axon. The action potential is then propagated along the axon.

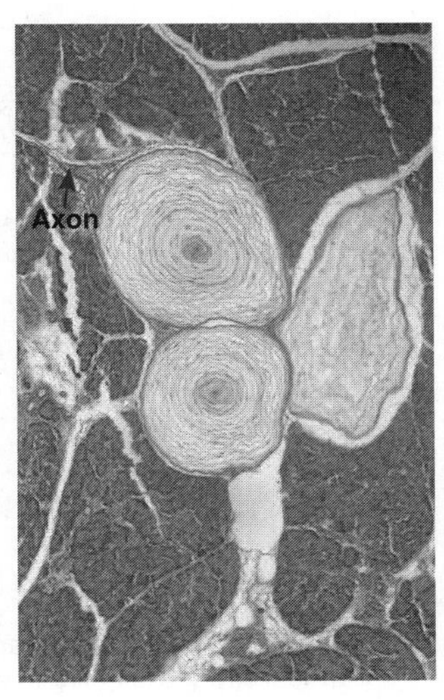

D. Fankhauser, University of Cincinnati, Clermont College

1. Explain why sensory receptors are termed 'biological transducers': ______________________

2. Identify one feature that all sensory receptors have in common: ______________________

3. Explain how a stimulus received by a sensory receptor is converted into an electrical response:

ISBN: 978-1-927309-20-9

47 Encoding Information

Key Idea: A sensory receptor communicates information about stimulus strength in the frequency of action potentials.
A receptor must do more than simply record a stimulus. It is important that it also provides information about the stimulus strength. Action potentials obey the 'all or none law' and are always the same size, so stimulus strength cannot be encoded by varying the amplitude of the action potentials. Instead, the frequency of impulses conveys information about the stimulus intensity; the higher the frequency of impulses, the stronger the stimulus. This encoding method is termed **frequency modulation**, and is the way that receptors inform the brain about stimulus strength. In the Pacinian corpuscle (below) frequency modulation is possible because a stronger pressure produces larger receptor potentials, which depolarise the spike generation zone to threshold more rapidly and produce a more rapid volley of action potentials. Sensory receptors also show **sensory adaptation** and will cease responding to a stimulus of the same intensity.

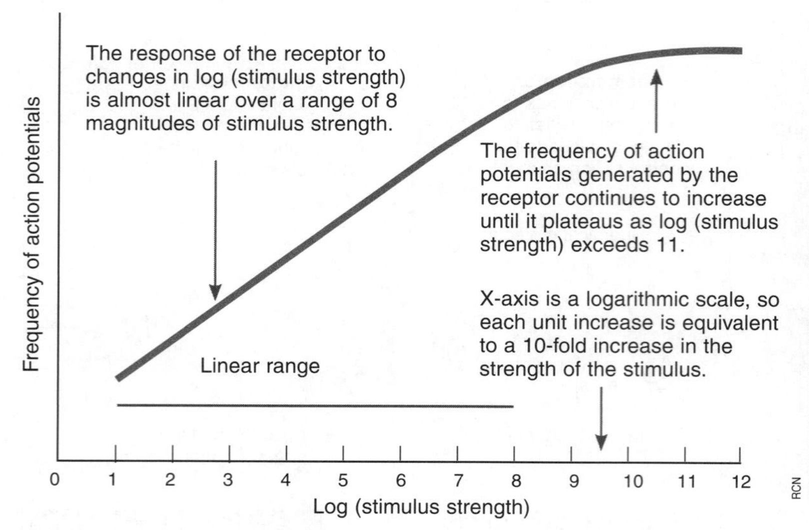

Receptors can use variation in action potential frequency to encode stimulus strengths that vary by nearly 11 orders of magnitude.

Layers of connective tissue deformed by pressure

Axon

RCN

Weak

Pressure

Pressure

Strong

A stronger stimulus (pressure) will produce a higher frequency of action potentials than a weaker stimulus.

Pacinian corpuscles are rapidly adapting receptors. They fire at the beginning and end of a stimulus, but do not respond to unchanging pressure.

1

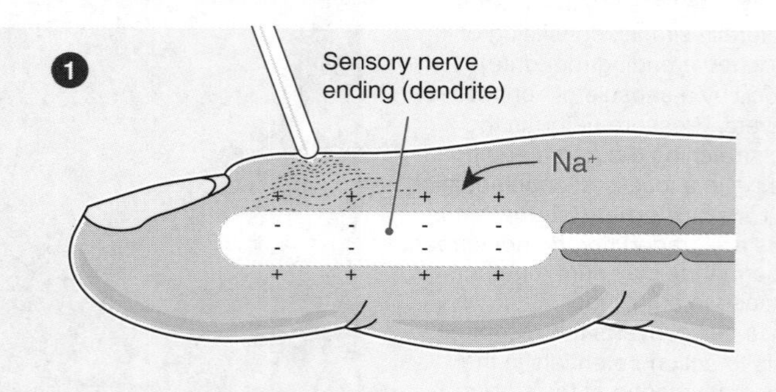

Deforming the corpuscle leads to an increase in the permeability of the nerve to sodium. Na^+ diffuses into the nerve ending creating a localised depolarisation. This depolarisation is called a **receptor potential**.

2

Depolarisation

Action potential

Na^+

Na^+

Axon (conducting region of cell)

The receptor potential spreads to the spike generation zone at the beginning of the axon. It depolarises the axon membrane to threshold and generates an action potential, which propagates along the axon.

1. (a) Explain how the strength of a stimulus is encoded by the nervous system: ______

   (b) Explain the significance of encoding information in this way: ______

2. Using the example of the Pacinian corpuscle, explain how stimulus strength is linked to frequency of action potentials: ______

3. Why is sensory adaptation important? ______

ISBN:978-1-927309-20-9

48 The Structure of the Eye

Key Idea: The eye is a sensory organ that converts light into nerve impulses resulting in the formation of a visual image.

The eye is a complex and highly sophisticated sense organ specialised to detect light. The adult eyeball is about 25 mm in diameter. Only the anterior one-sixth of its total surface area is exposed; the rest lies recessed and protected by the **orbit** into which it fits. The eyeball is protected and given shape by a fibrous tunic. The posterior part of this structure is the **sclera** (the white of the eye), while the anterior transparent portion is the **cornea**, which covers the coloured iris.

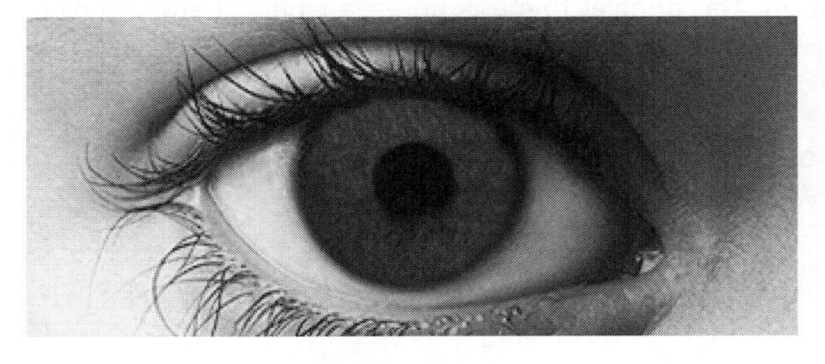

The structure and function of the human eye

The human eye is essentially a three layered structure comprising an outer fibrous layer (the sclera and cornea), a middle vascular layer (the choroid, ciliary body, and iris), and inner **retina** (neurones and **photoreceptor cells**). The shape of the eye is maintained by the fluid filled cavities (aqueous and vitreous humours), which also assist in light refraction. Eye colour is provided by the pigmented iris. The iris also regulates the entry of light into the eye through the contraction of circular and radial muscles.

Forming a visual image

Before light can reach the photoreceptor cells of the retina, it must pass through the cornea, aqueous humor, pupil, lens, and vitreous humour. For vision to occur, light reaching the photoreceptor cells must form an image on the retina. This requires **refraction** of the incoming light, **accommodation** of the lens, and **constriction** of the pupil.

The anterior of the eye is concerned mainly with **refracting** (bending) the incoming light rays so that they focus on the retina. Most refraction occurs at the cornea. The lens adjusts the degree of refraction to produce a sharp image. **Accommodation** adjusts the eye for near or far objects. Constriction of the pupil narrows the diameter of the hole through which light enters the eye, preventing light rays entering from the periphery.

The point at which the nerve fibres leave the eye as the optic nerve, is the **blind spot** (the point at which there are no photoreceptor cells). Nerve impulses travel along the optic nerves to the visual processing areas in the cerebral cortex. Images on the retina are inverted and reversed by the lens but the brain interprets the information it receives to correct for this image reversal.

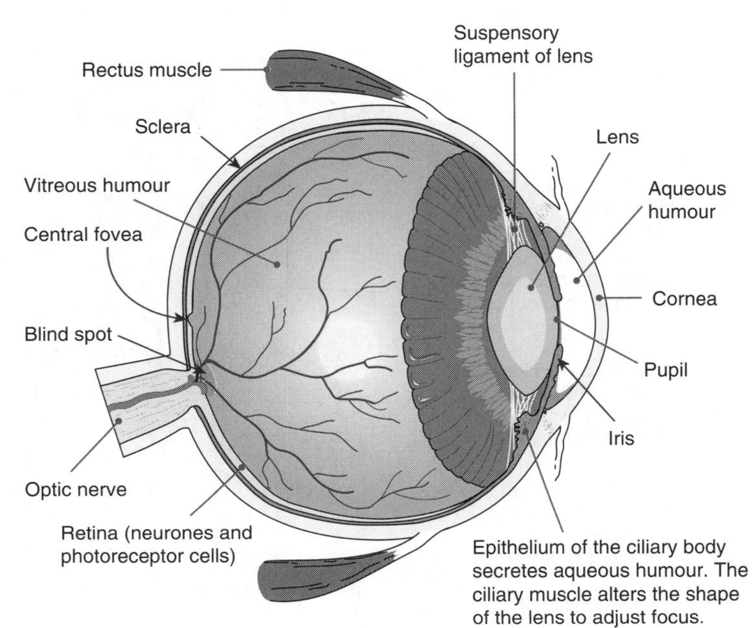

1. Identify the function of each of the structures of the eye listed below:

   (a) Cornea: ______

   (b) Ciliary body: ______

   (c) Iris: ______

2. (a) The first stage of vision involves forming an image on the retina. In simple terms, explain what this involves:

   ______

   (b) Explain how accommodation is achieved: ______

ISBN: 978-1-927309-20-9

49 The Physiology of Vision

Key Idea: The retina of the eye is a multilayered structure, which detects light and generates electrical responses. These are converted to action potentials in the optic nerve.

Vision in mammals is achieved by focussing light through a lens to form an image on the retina at the back of the eye. Light reaching the retina is absorbed by the photosensitive pigments associated with the membranes of the **photoreceptor cells** (the rods and cones). The pigment molecules are altered by the absorption of light and this causes them to produce electrical responses. These initial responses are converted to action potentials, which are transmitted via the **optic nerve** to the visual cortex of the brain, where the information is interpreted. The retina is not uniform. The **central fovea** is an area where there is a high density of cones and virtually no rods. It is the region of highest acuity.

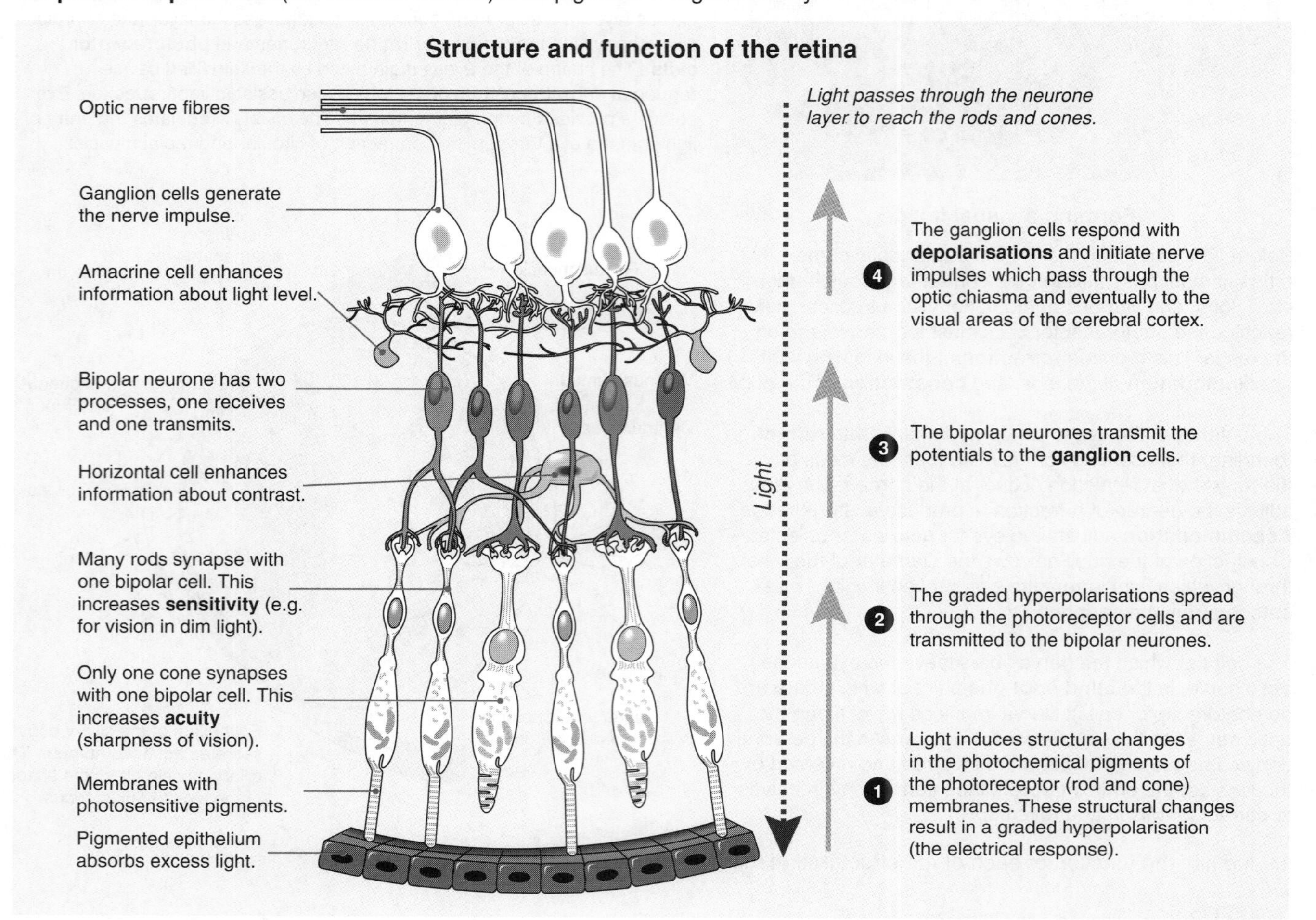

1. Describe the role of each of the following in human vision:

(a) Retina: ______

(b) Optic nerve: ______

(c) Central fovea: ______

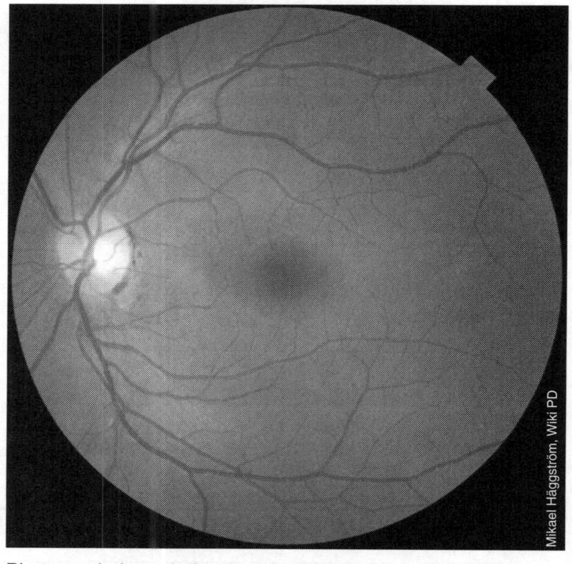

Photograph through the eye of a normal retina. The blind spot, where the where ganglion cell axons exit the eye to form the optic nerve, is seen as the bright area to the left of the image. The central fovea, where cone density is highest, is in the darker region at the centre of the image. Note the rich blood supply.

ISBN:978-1-927309-20-9

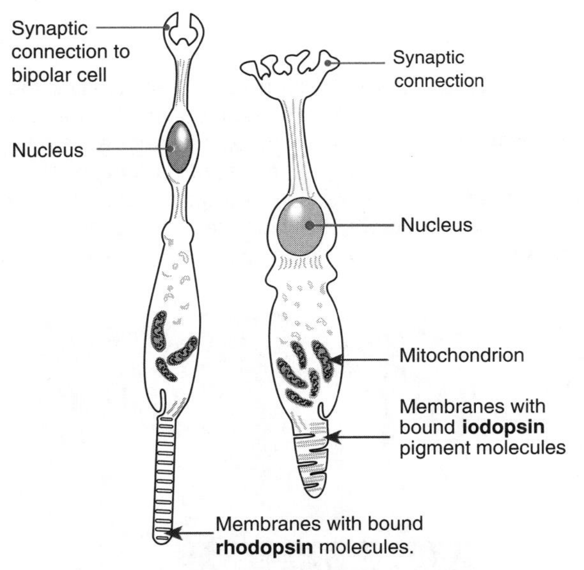

Structure of rod (left) and cone (right) photoreceptor cells

The basis of trichromatic vision

There are three classes of cones, each with a maximal response in either short (blue), intermediate (green) or long (yellow-green) wavelength light (below). The yellow-green cone is also sensitive to the red part of the spectrum and is often called the red cone. The differential responses of the cones to light of different wavelengths provides the basis of trichromatic colour vision.

Cone response to light wavelengths

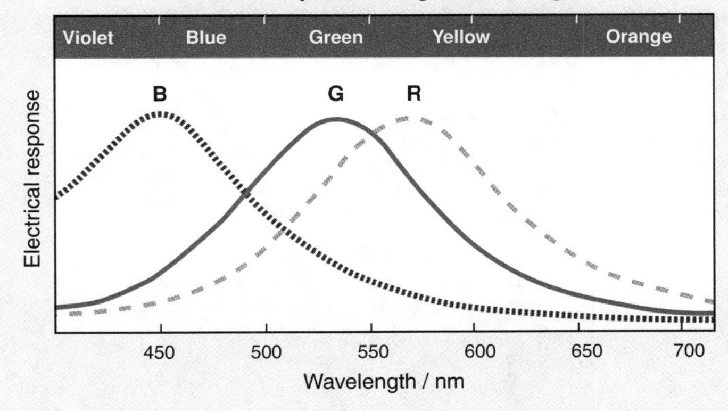

2. Complete the table below, comparing the features of rod and cone cells:

Feature	Rod cells	Cone cells
Visual pigment(s):		
Visual acuity:		
Overall function:		

3. Identify the three major types of neurone making up the retina and describe their basic function:

(a) ___

(b) ___

(c) ___

4. Identify two types of accessory neurones in the retina and describe their basic function:

(a) ___

(b) ___

5. Account for the differences in acuity (sharpness of vision) and sensitivity (to light level) between rod and cone cells:

___

___

6. (a) What is meant by the term **photochemical pigment** (photopigment)? ___

___

___

(b) Identify two photopigments and their location: ___

___

7. In your own words, explain how light is able to produce a nerve impulse in the ganglion cells: ___

___

___

___

___

ISBN: 978-1-927309-20-9

50 Sensitivity

Key Idea: The skin is an important sensory organ, with receptors for pain, pressure, touch, and temperature.

While some skin receptors are specialised receptor structures, many are simple unmyelinated nerves. Tactile (touch) and pressure receptors are **mechanoreceptors** and are stimulated by mechanical distortion. In the **Pacinian corpuscle**, the layers of tissue comprising the sensory structure are pushed together with pressure, stimulating the axon. Human skin is fairly uniform in structure, but the density and distribution of glands, hairs, and receptors varies according to the region of the body e.g. **Meissner's corpuscles**, are concentrated in areas sensitive to light touch and, in hairy skin, tactile receptors are clustered into specialised epithelial structures called touch domes or hair disks.

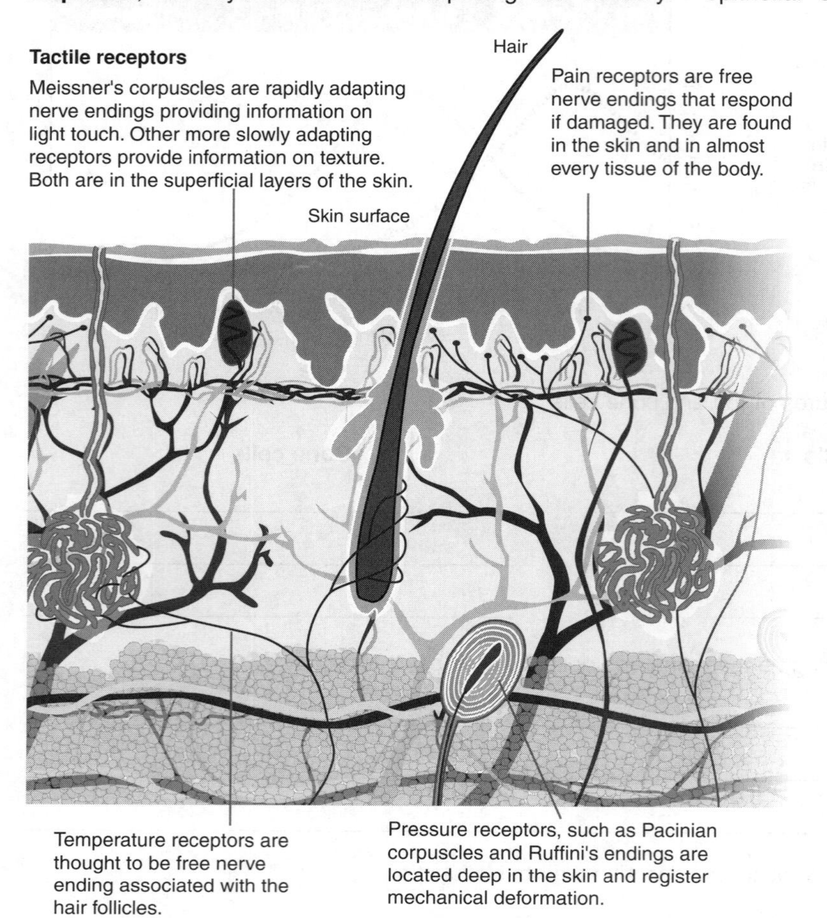

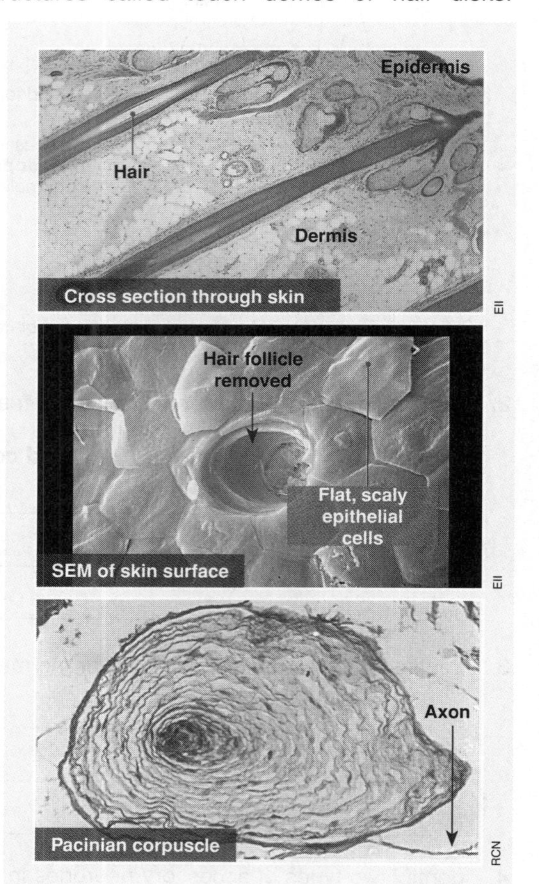

Testing the distribution of touch receptors

The receptors in the skin are more concentrated in some parts of the body than in others. The distribution of receptors can be tested by using the **two point touch test**. This involves finding the smallest distance at which someone can correctly distinguish two point stimuli.

Method

Repeatedly touch your lab partner's skin lightly with either one or two points of fine scissors or tweezers. Your partner's eyes should be closed. At each touch, they should report the sensation as "one" or "two", depending on whether they perceive one or two touches.

Begin with the scissor points far apart and gradually reduce the separation until only about 8 in 10 reports are correct.

This separation distance (in mm) is called the **two point threshold**. When the test subject can feel only one receptor (when there are two) it means that only one receptor is being stimulated. A large two point threshold indicates a low receptor density, a low one indicates a high receptor density.

Repeat this exercise for: the forearm, the back of the hand, the palm of the hand, the fingertip, and the lips, and then complete the table provide below:

Area of skin	Two point threshold / mm
Forearm	
Back of hand	
Palm of hand	
Fingertip	
Lips	

1. Name the region with the greatest number of touch receptors:

____

2. Name the region with the least number of touch receptors:

____

3. Explain why there is a difference between these two regions:

____

____

____

ISBN:978-1-927309-20-9

51 The Intrinsic Regulation of Heartbeat

Key Idea: Heartbeat is initiated by the sinoatrial node which acts as a pacemaker setting the basic heart rhythm.

The origin of the heart-beat is initiated by the heart (cardiac) muscle itself. The heartbeat is regulated by a conduction system consisting of the pacemaker (**sinoatrial node**) and a specialised conduction system of Purkyne tissue. The pacemaker sets the basic heart rhythm, but this rate can be influenced by hormones and by the cardiovascular control centre in the brainstem, which alters heart rate via parasympathetic and sympathetic nerves.

Generation of the heartbeat

The basic rhythmic heartbeat is **myogenic**. The nodal cells (SAN and atrioventricular node) spontaneously generate rhythmic action potentials without neural stimulation. The normal resting rate of self-excitation of the SAN is about 50 beats per minute.

The amount of blood ejected from the left ventricle per minute is called the **cardiac output**. It is determined by the **stroke volume** (the volume of blood ejected with each contraction) and the **heart rate** (number of beats per minute). Cardiac muscle responds to stretching by contracting more strongly. The greater the blood volume entering the ventricle, the greater the force of contraction. This relationship is important in regulating stroke volume in response to demand.

The hormone **epinephrine** (adrenaline) also influences cardiac output, increasing heart rate in preparation for vigorous activity. The sympathetic neurotransmitter **norepinephrine** (noradrenaline) has the same effect. Changing the rate and force of heart contraction is the main mechanism for controlling cardiac output in order to meet changing demands.

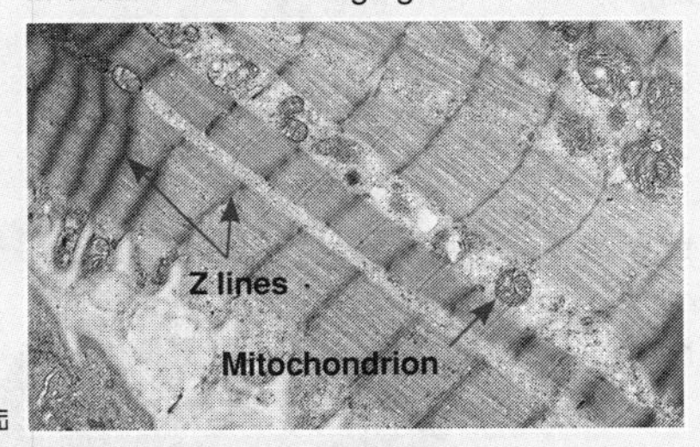

TEM of cardiac muscle showing striations in a fibre (muscle cell). The Z lines that delineate the contractile units of the rod-like units of the fibre. The fibres are joined by specialised electrical junctions called Intercalated discs, which allow impulses to spread rapidly through the heart muscle.

The **sinoatrial node** (SAN) or pacemaker is a small mass of specialised muscle cells on the wall of the right atrium, near the entry point of the superior vena cava. The pacemaker initiates the cardiac cycle, spontaneously generating **action potentials**, which cause the atria to contract. The SAN sets the basic heart rate, but this rate is influenced by hormones and impulses from the autonomic nervous system.

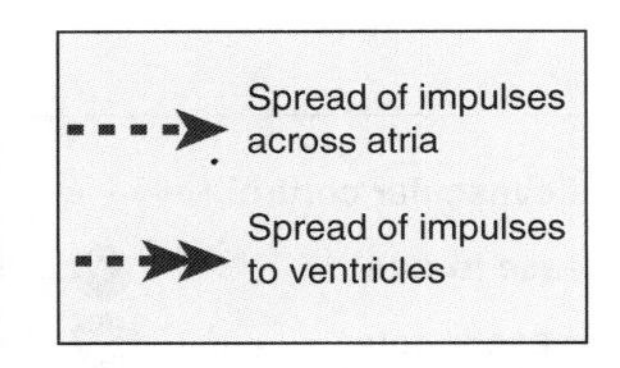

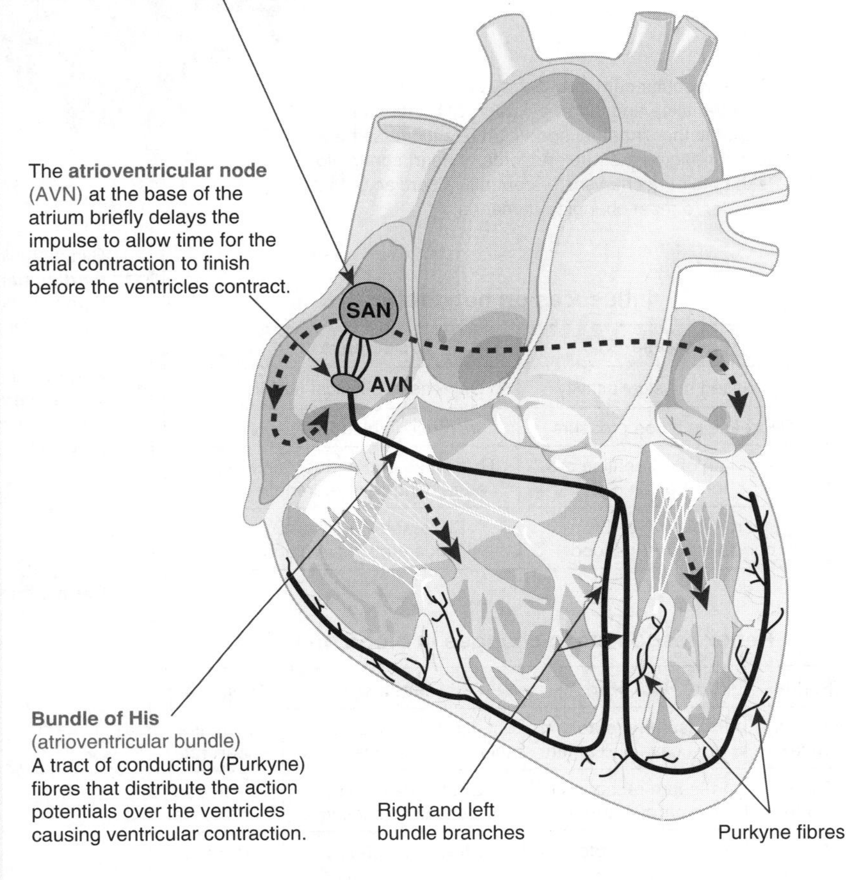

The **atrioventricular node** (AVN) at the base of the atrium briefly delays the impulse to allow time for the atrial contraction to finish before the ventricles contract.

Bundle of His (atrioventricular bundle) A tract of conducting (Purkyne) fibres that distribute the action potentials over the ventricles causing ventricular contraction.

Right and left bundle branches

Purkyne fibres

1. Describe the role of each of the following in heart activity:
   (a) The sinoatrial node: ____________________
   (b) The atrioventricular node: ____________________
   (c) The bundle of His: ____________________
   (d) Intercalated discs: ____________________

2. What is the significance of delaying the impulse at the AVN? ____________________

3. The heart-beat is intrinsic. Why is it important to be able to influence this basic rhythm via nerves and hormones?

4. (a) What is the effect of the hormone epinephrine on heart rate? ____________________
   (b) What sympathetic neurotransmitter has the same effect? ____________________

ISBN: 978-1-927309-20-9

52 Extrinsic Control of Heartbeat

Key Idea: The heart's basic rhythm is regulated via the cardiovascular control centre in response to hormones and input from sympathetic and parasympathetic nerves.

The pacemaker sets the basic rhythm of the heart, but this rate is influenced by the cardiovascular control centre, primarily in response to sensory information from pressure receptors in the walls of the blood vessels entering and leaving the heart. The main trigger for changing the basic rate of heart beat is change in blood pressure. The responses are mediated though simple reflexes.

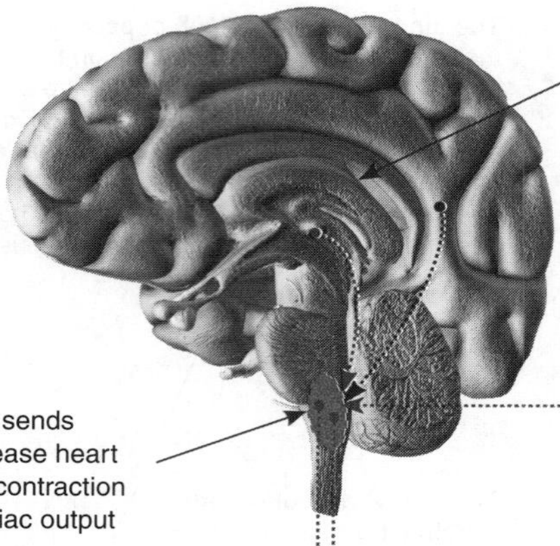

Cardiovascular control ....
Increase in rate +
Decrease in rate −

Higher brain centres influence the cardiovascular centre, e.g. excitement or anticipation of an event.

Baroreceptors in aorta, carotid arteries, and vena cava give feedback to cardiovascular centre on **blood pressure**. Blood pressure is directly related to the pumping action of the heart.

+ or −

Cardiovascular centre responds directly to noradrenaline and to low pH (high CO_2). It sends output to the sinoatrial node (SAN) to increase heart rate. Changing the rate and force of heart contraction is the main mechanism for controlling cardiac output in order to meet changing demands.

Sympathetic output to heart via **cardiac nerve** increases heart rate. Sympathetic output predominates during exercise or stress. +

Parasympathetic output to heart via **vagus nerve** decreases heart rate. Parasympathetic (vagal) output predominates during rest. −

Extrinsic input to SAN

Influences on heart rate

Increase	Decrease
Increased physical activity	Decreased physical activity
Decrease in blood pressure	Increase in blood pressure
Secretion of adrenaline or noradrenaline	Re-uptake and metabolism of adrenaline or noradrenaline
Increase in H^+ or CO_2 concentrations in blood	Decrease in H^+ or CO_2 concentrations in blood

Reflex responses to changes in blood pressure

Reflex	Receptor	Stimulus	Response
Bainbridge reflex	Pressure receptors in vena cava and atrium	Stretch caused by increased venous return	Increase heart rate
Carotid reflex	Pressure receptors in the carotid arteries	Stretch caused by increased arterial flow	Decrease heart rate
Aortic reflex	Pressure receptors in the aorta	Stretch caused by increased arterial flow	Decrease heart rate

Opposing actions keep blood pressure within narrow limits

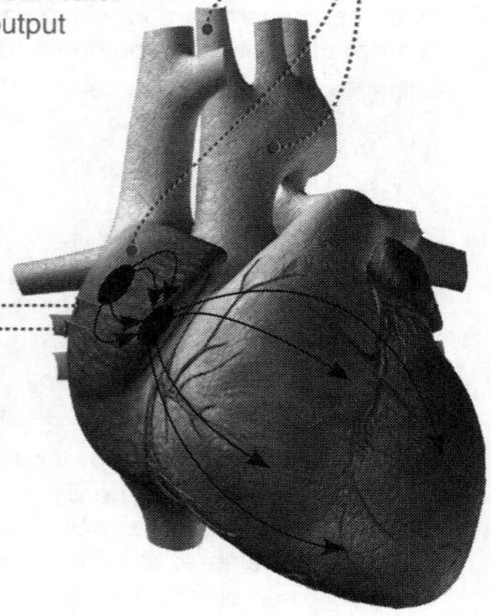

The intrinsic rhythm of the heart is influenced by the cardiovascular centre, which receives input from sensory neurones and hormones.

1. Explain how each of the following extrinsic factors influences the basic intrinsic rhythm of the heart:

   (a) Increased venous return: ____________________

   (b) Release of adrenaline in anticipation of an event: ____________________

   (c) Increase in blood CO_2: ____________________

2. How do these extrinsic factors bring about their effects? ____________________

   ____

3. What type of activity might cause increased venous return? ____________________

4. (a) Identify the nerve that brings about **increased** heart rate: ____________________

   (b) Identify the nerve that brings about **decreased** heart rate: ____________________

5. Account for the different responses to stretch in the vena cava and the aorta: ____________________

   ____

   ____

ISBN:978-1-927309-20-9

53 Investigating the Effect of Exercise

Key Idea: Breathing rate and heart rate both increase during exercise to meet the body's increased metabolic demands.

During exercise, the body's metabolic rate increases and the demand for oxygen increases. Oxygen is required for cellular respiration and ATP production. Increasing the rate of breathing delivers more oxygen to working tissues and enables them to make the ATP they need to keep working. An increased breathing rate also increases the rate at which carbon dioxide is expelled from the body. Heart rate also increases so blood can be moved around the body more quickly. This allows for faster delivery of oxygen and removal of carbon dioxide.

In this practical, you will work in groups of three to see how exercise affects breathing and heart rate. Choose one person to carry out the exercise and one person each to record heart rate and breathing rate.
Heart rate (beats per minute) is obtained by measuring the pulse (right) for 15 seconds and multiplying by four.
Breathing rate (breaths per minute) is measured by counting the number of breaths taken in 15 seconds and multiplying it by four.
CAUTION: The person exercising should have no known pre-existing heart or respiratory conditions.

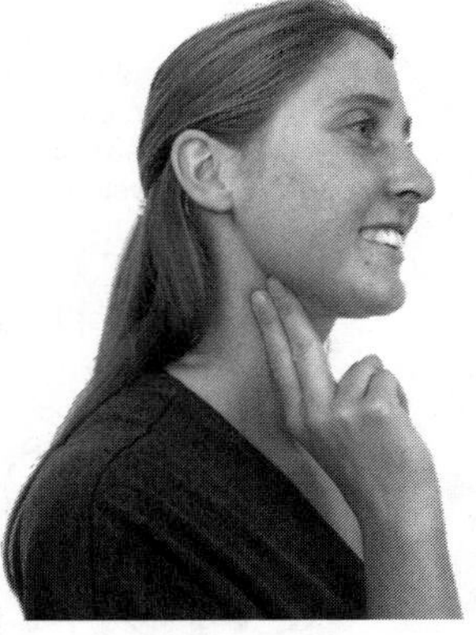

Gently press your index and middle fingers, not your thumb, against the carotid artery in the neck (just under the jaw) or the radial artery (on the wrist just under the thumb) until you feel a pulse.

Measuring the carotid pulse

Measuring the radial pulse

Procedure

Resting measurements

Have the person carrying out the exercise sit down on a chair for 5 minutes. They should try not to move. After 5 minutes of sitting, measure their heart rate and breathing rate. Record the resting data on the table (right).

Exercising measurements

Choose an exercise to perform. Some examples include step ups onto a chair, skipping rope, jumping jacks, and running in place.
Begin the exercise, and take measurements after 1, 2, 3, and 4 minutes of exercise. The person exercising should stop just long enough for the measurements to be taken. Record the results in the table.

Post exercise measurements

After the exercise period has finished, have the exerciser sit down in a chair. Take their measurements 1 and 5 minutes after finishing the exercise. Record the results on the table.

	Heart rate / beats minute^{-1}	Breathing rate / breaths minute^{-1}
Resting		
1 minute		
2 minutes		
3 minutes		
4 minutes		
1 minute after		
5 minutes after		

1. (a) Graph your results on separate piece of paper. You will need to use two vertical axes, one for heart rate and another for breathing rate. When you have finished answering the questions below, attach it to this page.

   (b) Analyse your graph and describe what happened to heart rate and breathing rate during exercise:

2. (a) Describe what happened to heart rate and breathing rate after exercise:

   (b) Why did this change occur?

ISBN: 978-1-927309-20-9

54 Neurones and Neurotransmitters

Key Idea: Neurones are electrically excitable cells that are specialised to process and transmit information via electrical and chemical signals. Neurotransmitters are chemicals that allow the transmission of signals between neurones.

Neurones are cells specialised to transmit information in the form of electrochemical signals from receptors (in the central nervous system) to effectors. Neurones consist of a cell body (soma) and long processes (dendrites and axons). Messages are transmitted between neurones by signalling molecules called neurotransmitters.

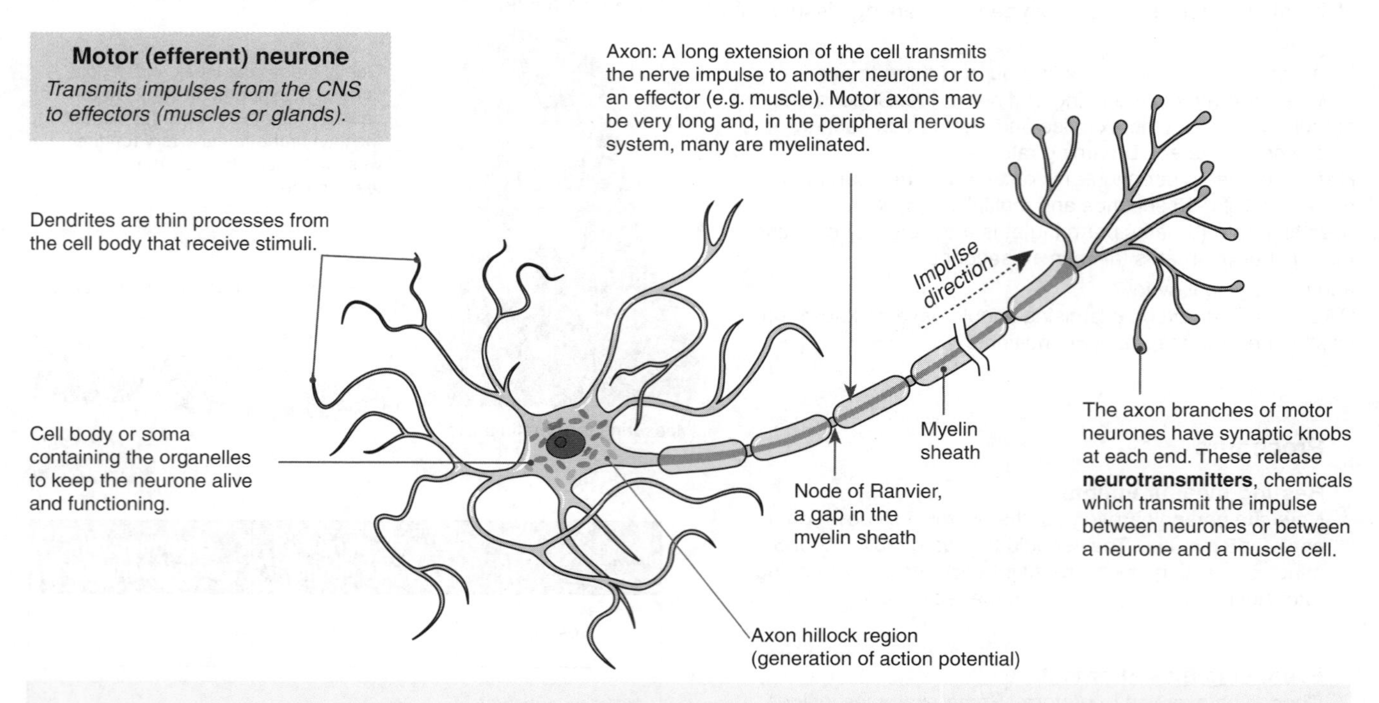

Neurotransmitters carry signals between neurones

Chemical signalling between neurones was first demonstrated in 1921 by Otto Loewi. In his experiment, the still beating hearts of two frogs were placed in connected flasks filled with saline solution. The vagus nerve (parasympathetic) of the first heart was still attached and was stimulated by electricity to reduce its rate of beating. After a delay, the rate of beating in the second heart also slowed. Increasing the beating rate in the first heart caused an increase in the beating rate in the second heart, showing electrical stimulus of the first heart caused it to release a chemical into the saline solution that then affected the heartbeat of the second heart. The chemical was found to be **acetylcholine**.

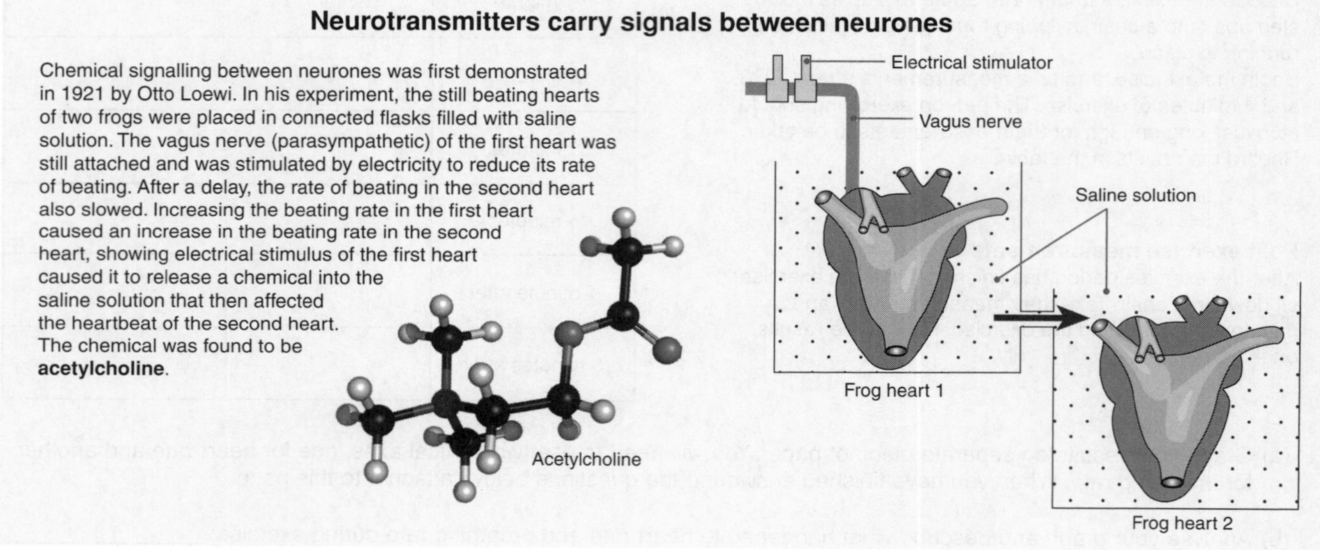

1. What is the function of a neurone? ______________________________

____

2. Describe the purpose of a neurotransmitter: ______________________________

____

3 (a) Explain why stimulating the first frog heart with electricity caused it to change its beating rate: ______________________________

____

____

(b) Explain why the second heart in the experiment reduced its beating rate after a delay: ______________________________

____

ISBN:978-1-927309-20-9

Where conduction speed is important, the axons of neurones are sheathed within a lipid and protein rich substance called **myelin**. Myelin is produced by **oligodendrocytes** in the central nervous system (CNS) and by **Schwann cells** in the peripheral nervous system (PNS). At intervals along the axons of myelinated neurones, there are gaps between neighbouring Schwann cells and their sheaths. These are called **nodes of Ranvier**. Myelin acts as an insulator, increasing the speed at which nerve impulses travel because it prevents ion flow across the neurone membrane and forces the current to "jump" along the axon from node to node.

Myelinated neurones

Diameter: 1-25 µm
Conduction speed: 6-120 ms^{-1}

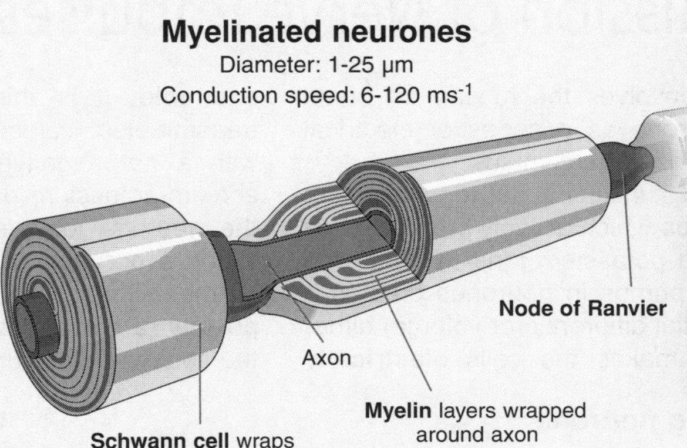

Myelin

Roadnottaken cc3.0

TEM cross section through a myelinated axon

Non-myelinated axons are relatively more common in the CNS where the distances travelled are less than in the PNS. Here, the axons are encased within the cytoplasmic extensions of oligodendrocytes or Schwann cells, rather than within a myelin sheath. **Impulses travel more slowly** because the nerve impulse is propagated along the entire axon membrane, rather than jumping from node to node as occurs in myelinated neurones.

Non-myelinated neurones

Diameter: <1 µm
Conduction speed: 0.2-0.5 ms^{-1}

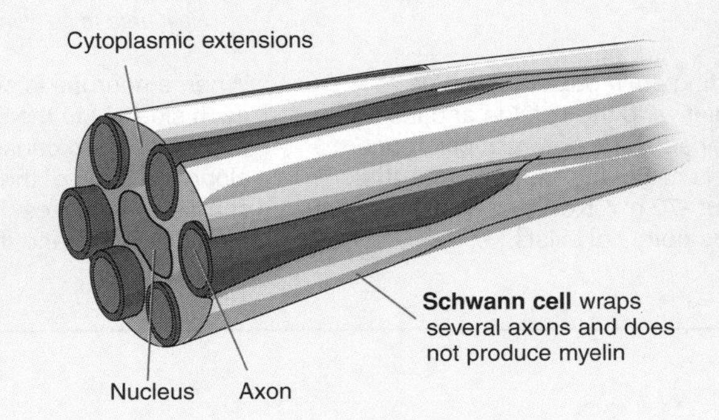

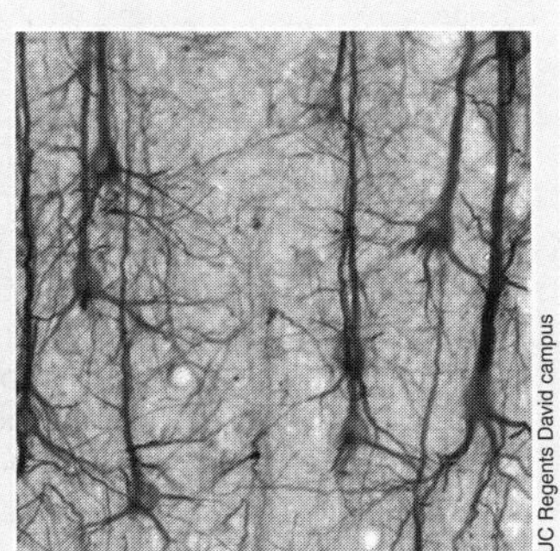

Unmyelinated pyramidal neurons of the cerebral cortex

Effect of temperature and diameter on conduction speed

Temperature and neurone diameter have a direct effect on the speed of signal conduction. Signal conduction increases as temperature increases (to a peak). Conduction also increases as axon diameter increases. In non-myelinated neurones, conduction velocity is proportional to the square root of neurone diameter.

Conduction speed and temperature

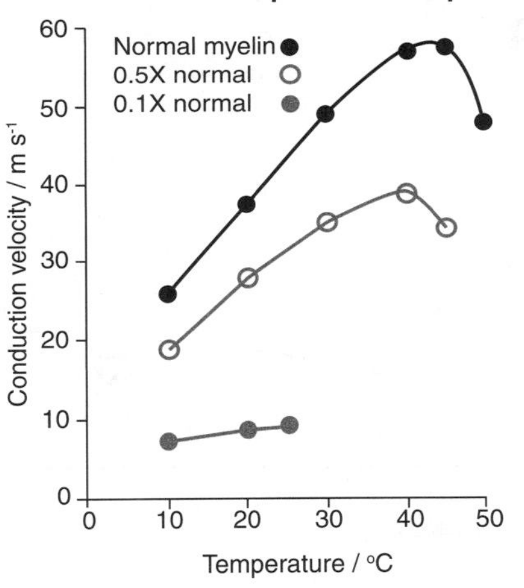

Conduction speed and axon diameter

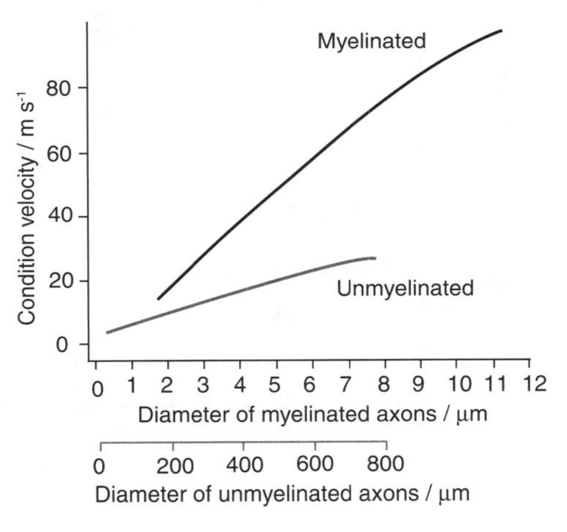

4. (a) What is the function of myelination in neurones? ______________________

   (b) How does myelination increase the speed of nerve impulse conduction? ______________________

   (c) Describe the adaptive advantage of faster conduction of nerve impulses: ______________________

   (d) For what diameter neurones is conduction in non-myelinated neurones faster than myelinated neurones?

   ____

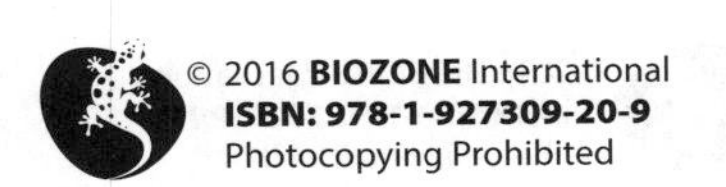

ISBN: 978-1-927309-20-9

55 Transmission of Nerve Impulses

Key Idea: A nerve impulse involves the movement of an action potential along a neurone as a series of electrical depolarisation events in response to a stimulus.

The plasma membranes of cells, including neurones, contain **sodium-potassium ion pumps** which actively pump sodium ions (Na^+) out of the cell and potassium ions (K^+) into the cell. The action of these ion pumps in neurones creates a separation of charge (a potential difference or voltage) either side of the membrane and makes the cells **electrically excitable**. It is this property that enables neurones to transmit electrical impulses. The **resting state** of a neurone, with a net negative charge inside, is maintained by the sodium-potassium pumps, which actively move two K^+ into the neurone for every three Na^+ moved out (below left). When a nerve is stimulated, a brief increase in membrane permeability to Na^+ temporarily reverses the membrane polarity (a **depolarisation**). After the nerve impulse passes, the sodium-potassium pump restores the resting potential.

The resting neurone

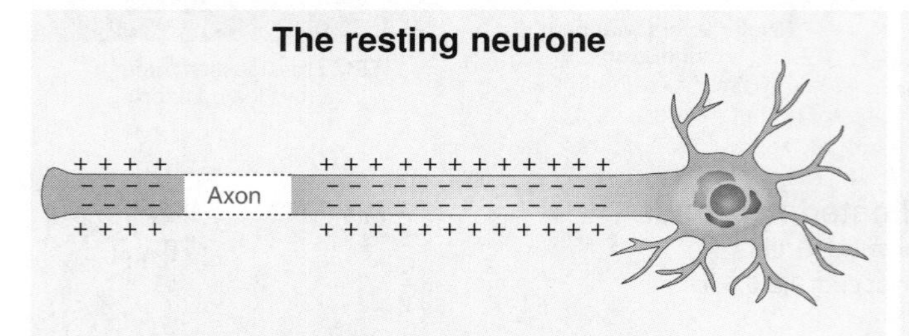

When a neurone is not transmitting an impulse, the inside of the cell is negatively charged relative to the outside and the cell is said to be electrically polarised. The potential difference (voltage) across the membrane is called the **resting potential**. For most nerve cells this is about -70 mV. Nerve transmission is possible because this membrane potential exists.

The nerve impulse

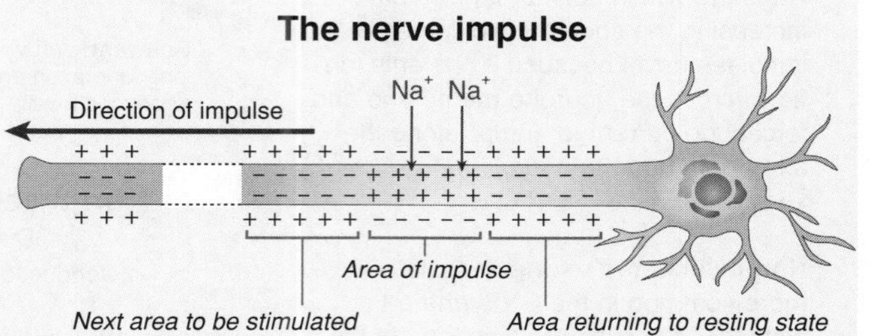

When a neurone is stimulated, the distribution of charges on each side of the membrane briefly reverses. This process of **depolarisation** causes a burst of electrical activity to pass along the axon of the neurone as an **action potential**. As the charge reversal reaches one region, local currents depolarise the next region and the impulse spreads along the axon.

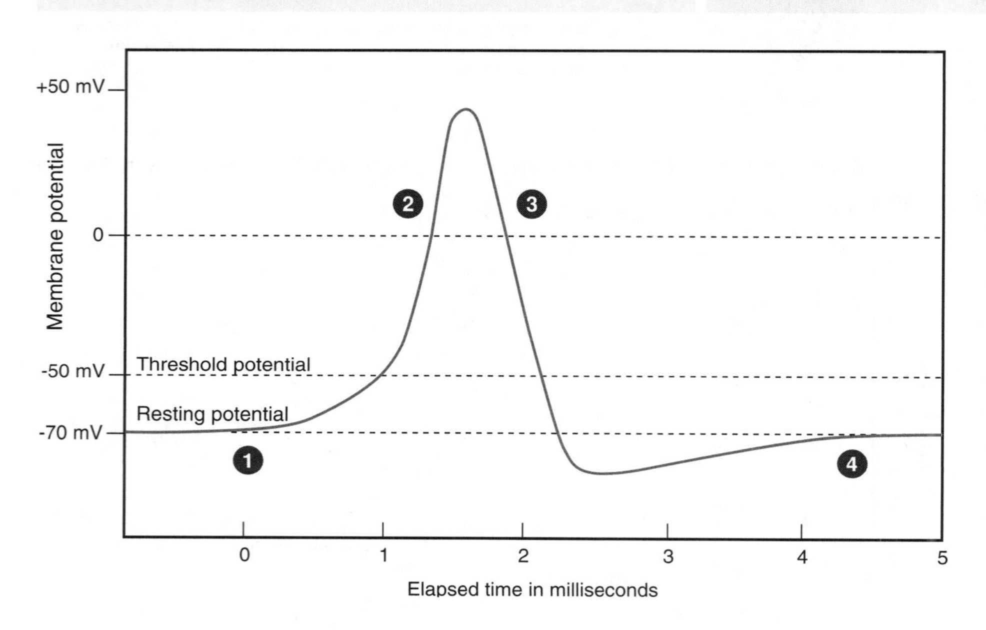

The depolarisation in an axon can be shown as a change in membrane potential (in millivolts). A stimulus must be strong enough to reach the **threshold potential** before an action potential is generated. This is the voltage at which the depolarisation of the membrane becomes unstoppable.

The action potential is **all or nothing** in its generation and because of this, impulses (once generated) always reach threshold and move along the axon without attenuation. The resting potential is restored by the movement of potassium ions (K^+) out of the cell. During this **refractory period**, the nerve cannot respond, so nerve impulses are discrete.

Voltage-gated ion channels and the course of an action potential

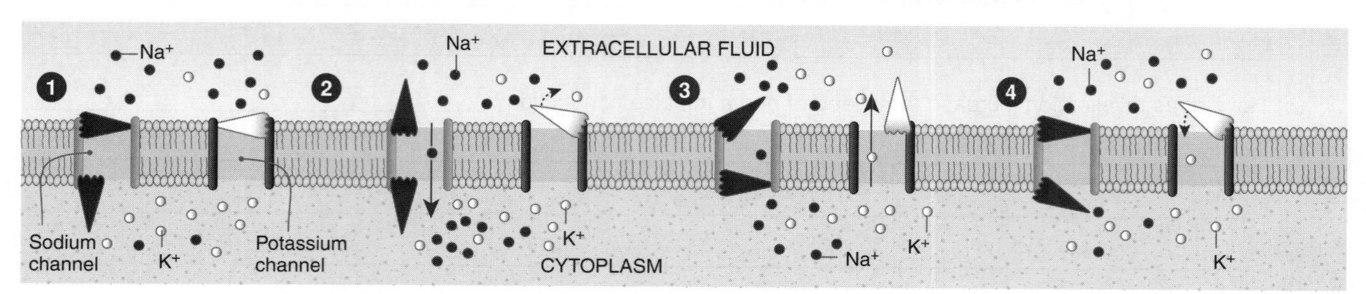

Resting state:
Voltage activated Na^+ and K^+ channels are closed.

Depolarisation:
Voltage activated Na^+ channels open and there is a rapid influx of Na^+ ions. The interior of the neurone becomes positive relative to the outside.

Repolarisation:
Voltage activated Na^+ channels close and the K^+ channels open; K^+ moves out of the cell, restoring the negative charge to the cell interior.

Returning to resting state:
Voltage activated Na^+ and K^+ channels close to return the neurone to the resting state.

ISBN:978-1-927309-20-9

Axon myelination is a feature of vertebrate nervous systems and it enables them to achieve very rapid speeds of nerve conduction. Myelinated neurones conduct impulses by **saltatory conduction**, a term that describes how the impulse jumps along the fibre. In a myelinated neurone, **action potentials are generated only at the nodes**, which is where the voltage gated channels occur. The axon is insulated so the action potential at one node is sufficient to trigger an action potential in the next node and the impulse jumps along the fibre. Contrast this with a non-myelinated neurone in which voltage-gated channels occur along the entire length of the axon.

As well as increasing the speed of conduction, the myelin sheath reduces energy expenditure because the area over which depolarisation occurs is less (and therefore also the number of sodium and potassium ions that need to be pumped to restore the resting potential).

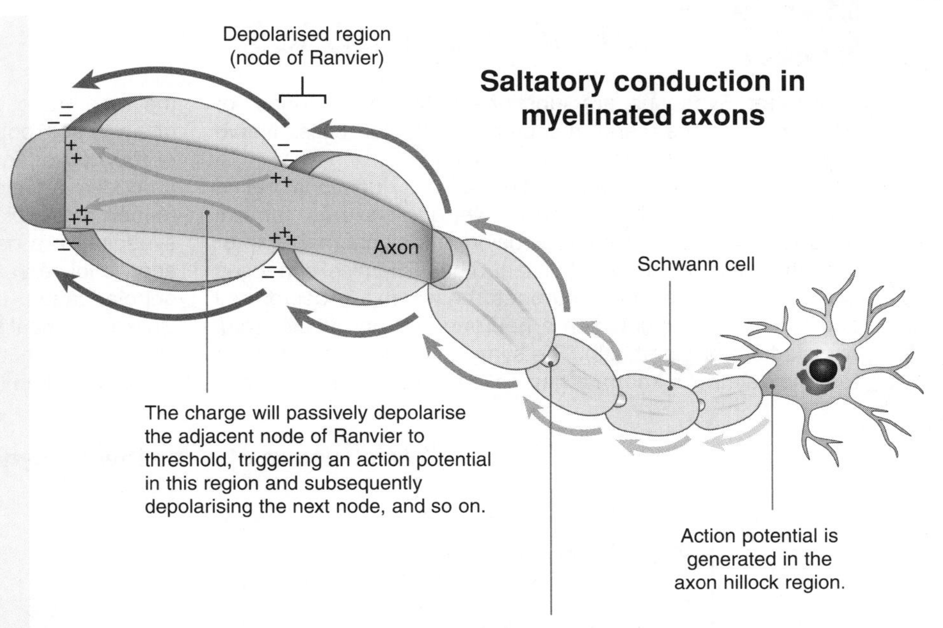

1. In your own words, define what an **action potential** is: ____________________

____

2. (a) Identify the defining **functional feature** of neurones: ____________________

____

(b) How does this differ from the supporting tissue (e.g. Schwann cells) of the nervous system? ____________________

____

3. Describe the movement of voltage-gated channels and ions associated with:

(a) Depolarisation of the neurone: ____________________

(b) Repolarisation of the neurone: ____________________

4. Summarise the sequence of events in a neurone when it receives a stimulus sufficient to reach threshold:

____

____

____

____

____

5. How is the resting potential restored in a neurone after an action potential has passed? ____________________

____

6. (a) Explain how an action potential travels in a **myelinated neurone**: ____________________

____

(b) How does this differ from its travel in a **non-myelinated neurone**? ____________________

____

7. Explain how the **refractory period** influences the direction in which an impulse will travel: ____________________

____

____

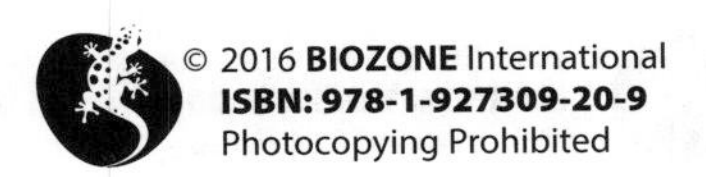

ISBN: 978-1-927309-20-9

56 Chemical Synapses

Key Idea: Synapses are junctions between neurones, or between neurones and receptor or effector cells. Nerve impulses are transmitted across synapses.

Action potentials are transmitted across junctions called synapses. Almost all synapses in vertebrates are chemical synapses, which involve the diffusion of a signal molecule or neurotransmitter from one cell to another. Chemical synapses can occur between two neurones, between a receptor cell and a neurone, or between a neurone and an effector (e.g. muscle fibre or gland cell). The synapse consists of the axon terminal (synaptic knob), a gap called the synaptic cleft, and the membrane of the post-synaptic (receiving) cell. Arrival of an action potential at the axon terminal causes release of the neurotransmitter, which diffuses across the cleft and produces an electrical response in the post-synaptic cell (either a depolarisation or hyperpolarisation). Cholinergic synapses are named for the neurotransmitter they release, acetylcholine (ACh). In the example below, ACh results in depolarisation (excitation) of the post-synaptic neurone. Unlike electrical synapses, in which transmission can occur in either direction, transmission at chemical synapses is always in one direction (unidirectional).

The structure of a cholinergic synapse

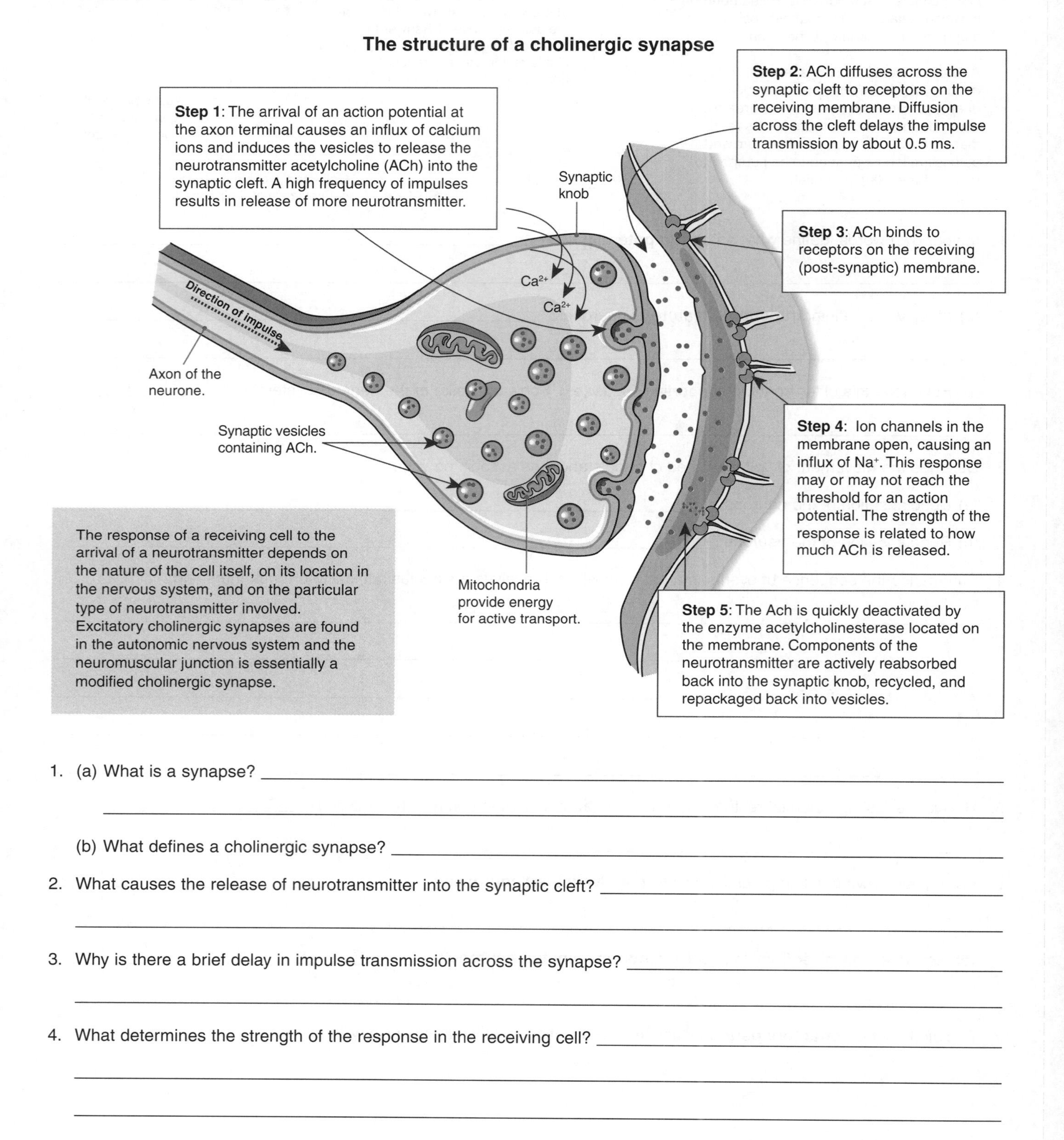

The response of a receiving cell to the arrival of a neurotransmitter depends on the nature of the cell itself, on its location in the nervous system, and on the particular type of neurotransmitter involved. Excitatory cholinergic synapses are found in the autonomic nervous system and the neuromuscular junction is essentially a modified cholinergic synapse.

1. (a) What is a synapse? ______

   (b) What defines a cholinergic synapse? ______

2. What causes the release of neurotransmitter into the synaptic cleft? ______

3. Why is there a brief delay in impulse transmission across the synapse? ______

4. What determines the strength of the response in the receiving cell? ______

KNOW WEB 56 LINK 57 LINK 58

ISBN:978-1-927309-20-9

The neuromuscular junction

The neuromuscular junction is a specialised cholinergic synapse between a motor neurone and a muscle fibre. Functionally, they operate in the same way as the excitatory cholinergic synapse pictured opposite.

- Arrival of an action potential at the neuromuscular junction results in depolarisation of the muscle fibre membrane (the sarcolemma) and this results in contraction of the muscle fibre.
- For a muscle fibre to contract, it must receive a threshold stimulus in the form of an action potential. Action potentials are carried by motor neurones from the central nervous system to the muscle fibres they supply. The arrival of an action potential at the neuromuscular junction results in release of the neurotransmitter acetylcholine and contraction of the fibre.
- The response of a single muscle fibre is **all-or-none**, meaning it contracts maximally or not at all.

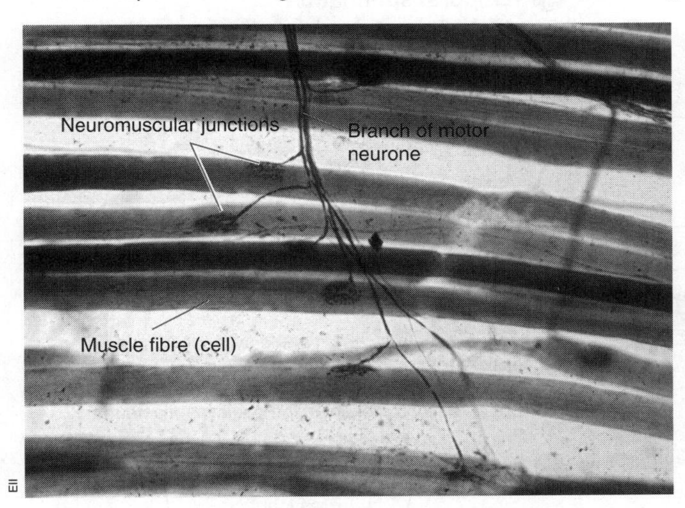

Axon terminals of a motor neurone supplying a muscle. Axon branches end on the sarcolemma (plasma membrane) of a muscle fibre at regions called neuromuscular junctions. Each fibre receives a branch of an axon, but one axon may supply many muscle fibres. A motor neurone and all the fibres it innervates is called a motor unit.

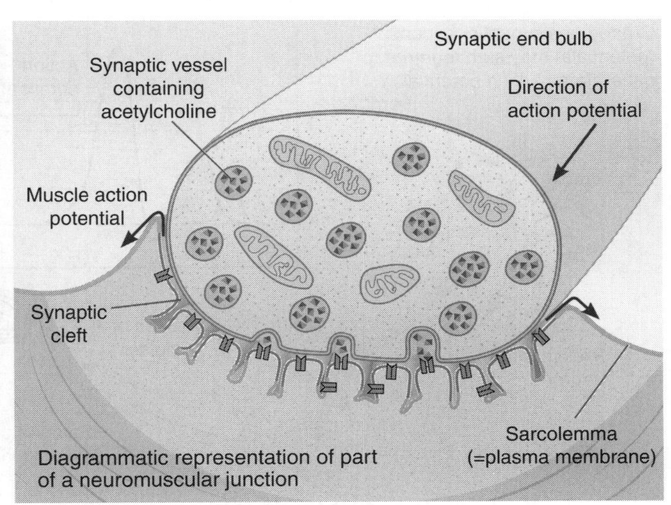

When an action potential arrives at the neuromuscular junction, it causes release of acetylcholine, which diffuses across the synaptic cleft to stimulate an action potential in the sarcolemma. The action potential travels throughout the muscle fibre causing muscle contraction.

5. What factors determine the response of the post-synaptic cell? ______________________

____

____

6. (a) How is the neurotransmitter is deactivated? ______________________

____

(b) Why do you think it is important for the neurotransmitter to be deactivated soon after its release? ______________________

____

(c) Why is transmission at chemical synapses unidirectional and what is the significance of this? ______________________

____

____

____

7. (a) In what way is the neuromuscular junction (above) similar to the cholinergic synapse described opposite: ______________________

____

____

(b) In what ways are these two synaptic junctions different? ______________________

____

____

____

ISBN: 978-1-927309-20-9

57 Integration at Synapses

Key Idea: Synapses play a pivotal role in the ability of the nervous system to respond appropriately to stimulation and to adapt to change by integrating all inputs.

The nature of synaptic transmission in the nervous system allows the **integration** (interpretation and coordination) of inputs from many sources. These inputs can be excitatory (causing depolarisation) or inhibitory (making an action potential less likely). It is the sum of all excitatory and inhibitory inputs that leads to the final response in a post-synaptic cell. Synaptic integration is behind all the various responses we have to stimuli. It is also the most probable mechanism by which learning and memory are achieved.

Summation at synapses

Graded postsynaptic responses (potentials) may sum together to generate an action potential.

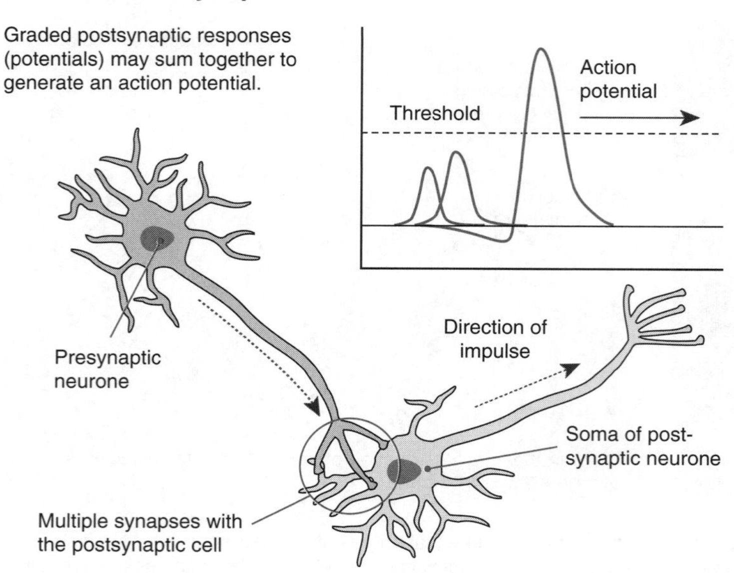

Nerve transmission across chemical synapses has several advantages, despite the delay caused by neurotransmitter diffusion. Chemical synapses transmit impulses in one direction to a precise location and, because they rely on a limited supply of neurotransmitter, they are subject to fatigue (inability to respond to repeated stimulation). This protects the system against overstimulation.

Synapses also act as centres for the **integration** of inputs from many sources. The response of a postsynaptic cell is often not strong enough on its own to generate an action potential. However, because the strength of the response is related to the amount of neurotransmitter released, subthreshold responses can sum together to produce a response in the post-synaptic cell. This additive effect is termed **summation**. Summation can be **temporal** or **spatial** (right).

1 Temporal summation

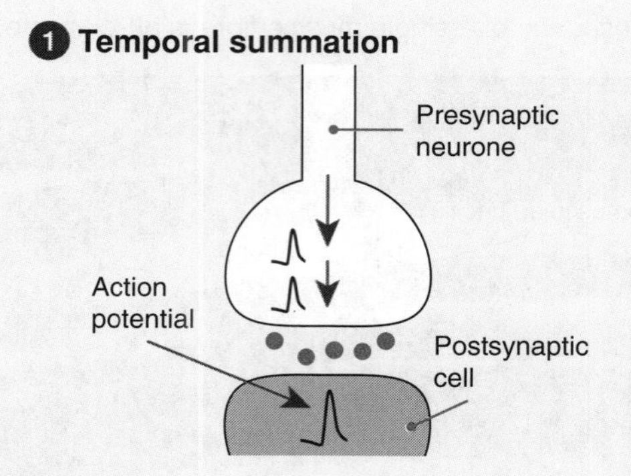

Several impulses may arrive at the synapse in quick succession from a single axon. The individual responses are so close in time that they sum to reach threshold and produce an action potential in the postsynaptic neurone.

2 Spatial summation

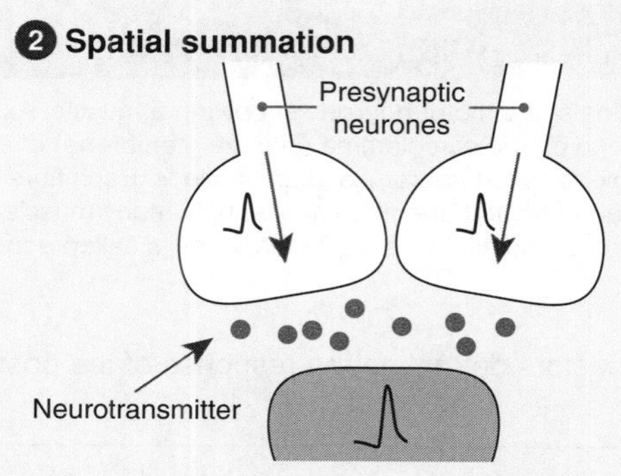

Individual impulses from spatially separated axon terminals may arrive simultaneously at different regions of the same postsynaptic neurone. The responses from the different places sum to produce an action potential.

1. Explain the purpose of nervous system integration: ______

2. Describe two advantages of chemical synapses:

   (a) ______

   (b) ______

3. (a) Explain what is meant by **summation**: ______

   (b) In simple terms, distinguish between temporal and spatial summation: ______

ISBN:978-1-927309-20-9

58 Drugs at Synapses

Key Idea: Drugs may increase or decrease the effect of neurotransmitters at synapses.

Drugs may act at synapses either mimicking or blocking the usual effect of a neurotransmitter (whether it be excitatory or inhibitory). Drugs that increase the usual effect of a neurotransmitter are called **agonists** while those that decrease their effect are called **antagonists**. Many recreational and therapeutic drugs work through their action at synapses, controlling the response of the receiving cell to incoming action potentials. The diagram below shows the effect of drugs at cholinergic synapses (those with receptors for acetylcholine). Acetylcholine receptors are classified as nicotinic or muscarinic according to their response to the chemicals nicotine or muscarine (a fungal toxin).

Drugs at cholinergic synapses

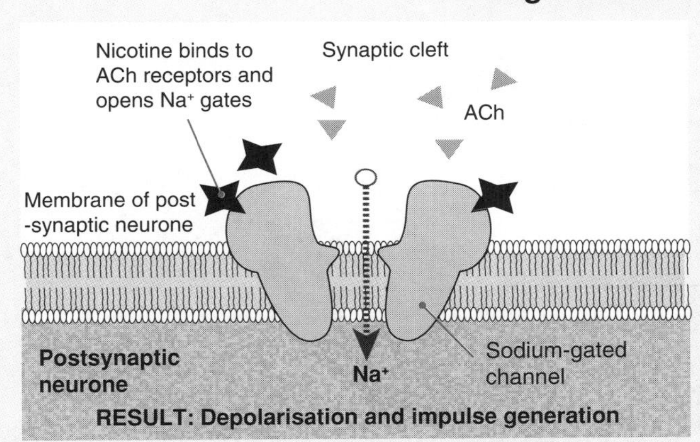

Agonistic drugs increase the usual effect of a neurotransmitter. If a neurotransmitter is inhibitory, an agonist will increase the inhibitory effect. If it is excitatory, an agonist will increase the excitatory effect. Agonists enhance the usual effect of a neurotransmitter by mimicking its action at the synapse, by preventing breakdown of the neurotransmitter, or by increasing neurotransmitter production. For example **nicotine** acts as an agonist by binding to and activating nicotinic acetylcholine (Ach) receptors on the postsynaptic membrane. This opens sodium gates, leading to a sodium influx and membrane depolarisation.

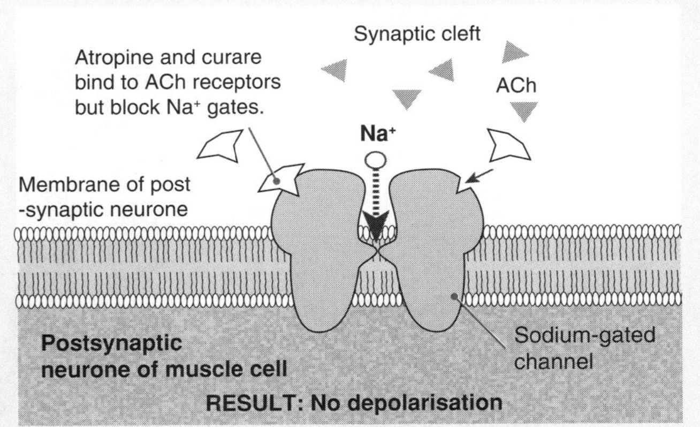

Antagonistic drugs decrease the usual effect of neurotransmitters. If a neurotransmitter is inhibitory, an antagonist will decrease the inhibitory effect. If it is excitatory, an antagonist will decrease the excitatory effect. Antagonists decrease the usual effect of a neurotransmitter by blocking receptor sites, increasing the rate of neurotransmitter removal, or decreasing neurotransmitter production. **Atropine** and **curare** both act as antagonists at some cholinergic synapses. They compete with Ach for binding sites on the postsynaptic membrane, and block sodium influx so that impulses are not generated. In muscles, contraction is prevented.

Nicotine is the highly addictive substance in cigarettes. It acts on nicotinic acetylcholine receptors, and indirectly affects the levels of several neurotransmitters, including dopamine. Dopamine produces feelings of euphoria and relaxation, which create nicotine addiction.

Fly agaric (*Amanita muscaria*)

Onderwijsgek at nl.wikipedia cc 3.0

Muscarine, a compound found in several types of mushrooms, binds to muscarinic acetylcholine receptors. Muscarine is used to treat a number of medical conditions (e.g. glaucoma), but consumption of the mushrooms can deliver a fatal overdose of muscarine.

Eastern green mamba

Mamba snake venom contains a number of neurotoxins including **dendrotoxins**. These small peptide molecules act as acetylcholine receptor antagonists (blocking muscarinic receptors). They have many effects including disrupting muscle contraction.

1. Describe the action of an:

   (a) Antagonistic drug: ______

   (b) Agonistic drug: ______

2. Describe the effect of an antagonistic drug on a neurotransmitter with an inhibitory effect: ______

3. Explain why atropine and curare are described as direct antagonists: ______

4. How would an agonist that prevents neurotransmitter breakdown exert its effect? ______

ISBN: 978-1-927309-20-9

59 Antagonistic Muscles

Key Idea: Antagonistic muscles are muscle pairs that have opposite actions to each other. Together, their opposing actions bring about movement of body parts.

In both vertebrates and invertebrates, muscle provide the contractile force to move body parts. Muscles create movement of body parts when they contract across joints. Because muscles can only pull and not push, most body movements are achieved through the action of opposing sets of muscles called **antagonistic muscles**. Antagonistic muscles function by producing opposite movements, as one muscle contracts (shortens), the other relaxes (lengthens). Skeletal muscles are attached to the skeleton by tough connective tissue structures (**tendons** in vertebrates or attachment fibres in insects). They always have at least two attachments: an origin and an insertion. Body parts move when a muscle contracts across a joint. The type and degree of movement depends on how much movement the joint allows and where the muscle is located in relation to the joint.

Muscles of the upper arm

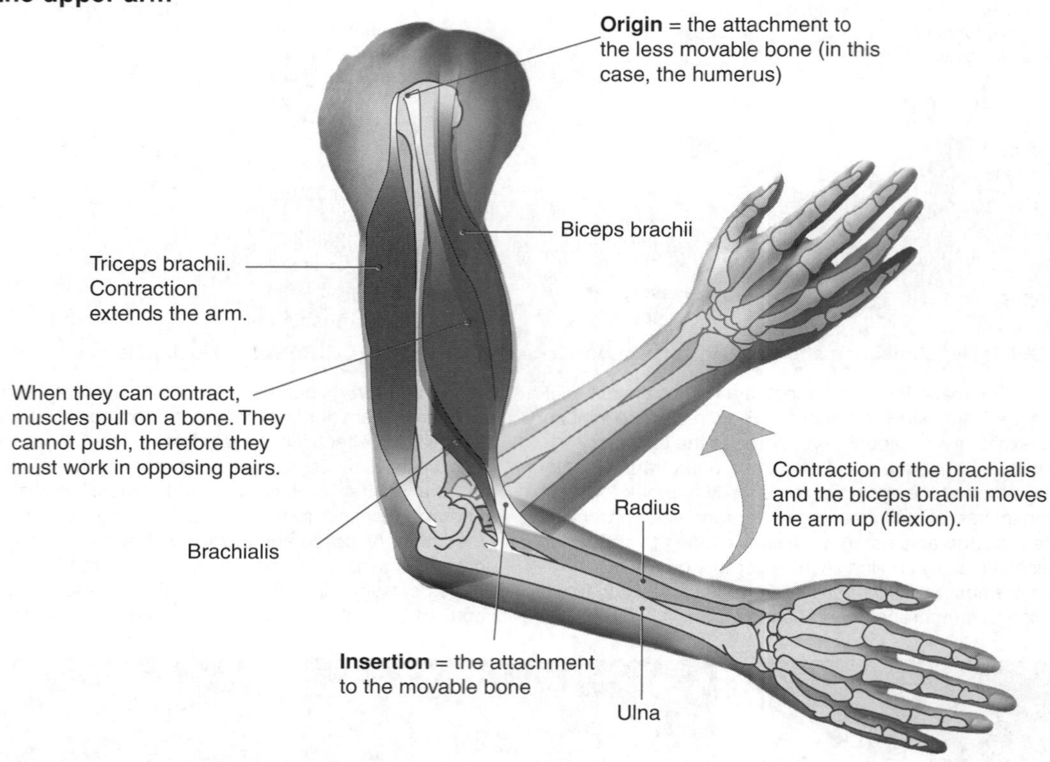

Opposing movements require opposing muscles

The **flexion** (bending) and **extension** (unbending) of limbs is caused by the action of **antagonistic muscles**. Antagonistic muscles work in pairs and their actions oppose each other. During movement of a limb, muscles other than those primarily responsible for the movement may be involved to fine tune the movement.

Every coordinated movement in the body requires the application of muscle force. This is accomplished by the action of agonists, antagonists, and synergists. The opposing action of agonists and antagonists (working constantly at a low level) also produces muscle tone. Note that either muscle in an antagonistic pair can act as the agonist or prime mover, depending on the particular movement (for example, flexion or extension).

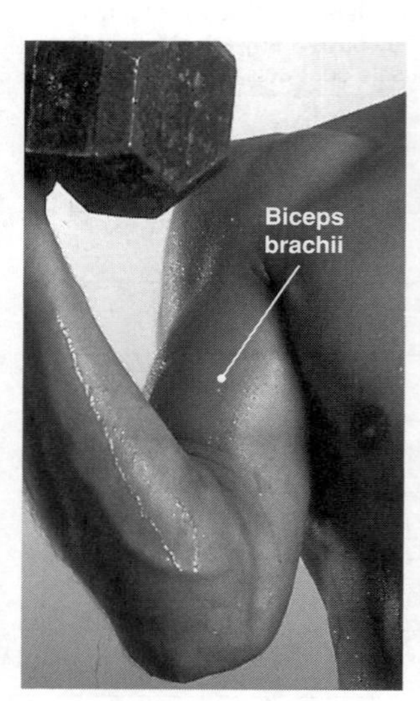

Agonists or prime movers: muscles that are primarily responsible for the movement and produce most of the force required.

Antagonists: muscles that oppose the prime mover. They may also play a protective role by preventing over-stretching of the prime mover.

Synergists: muscles that assist the prime movers and may be involved in fine-tuning the direction of the movement.

During flexion of the forearm (left) the **brachialis** muscle acts as the prime mover and the **biceps brachii** is the synergist. The antagonist, the **triceps brachii** at the back of the arm, is relaxed. During extension, their roles are reversed.

Movement of the upper leg is achieved through the action of several large groups of muscles, collectively called the **quadriceps** and the **hamstrings**.

The hamstrings are actually a collection of three muscles, which act together to flex the leg.

The quadriceps at the front of the thigh (a collection of four large muscles) opposes the motion of the hamstrings and extends the leg.

When the prime mover contracts forcefully, the antagonist also contracts very slightly. This stops over-stretching and allows greater control over thigh movement.

ISBN:978-1-927309-20-9

Levers

The skeleton works as a system of levers, with the muscles providing the force required to move these levers. The joint acts as the fulcrum (F), the muscles exert the effort force (E) and the weight of the bone being moved represents the load (L). There are three classes of levers, examples in the body are shown below:

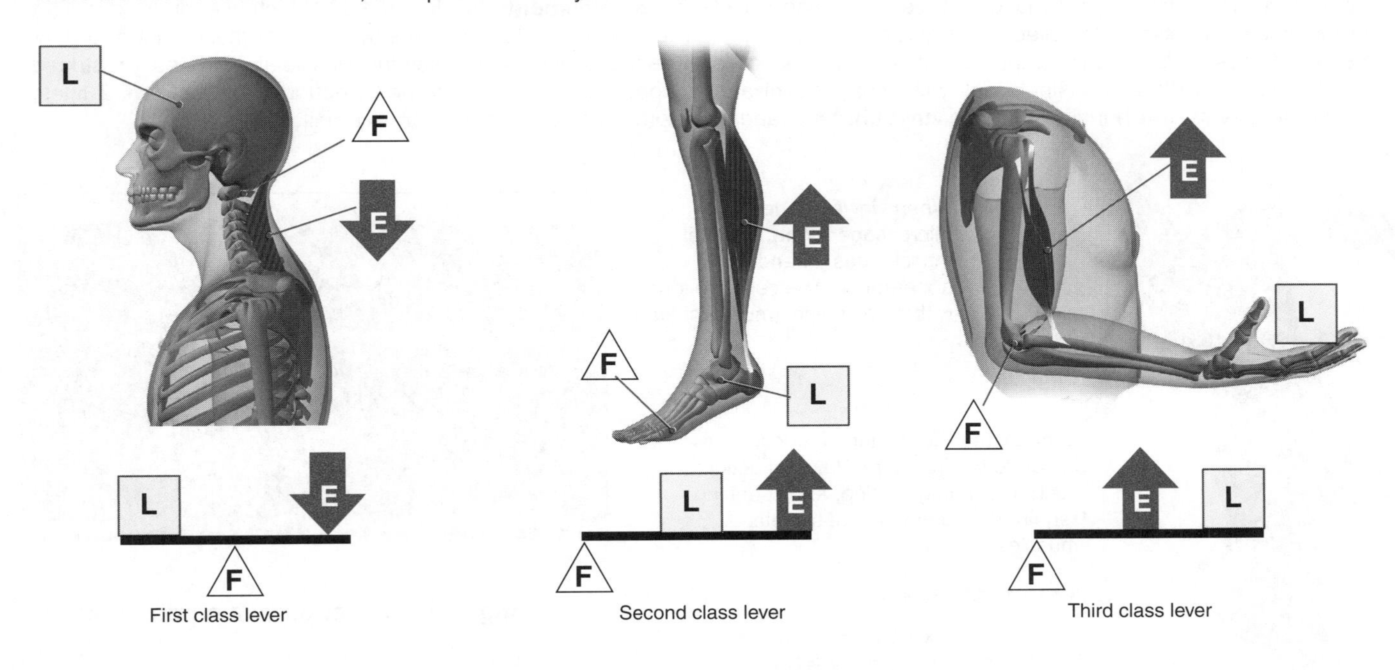

First class lever

Second class lever

Third class lever

1. Explain why the muscles that cause movement of body parts tend to operate as antagonistic pairs:

2. Describe the relationship between muscles and joints. Using appropriate terminology, explain how antagonistic muscles act together to raise and lower a limb:

3. Describe the role of each of the following muscles in moving a limb:

   (a) Prime mover:

   (b) Antagonist:

   (c) Synergist:

4. Explain the role of joints in the movement of body parts:

5. (a) Identify the insertion for the biceps brachii during flexion of the forearm:

   (b) Identify the insertion of the brachialis muscle during flexion of the forearm:

   (c) Identify the antagonist during flexion of the forearm:

   (d) Given its insertion, describe the forearm movement during which the biceps brachii is the prime mover:

6. What class of lever is involved during flexion of the knee joint (raising the heel towards the buttocks)?

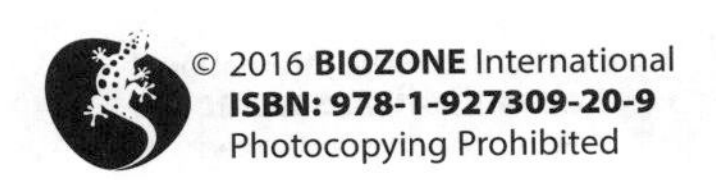

ISBN: 978-1-927309-20-9

60 Skeletal Muscle Structure and Function

Key Idea: Skeletal muscle is organised into bundles of muscle cells or fibres. The muscle fibres are made up of repeating contractile units called sarcomeres.

Skeletal muscle is organised into bundles of muscle cells or fibres. Each **fibre** is a single cell with many nuclei and each fibre is itself a bundle of smaller **myofibrils** arranged lengthwise. Each myofibril is in turn composed of two kinds of **myofilaments** (thick and thin), which overlap to form light and dark bands. It is the alternation of these light and dark bands which gives skeletal muscle its striated or striped appearance. The **sarcomere**, bounded by the dark Z lines, forms one complete contractile unit.

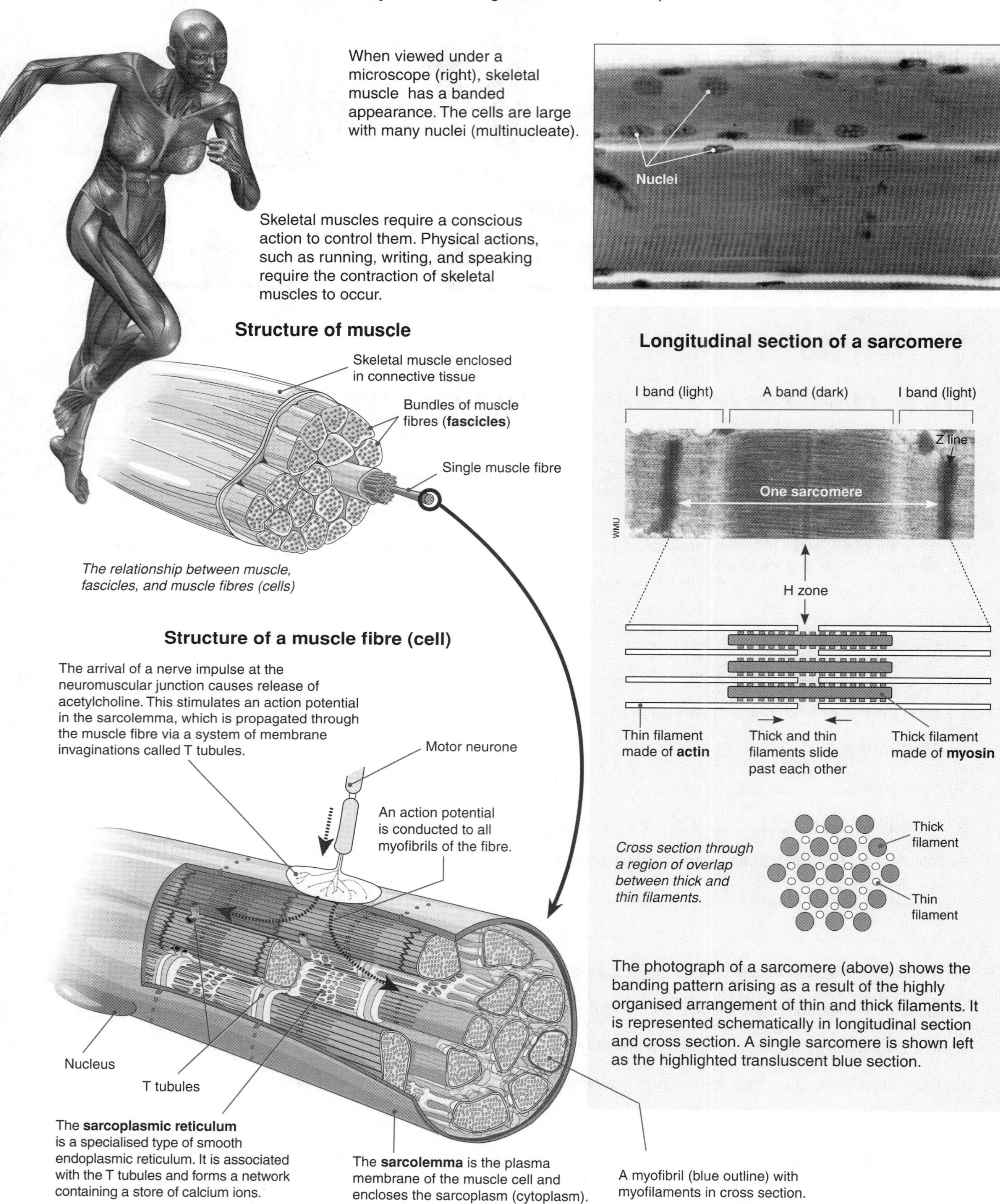

The photograph of a sarcomere (above) shows the banding pattern arising as a result of the highly organised arrangement of thin and thick filaments. It is represented schematically in longitudinal section and cross section. A single sarcomere is shown left as the highlighted transluscent blue section.

ISBN:978-1-927309-20-9

The banding pattern of myofibrils

Within a myofibril, the thin filaments, held together by the **Z lines**, project in both directions. The arrival of an action potential sets in motion a series of events that cause the thick and thin filaments to slide past each other. This is called **contraction** and it results in shortening of the muscle fibre and is accompanied by a visible change in the appearance of the myofibril: the I band and the sarcomere shorten and H zone shortens or disappears (below).

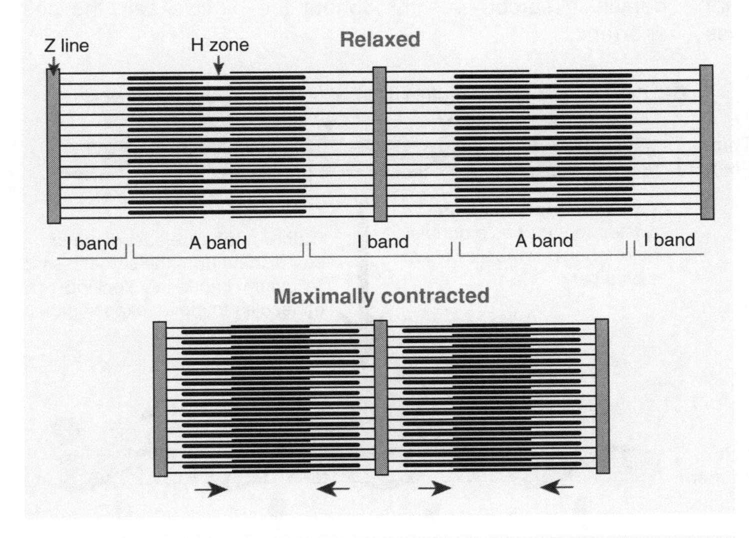

The response of a single muscle fibre to stimulation is to contract maximally or not at all; its response is referred to as the **all-or-none law** of muscle contraction. If the stimulus is not strong enough to produce an action potential, the muscle fibre will not respond. However skeletal muscles as a whole are able to produce varying levels of contractile force. These are called **graded responses** (right).

Muscles have graded responses

Muscle fibres respond to an action potential by contracting maximally, yet skeletal muscles as a whole can produce **contractions of varying force**. This is achieved by changing the frequency of stimulation (more rapid arrival of action potentials) and by changing the number of fibres active at any one time. A stronger muscle contraction is produced when a large number of muscle fibres are recruited (below left), whereas less strenuous movements, such as picking up a pen, require fewer active fibres (below right).

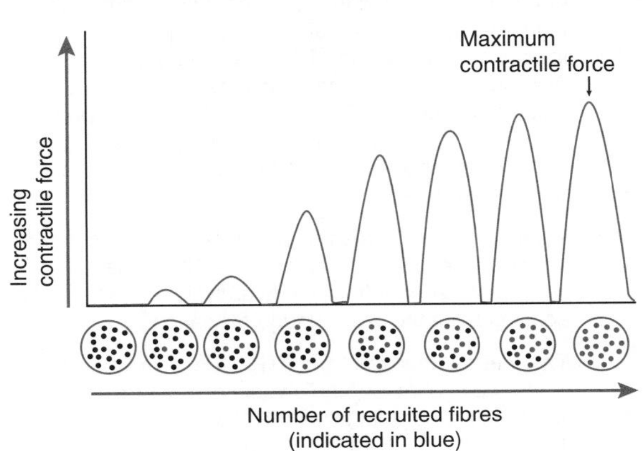

1. (a) Explain the cause of the banding pattern visible in striated muscle: ____________________

   ____

   (b) Explain the change in appearance of a myofibril during contraction with reference to the following:

   The I band: ____________________

   The H zone: ____________________

   The sarcomere: ____________________

2. Study the electron micrograph of the sarcomere (opposite).

   (a) Is it in a contracted or relaxed state (use the diagram, top left to help you decide): ____________________

   (b) Explain your answer: ____________________

3. What is meant by the all-or-none response of a muscle fibre? ____________________

4. Name two ways in which a muscle as a whole can produce contractions of varying force:

   (a) ____________________

   (b) ____________________

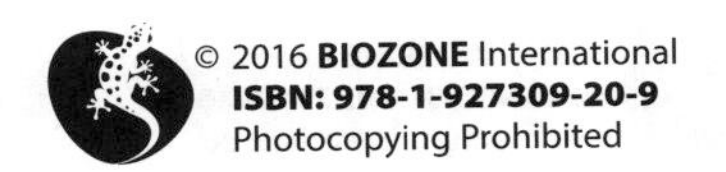

ISBN: 978-1-927309-20-9

61 The Sliding Filament Theory

Key Idea: The sliding filament theory describes how muscle contraction occurs when the thick and thin myofibrils of a muscle fibre slide past one another. Calcium ions and ATP are required.

The structure and arrangement of the thick and thin filaments in a muscle fibre make it possible for them to slide past each other and cause shortening (contraction) of the muscle. The ends of the thick myosin filaments have cross bridges that can link to adjacent thin actin filaments. When the cross bridges of the thick filaments connect to the thin filaments, a shape change moves one filament past the other. Two things are necessary for cross bridge formation: calcium ions, which are released from the sarcoplasmic reticulum when the muscle receives an action potential, and ATP, which is present in the muscle fibre and is hydrolysed by ATPase enzymes on the myosin. When cross bridges attach and detach in sarcomeres throughout the muscle cell, the cell shortens.

The sliding filament theory

Muscle contraction requires calcium ions (Ca^{2+}) and energy (in the form of ATP) in order for the thick and thin filaments to slide past each other. The steps are:

1. The binding sites on the **actin** molecule (to which myosin 'heads' will locate) are blocked by a complex of two protein molecules: **tropomyosin** and **troponin**.
2. Prior to muscle contraction, ATP binds to the heads of the myosin molecules, priming them in an erect high energy state. Arrival of an action potential is transmitted along the T tubules and causes a release of Ca^{2+} from the sarcoplasmic reticulum into the sarcoplasm. The Ca^{2+} binds to the troponin and causes the blocking complex to move so that the myosin binding sites on the actin filament become exposed.
3. The heads of the cross-bridging myosin molecules attach to the binding sites on the actin filament. Release of energy from the hydrolysis of ATP accompanies the cross bridge formation.
4. The energy released from ATP hydrolysis causes a change in shape of the myosin **cross bridge**, resulting in a bending action (*the power stroke*). This causes the actin filaments to slide past the myosin filaments towards the centre of the sarcomere.
5. (Not illustrated). Fresh ATP attaches to the myosin molecules, releasing them from the binding sites and repriming them for a repeat movement. They become attached further along the actin chain as long as ATP and Ca^{2+} are available.

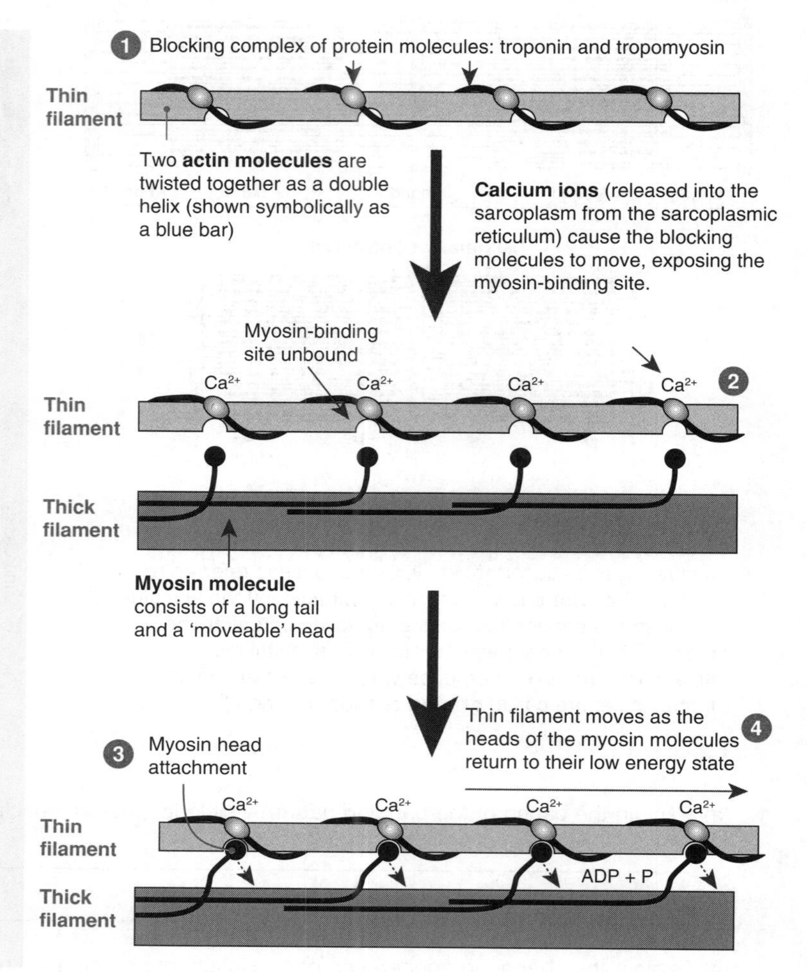

1. Match the following chemicals with their functional role in muscle movement (draw a line between matching pairs):

(a) Myosin • Bind to the actin molecule in a way that prevents myosin head from forming a cross bridge

(b) Actin • Supplies energy for the flexing of the myosin 'head' (power stroke)

(c) Calcium ions • Has a moveable head that provides a power stroke when activated

(d) Troponin-tropomyosin • Two protein molecules twisted in a helix shape that form the thin filament of a myofibril

(e) ATP • Bind to the blocking molecules, causing them to move and expose the myosin binding site

2. (a) Identify the two things necessary for cross bridge formation: ______

(b) Explain where each of these comes from: ______

______

______

3. Why are there abundant mitochondria in a muscle fibre? ______

______

ISBN:978-1-927309-20-9

62 Energy for Muscle Contraction

Key Idea: Three energy systems supply energy (ATP) to carry out muscle contraction: the ATP-CP system, the glycolytic system, and the oxidative system.

During exercise, the energy demands of skeletal muscle can increase up to 20 times. In order to continue to contract during exercise, energy in the form of ATP must be supplied. Three energy systems do this: the ATP-CP system, the glycolytic system, and the oxidative system. The ultimate sources of energy for ATP generation in muscle via these systems are glucose, and stores of glycogen and triglycerides. Prolonged exercise utilises the oxidative system and relies on a constant supply of oxygen to the tissues. Anaerobic pathways provide lower yields of ATP and are important for brief periods of high intensity, but unsustained exercise.

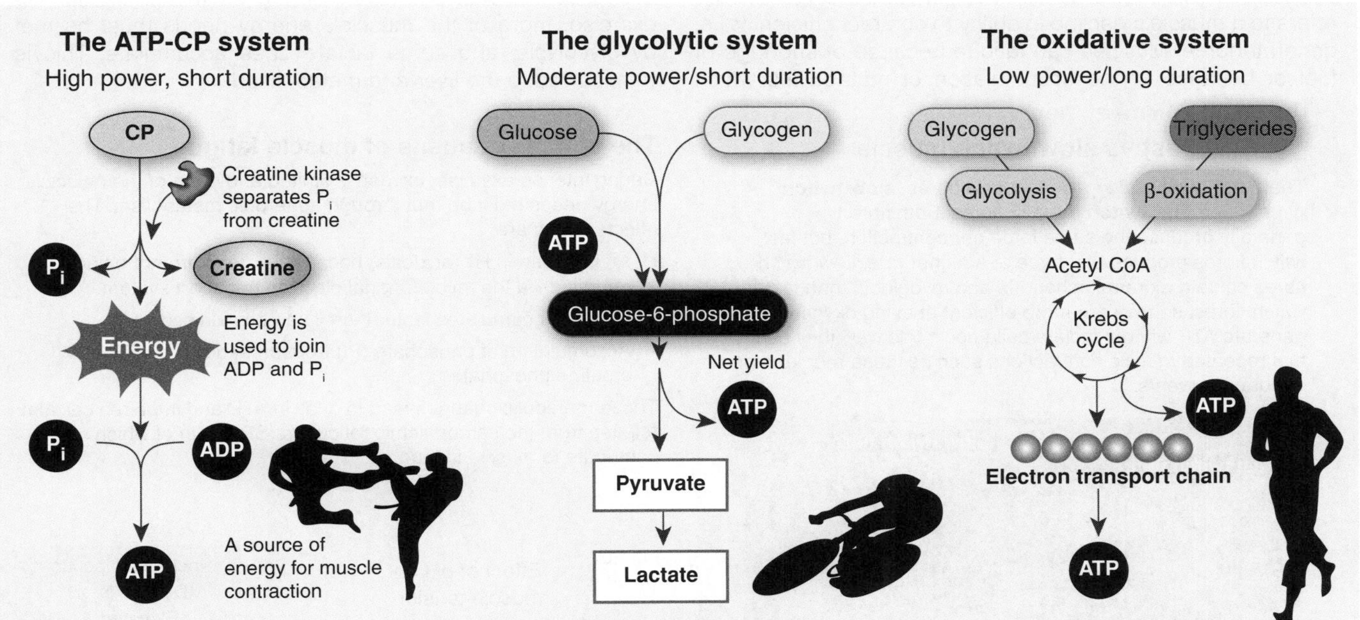

The ATP-CP system is the simplest of the systems supplying energy for muscle contraction. CP or **creatine phosphate** is a high energy compound that stores enough energy for brief periods of muscular effort. Energy released from the breakdown of CP is not used directly to accomplish cellular work. Instead it rebuilds ATP to maintain a relatively constant supply. This process is anaerobic and occurs rapidly.

CP levels decline steadily as it is used to replenish depleted ATP levels. The ATP-CP system maintains energy levels for 3-15 s of maximum effort. Beyond this, the muscle must rely on other energy systems.

ATP is also be provided by **glycolysis**. The ATP yield from glycolysis is low (only net 2 ATP per molecule of glucose), but it produces ATP rapidly and does not require oxygen. The fuel for the glycolytic system is glucose in the blood, or glycogen, which is stored in the muscle or liver and broken down to produce glucose-6-phosphate. Pyruvate is reduced to lactate, regenerating NAD^+ and allowing further glycolysis.

Glycolysis provides ATP for a few minutes of muscle contraction. It causes an accumulation of H^+ (because protons are not removed via mitochondrial respiration) and lactate in the tissues contributing to reduced muscle function.

In the oxidative system, glucose is completely broken down to yield around 36 molecules of ATP. This process uses oxygen and occurs in the mitochondria. Aerobic metabolism has a high energy yield and is the primary method of energy production during sustained activity.

Oxidative metabolism relies on a continued supply of oxygen and on glucose, stored glycogen, or stored triglycerides for fuel. Triglycerides provide free fatty acids, which are oxidised in the mitochondria by the successive removal of two-carbon fragments (a process called beta-oxidation). These two carbon units enter the Krebs cycle as acetyl coenzyme A (acetyl CoA).

1. Summarise the features of the three energy systems in the table below:

	ATP-CP system	Glycolytic system	Oxidative system
ATP supplied by:			
Duration of ATP supply:			

2. Why is the supply of energy through the glycolytic system is limited: ______

3. (a) What does the oxidative system rely on in order to supply energy for muscle contraction? ______

(b) How is this provided as exercise intensity increases? ______

ISBN: 978-1-927309-20-9

LINK 63 LINK 61 KNOW

63 Muscle Fatigue

Key Idea: Muscle fatigue refers to the decline in a muscle's ability to generate force in a prolonged or repeated contraction. Fast twitch muscles fatigue faster than slow twitch muscles.

Muscle fibres are primarily of two types: fast twitch (FT) or slow twitch (ST). Fast twitch fibres predominate during anaerobic, explosive activity, whereas slow twitch fibres predominate during endurance activity. Long or intense periods of vigorous activity can result in muscle fatigue, which refers to a muscle's decline in ability to contract efficiently, i.e. generate force. Muscles can fatigue because of shortage of fuel or because of the accumulation of metabolites which interfere with the activity of calcium in the muscle. Training can increase the length of time it takes for muscles to fatigue. Contrary to older thinking, muscle fatigue is not caused by the toxic effects of lactic acid accumulation in oxygen-starved muscle. In fact, lactate formed during exercise is an important source of fuel (through conversion to glucose) and delays fatigue and metabolic acidosis during moderate activity by acting as a buffer. However, during sustained exhausting exercise, more of the muscle's energy needs must be met by glycolysis, and some lactate does accumulate. This is transported to the liver and metabolised.

Fast vs slow twitch muscle

There are two basic types of muscle fibres: **slow twitch** (type I) and **fast twitch** (type II) fibres. Both fibre types generally produce the same force per contraction, but fast twitch fibres produce that force at a higher rate. Low twitch fibres contain more mitochondria and myoglobin than fast twitch fibres, so they are more efficient at using oxygen to generate ATP without lactate build up. In this way, they can fuel repeated muscle contractions such as those required for endurance events.

Type II fast twitch fibers are classified further according to their metabolism:
- Type IIa (intermediate) = some oxidative capacity
- Type IIb = fast glycolytic only

Slow twitch fibers appear light coloured when stained with a myofibrillar ATPase stain.

Feature	Fast twitch	Slow twitch
Colour	White	Red
Diameter	Large	Small
Contraction rate	Fast	Slow
ATP production	Fast	Slow
Metabolism	Anaerobic	Aerobic
Rate of fatigue	Fast	Slow
Power	High	Low

The complex causes of muscle fatigue

During intense exercise, oxygen is limited and more of the muscle's energy needs must be met through anaerobic metabolism. The effects of this are:

- An increase in H^+ (acidosis) because protons are not being removed via the mitochondrial electron transport system.
- Lactate accumulates faster than it can be oxidised.
- Accumulation of phosphate (*Pi*) from breakdown of ATP and creatine phosphate

These metabolic changes lead to a fall in ATP and impaired calcium release from the sarcoplasmic reticulum (SR), both of which contribute to muscle fatigue.

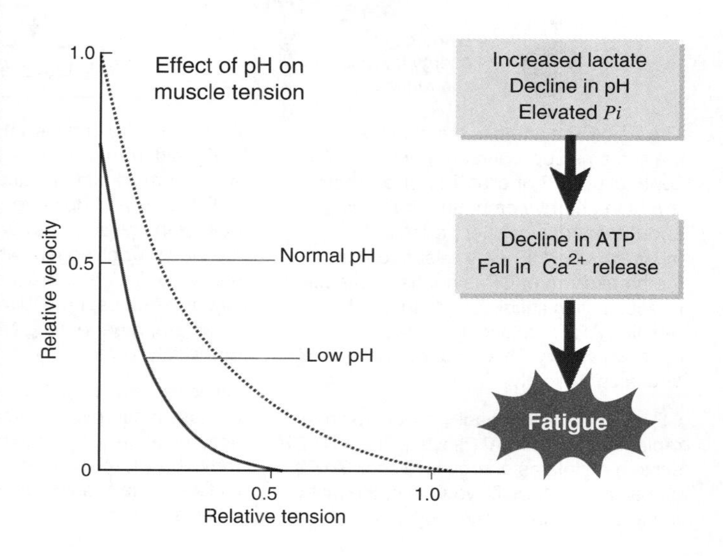

1. Contrast fast and slow twitch muscles concerning their rate of fatigue, rate of contraction, and ability to produce power:

2. (a) Name an important limiting factor during intense exercise:

   (b) How does the lack of this factor cause fatigue?

3. What is the maximum tension a muscle can produce under low pH compared to normal conditions?

KNOW LINK 64

ISBN:978-1-927309-20-9

64 Investigating Muscle Fatigue

Key Idea: A simple experiment can be used to show how muscles fatigue in response to prolonged work.

When skeletal muscle undergoes prolonged or repetitive work, it becomes fatigued, meaning it loses its ability to produce contractile force. Muscle fatigue can be measured electronically by recording the force of the contraction over time, or can be studied more simply by measuring the number of repetitive contractions that can take place over a set time. In the activity below you will work in a group of three to test the effect of muscle fatigue in your fingers.

In this practical, you will demonstrate the effects of muscle fatigue in fingers by opening and closing a spring-loaded peg over ten 10 second intervals. You will need to work in a group of three for this experiment.

Test subject: The person who opens and closes the peg.
Time keeper: Calls out the time in 10 second intervals.
Recorder: Records the number of times the peg is opened in each 10 second interval.

The test subject holds the clothes peg comfortably with the thumb and forefinger of their dominant hand (the hand they write with). They should practise opening and closing the peg fully several times before beginning the experiment.

When the timekeeper says go, the test subject opens and closes the peg fully as many times as possible for the duration of the experiment. The timekeeper calls out each 10 second interval so that the recorder can accurately record the data in the chart right. Switch roles until everyone in your group has completed the experiment.

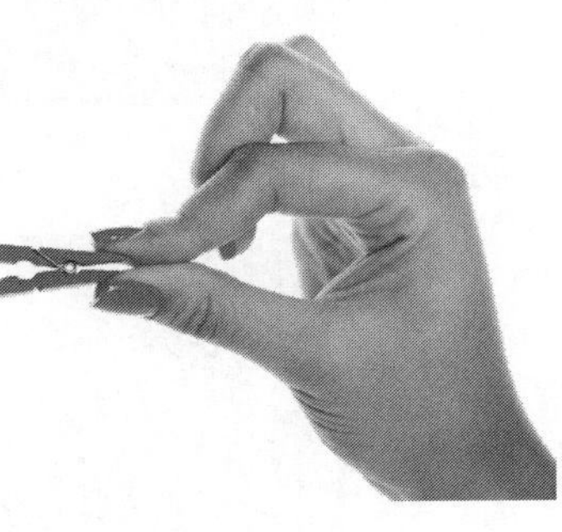

	Student 1	Student 2	Student 3
1st 10 sec			
2nd 10 sec			
3rd 10 sec			
4th 10 sec			
5th 10 sec			
6th 10 sec			
7th 10 sec			
8th 10 sec			
9th 10 sec			
10th 10 sec			

1. (a) On the grid (right) plot the data for all three individuals:

   (b) Describe the results: ______________________

   ____

   ____

   ____

   ____

   ____

   ____

   (c) Are these results what you expected? Why or why not?

   ____

   ____

   ____

   ____

   (d) Predict what would happen if the experiment was repeated using the non-dominant hand:

   ____

   ____

ISBN: 978-1-927309-20-9

65 Homeostasis

Key Idea: Homeostasis refers to the (relatively) constant physiological state of the body despite fluctuations in the external environment.

Organisms maintain a relatively constant physiological state, called **homeostasis**, despite changes in their environment. Any change in the environment to which an organism responds is called a **stimulus** and, because environmental stimuli are not static, organisms must also adjust their behaviour and physiology constantly to maintain homeostasis. This requires the coordinated activity of the body's organ systems. Homeostatic mechanisms prevent deviations from the steady state and keep the body's internal conditions within strict limits. Deviations from these limits can be harmful.

An example of homeostasis occurs when you exercise (right). Your body must keep your body temperature constant at about 37.0°C despite the increased heat generated by activity. Similarly, you must regulate blood sugar levels and blood pH, water and electrolyte balance, and blood pressure. Your body's organ systems carry out these tasks.

To maintain homeostasis, the body must detect stimuli through receptors, process this sensory information, and respond to it appropriately via effectors. The responses provide new feedback to the receptor. These three components are illustrated below.

How homeostasis is maintained

Muscles and glands

Sense organ (e.g. eye)

Receptor

Detects change and sends a message to the control centre.

Effector

Responds to the output from the control centre.

Brain and spinal cord

Control centre

Receives the message and coordinates a response. Sends an output message to an effector.

The analogy of a thermostat on a heater is a good way to understand how homeostasis is maintained. A heater has sensors (a receptor) to monitor room temperature. It also has a control centre to receive and process the data from the sensors. Depending on the data it receives, the control centre activates the effector (heating unit), switching it on or off. When the room is too cold, the heater switches on. When it is too hot, the heater switches off. This maintains a constant temperature.

1. What is homeostasis? ____________________

____

2. What is the role of the following components in maintaining homeostasis:

(a) Receptor: ____________________

____

(b) Control centre: ____________________

____

(c) Effector: ____________________

____

ISBN:978-1-927309-20-9

66 Negative Feedback

Key Idea: Negative feedback mechanisms detect departures from a set point norm and act to restore the steady state.

Most physiological systems achieve homeostasis through negative feedback. In negative feedback systems, movement away from a steady state is detected and triggers a mechanism to counteract that change. **Negative feedback** has a stabilising effect, dampening variations from a set point and returning internal conditions to a steady state. This steady state provides the specific conditions of temperature, pH, and osmolarity required for the billions of enzyme-catalysed reactions that constitute metabolism. Enzymes work correctly within very narrow limits. Outside those limits they break down or denature and are unable to catalyse reactions in the body. For example, most enzymes in the body denature above 40°C. If the body's internal environment reaches this temperature for any prolonged time it may result in death.

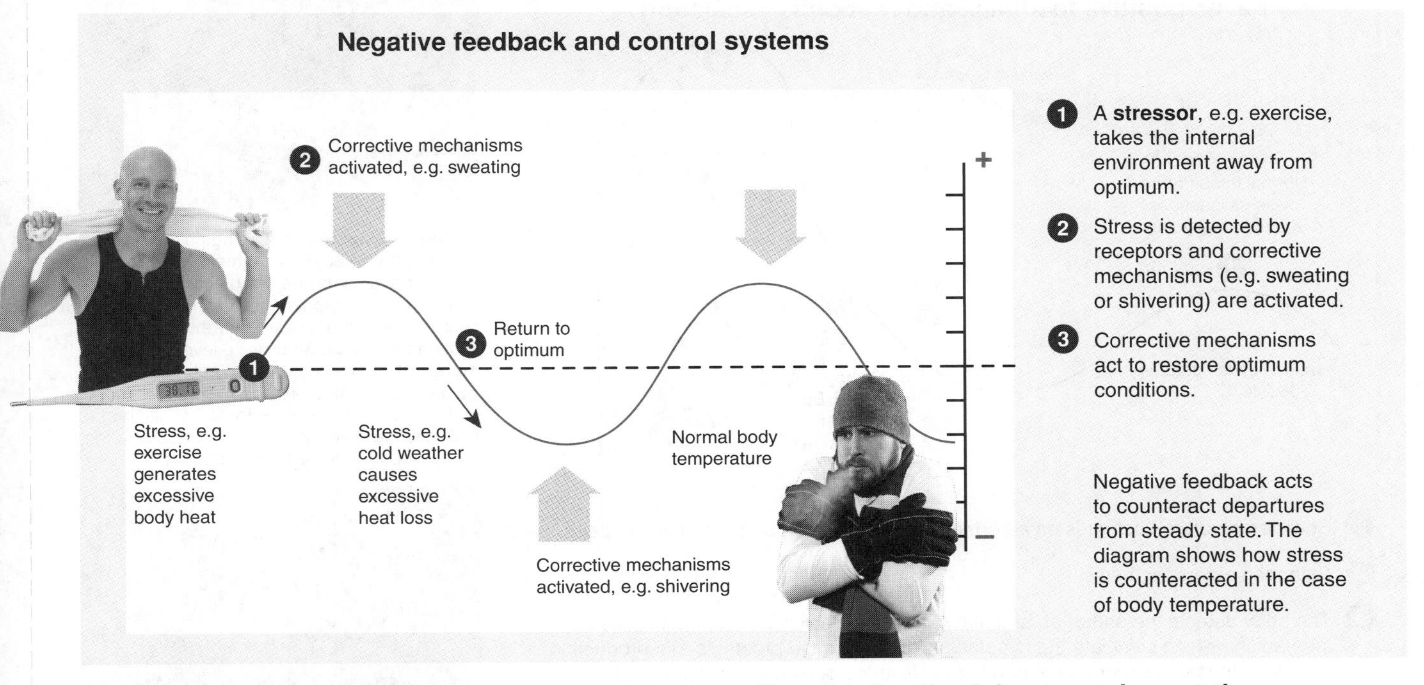

Negative feedback in blood pH

Regulation of ventilation rate helps to maintain blood pH between 7.35 and 7.45. Low blood pH stimulates increased breathing rate, which reduces H^+ via exhalation. This reduces sensory input to the medulla and breathing returns to normal.

High H^+ (low pH) in blood

Chemoreceptors in the medulla oblongata detect changes in H^+ and send impulses to the lungs.

Lung ventilation rate increases.

More CO_2 is exhaled, reducing H_2CO_3 in the blood and therefore reducing H^+.

H^+ level in blood falls

Negative feedback in stomach emptying

Empty stomach. Stomach wall is relaxed.

Stretch receptors are deactivated

A

Food is eaten

Food enters the stomach, stretching the stomach wall.

Stretch receptors are activated

Smooth muscle in the stomach wall contracts. Food is mixed and emptied from the stomach.

B

1. How do negative feedback mechanisms maintain homeostasis in a variable environment? ______________________

____

____

2. On the diagram of stomach emptying, state:

The stimulus at: A: ______________________ The response at B: ______________________

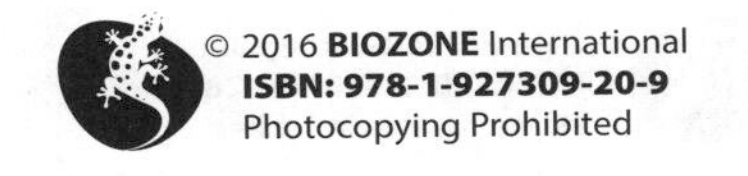

ISBN: 978-1-927309-20-9

67 Positive Feedback

Key Idea: Positive feedback results in the escalation of a response to a stimulus. It causes system instability and occurs when a particular outcome or resolution is required.

Positive feedback mechanisms amplify a physiological response in order to achieve a particular result. Labour, fever, blood clotting, and fruit ripening all involve positive feedback. Normally, a positive feedback loop is ended when the natural resolution is reached (e.g. baby is born, pathogen is destroyed). Positive feedback is relatively rare because such mechanisms are unstable and potentially damaging.

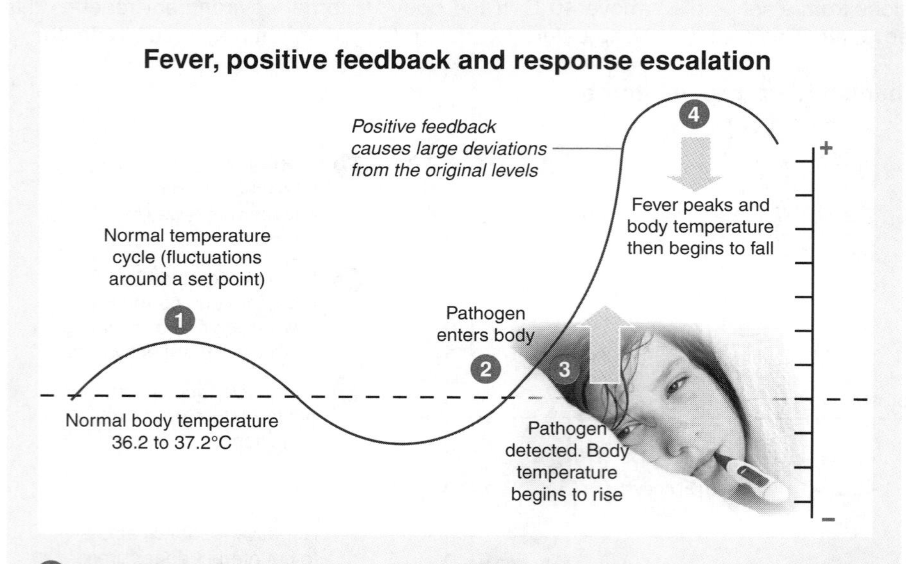

1. Body temperature fluctuates on a normal, regular basis around a narrow set point.
2. Pathogen enters the body.
3. The body detects the pathogen and macrophages attack it. Macrophages release interleukins which stimulate the hypothalamus to increase prostaglandin production and reset the body's thermostat to a higher 'fever' level by shivering (the chill phase).
4. The fever breaks when the infection subsides. Levels of circulating interleukins (and other fever-associated chemicals) fall, and the body's thermostat is reset to normal. This ends the positive feedback escalation and normal controls resume. If the infection persists, the escalation may continue, and the fever may intensify. Body temperatures in excess of 43°C are often fatal or result in brain damage.

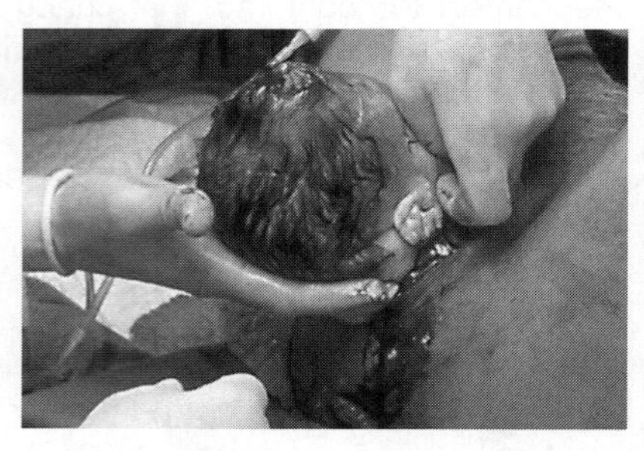

Labour and lactation: During childbirth (above), the release of oxytocin intensifies the contractions of the uterus so that labour proceeds to its conclusion. The birth itself restores the system by removing the initiating stimulus. After birth, levels of the milk-production hormone prolactin increase. Suckling maintains prolactin secretion and causes the release of oxytocin, resulting in milk release. The more an infant suckles, the more these hormones are produced.

Ethylene is a gaseous plant hormone involved in fruit ripening. It accelerates the ripening of fruit in its vicinity so nearby fruit also ripens, releasing more ethylene. Over-exposure to ethylene causes fruit to over-ripen (rot).

1. (a) What is the biological role of positive feedback loops? Describe an example: ______________________

____

____

____

(b) Why is positive feedback inherently unstable (contrast with negative feedback)? ______________________

____

____

____

(c) How is a positive feedback loop normally stopped? ______________________

____

(d) Describe a situation in which this might not happen. What would be the result? ______________________

____

____

____

ISBN:978-1-927309-20-9

68 Cell Signalling

Key Idea: Cells use signals (chemical messengers) to communicate and to gather information about, and respond to, changes in their cellular environment.

Cells communicate and bring about responses by producing and reacting to signal molecules. Three main pathways for cell signalling exist. The endocrine pathway involves the transport of hormones in the blood. In paracrine signalling, the signal travels an intermediate distance to act upon neighbouring cells. Autocrine signalling involves a cell producing and reacting to its own signal.

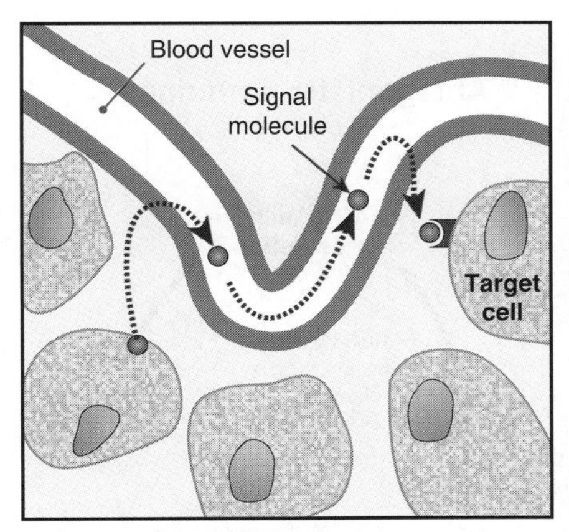

Endocrine signalling: Hormone signals are released by ductless endocrine glands and are carried by the circulatory system through the body to the target cells. Examples include sex hormones, growth factors and neurohormones such as dopamine.

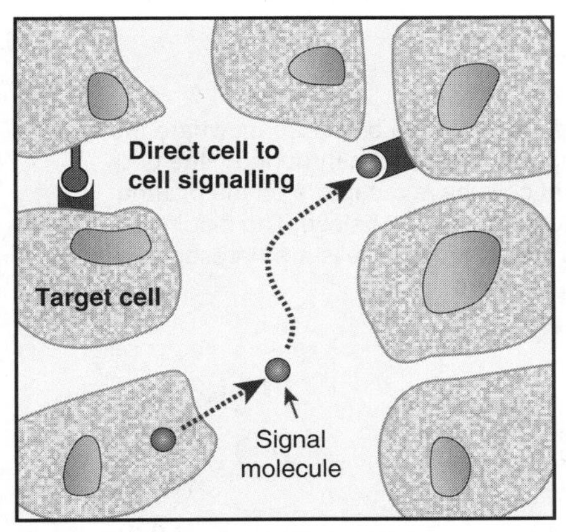

Paracrine signalling: Signals released from a cell act on target cells close by. The messenger can be transferred through the extracellular fluid (e.g. at synapses) or directly between cells. Examples include neurotransmitters and prostaglandins.

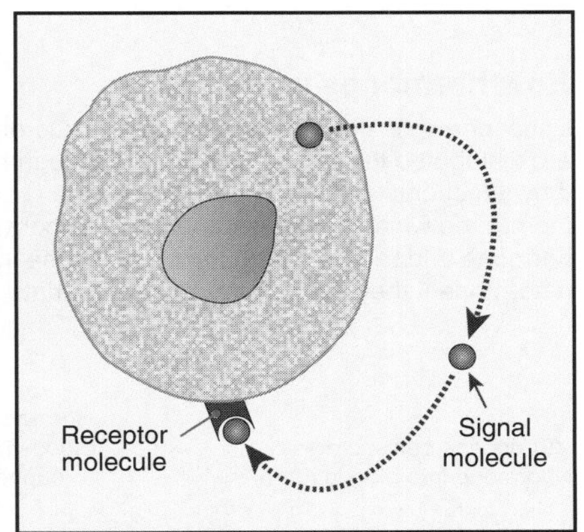

Autocrine signalling: Cells produce and react to their own signals. In vertebrates, when a foreign antibody enters the body, some T-cells produce a growth factor to stimulate their own production. The increased number of T-cells helps to fight the infection.

Signalling receptors and signalling molecules

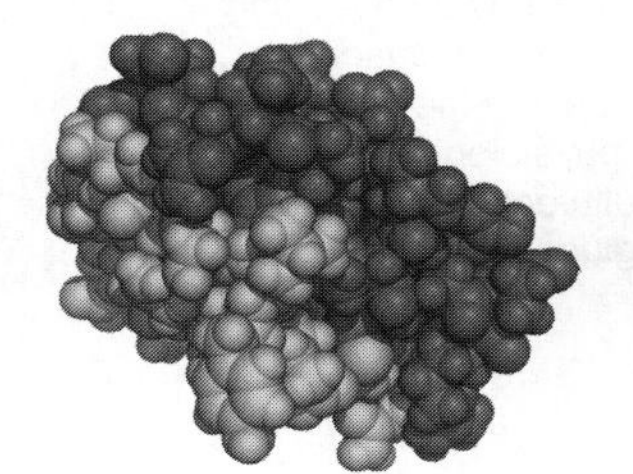

Insulin is an endocrine signalling molecule (hormone) which regulates cellular uptake of glucose.

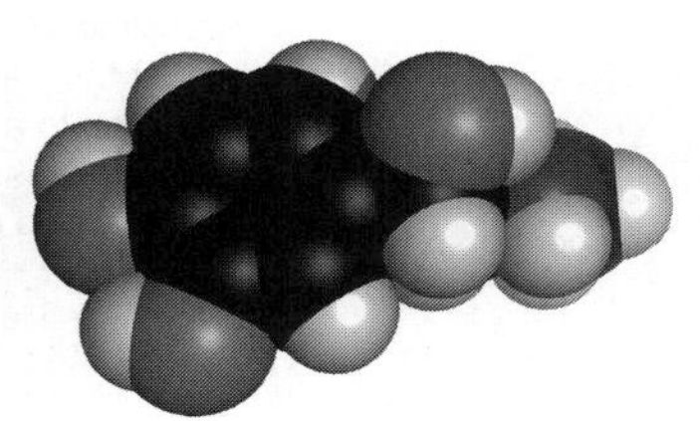

Adrenaline can behave as a hormone or a neurotransmitter. It acts to increase blood glucose.

The binding sites of cell receptors are specific only to certain **ligands** (signal molecules). This stops them reacting to every signal the cell encounters. Receptors generally fall into two main categories:

- **Cytoplasmic receptors**

Cytoplasmic receptors, located within the cell cytoplasm, bind ligands which are able to cross the plasma membrane unaided.

- **Transmembrane receptors**

These span the cell membrane and bind ligands which cannot cross the plasma membrane on their own. They have an extra-cellular domain outside the cell, and an intracellular domain within the cell cytosol.

Ion channels, protein kinases and G-protein linked receptors are examples of transmembrane receptors (see diagram on right).

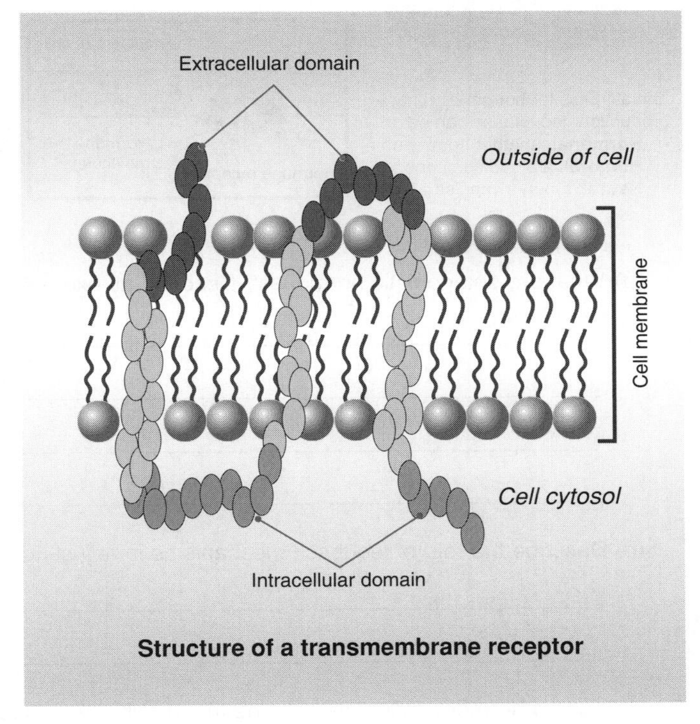

Structure of a transmembrane receptor

1. Briefly describe the three types of cell signalling:

(a)___

(b)___

(c)___

2. Identify the components that all three cell signalling types have in common: ___

ISBN: 978-1-927309-20-9

69 Hormonal Regulatory Systems

Key Idea: The endocrine system regulates physiological processes by releasing blood borne chemical messengers (called hormones) which interact with target cells.

The endocrine system is made up of endocrine cells (organised into endocrine glands) and the hormones they produce. Hormones are potent chemical regulators. They are produced in very small quantities but can exert a very large effect on metabolism. Endocrine glands secrete hormones directly into the bloodstream rather than through a duct or tube. The basis of hormonal control and the role of negative feedback mechanisms in regulating hormone levels are described below.

How hormones work

Endocrine cells produce hormones and secrete them into the bloodstream where they are distributed throughout the body. Although hormones are sent throughout the body, they affect only specific target cells. These target cells have receptors on the plasma membrane which recognise and bind the hormone (see inset, below). The binding of hormone and receptor triggers the response in the target cell. Cells are unresponsive to a hormone if they do not have the appropriate receptors.

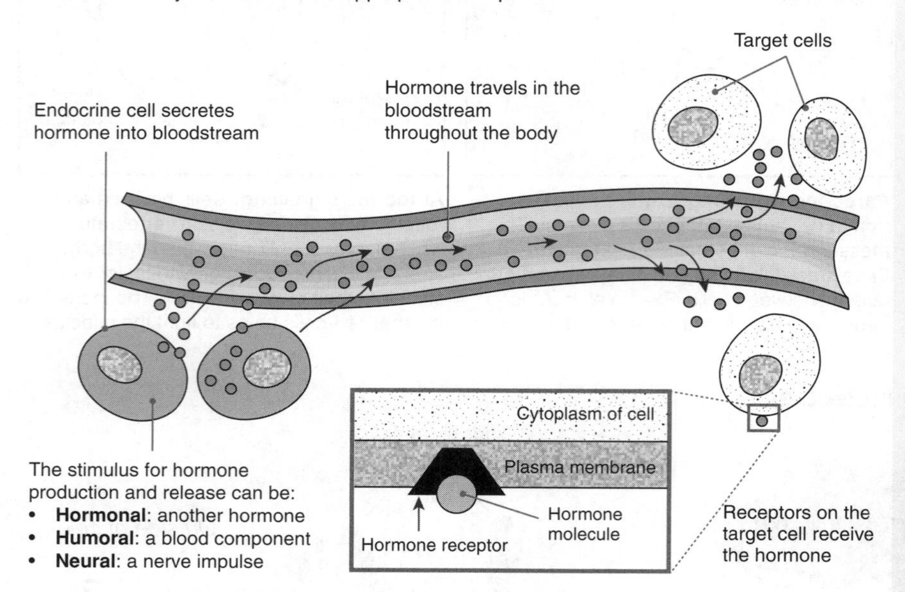

Antagonistic hormones

Insulin secretion

Blood glucose rises: insulin is released

Lowers blood glucose level

Blood glucose falls: glucagon is released

Glucagon secretion

Raises blood glucose level

The effects of one hormone are often counteracted by an opposing hormone. Feedback mechanisms adjust the balance of the two hormones to maintain a physiological function. Example: insulin acts to decrease blood glucose and glucagon acts to raise it.

1. (a) What are antagonistic hormones? Describe an example of how two such hormones operate: ______________________

(b) Describe the role of feedback mechanisms in adjusting hormone levels (explain using an example if this is helpful): ______________________

2. How can a hormone influence only the target cells even though all cells may receive the hormone? ______________________

3. Explain why hormonal control differs from nervous system control with respect to the following:

(a) The speed of hormonal responses is slower: ______________________

(b) Hormonal responses are generally longer lasting: ______________________

ISBN:978-1-927309-20-9

70 Control of Blood Glucose

Key Idea: The endocrine part of the pancreas (the α and β cells of the islets of Langerhans) produces two hormones, glucagon and insulin, which maintain blood glucose homeostasis through negative feedback.

Insulin promotes a decrease in blood glucose by promoting cellular uptake of glucose and synthesis of glycogen. Glucagon promotes an increase in blood glucose through the breakdown of glycogen and the synthesis of glucose from amino acids. Negative feedback stops hormone secretion when normal blood glucose levels are restored. Blood glucose homeostasis allows energy to be available to cells as needed. Extra energy is stored as glycogen or fat. These storage molecules are converted to glucose when energy is needed. The liver has a central role in these carbohydrate conversions. One of the consequences of a disruption to the insulin-glucagon system is the disease **diabetes mellitus**.

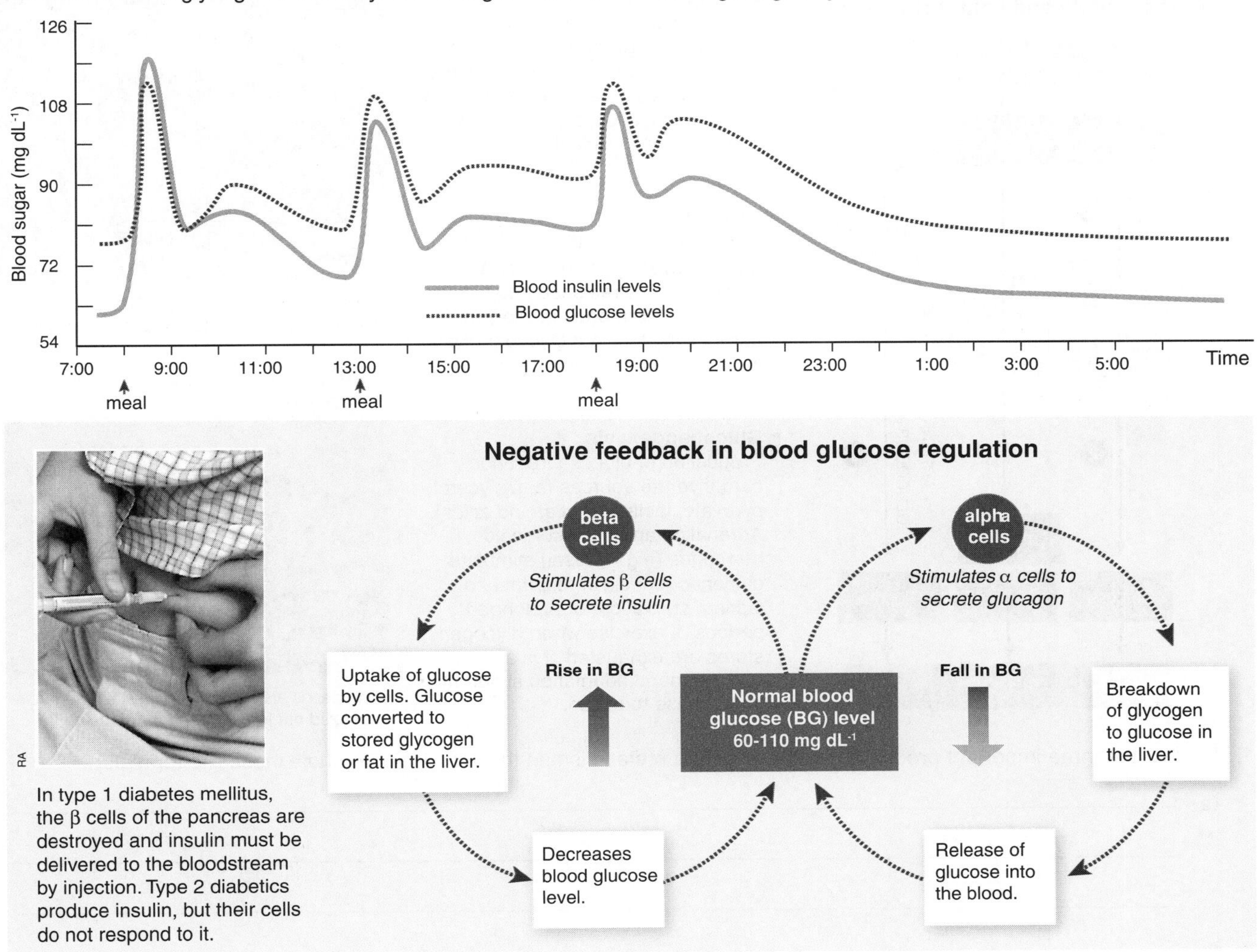

In type 1 diabetes mellitus, the β cells of the pancreas are destroyed and insulin must be delivered to the bloodstream by injection. Type 2 diabetics produce insulin, but their cells do not respond to it.

1. (a) Identify the stimulus for the release of insulin: ______

   (b) Identify the stimulus for the release of glucagon: ______

   (c) Explain how glucagon brings about an increase in blood glucose level: ______

   (d) Explain how insulin brings about a decrease in blood glucose level: ______

2. Explain the pattern of fluctuations in blood glucose and blood insulin levels in the graph above:

3. The stimulus for the production and release of insulin and glucagon is: hormonal / humoral / neural (circle one):

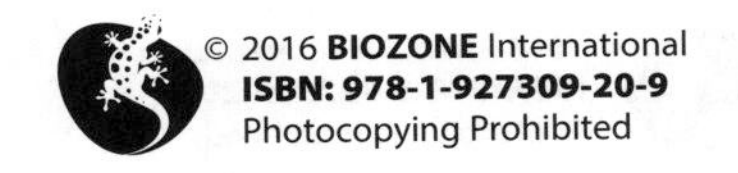

ISBN: 978-1-927309-20-9

71 The Liver's Role in Carbohydrate Metabolism

Key Idea: The interconversion of glycogen and glucose occurs in the liver in response insulin, glucagon and adrenaline.

Insulin and glucagon are antagonistic hormones secreted by α and β cells of the pancreas. The liver has a central role in the body's carbohydrate metabolism. In the liver, insulin stimulates the conversion of glucose to glycogen whereas glucagon stimulates the release of glucose from stored glycogen. In addition, adrenaline, released by the adrenal glands, also plays a role in the release of glucose into the blood, although it targets muscle cells more the liver cells.

Overview of carbohydrate metabolism in the liver

Carbohydrate and lipid metabolism

SMALL INTESTINE
Hexose sugars
Lipids
+ INSULIN
1
Glycerol + amino acids
Glycogen
Fats
+ GLUCAGON
2
+ ADRENALINE GLUCOCORTICOIDS
3
Fatty acids
Glucose
Glycerol
Glucose
BLOOD

▶ **Glycogenesis**
Excess glucose in the blood is converted to **glycogen** (a glucose polysaccharide). **Insulin** stimulates glycogenesis in response to high blood glucose. Glycogen is stored in the liver and muscle tissue.

▶ **Glycogenolysis**
Conversion of stored glycogen to glucose (glycogen breakdown). The free glucose is released into the blood. The hormones **glucagon** and adrenaline stimulate glycogenolysis in response to low blood glucose.

▶ **Gluconeogenesis**
Production of glucose from non-carbohydrate sources (e.g. glycerol, pyruvate, lactate, and amino acids). Adrenaline and glucocorticoid hormones (e.g. cortisol) stimulate gluconeogenesis in response to fasting, starvation, or prolonged periods of exercise when glycogen stores are exhausted. It is also part of the general adaptation syndrome in response to stress.

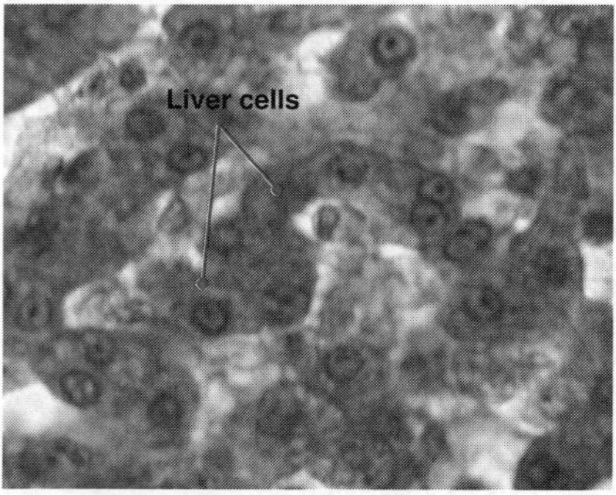

Glycogen is stored within the liver cells. Glucagon stimulates its conversion to glucose.

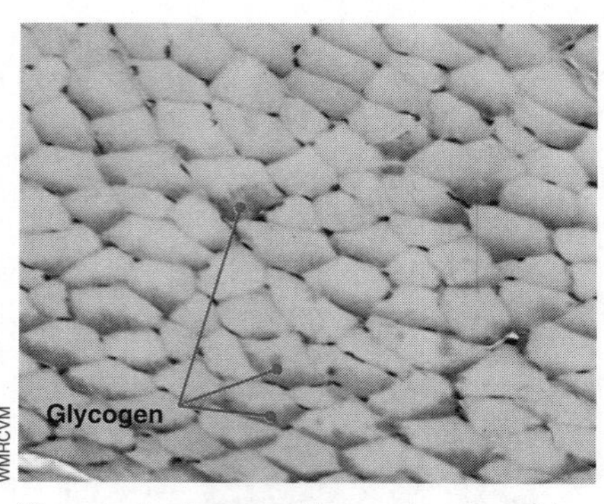

Glycogen is also stored in muscle, where it is squeezed out to the periphery of the cells.

1. Explain the three important processes of carbohydrate metabolism in the liver, including how these are regulated:

   (a) ______________________________

   (b) ______________________________

   (c) ______________________________

2. Identify the processes occurring at each numbered stage on the diagram above, right:

   (a) Process occurring at point 1: ______________________________

   (b) Process occurring at point 2: ______________________________

   (c) Process occurring at point 3: ______________________________

3. Explain why it is important that the body can readily convert and produce different forms of carbohydrates:

   ____

ISBN:978-1-927309-20-9

72 Insulin and Glucose Uptake

Key Idea: Activation of the insulin receptor by insulin causes a signal cascade that results in cellular glucose uptake.

Insulin is a peptide hormone secreted by the pancreas. Its primary role is in maintaining blood glucose homeostasis. Insulin production is tightly regulated because over- or underproduction has serious physiological consequences (including death). Too little insulin results in elevated blood glucose levels (hyperglycaemia), whereas too much insulin results in low blood glucose levels (hypoglycaemia). The insulin receptors are surface-bound protein kinase receptors. They facilitate a cellular response by catalysing the transfer of a phosphate group from ATP to a target protein.

Glucose uptake pathway

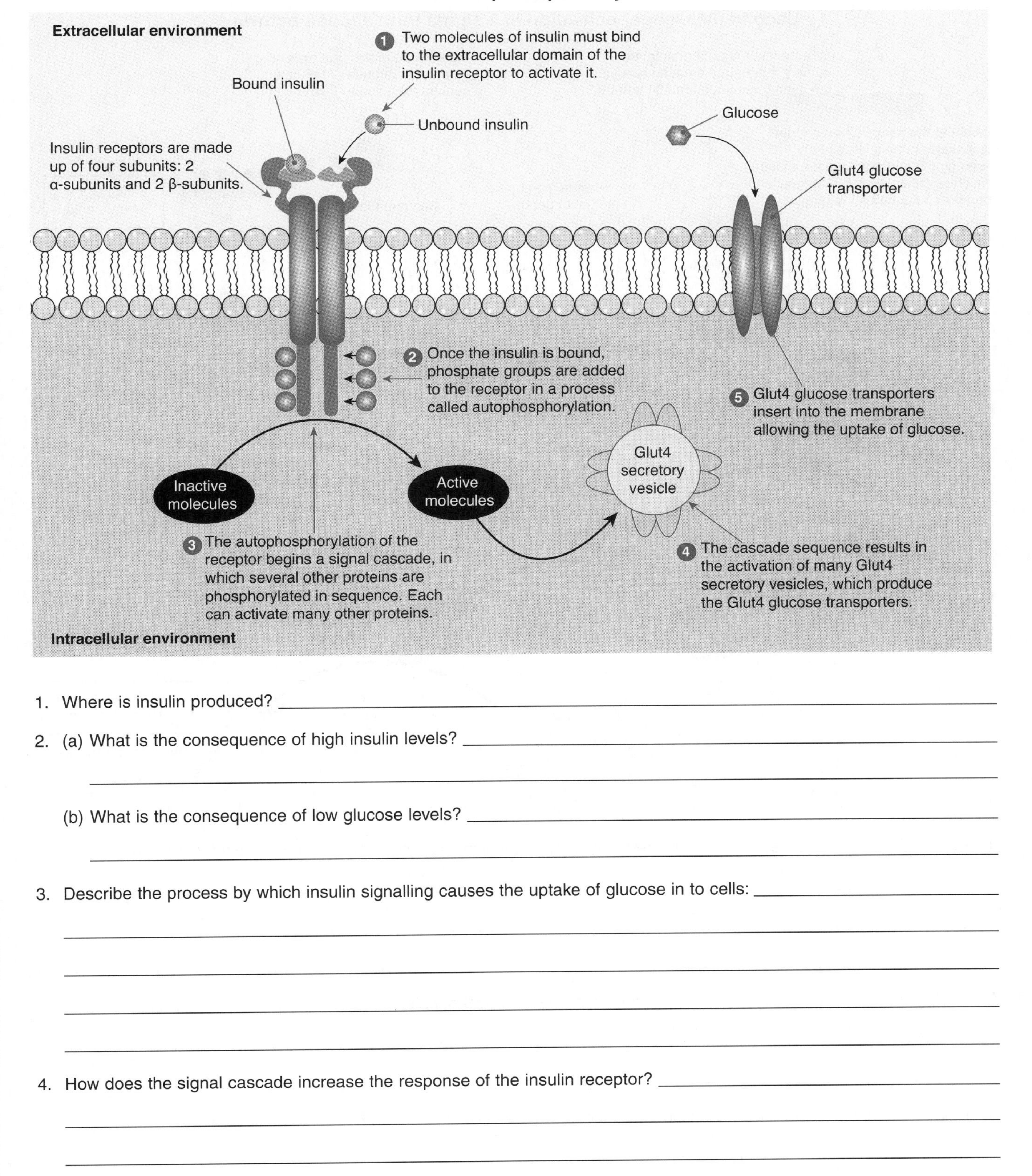

1. Where is insulin produced? ______

2. (a) What is the consequence of high insulin levels? ______

   (b) What is the consequence of low glucose levels? ______

3. Describe the process by which insulin signalling causes the uptake of glucose in to cells: ______

4. How does the signal cascade increase the response of the insulin receptor? ______

ISBN: 978-1-927309-20-9

73 Adrenaline and Glucose Metabolism

Key Idea: Adrenaline and glucagon signal a cell to convert glycogen to glucose molecules.

Adrenaline and glucagon (as well as insulin) act as signal molecules in a second messenger system. An extracellular signal brings about an intracellular response through a **signal transduction pathway**. The signal molecule (e.g. adrenaline) binds to the specific receptor of a target cell and starts a cascade of reactions that amplifies the original signal and brings about a response (e.g. enzyme activation). The second messenger in the system below is cyclic AMP. Adrenaline binds to the β-adrenergic receptor in muscle fibres whereas glucagon binds to the glucagon receptor. Adrenaline and glucagon produce the same response in cells, i.e. the production of glucose from glycogen.

Second messenger activation in a signal transduction pathway

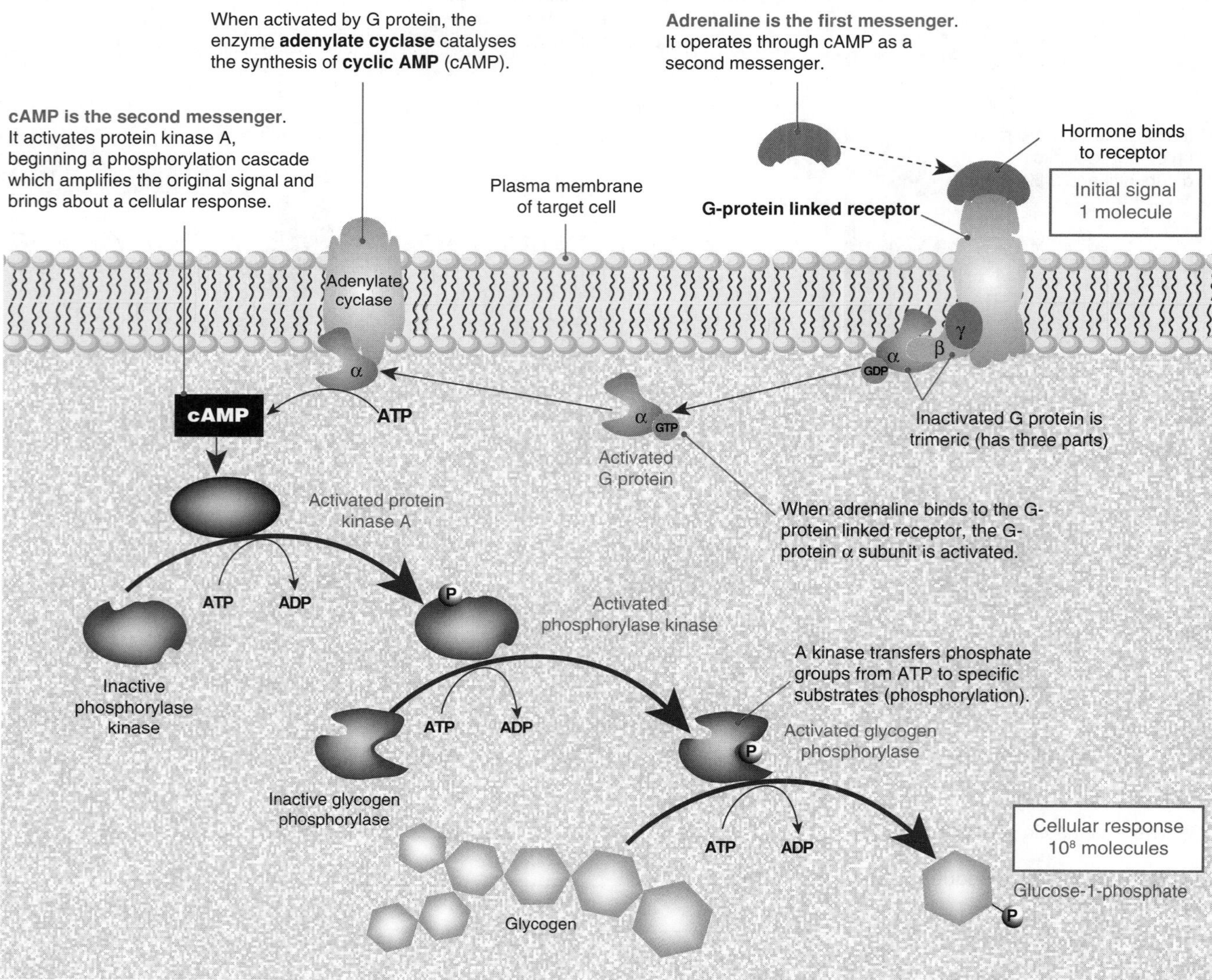

1. Explain why a second messenger is needed to convey a signal inside a cell from a water soluble first messenger:

2. Explain why each molecule in the cAMP signal cascade is phosphorylated:

3. What is the role of G proteins in coupling the receptor to the cellular response?

ISBN:978-1-927309-20-9

74 Type 1 Diabetes Mellitus

Key Idea: In type 1 diabetes, the insulin-producing cells of the pancreas are destroyed and insulin cannot be made.

When the insulin-glucagon system is disrupted the disease **diabetes mellitus** occurs. Diabetes mellitus is characterised by **hyperglycaemia** (high blood sugar). **Type 1 diabetes** is characterised by **absolute insulin deficiency**. It usually begins in childhood as a result of autoimmune destruction of the insulin-producing cells of the pancreas. For this reason, it was once called juvenile-onset diabetes. It is a severe, incurable condition, and is treated with insulin injections.

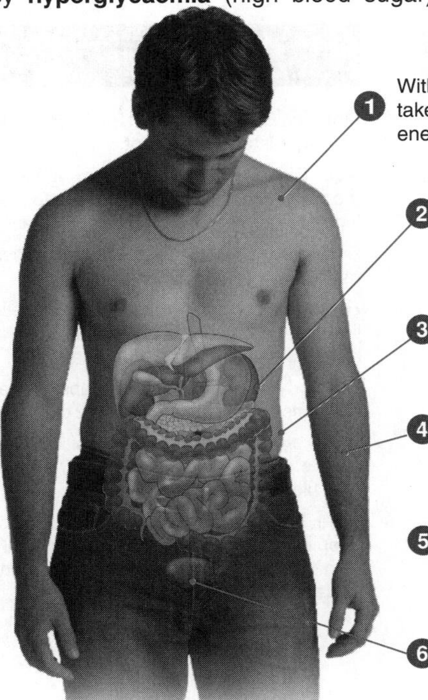

1. Without insulin, cells cannot take up glucose and so lack an energy source for metabolism.
2. Production of urine from the kidneys increases to clear the body of excess blood glucose. Glucose is present in the urine.
3. There is constant thirst. Weight is lost despite hunger and overeating.
4. Inability to utilise glucose leads to muscle weakness and fatigue.
5. Fats are metabolised for energy leading to a fall in blood pH (ketosis). This is potentially fatal.
6. High sugar levels in blood and urine promote bacterial and fungal infections of the bladder and urinogenital tract.

Cause of type 1 diabetes mellitus

Incidence: About 10-15% of all diabetics.

Age at onset: Early; often in childhood.

Symptoms: Hyperglycaemia (high blood sugar), excretion of glucose in the urine (glucosuria), increased urine production, excessive thirst and hunger, weight loss, and ketosis.

Cause: Absolute deficiency of insulin due to lack of insulin production (pancreatic beta cells are destroyed in an autoimmune reaction). There is a genetic component but usually a childhood viral infection triggers the development of the disease. Mumps, coxsackie, and rubella are implicated.

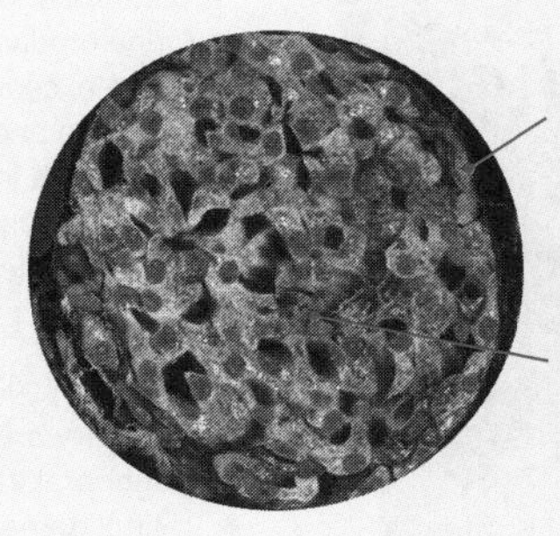

α **cells** produce glucagon, which promotes glucose release from the liver.

β **cells** (most of the cells in this field of view) produce insulin, the hormone promoting cellular uptake of glucose. Cells are destroyed in type 1 diabetes mellitus.

Cell types in the endocrine region of a normal pancreas

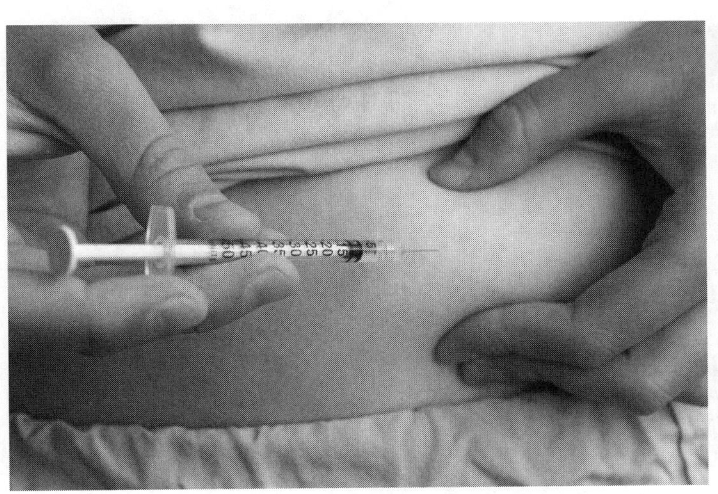

Treatments of type 1 diabetes mellitus

Present treatments: Regular insulin injections are combined with dietary management to keep blood sugar levels stable. Blood glucose levels are monitored regularly with testing kits to guard against sudden, potentially fatal, falls in blood glucose (hypoglycaemia).

Insulin was once extracted from dead animals, but animal-derived insulin produces many side effects. Genetically engineered microbes now provide low cost human insulin, without the side effects associated with animal insulin.

Newer treatments: Cell therapy involves transplanting islet cells into the patient where they produce insulin and regulate blood sugar levels. The islet cells may be derived from stem cells or from the pancreatic tissue of pigs. New technology developed by New Zealand company Living Cell Technologies Ltd, encapsulates the pig islet cells within microspheres so they are protected from destruction by the patient's immune system.

Cell therapy promises to be an effective way to provide sustained relief for diabetes.

1. Describe the symptoms of type 1 diabetes mellitus and relate these to the physiological cause of the disease:

2. Explain how regular insulin injections assist the type 1 diabetic to maintain their blood glucose homeostasis:

ISBN: 978-1-927309-20-9

LINK 75 LINK 72 WEB 74 KNOW

75 Type 2 Diabetes Mellitus

Key Idea: In type 2 diabetes, the pancreas produces insulin, but the body does not respond to it appropriately.

In type 2 diabetes, the pancreas produces insulin, but the quantities are insufficient or the body's cells do not react to it, so blood glucose levels remain high. For this reason, **type 2 diabetes** is sometimes called **insulin resistance diabetes**. Type 2 diabetes is a chronic, progressive disease, and gets worse with age if not managed. The long-term effects of high blood sugar include heart disease, strokes, loss of vision, and kidney failure.

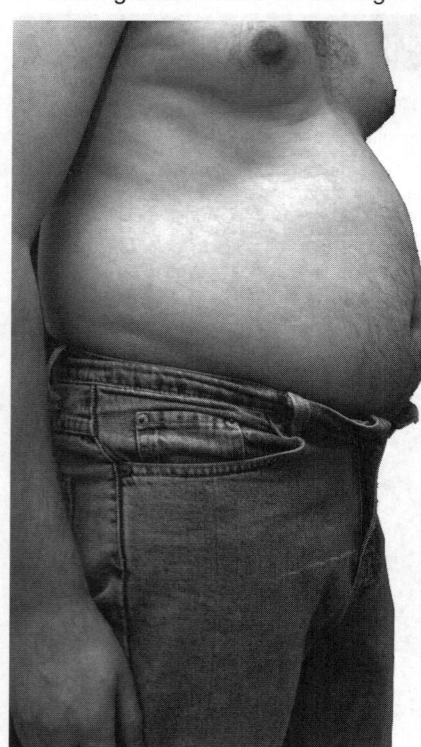

Symptoms of type 2 diabetes mellitus

a Symptoms may be mild at first. The body's cells do not respond appropriately to the insulin that is present and blood glucose levels become elevated. Normal blood glucose level is 60-110 mg dL^{-1}. In diabetics, fasting blood glucose level is 126 mg dL^{-1} or higher.

b Symptoms occur with varying degrees of severity:

- Cells are starved of fuel. This can lead to increased appetite and overeating and may contribute to an existing obesity problem.
- Urine production increases to rid the body of the excess glucose. Glucose is present in the urine and patients are frequently very thirsty.
- The body's inability to use glucose properly leads to muscle weakness and fatigue, irritability, frequent infections, and poor wound healing.

c Uncontrolled elevated blood glucose eventually results in damage to the blood vessels and leads to:

- coronary artery disease
- peripheral vascular disease
- retinal damage, blurred vision and blindness
- kidney damage and renal failure
- persistent ulcers and gangrene

Risk factors

Obesity: BMI greater than 27. Distribution of weight is also important.

Age: Risk increases with age, although the incidence of type 2 diabetes is increasingly reported in obese children.

Sedentary lifestyle: Inactivity increases risk through its effects on bodyweight.

Family history: There is a strong genetic link for type 2 diabetes. Those with a family history of the disease are at greater risk.

Ethnicity: Certain ethnic groups are at higher risk of developing of type 2 diabetes.

High blood pressure: Up to 60% of people with undiagnosed diabetes have high blood pressure.

High blood lipids: More than 40% of people with diabetes have abnormally high levels of cholesterol and similar lipids in the blood.

Treating Type 2 Diabetes

Diabetes is not curable but can be managed to minimise the health effects:

- Regularly check blood glucose level
- Manage diet to reduce fluctuations in blood glucose level
- Exercise regularly
- Reduce weight
- Reduce blood pressure
- Reduce or stop smoking
- Take prescribed anti-diabetic drugs
- Insulin therapy may be required

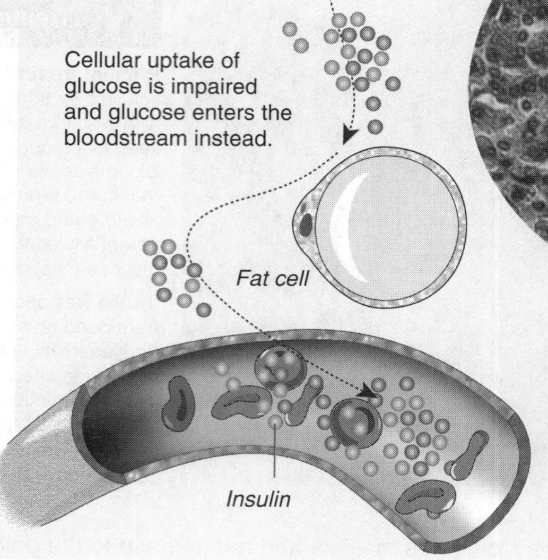

The **beta cells** of the pancreatic islets (above) produce insulin, the hormone responsible for the cellular uptake of glucose. In type 2 diabetes, the body's cells do not utilise the insulin properly.

1. Distinguish between type 1 and type 2 diabetes, relating the differences to the different methods of treatment:

2. Why is the increase in type 2 diabetes considered epidemic in the developed world?

ISBN:978-1-927309-20-9

76 Measuring Glucose in a Urine Sample

Key Idea: The level of glucose in urine can be measured using colorimetry.

During initial filtration in the kidney nephron, glucose is lost to the filtrate. However, urine ordinarily contains no glucose because it is reabsorbed by the proximal tubules of the kidney. If the amount of glucose in the blood is very high (160-180 mg dL^{-1}) so much glucose is lost to the filtrate that the proximal tubules can not return all of it to the blood and glucose is lost to the urine. The presence of this 'overflow' glucose can be measured using the Benedict's test for reducing sugars. The presence of glucose in the urine can indicate diabetes mellitus.

Aim:

To find the concentration of glucose in three unknown urine samples.

Method:

A calibration curve was prepared using prepared artificial 'urine' containing known concentrations of glucose (right). A Benedict's test was performed on each and the absorbance of the resulting solution was measured using a colorimeter. The Benedict's test was then performed on unknown urine samples and the absorbance of the resulting solutions measured and compared to the calibration curve.

Benedict's test for reducing sugars:

Benedict's solution is added, and the sample is placed in a water bath at 90°C or heated over a Bunsen burner.

Results:

Sample 1 absorbance: 0.10
Sample 2 absorbance: 0.05
Sample 3 absorbance: 0.82

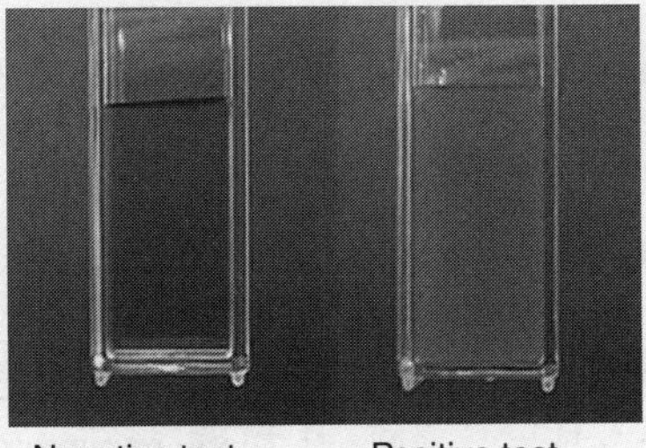

Colorimetric analysis of glucose

Prepare glucose standards

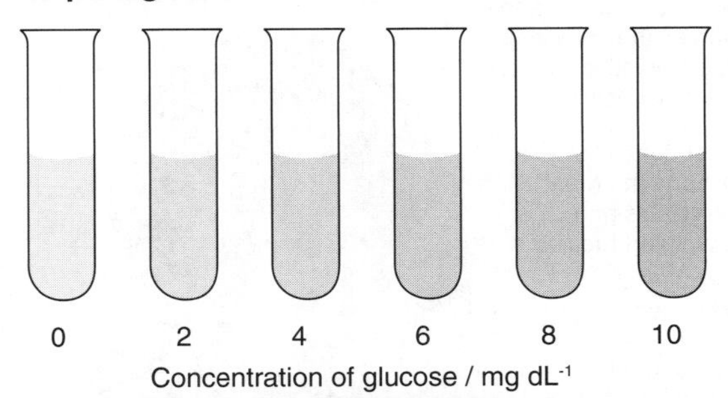

Produce the calibration curve

Cool and filter samples as required. Using a red filter, measure the absorbance (at 735 nm) for each of the known dilutions and use these values to produce a calibration curve for glucose.

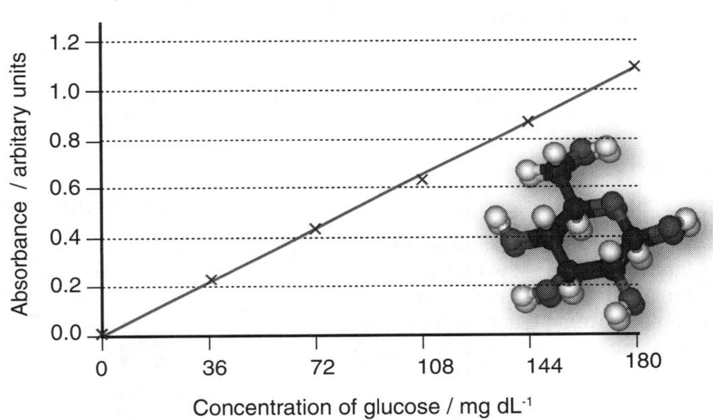

1. Use the calibration curve provided to estimate the glucose content of the urine samples:

   (a) Sample 1: ______

   (b) Sample 2: ______

   (c) Sample 3: ______

2. (a) Which of the samples gives the most cause for concern? ______

   (b) Explain why: ______

3. What is the purpose of the 0 mg dL^{-1} tube? ______

4. What would you do if the absorbance values you obtained for most of your 'unknowns' were outside the range of your calibration curve?

______

ISBN: 978-1-927309-20-9

77 The Physiology of the Kidney

Key Idea: The functional unit of the kidney is the nephron. It is a selective filter element, comprising a renal corpuscle and its associated tubules and ducts.

Kidneys are the organs responsible for excretion of wastes and for regulating fluid and ion balance. They are made up of millions of selective filtering units called nephrons, which produce an excretory fluid called urine. Ultrafiltration, i.e. forcing fluid and dissolved substances through a membrane by pressure, occurs in the first part of the nephron and creates an initial filtrate of the blood. This filtrate is modified by secretion and reabsorption to create the final urine. The formation of the glomerular filtrate depends on the pressure of the blood entering the nephron (below). If it increases, filtration rate increases; when it falls, glomerular filtration rate also falls. This process is precisely regulated so that glomerular filtration rate per day stays constant.

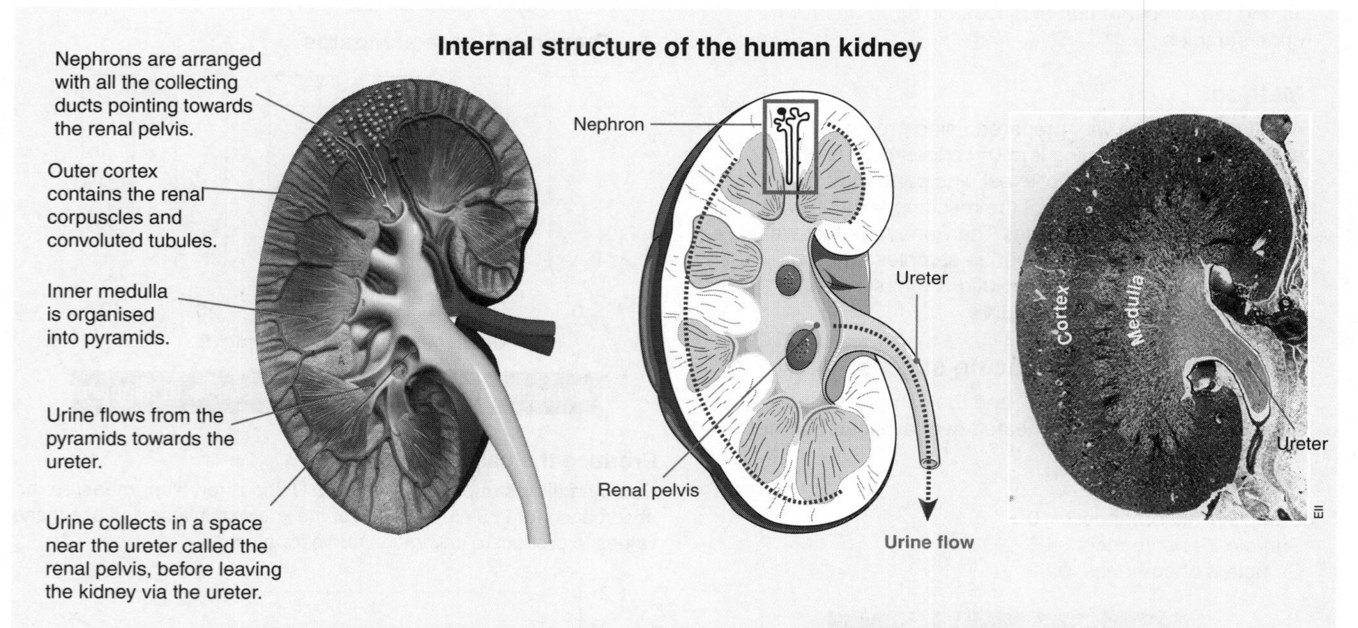

The inside of the kidney appears striped because the nephrons are all aligned so that urine is concentrated as it flows towards the ureter. The outer cortex and inner medulla can be seen in the low power LM of the kidney, far right.

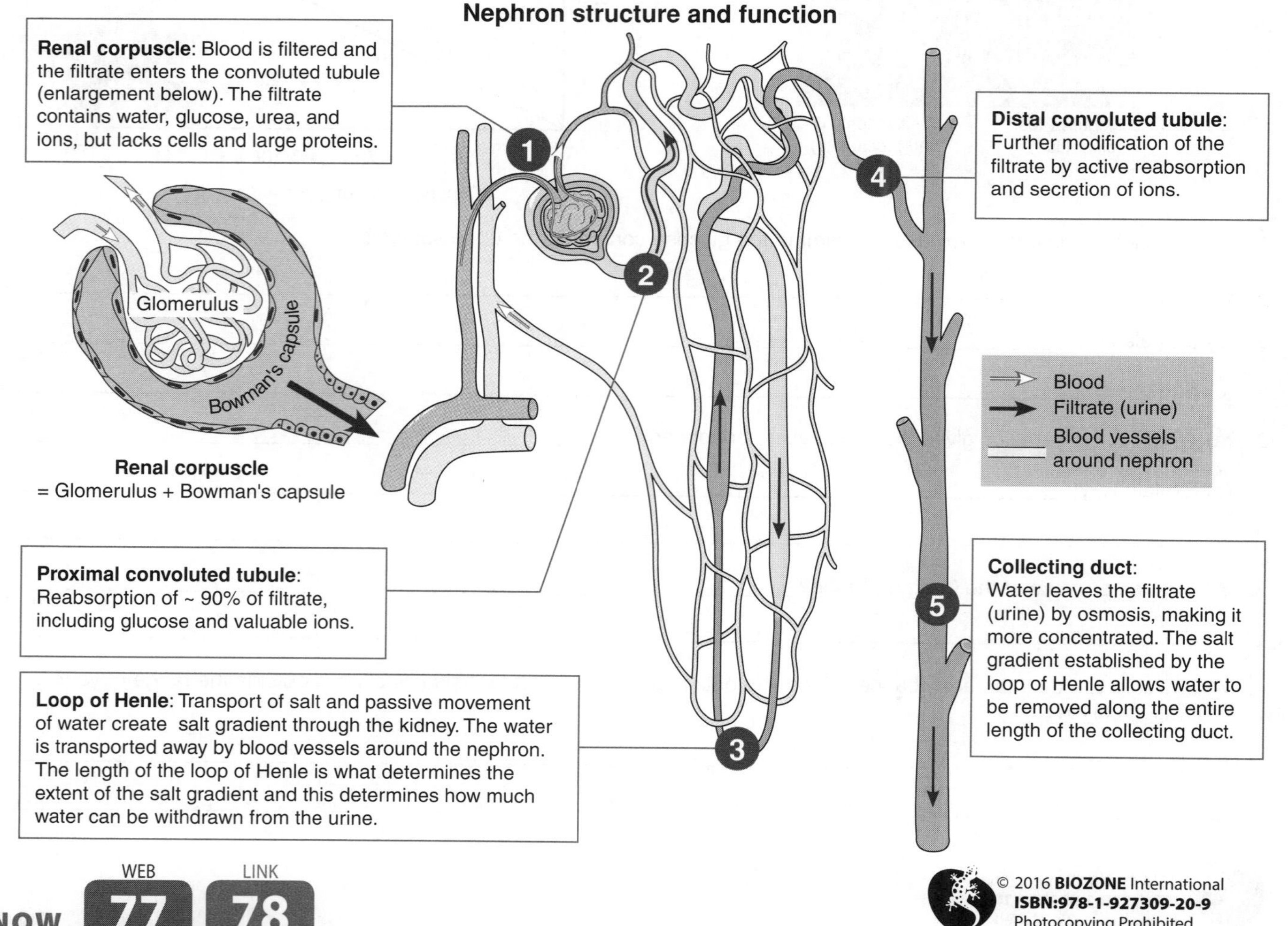

ISBN:978-1-927309-20-9

1. Summarise the main activities in each of the five regions of the nephron:

   (a) Renal corpuscle:

   (b) Proximal (near) convoluted tubule:

   (c) Loop of Henle:

   (d) Distal (far) convoluted tubule:

   (e) Collecting duct:

2. Why are the nephrons all aligned in the same way in the kidney:

3. Why does the kidney receive blood at a higher pressure than other organs?

4. Explain the importance of the following in the production of urine in the kidney nephron:

   (a) Filtration of the blood at the glomerulus:

   (b) Active secretion:

   (c) Reabsorption:

5. (a) What is the purpose of the salt gradient in the kidney?

   (b) How is this salt gradient produced?

6. The graph below shows the volume of urine collected from a subject after drinking 1000 cm^3 of distilled water. The subject's urine was collected at 25 minute intervals over a number of hours.

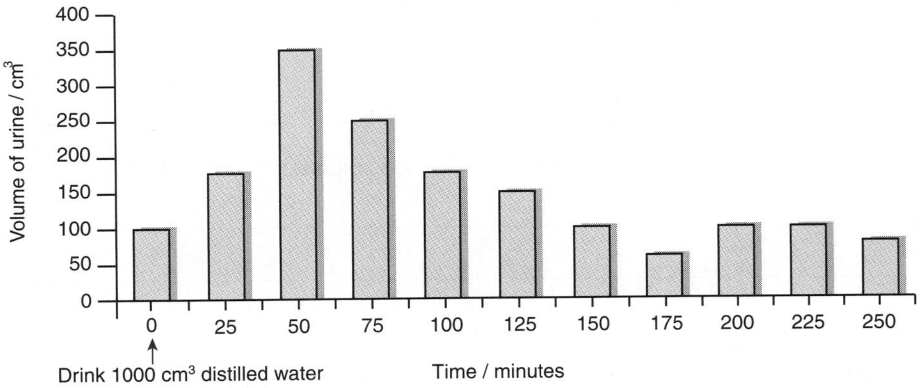

Volume of urine / cm^3
400
350
300
250
200
150
100
50
0
0
25
50
75
100
125
150
175
200
225
250
Drink 1000 cm^3 distilled water
Time / minutes

   (a) Describe the changes in urine output during the experiment:

   (b) Explain the difference in the volume of urine collected at 25 minutes and 50 minutes:

ISBN: 978-1-927309-20-9

78 Osmoregulation

Key Idea: Antidiuretic hormone (ADH) helps maintain water balance by regulating water absorption by the kidneys.

The body regulates water balance in response to fluctuations in fluid gains and losses. One mechanism by which fluid balance is maintained is by varying the volume of water absorbed by the kidneys and thereby the concentration of the urine produced. This involves a hormone called **antidiuretic hormone** (ADH). Osmoreceptors in the hypothalamus monitor blood osmolarity (water content) and send messages to the pituitary gland which regulates the amount of ADH released. ADH promotes the reabsorption of water from the kidney collecting ducts, producing a less or more concentrated urine.

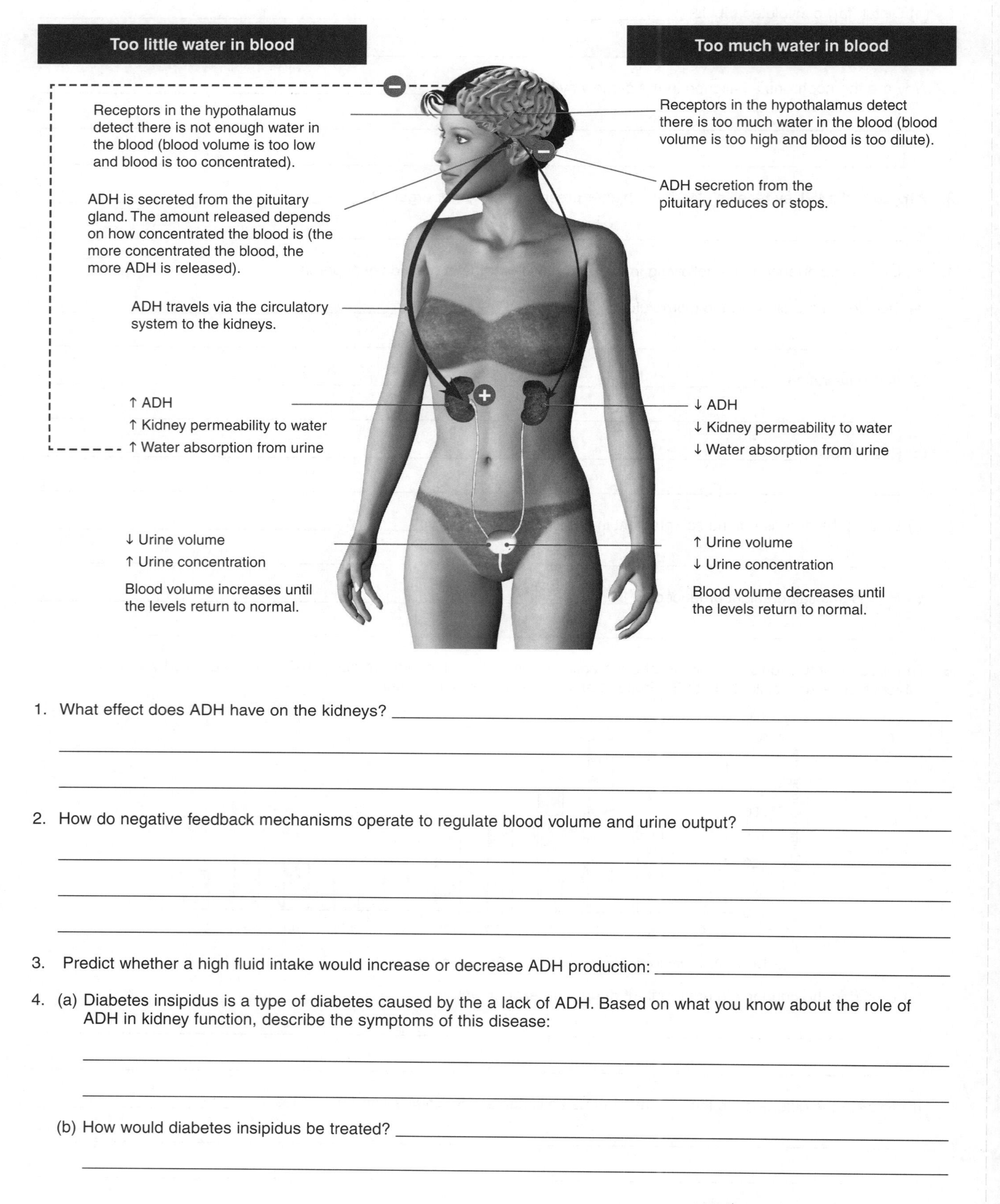

1. What effect does ADH have on the kidneys? ______________________________

____

____

2. How do negative feedback mechanisms operate to regulate blood volume and urine output? ______________________________

____

____

____

3. Predict whether a high fluid intake would increase or decrease ADH production: ______________________________

4. (a) Diabetes insipidus is a type of diabetes caused by the a lack of ADH. Based on what you know about the role of ADH in kidney function, describe the symptoms of this disease:

____

____

(b) How would diabetes insipidus be treated? ______________________________

____

ISBN:978-1-927309-20-9

79 Chapter Review

Summarise what you know about this topic under the headings and sub-headings provided. You can draw diagrams or mind maps, or write short notes to organise your thoughts. Use the images and hints to help you and refer back to the introduction to check the points covered:

Plant and animal responses
HINT: Taxes, kineses, and tropisms.

Sensitivity
HINT: Sensing the environment and vision.

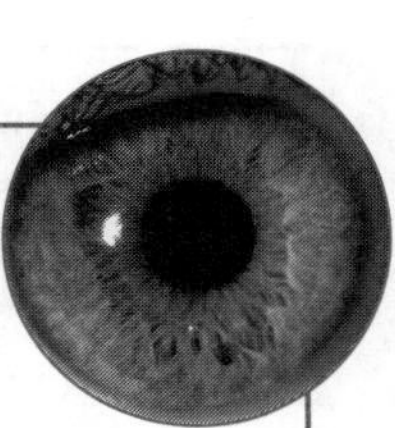

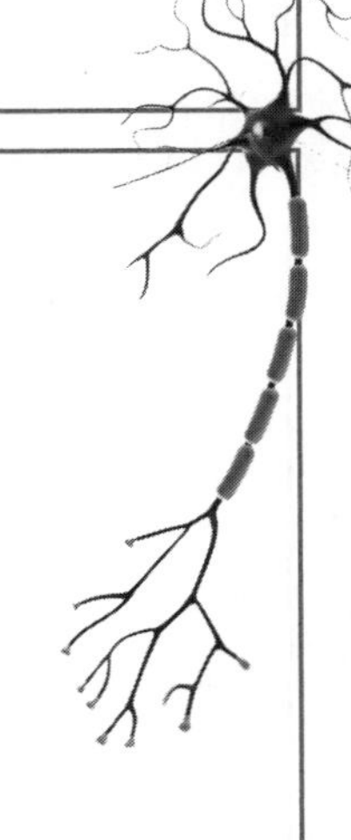

Neurones and signal transmission
HINT: Neurones, neurotransmitters, and synapses.

ISBN: 978-1-927309-20-9

Muscles and movement

HINT: Muscle structure and function.

Homeostasis

HINT: Negative and positive feedback, blood glucose, osmoregulation.

ISBN:978-1-927309-20-9

80 KEY TERMS AND IDEAS: Did You Get It?

1. (a) What is the name given to a plant growth response to directional light? ______

   (b) What is the name given to a plant growth response to gravity? ______

   (c) What is the name given to a plant response that is independent of stimulus direction? ______

   (d) What plant hormone is principally responsible for the phototropic effect? ______

2. (a) What responses are being shown by the orchid in the photo (right):

   ______

   ______

   ______

   ______

   (b) What is the stimulus involved? ______

   ______

3. (a) Label the components of this neurone (right) using the following word list: *cell body, axon, dendrites, node of Ranvier.*

   (b) Is this neurone myelinated or unmyelinated?(delete one)

   (c) Explain your answer: ______

   ______

   ______

   (d) In what form do electrical signals travel in this cell?

   ______

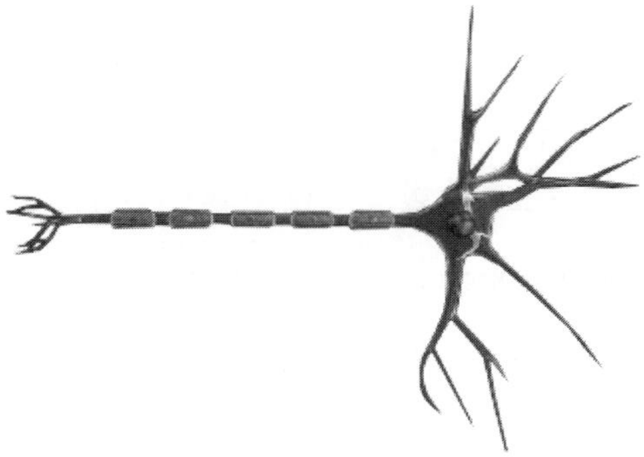

4. The graph below shows a real recording of the changes in membrane potential in an axon during transmission of an action potential (rather than an idealised schematic). Match each stage (A-E) to the correct summary provided below.

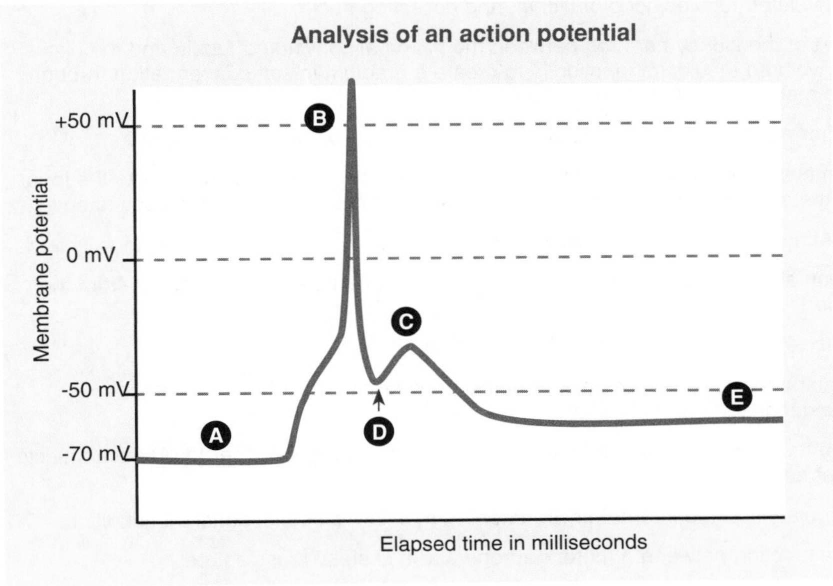

Analysis of an action potential

Membrane potential: +50 mV, 0 mV, -50 mV, -70 mV

Stages: A, B, C, D, E

Elapsed time in milliseconds

- ☐ Membrane depolarisation (due to rapid Na^+ entry across the axon membrane).
- ☐ Hyperpolarisation (an overshoot caused by the delay in closing of the K^+ channels).
- ☐ Return to resting potential after the stimulus has passed.
- ☐ Repolarisation as the Na^+ channels close and slower K^+ channels begin to open.
- ☐ The membrane's resting potential.

5. (a) Name the excretory organ of vertebrates: ______

   (b) Name the selective filtering element of the kidney: ______

   (c) The length of this structure is directly related to the ability of an organism to concentrate urine: ______

   (d) Name the hormone involved in controlling urine output: ______

ISBN: 978-1-927309-20-9

6. Test your knowledge about feedback mechanisms by studying the two graphs below, and answering the questions about them. In your answers, use biological terms appropriately to show your understanding.

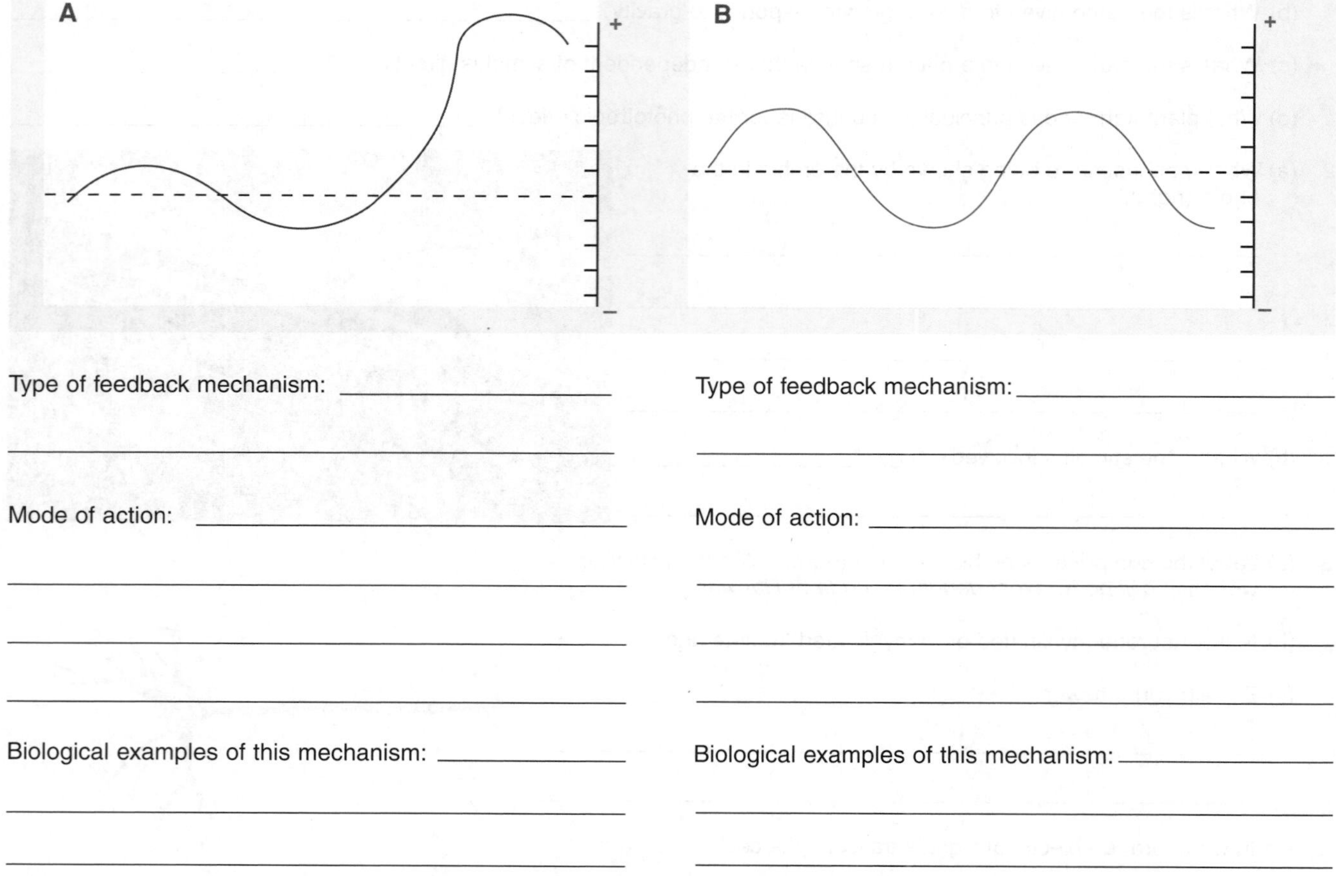

Type of feedback mechanism: ______________

Mode of action: ______________

Biological examples of this mechanism: ______________

Type of feedback mechanism: ______________

Mode of action: ______________

Biological examples of this mechanism: ______________

7. Test your vocabulary by matching each term to its definition, as identified by its preceding letter code.

auxin ....

excretion ....

homeostasis ....

insulin ....

kidney ....

loop of Henle ....

muscle fibre ....

negative feedback ....

nephron ....

neuromuscular junction ....

positive feedback ....

reflex ....

sliding filament hypothesis ....

synapse ....

A The functional unit of the kidney comprising the glomerulus, Bowman's capsule, convoluted tubules, loop of Henle, and collecting duct.

B Part of the kidney nephron between the proximal convoluted tubule and the distal convoluted tubule. Its function is to create a gradient in salt concentration through the medullary region of the kidney.

C A hormone, secreted by the pancreas, that lowers blood glucose levels.

D A mechanism in which the output of a system acts to oppose changes to the input of the system. The net effect is to stabilise the system and dampen fluctuations.

E Elimination of the waste products of metabolism.

F Bean shaped organ which removes and concentrates metabolic wastes from the blood.

G Muscle cell containing a bundle of myofibrils.

H A plant hormone responsible for apical dominance, phototropism, and cell elongation.

I The theory of how thin and thick filaments slide past each other to produce muscle contraction.

J Regulation of the internal environment to maintain a stable, constant condition.

K The junction between a motor neurone and a skeletal muscle fibre.

L The gap between neighbouring neurones or between a neurone and an effector.

M A destabilising mechanism in which the output of the system causes an escalation in the initial response.

N An automatic response to a stimulus involving a small number of neurones and a central nervous system (CNS) processing point.

ISBN:978-1-927309-20-9

Topic 7

Genetics, populations, evolution, and ecosystems

Key terms

adaptation
allele
allele frequency
allopatric speciation
carrying capacity
codominant allele
community
competition
conservation
dihybrid cross
diploid
dominant allele
ecological succession
ecosystem
epistasis
evolution
fitness
founder effect
gene flow (=migration)
gene pool
genetic bottleneck
genetic drift
genetic equilibrium
genotype
haploid
Hardy-Weinberg principle
heterozygous
homologous chromosomes
homozygous
linkage (linked genes)
locus
mark and recapture
meiosis
monohybrid cross
multiple alleles
multiple genes
mutation
natural selection
niche
phenotype
polygenes
polyploidy
population
predation
quadrat
recessive allele
reproductive isolation
sex linkage
species
sympatric speciation
trait
transect

7.1 Inheritance

Learning outcomes

	Learning outcome	Activity number
☐	1 Define and distinguish between genotype and phenotype.	81
☐	2 Distinguish between genes and alleles. Explain what it means when alleles are dominant, recessive, or codominant.	82
☐	3 Define the term locus. Distinguish between the terms homozygous and heterozygous in relation to alleles in diploid organisms.	82
☐	4 Demonstrate appropriate use of the terms used in studying inheritance: allele, locus, trait, heterozygous, homozygous, genotype, phenotype, cross, test cross, back cross, carrier, F_1/F_2 generation.	87 96
☐	5 Use genetic diagrams to interpret or predict the results of monohybrid and dihybrid crosses involving dominant, recessive, codominant, and multiple alleles. Distinguish between multiple alleles and multiple genes.	83 84 85 86 87 90 96 99
☐	6 Use labelled genetic diagrams to interpret or predict the results of genetic crosses involving sex linkage, autosomal linkage, and gene interactions, including epistasis and polygenic inheritance (multiple genes).	88 89 91 92 93 97 98 99
☐	7 **AT** Investigate genetic ratios using crosses of *Drosophila* or *Brassica rapa*.	91 92 94
☐	8 Use the chi-squared test (for goodness of fit) to test the significance of differences between the observed and expected results of genetic crosses.	94 95

7.2 Populations

Learning outcomes

	Learning outcome	Activity number
☐	9 Understand what is meant by a population and know that species may exist as one or more populations.	101 120
☐	10 **AT** Collect and analyse data about the frequency of observable phenotypes in a single population.	99
☐	11 Understand the concept of the gene pool and explain how allele frequencies are expressed for populations. Explain the effect of mutation, gene flow (migration), natural selection, and genetic drift on the allele frequencies of populations.	101 102 103 104 107-111
☐	12 State the conditions required for genetic equilibrium in a population and explain the consequences of these conditions rarely, if ever, being satisfied. Understand the Hardy-Weinberg principle as a mathematical model for genetic equilibrium.	102 105
☐	13 Use the Hardy-Weinberg equation to calculate and analyse the frequencies of alleles, genotypes, and phenotypes in a population.	105 106

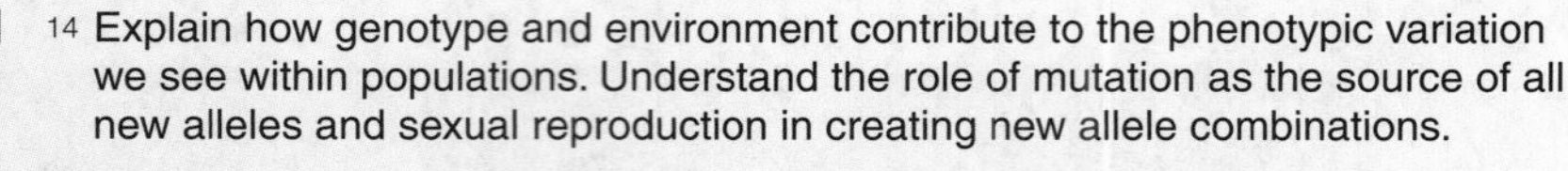

7.3 Evolution may lead to speciation

Learning outcomes

		Learning outcome	Activity number
☐	14	Explain how genotype and environment contribute to the phenotypic variation we see within populations. Understand the role of mutation as the source of all new alleles and sexual reproduction in creating new allele combinations.	81 100
☐	15	Explain how sexual reproduction contributes to genetic variation within a species, including reference to linkage and recombination in meiosis and random fusion of gametes at fertilisation.	100
☐	16	Recall the principles by which natural selection sorts variation and establishes adaptive phenotypes. Explain the role of predation, disease, and competition in natural selection (differential survival and reproduction).	101 107-110
☐	17	Using appropriate examples, describe and explain the effect of natural selection on the allele frequencies in gene pools. Examples could include:	103 104
☐	i	Stabilising selection, e.g. for human birth weight and regional skin colour.	110
☐	ii	Directional selection, e.g. melanism in rock pocket mice or peppered moths.	107 108
☐	iii	Disruptive selection, e.g. selection for different beak sizes in Galápagos finches.	109
☐	18	Describe evolution as the change in the allele frequencies of a population. Explain how geographic separation of a species into two isolates can result in increasing genetic divergence and formation of a new species (allopatric speciation).	101 114 116-118
☐	19	Explain the role of reproductive isolation in establishing and maintaining the integrity of new and existing species. Describe reproductive isolating mechanisms operating before and after fertilisation.	114 115
☐	20	Distinguish allopatric and sympatric speciation and explain why sympatric speciation is typically more common in plants.	116 119
☐	21	Recognise genetic drift as an important process in the evolution of small populations. Describe its consequences (e.g. fixation or loss of alleles) and the conditions under which it is important, e.g. after a population (genetic) bottleneck or in founder populations (the founder effect).	111-113
☐	22	Explain how evolutionary change over a long period of time has resulted in a great diversity of species.	118

7.4 Populations in ecosystems

Learning outcomes

		Learning outcome	Activity number
☐	23	Define the terms ecosystem, community, population, species, and environment With reference to specific examples, explain that ecosystems exist on different scales (from small to very large). Categorise the components of an ecosystem as biotic or abiotic factors and explain how these influence species distribution.	120-122
☐	24	Explain how, within a habitat, species occupy a niche (functional position) governed by adaptation to biotic and abiotic conditions.	123 124
☐	25	**PR-12** Investigate the effect of a named environmental factor on the distribution of a species.	140 142 143
☐	26	Understand the factors determining population size. Distinguish between exponential and logistic growth in populations. Define carrying capacity.	125 126
☐	27	**AT** Use turbidity measurements to investigate the growth rate of a microbial culture. Represent that growth on a logarithmic scale.	127 128
☐	28	Explain how the carrying capacity of an ecosystem is affected by abiotic factors and by intraspecific and interspecific interactions, e.g. predation and competition.	129-135
☐	29	Describe how population size can be estimated using sampling methods that are appropriate for the species involved and the environment, e.g. quadrats along a belt transect (e.g. for vegetation) and mark and recapture (for motile organisms).	136-139 141
☐	30	**AT** Estimate population size of a motile species from mark and capture data using the Lincoln Index.	141
☐	31	Recognise ecosystems as resilient, dynamic systems. Describe succession from colonisation by pioneer species to a climax community, including the role of earlier successional species in modifying the environment for later species. Explain how the changes may result in a change in biodiversity.	144-148
☐	32	Explain how conservation of some habitats, e.g. meadows, may involve management of a natural succession.	149
☐	33	Explain how the conflict between human needs and conservation must be managed to ensure the sustainability of resources.	150-153

81 Genotype and Phenotype

Key Idea: An organism's phenotype is influenced by the effects of the environment during and after development, even though the genotype remains unaffected.

The phenotype encoded by genes is a product not only of the genes themselves, but of their internal and external environment and the variations in the way those genes are controlled (epigenetics). Even identical twins have minor differences in their appearance due to epigenetic and environmental factors such as diet and intrauterine environment. Genes, together with epigenetic and environmental factors determine the unique phenotype that is produced.

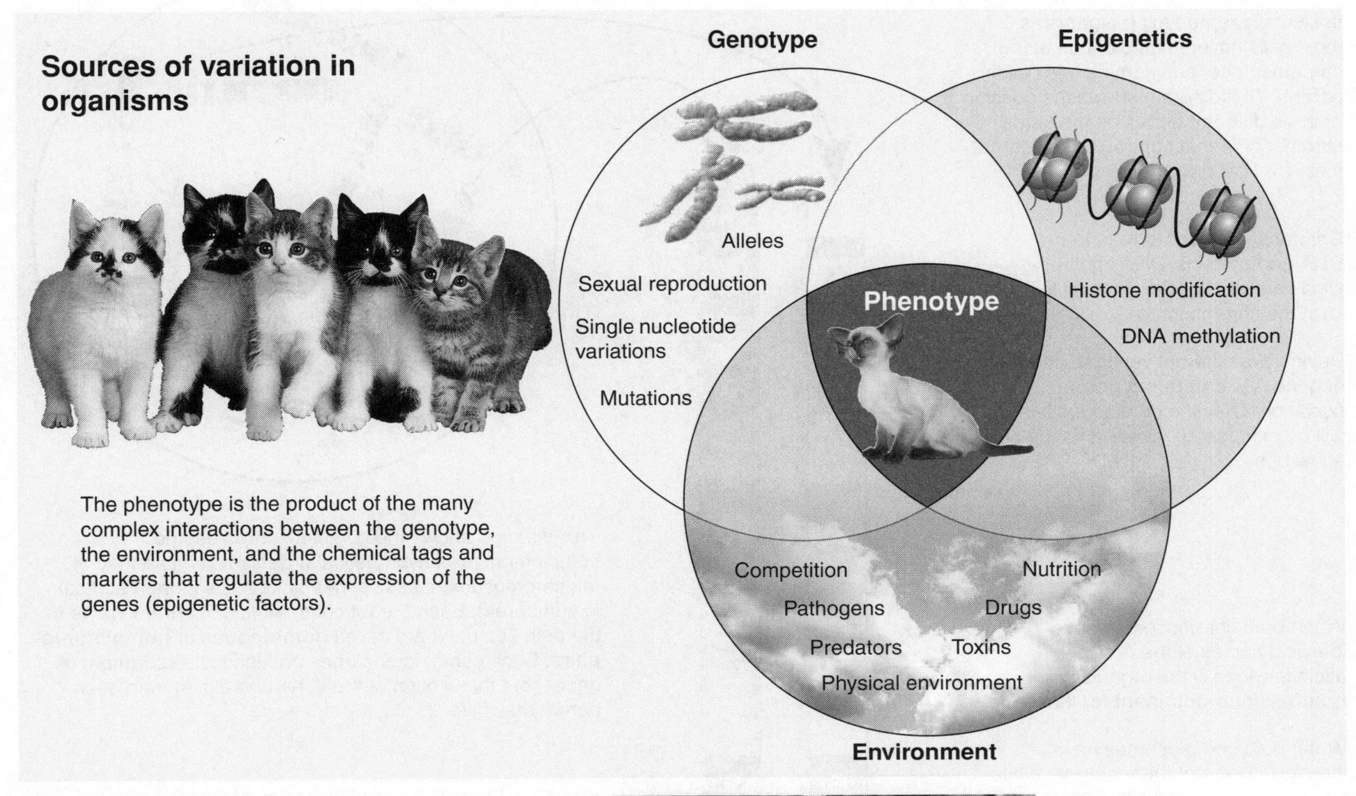

The phenotype is the product of the many complex interactions between the genotype, the environment, and the chemical tags and markers that regulate the expression of the genes (epigenetic factors).

Plasticity and polyphenism

Polyphenism is the expression of different phenotypes in a species due to environmental influences. Examples include sex determination in reptiles and changes in pigmentation in the wings of some butterfly species as the seasons change. The amount of change in a phenotype due environmental influences is called its phenotypic plasticity. Plants show a high amount of phenotypic plasticity because they are unable to move and must therefore adjust to a wide range of environmental changes throughout their lives. Animals generally show a much lower level of phenotypic plasticity, although plasticity in behaviour is often very high.

All worker bees and the queen bee (circled left) in a hive have the same genome, yet the queen looks and behaves very differently from the workers. Bee larvae fed a substance called royal jelly develop into queens. Without the royal jelly, the larvae develop into workers.

1. (a) What are some sources of genetically induced variation? ____________________

   (b) What are some sources of environmentally induced variation? ____________________

2. Explain why genetically identical twins are not always phenotypically identical: ____________________

3. Explain why behaviour is often highly plastic: ____________________

4. Give an example in which an organism's environment produces a marked phenotypic change: ____________________

ISBN: 978-1-927309-20-9

82 Alleles

Key Idea: Eukaryotes generally have paired chromosomes. Each chromosome contains many genes and each gene may have a number of versions called alleles.

Sexually reproducing organisms usually have paired sets of chromosomes, one set from each parent. The equivalent chromosomes that form a pair are termed **homologues**. They carry equivalent sets of genes, but there is the potential for different versions of a gene (**alleles**) to exist in a population.

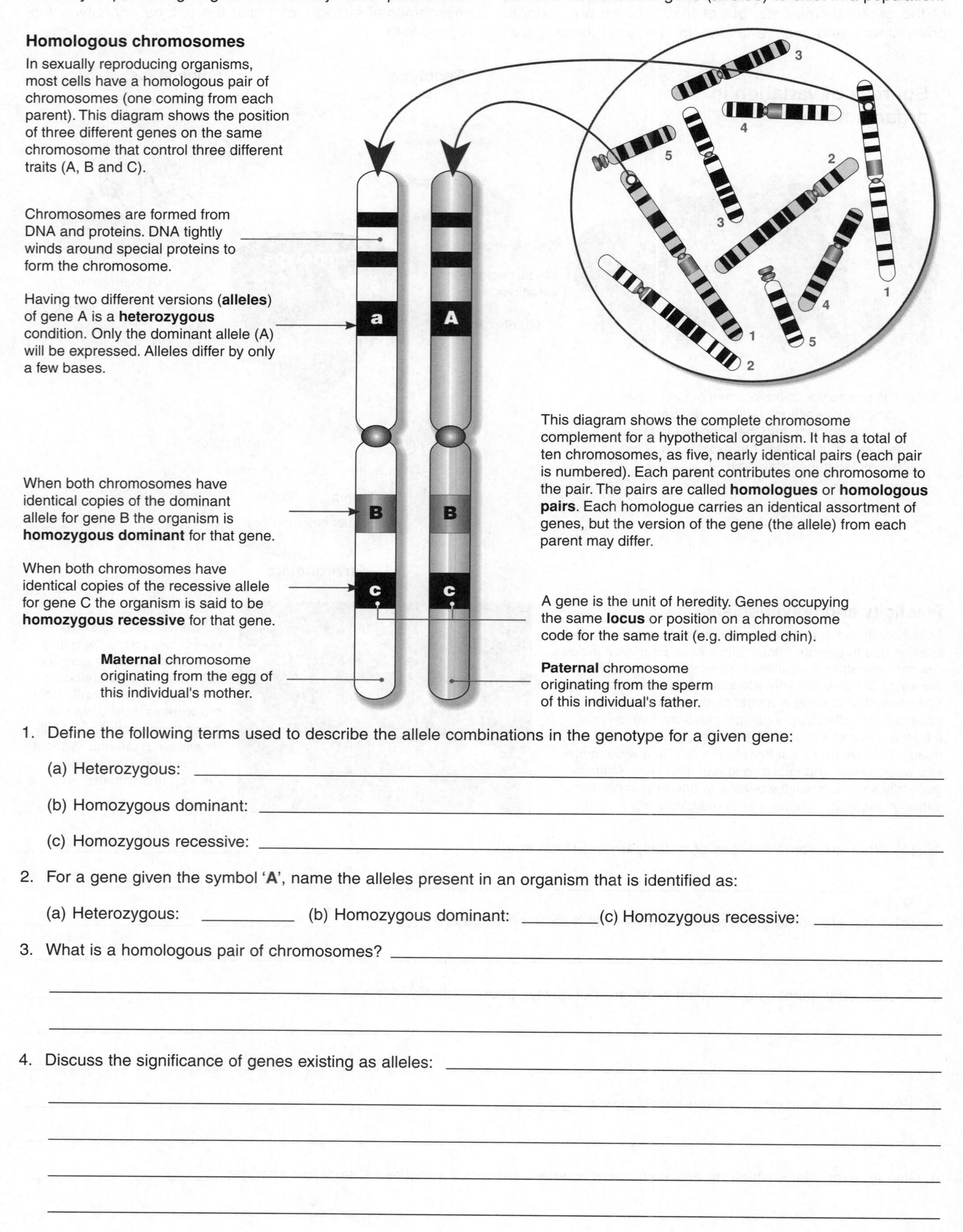

1. Define the following terms used to describe the allele combinations in the genotype for a given gene:

   (a) Heterozygous: ______________________________

   (b) Homozygous dominant: ______________________________

   (c) Homozygous recessive: ______________________________

2. For a gene given the symbol '**A**', name the alleles present in an organism that is identified as:

   (a) Heterozygous: __________ (b) Homozygous dominant: __________ (c) Homozygous recessive: __________

3. What is a homologous pair of chromosomes? ______________________________

   ____

   ____

4. Discuss the significance of genes existing as alleles: ______________________________

   ____

   ____

   ____

   ____

ISBN:978-1-927309-20-9

83 The Monohybrid Cross

Key Idea: The outcome of a cross depends on the parental genotypes. A true breeding parent is homozygous for the gene involved.

Examine the diagrams depicting monohybrid (single gene) inheritance. The F_1 generation by definition describes the offspring of a cross between distinctly different, **true-breeding** (homozygous) parents. A **back cross** refers to any cross between an offspring and one of its parents. If the back cross is to a homozygous recessive, it is diagnostic, and is therefore called a test cross.

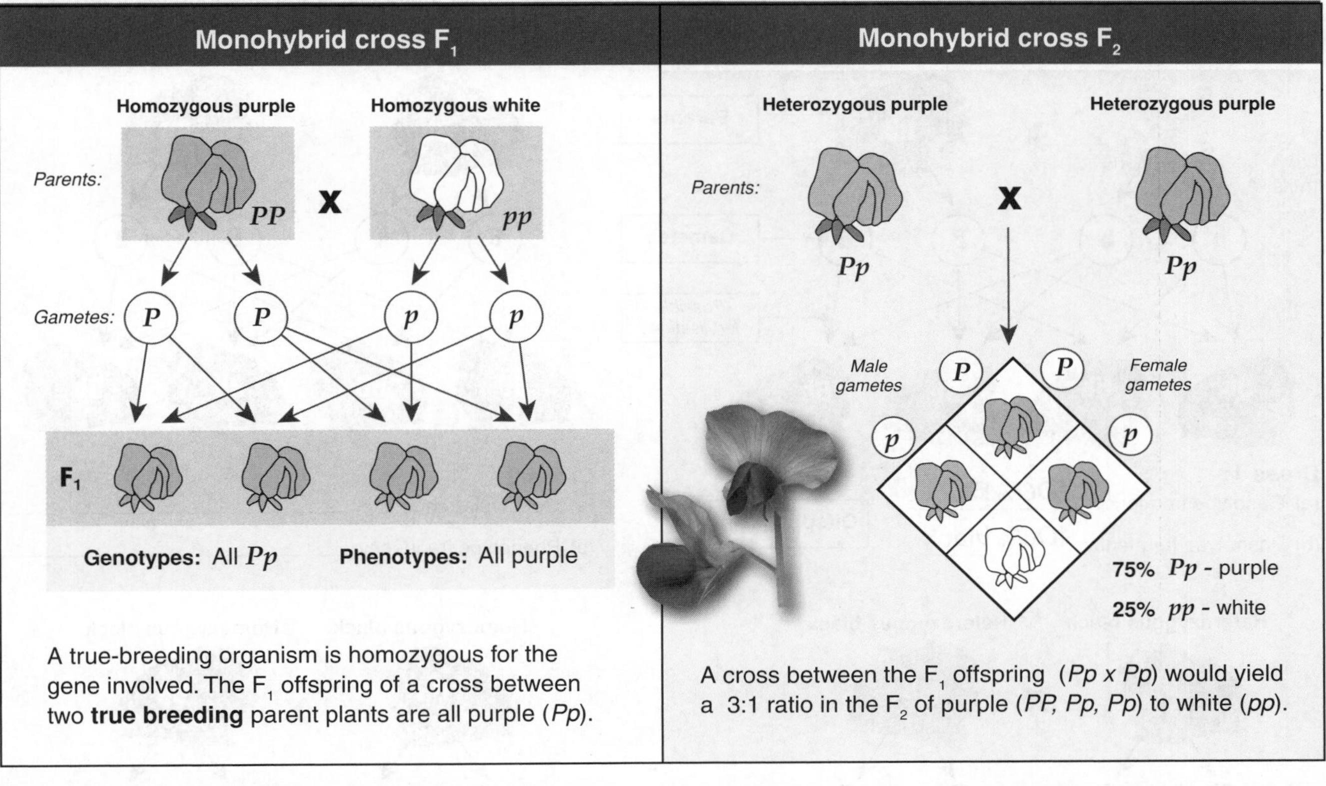

A true-breeding organism is homozygous for the gene involved. The F_1 offspring of a cross between two **true breeding** parent plants are all purple (*Pp*).

A cross between the F_1 offspring (*Pp x Pp*) would yield a 3:1 ratio in the F_2 of purple (*PP, Pp, Pp*) to white (*pp*).

1. Study the diagrams above and explain why white flower colour does not appear in the F_1 generation but reappears in the F_2 generation:

____

____

____

____

2. Complete the crosses below:

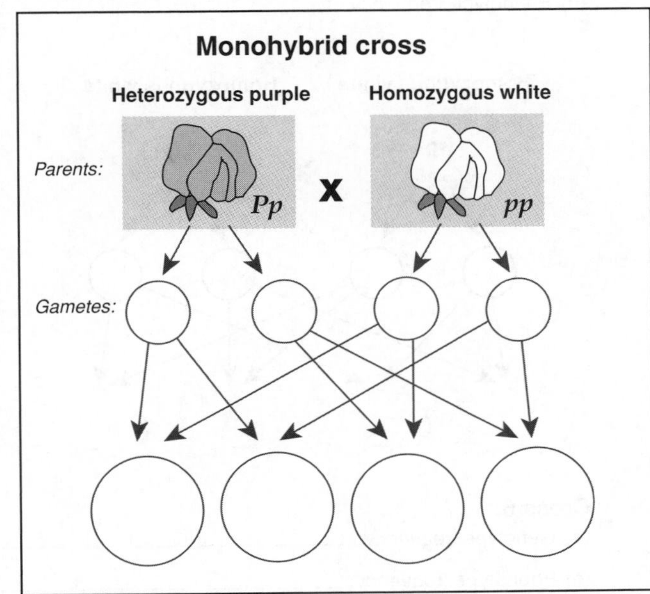

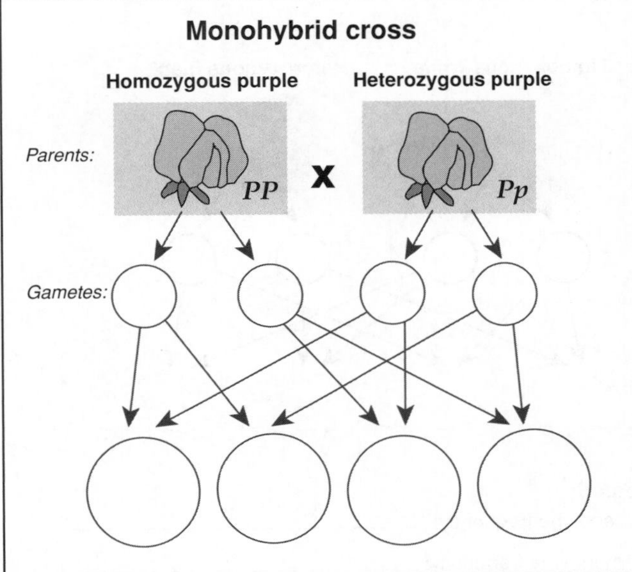

ISBN: 978-1-927309-20-9

84 Practising Monohybrid Crosses

Key Idea: A monohybrid cross studies the inheritance pattern of one gene. The offspring of these crosses occur in predictable ratios.

In this activity, you will examine six types of matings possible for a pair of alleles governing coat colour in guinea pigs. A dominant allele (**B**) produces **black** hair and its recessive allele (**b**), produces white. Each parent can produce two types of gamete by meiosis. Determine the **genotype** and **phenotype frequencies** for the crosses below. For crosses 3 to 6, also determine the gametes produced by each parent (write these in the circles) and offspring genotypes and phenotypes (write these inside the offspring shapes).

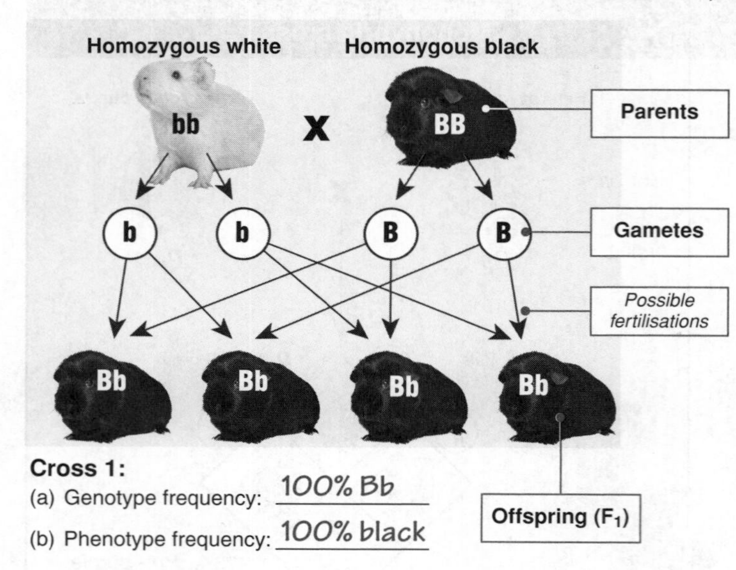

Cross 1:

(a) Genotype frequency: 100% Bb

(b) Phenotype frequency: 100% black

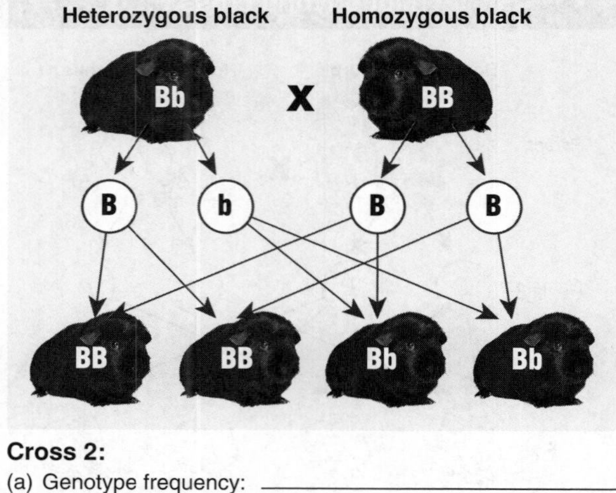

Cross 2:

(a) Genotype frequency: ______________________

(b) Phenotype frequency: ______________________

Cross 3:

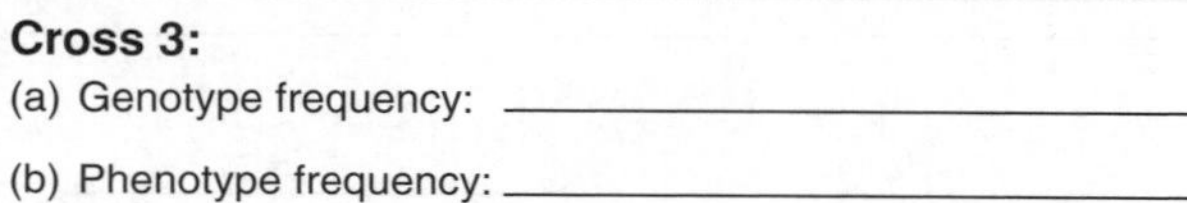

(a) Genotype frequency: ______________________

(b) Phenotype frequency: ______________________

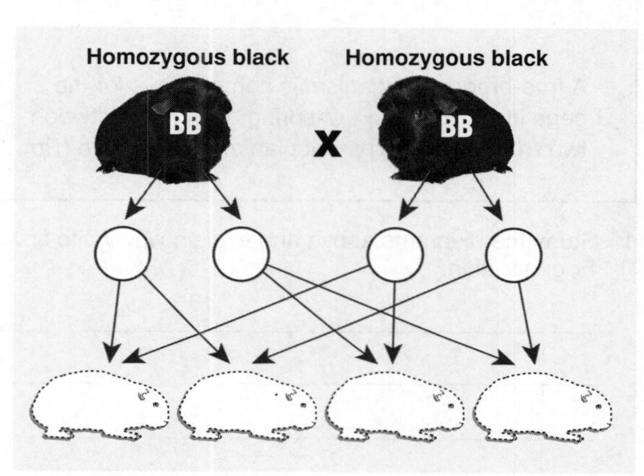

Cross 4:

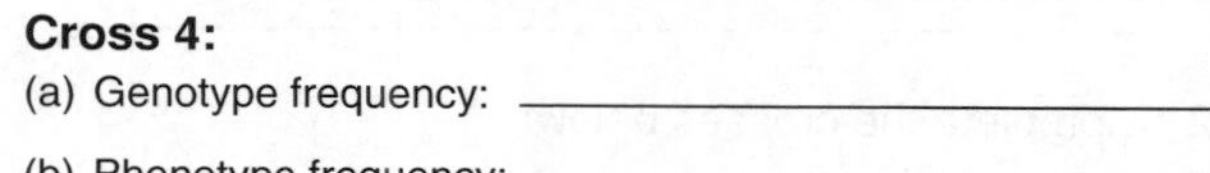

(a) Genotype frequency: ______________________

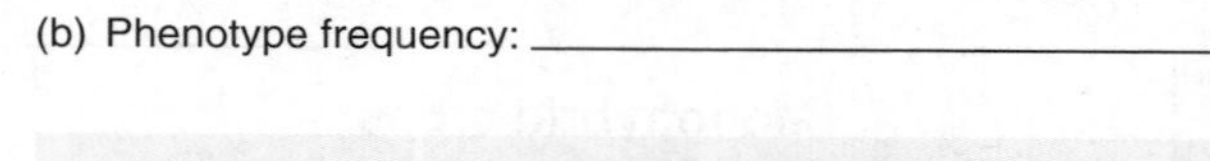

(b) Phenotype frequency: ______________________

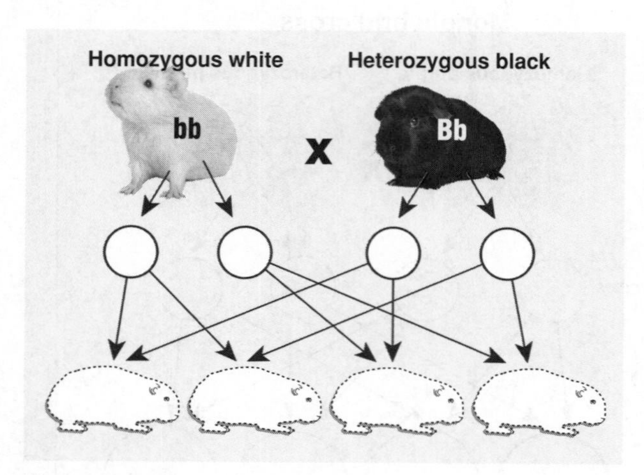

Cross 5:

(a) Genotype frequency: ______________________

(b) Phenotype frequency: ______________________

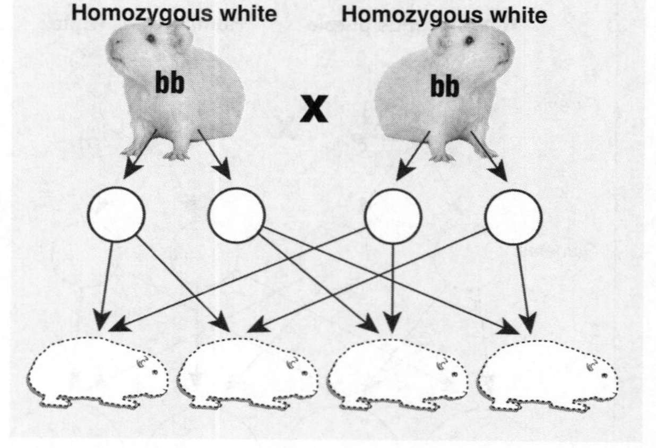

Cross 6:

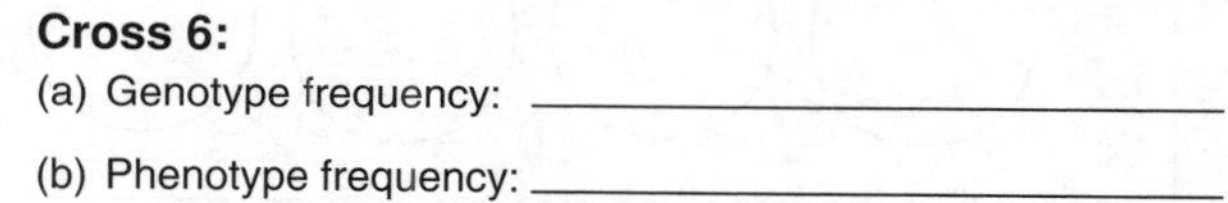

(a) Genotype frequency: ______________________

(b) Phenotype frequency: ______________________

ISBN:978-1-927309-20-9

85 Codominance of Alleles

Key Idea: In inheritance involving codominant alleles, neither allele is recessive and both alleles are equally and independently expressed in the heterozygote.

Codominance is an inheritance pattern in which both alleles in a heterozygote contribute to the phenotype and both alleles are **independently** and **equally expressed**. Examples include the human blood group AB and certain coat colours in horses and cattle. Reddish coat colour is equally dominant with white. Animals that have both alleles have coats that are roan (both red and white hairs are present).

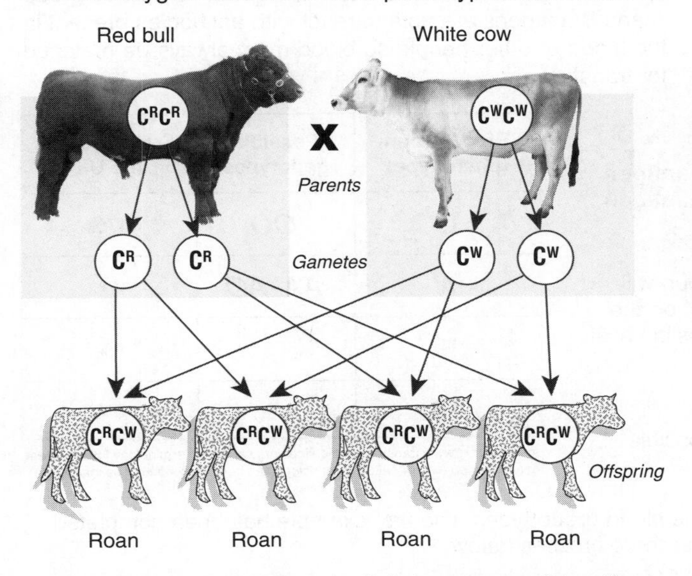

A roan shorthorn heifer

In the shorthorn cattle breed, coat colour is inherited. White shorthorn parents always produce calves with white coats. Red parents always produce red calves. However, when a red parent mates with a white one, the calves have a coat colour that is different from either parent; a mixture of red and white hairs, called roan. Use the example (left) to help you to solve the problems below.

1. Explain how codominance of alleles can result in offspring with a phenotype that is different from either parent:

2. A white bull is mated with a roan cow (right):

   (a) Fill in the spaces to show the genotypes and phenotypes for parents and calves:

   (b) What is the phenotype ratio for this cross?

   (c) How could a cattle farmer control the breeding so that the herd ultimately consisted of only red cattle:

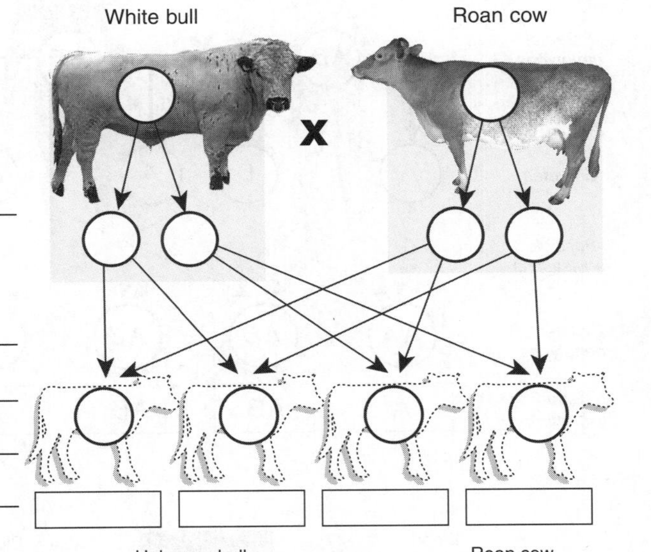

3. A farmer has only roan cattle on his farm. He suspects that one of the neighbours' bulls may have jumped the fence to mate with his cows earlier in the year because half the calves born were red and half were roan. One neighbour has a red bull, the other has a roan.

   (a) Fill in the spaces (right) to show the genotype and phenotype for parents and calves.

   (b) Which bull serviced the cows? **red** or **roan** (*delete one*)

4. Describe the classical phenotypic ratio for a codominant gene resulting from the cross of two heterozygous parents (e.g. a cross between two roan cattle):

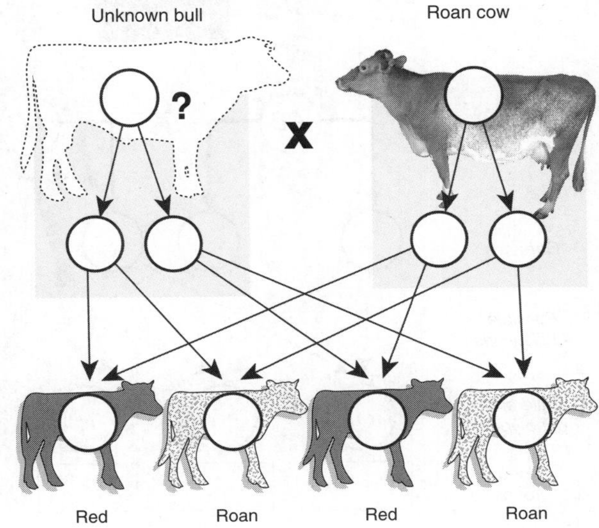

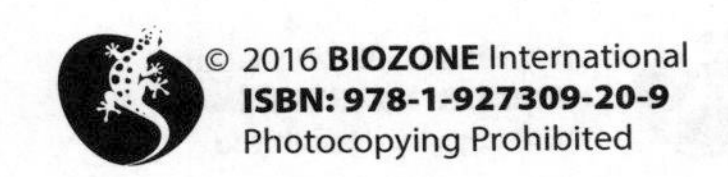

ISBN: 978-1-927309-20-9

86 Codominance in Multiple Allele Systems

Key Idea: The human ABO blood group system is a multiple allele system involving the codominant alleles *A* and *B* and the recessive allele *O*.

The four common blood groups of the human 'ABO blood group system' are determined by three alleles: ***A***, ***B***, and ***O***. The ABO antigens consist of sugars attached to the surface of red blood cells. The alleles code for enzymes (proteins) that join these sugars together. The allele O produces a non-functioning enzyme that is unable to make any changes to the basic antigen (sugar) molecule. The other two alleles (***A***, ***B***) are **codominant** and are expressed equally. They each produce a different functional enzyme that adds a different, specific sugar to the basic sugar molecule. The blood group A and B antigens are able to react with antibodies present in the blood of other people so blood must always be matched for transfusion.

Recessive allele: ***O*** produces a non-functioning protein
Dominant allele: ***A*** produces an enzyme which forms **A antigen**
Dominant allele: B produces an enzyme which forms **B antigen**

If a person has the ***AO*** allele combination then their blood group will be group **A**. The presence of the recessive allele has no effect on the blood group in the presence of a dominant allele. Another possible allele combination that can create the same blood group is ***AA***.

Blood group (phenotype)	Possible genotypes	Frequency in the UK
O	*OO*	47%
A	*AA, AO*	42%
B		8%
AB		3%

Source: http://www.transfusionguidelines.org.uk/ Allele terminology follows latest recommended (use of *I* allele terminology has been discontinued as inaccurate)

1. Use the information above to complete the table for the possible genotypes for blood group B and group AB.

2. Below are four crosses possible between couples of various blood group types. The first example has been completed for you. Complete the genotype and phenotype for the other three crosses below:

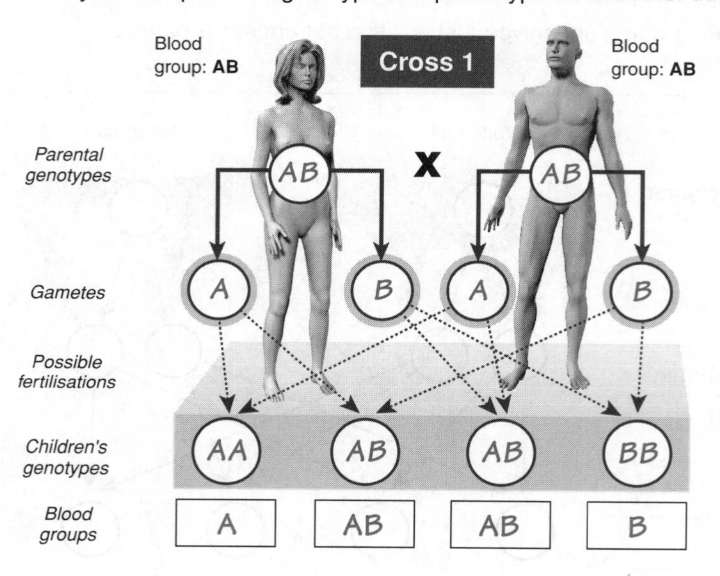

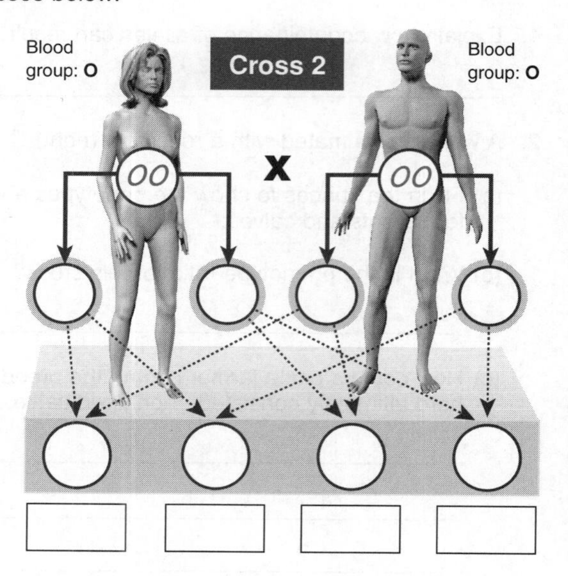

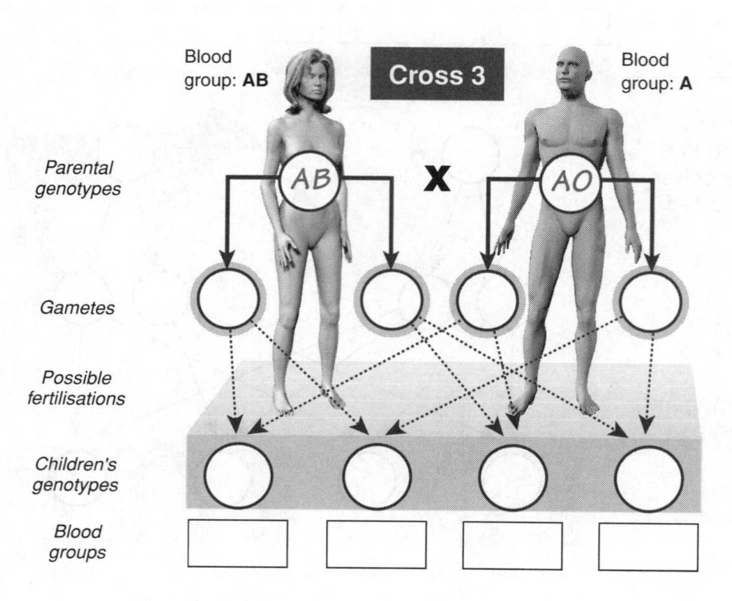

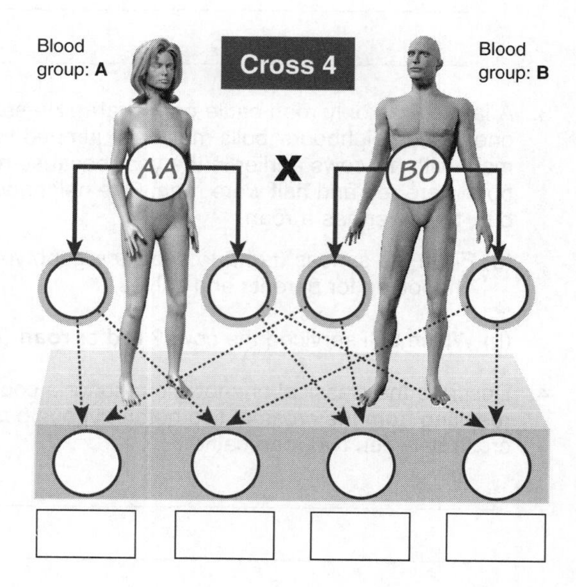

ISBN:978-1-927309-20-9

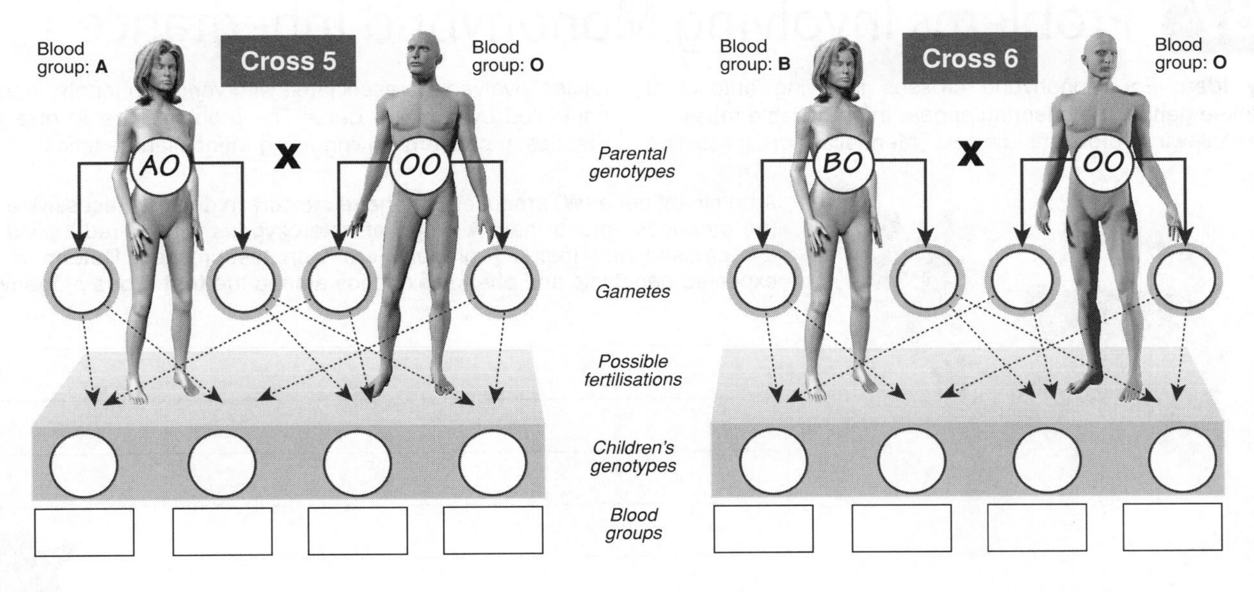

3. A wife is heterozygous for blood group **A** and the husband has blood group **O**.

   (a) Give the genotypes of each parent (fill in spaces on the diagram on the right).

   Determine the probability of:

   (b) One child having blood group **O**:

   ____

   (c) One child having blood group **A**:

   ____

   (d) One child having blood group **AB**:

   ____

4. In a court case involving a paternity dispute (i.e. who is the father of a child) a man claims that a male child (blood group **B**) born to a woman is his son and wants custody. The woman claims that he is not the father.

   (a) If the man has a blood group **O** and the woman has a blood group **A**, could the child be his son? Use the diagram on the right to illustrate the genotypes of the three people involved.

   (b) State with reasons whether the man can be correct in his claim:

   ____

   ____

   ____

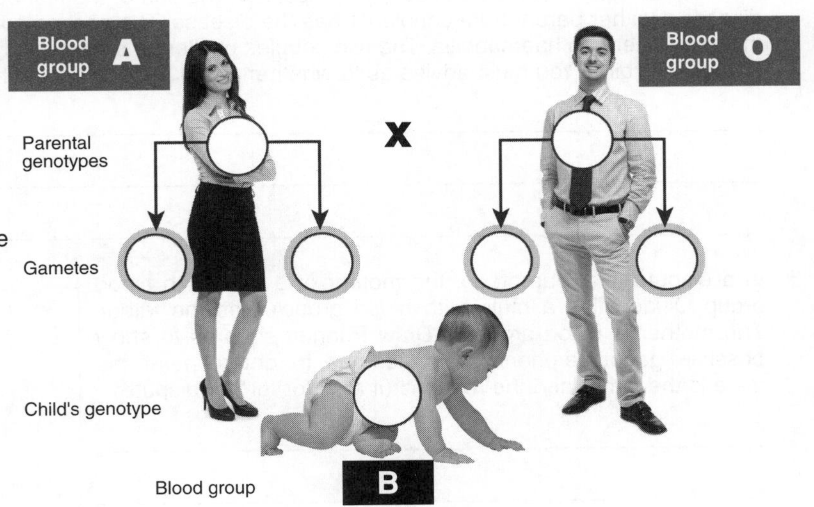

5. Give the blood groups which are possible for children of the following parents (remember that in some cases you don't know if the parent is homozygous or heterozygous).

   (a) Mother is group **AB** and father is group **O**: ______________________________

   (b) Father is group **B** and mother is group **A**: ______________________________

87 Problems Involving Monohybrid Inheritance

Key Idea: For monohybrid crosses involving autosomal unlinked genes, the offspring appear in predictable ratios.

The following problems involve Mendelian crosses. The alleles involved are associated with various phenotypic traits controlled by a single gene. The problems are to give you practise in problem solving using Mendelian genetics.

1. A dominant gene (**W**) produces wire-haired texture in dogs; its recessive allele (**w**) produces smooth hair. A group of heterozygous wire-haired individuals are crossed and their F_1 progeny are then test-crossed. Determine the expected genotypic and phenotypic ratios among the **test cross** progeny:

2. In sheep, black wool is due to a recessive allele (**b**) and white wool to its dominant allele (**B**). A white ram is crossed to a white ewe. Both animals carry the black allele (b). They produce a white ram lamb, which is then back crossed to the female parent. Determine the probability of the **back cross** offspring being black:

3. A homozygous recessive allele, **aa**, is responsible for albinism. Humans can exhibit this phenotype. In each of the following cases, determine the possible genotypes of the mother and father, and their children:

   (a) Both parents have normal phenotypes; some of their children are albino and others are unaffected:

   (b) Both parents are albino and have only albino children:

   (c) The woman is unaffected, the man is albino, and they have one albino child and three unaffected children:

4. Two mothers give birth to sons at a busy hospital. The son of the first couple has haemophilia, a recessive, X-linked disease. Neither parent from couple #1 has the disease. The second couple has an unaffected son, despite the fact that the father has haemophilia. The two couples challenge the hospital in court, claiming their babies must have been swapped at birth. You must advise as to whether or not the sons could have been swapped. What would you say?

5. In a dispute over parentage, the mother of a child with blood group O identifies a male with blood group A as the father. The mother is blood group B. Draw Punnett squares to show possible genotype/phenotype outcomes to determine if the male is the father and the reasons (if any) for further dispute:

ISBN:978-1-927309-20-9

88 Sex Linked Genes

Key Idea: Many genes on the X chromosome do not have a match on the Y chromosome. In males, a recessive allele on the X chromosome cannot therefore be masked by a dominant allele.

Sex linkage occurs when a gene is located on a sex chromosome (usually the X). The result of this is that the character encoded by the gene is usually seen only in one sex (the heterogametic sex). In humans, recessive sex linked genes cause a number of heritable disorders in males, e.g. haemophilia. Women who have a recessive allele are said to be carriers. One of the gene loci controlling coat colour in cats is sex-linked. The two alleles, red and non-red (or black), are found only on the X-chromosome.

Allele types

Xo = Non-red (=black)
XO = Red

Genotypes		**Phenotypes**
XoXo, XoY	=	Black coated female, male
XOXO, XOY	=	Orange coated female, male
X_OX_o	=	Tortoiseshell (intermingled black and orange in fur) in female cats only

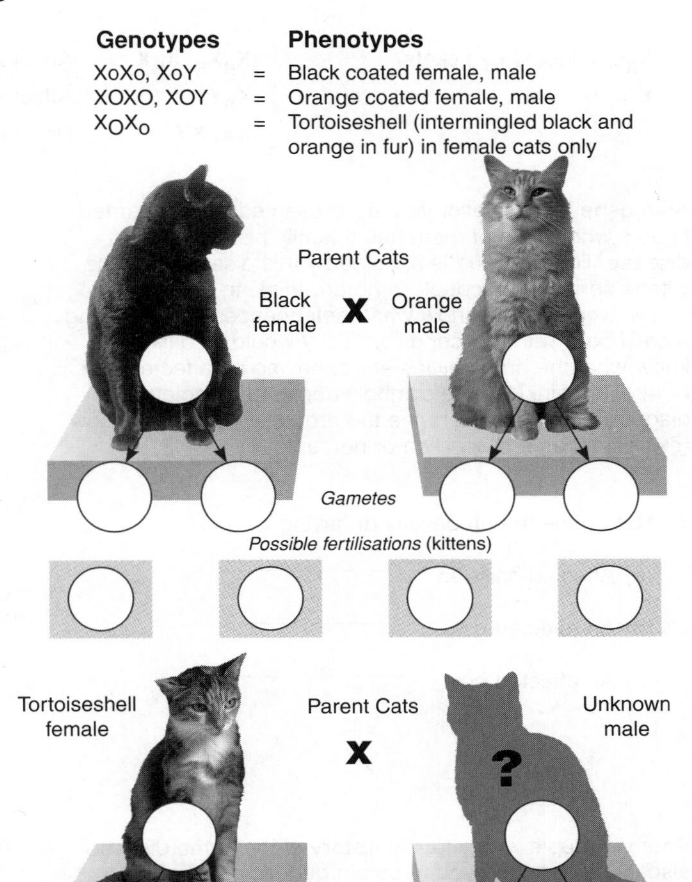

1. An owner of a cat is thinking of mating her black female cat with an orange male cat. Before she does this, she would like to know what possible coat colours could result from such a cross. Use the symbols above to fill in the diagram on the right. Summarise the possible genotypes and phenotypes of the kittens in the tables below.

	Genotypes	Phenotypes
Male kittens		

Female kittens		

2. A female tortoiseshell cat mated with an unknown male cat in the neighbourhood and has given birth to a litter of six kittens. The owner of this female cat wants to know what the appearance and the genotype of the father was of these kittens. Use the symbols above to fill in the diagram on the right. Also show the possible fertilisations by placing appropriate arrows.

Describe the father cat's:

(a) Genotype: ____________________

(b) Phenotype: ____________________

Tortoiseshell female — Parent Cats — X — Unknown male ?

Gametes

Possible fertilisations (kittens)

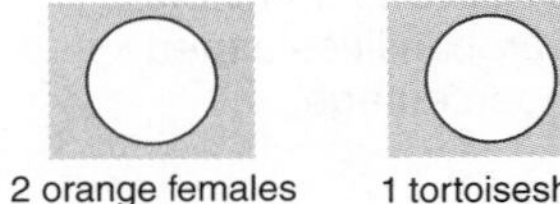

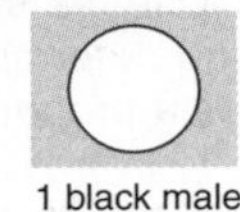

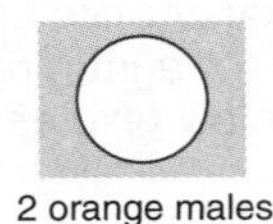

3. The owner of another cat, a black female, also wants to know which cat fathered her two tortoiseshell female and two black male kittens. Use the symbols above to fill in the diagram on the right. Show the possible fertilisations by placing appropriate arrows.

Describe the father cat's:

(a) Genotype: ____________________

(b) Phenotype: ____________________

(c) Was it the same male cat that fathered both this litter and the one above?

YES / NO (*delete one*)

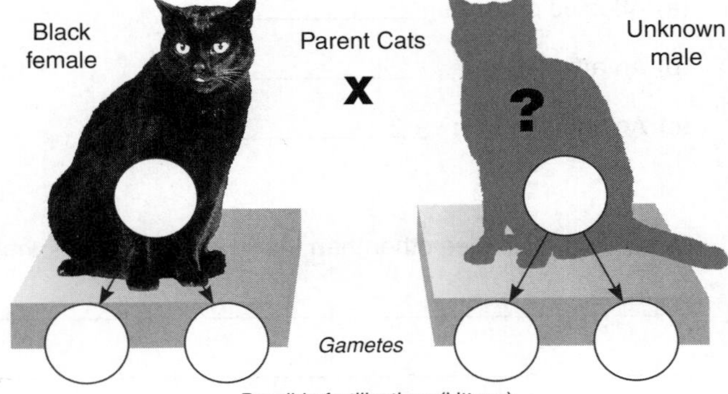

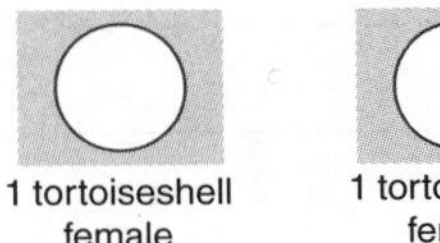

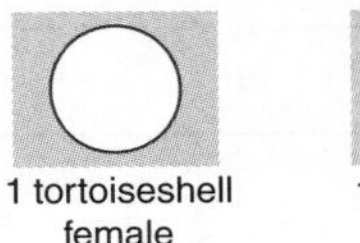

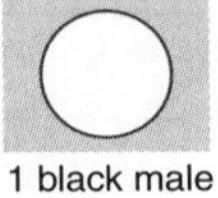

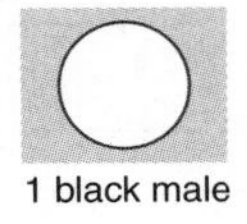

ISBN: 978-1-927309-20-9

Dominant allele in humans

A rare form of rickets in humans is determined by a **dominant** allele of a gene on the **X chromosome** (it is not found on the Y chromosome). This condition is not successfully treated with vitamin D therapy. The allele types, genotypes, and phenotypes are as follows:

Allele types	Genotypes		Phenotypes
X_R = affected by rickets	X_RX_R, X_RX	=	Affected female
X = normal	X_RY	=	Affected male
	XX, XY	=	Normal female, male

As a genetic counsellor you are presented with a married couple where one of them has a family history of this disease. The husband is affected by this disease and the wife is normal. The couple, who are thinking of starting a family, would like to know what their chances are of having a child born with this condition. They would also like to know what the probabilities are of having an affected boy or affected girl. Use the symbols above to complete the diagram right and determine the probabilities stated below (expressed as a proportion or percentage).

Normal wife | Affected husband

Parents X

Gametes

Possible fertilisations

Children

4. Determine the probability of having:

(a) Affected children: ____________________

(b) An affected girl: ____________________

(c) An affected boy: ____________________

Another couple with a family history of the same disease also come in to see you to obtain genetic counselling. In this case, the husband is normal and the wife is affected. The wife's father was not affected by this disease. Determine what their chances are of having a child born with this condition. They would also like to know what the probabilities are of having an affected boy or affected girl. Use the symbols above to complete the diagram right and determine the probabilities stated below (expressed as a proportion or percentage).

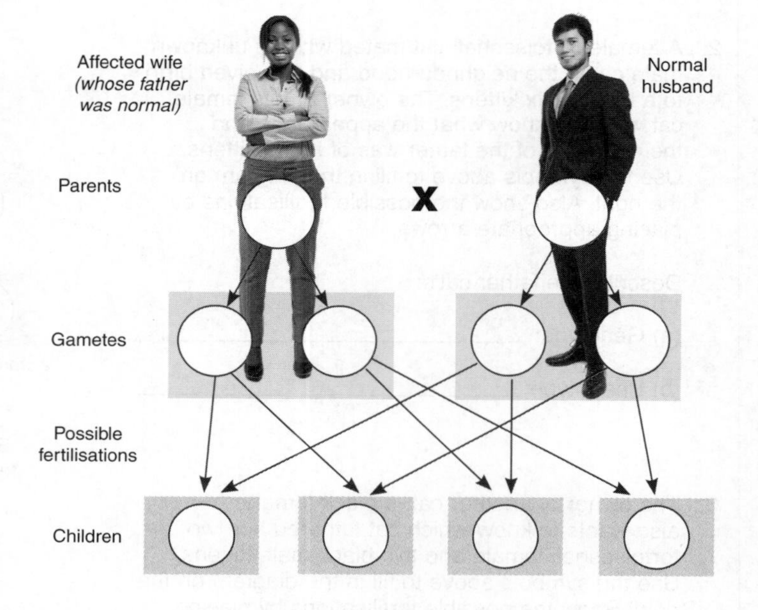

5. Determine the probability of having:

(a) Affected children: ____________________

(b) An affected girl: ____________________

(c) An affected boy: ____________________

6. Describing examples other than those above, discuss the role of **sex linkage** in the inheritance of genetic disorders:

____

____

____

____

ISBN:978-1-927309-20-9

89 Inheritance Patterns

Key Idea: Sex-linked traits and autosomal traits have different inheritance patterns.

Complete the following monohybrid crosses for different types of inheritance patterns in humans: autosomal recessive, autosomal dominant, sex linked recessive, and sex linked dominant inheritance.

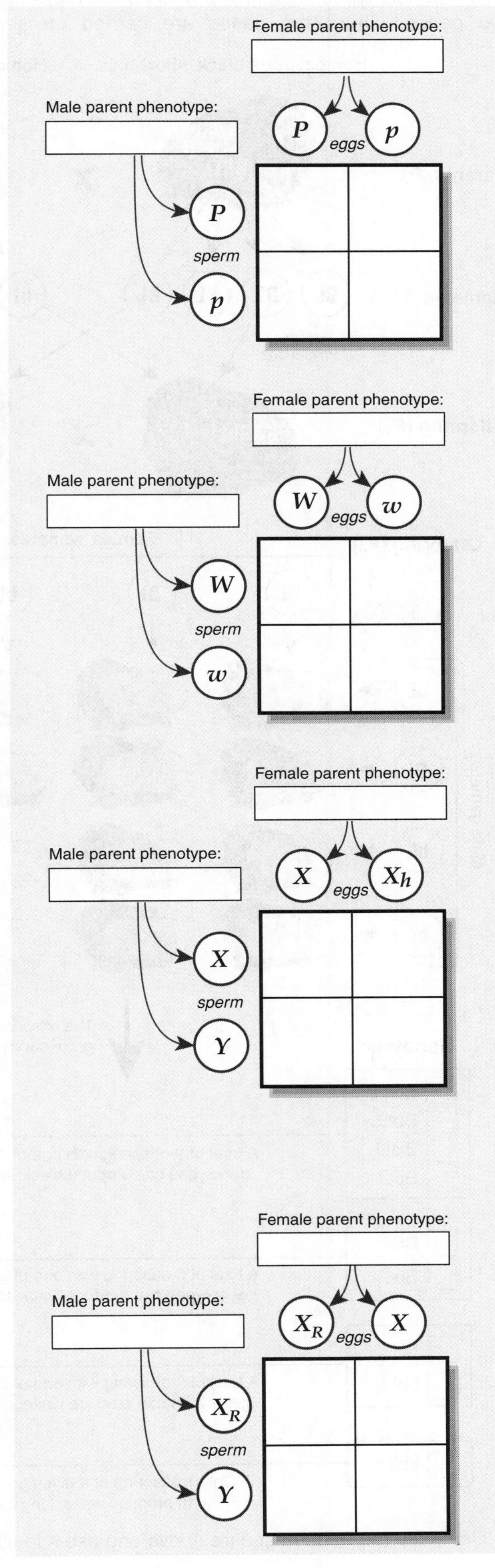

1. **Inheritance of autosomal recessive traits**
   Example: *Albinism*

   Albinism (lack of pigment in hair, eyes and skin) is inherited as an autosomal recessive allele (not sex-linked).

   Using the codes: **PP** (normal) **Pp** (carrier)
   pp (albino)

   (a) Enter the parent phenotypes and complete the Punnett square for a cross between two carrier genotypes.

   (b) Give the ratios for the phenotypes from this cross.

   Phenotype ratios: ______________________________

   ____

2. **Inheritance of autosomal dominant traits**
   Example: *Woolly hair*

   Woolly hair is inherited as an autosomal dominant allele. Each affected individual will have at least one affected parent.

   Using the codes: **WW** (woolly hair)
   Ww (woolly hair, heterozygous)
   ww (normal hair)

   (a) Enter the parent phenotypes and complete the Punnett square for a cross between two heterozygous individuals.

   (b) Give the ratios for the phenotypes from this cross.

   Phenotype ratios: ______________________________

   ____

3. **Inheritance of sex linked recessive traits**
   Example: *Haemophilia*

   Inheritance of haemophilia is sex linked. Males with the recessive (haemophilia) allele, are affected. Females can be carriers.

   Using the codes: **XX** (normal female)
   XX_h (carrier female)
   X_hX_h (haemophiliac female)
   XY (normal male)
   X_hY (haemophiliac male)

   (a) Enter the parent phenotypes and complete the Punnett square for a cross between a normal male and a carrier female.

   (b) Give the ratios for the phenotypes from this cross.

   Phenotype ratios: ______________________________

   ____

4. **Inheritance of sex linked dominant traits**
   Example: *Sex linked form of rickets*

   A rare form of rickets is inherited on the X chromosome.

   Using the codes: **XX** (normal female); **XY** (normal male)
   X_RX (affected heterozygote female)
   X_RX_R (affected female)
   X_RY (affected male)

   (a) Enter the parent phenotypes and complete the Punnett square for a cross between an affected male and heterozygous female.

   (b) Give the ratios for the phenotypes from this cross.

   Phenotype ratios: ______________________________

   ____

ISBN: 978-1-927309-20-9

90 Dihybrid Cross

Key Idea: A dihybrid cross studies the inheritance pattern of two genes. In crosses involving unlinked autosomal genes, the offspring occur in predictable ratios.
There are four types of gamete produced in a cross involving two genes, where the genes are carried on separate chromosomes and are sorted independently of each other during meiosis. The two genes in the example below are on separate chromosomes and control two unrelated characteristics, **hair colour** and **coat length**. Black (B) and short (L) are dominant to white and long.

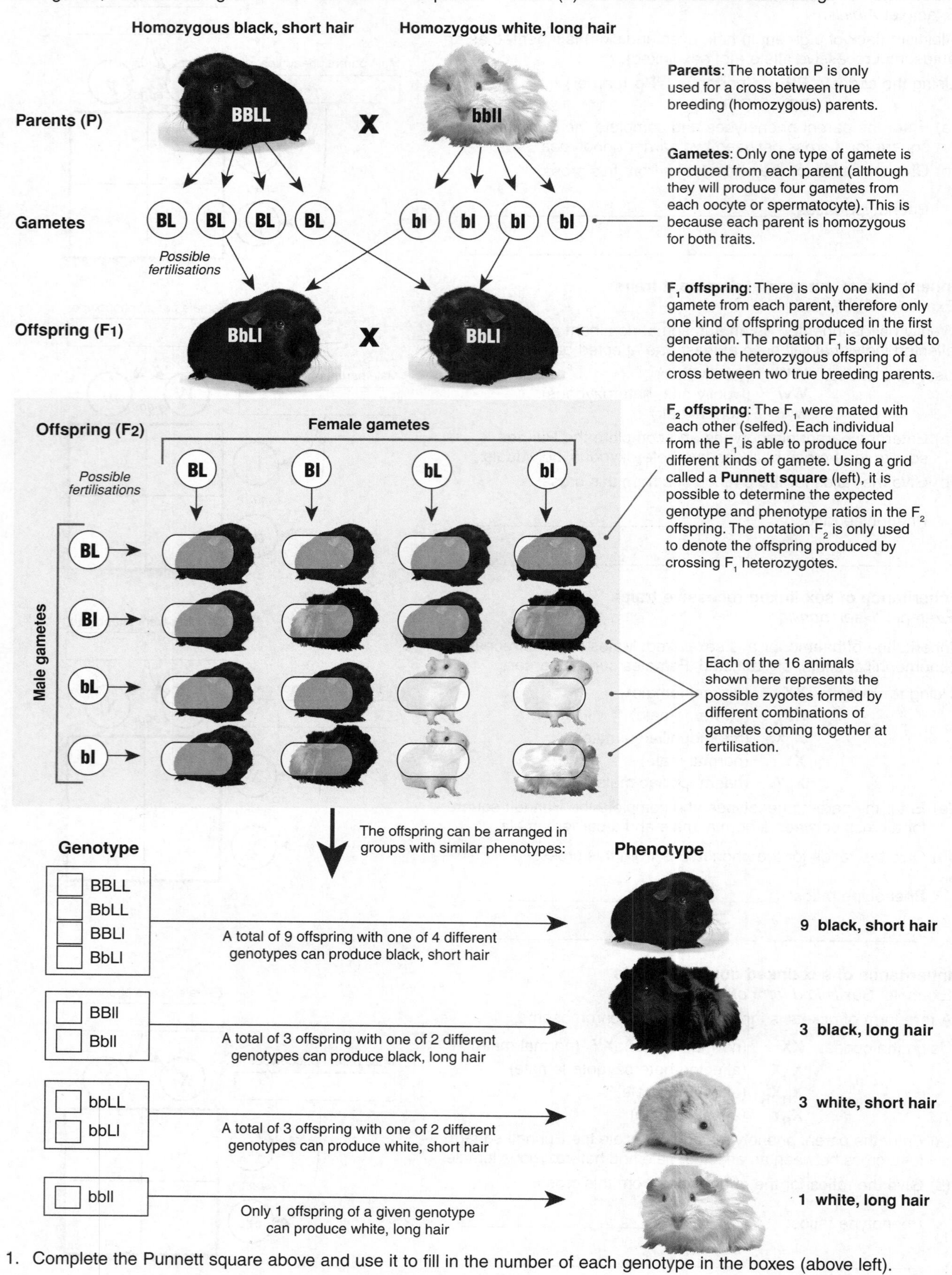

1. Complete the Punnett square above and use it to fill in the number of each genotype in the boxes (above left).

ISBN:978-1-927309-20-9

91 Inheritance of Linked Genes

Key Idea: Linked genes are genes found on the same chromosome and tend to be inherited together. Linkage reduces the genetic variation in the offspring.

Genes are **linked** when they are on the same chromosome. Linked genes tend to be inherited together and the extent of crossing over depends on how close together they are on the chromosome. In genetic crosses, linkage is indicated when a greater proportion of the offspring from a cross are of the parental type (than would be expected if the alleles were on separate chromosomes and assorting independently). Linkage reduces the genetic variation that can be produced in the offspring.

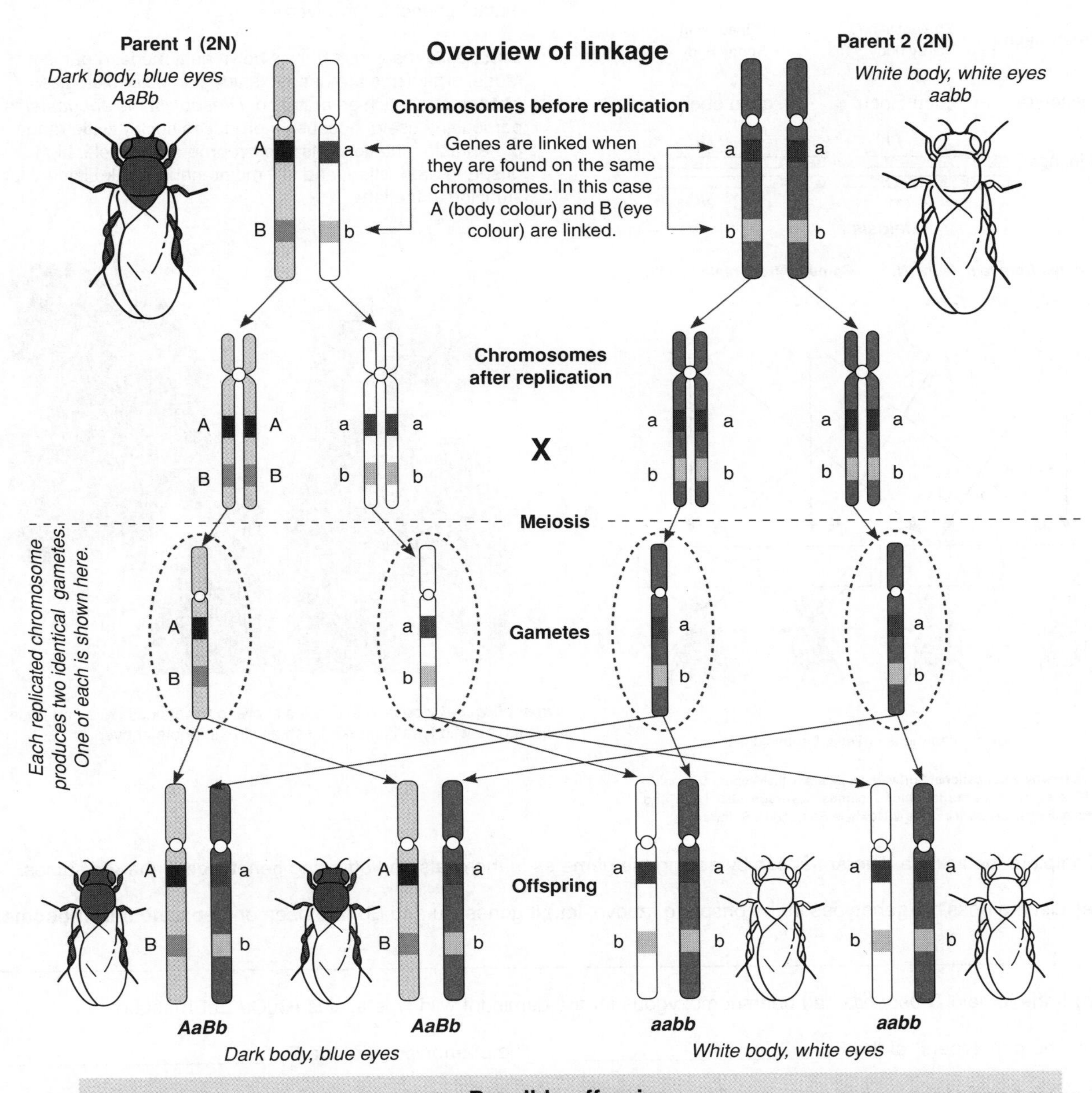

Possible offspring

Only two kinds of genotype combinations are possible. They are they same as the parent genotype.

1. What is the effect of **linkage** on the inheritance of genes? ______________________

2. Explain how linkage decreases the amount of genetic variation in the offspring: ______________________

ISBN: 978-1-927309-20-9

An example of linked genes in *Drosophila*

The genes for wing shape and body colour are linked (they are on the same chromosome).

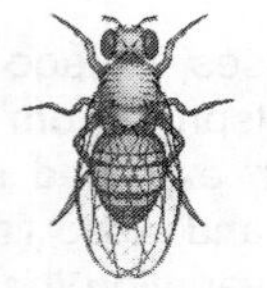

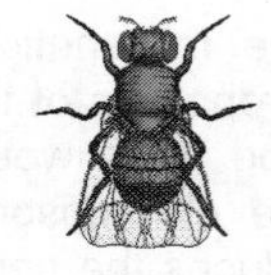

Parent	**Wild type female**	**Mutant male**
Phenotype	Straight wing Grey body	Curled wing Ebony body
Genotype	Cucu Ebeb	cucu ebeb

Linkage

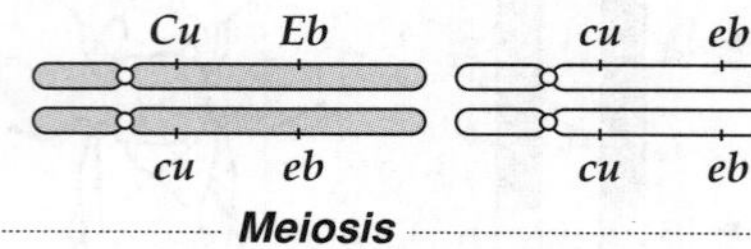

Meiosis

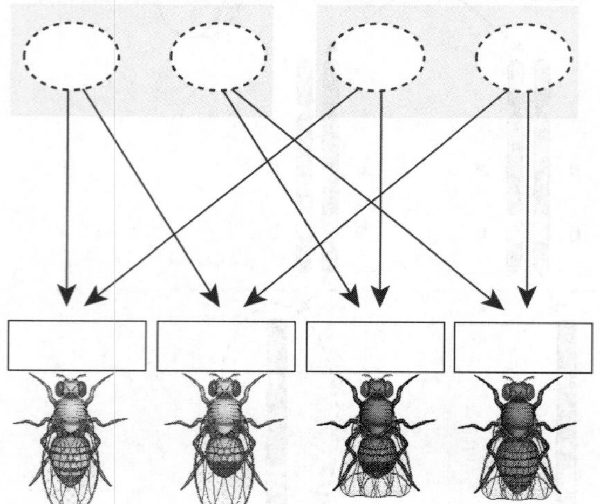

Sex of offspring is irrelevant in this case

Contact **Newbyte Educational Software** for details of their superb *Drosophila Genetics* software package which includes coverage of linkage and recombination. *Drosophila* images © Newbyte Educational Software.

Drosophila and linked genes

In the example shown left, wild type alleles are dominant and are given an upper case symbol of the mutant phenotype (Cu or Eb). This notation used for *Drosophila* departs from the convention of using the dominant gene to provide the symbol. This is necessary because there are many mutant alternative phenotypes to the wild type (e.g. curled and vestigial wings). A lower case symbol of the wild type (e.g. ss for straight wing) would not indicate the mutant phenotype involved.

Drosophila melanogaster is known as a model organism. Model organisms are used to study particular biological phenomena, such as mutation. *Drosophila melanogaster* is particularly useful because it produces such a wide range of heritable mutations. Its short reproduction cycle, high offspring production, and low maintenance make it ideal for studying in the lab.

Drosophila melanogaster examples showing variations in eye and body colour. The wild type is marked with a w in the photo above.

3. Complete the linkage diagram above by adding the gametes in the ovals and offspring genotypes in the rectangles.

4. (a) List the possible genotypes in the offspring (above, left) if genes Cu and Eb had been on **separate chromosomes**:

   ____

   (b) If the female *Drosophila* had been homozygous for the dominant wild type alleles (CuCu EbEb), state:

   The genotype(s) of the F_1: ____________________ The phenotype(s) of the F_1: ____________________

5. A second pair of *Drosophila* are mated. The female genotype is Vgvg EbEb (straight wings, grey body), while the male genotype is vgvg ebeb (vestigial wings, ebony body). Assuming the genes are linked, carry out the cross and list the genotypes and phenotypes of the offspring. Note vg = vestigial (no) wings:

   ____

   ____

   ____

   The genotype(s) of the F_1: ____________________ The phenotype(s) of the F_1: ____________________

6. Explain why *Drosophila* are often used as model organisms in the study of genetics: ____________________

   ____

   ____

ISBN:978-1-927309-20-9

92 Recombination and Dihybrid Inheritance

Key Idea: Recombination is the exchange of alleles between homologous chromosomes as a result of crossing over. Recombination increases the genetic variation in the offspring. The alleles of parental linkage groups separate and new associations of alleles are formed in the gametes. Offspring formed from these gametes are called **recombinants** and show combinations of characteristics not seen in the parents.

In contrast to linkage, recombination increases genetic variation in the offspring. Recombination between the alleles of parental linkage groups is indicated by the appearance of non-parental types in the offspring, although not in the numbers that would be expected had the alleles been on separate chromosomes (independent assortment).

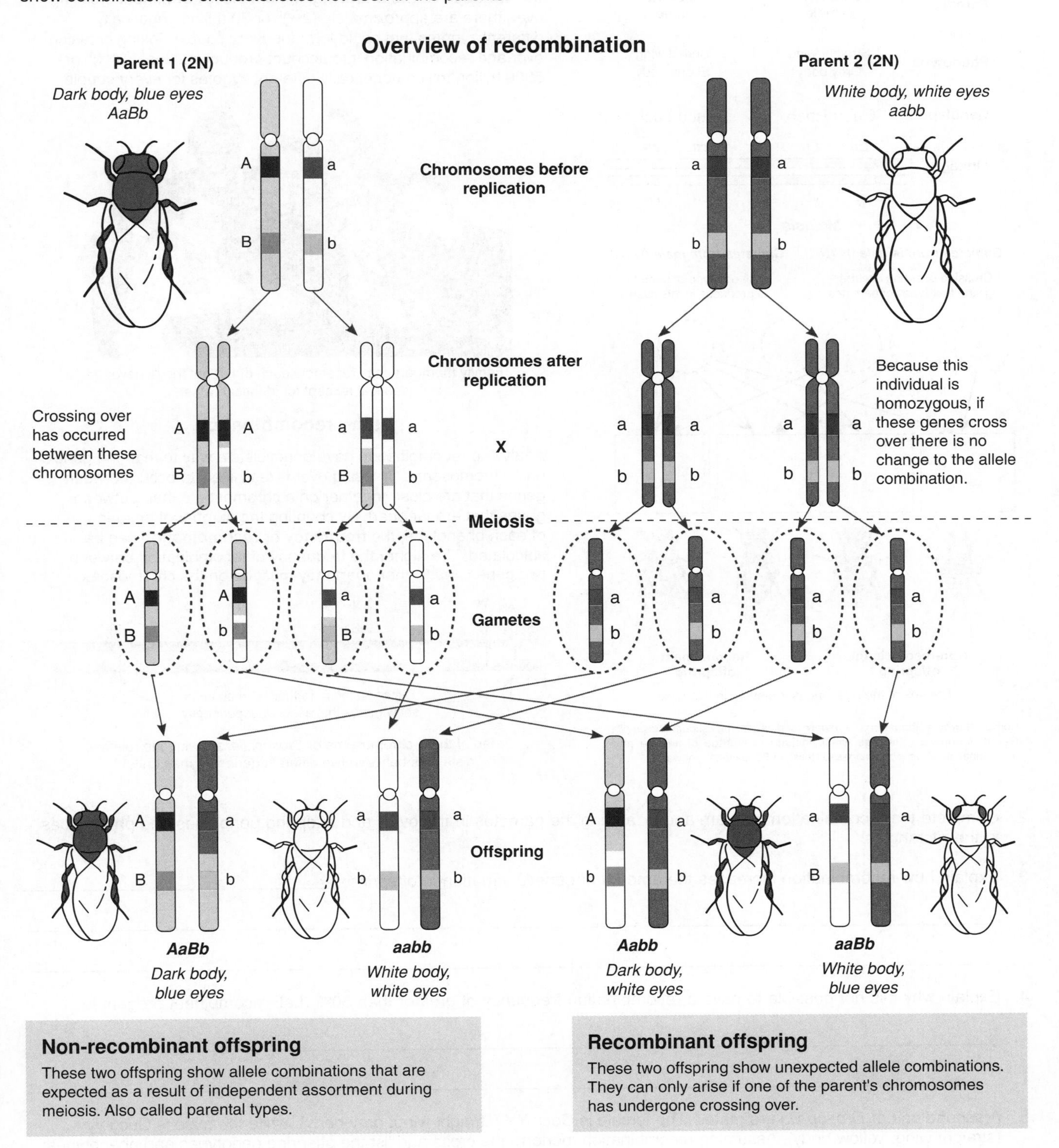

Non-recombinant offspring

These two offspring show allele combinations that are expected as a result of independent assortment during meiosis. Also called parental types.

Recombinant offspring

These two offspring show unexpected allele combinations. They can only arise if one of the parent's chromosomes has undergone crossing over.

1. Describe the effect of **recombination** on the inheritance of genes: ______________________________

____

____

ISBN: 978-1-927309-20-9

An example of recombination

In the female parent, crossing over occurs between the linked genes for wing shape and body colour

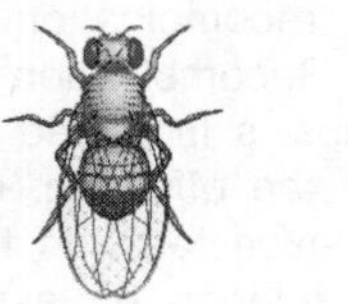

	Wild type female	Mutant male
Parent		
Phenotype	Straight wing Grey body	Curled wing Ebony body
Genotype	Cucu Ebeb	cucu ebeb
Linkage	*Cu Eb* / *cu eb*	*cu eb* / *cu eb*

Meiosis

Gametes from female fly (N)

Crossing over has occurred, giving four types of gametes

Gametes from male fly (N)

Only one type of gamete is produced in this case

Non-recombinant offspring

Recombinant offspring

The sex of the offspring is irrelevant in this case

Contact **Newbyte Educational Software** for details of their superb *Drosophila Genetics* software package which includes coverage of linkage and recombination. *Drosophila* images © Newbyte Educational Software.

The cross (left) uses the same genotypes as the previous activity but, in this case, crossing over occurs between the alleles in a linkage group in one parent. The symbology used is the same.

Recombination produces variation

If crossing over does not occur, the possible combinations in the gametes is limited. **Crossing over and recombination increase the variation in the offspring**. In humans, even without crossing over, there are approximately $(2^{23})^2$ or 70 trillion genetically different zygotes that could form for every couple. Taking crossing over and recombination into account produces at least $(4^{23})^2$ or 5000 trillion trillion genetically different zygotes for every couple.

Family members may resemble each other, but they'll never be identical (except for identical twins).

Using recombination

Analysing recombination gave geneticists a way to map the genes on a chromosome. Crossing over is less likely to occur between genes that are close together on a chromosome than between genes that are far apart. By counting the number of offspring of each phenotype, the **frequency of recombination** can be calculated. The higher the frequency of recombination between two genes, the further apart they must be on the chromosome.

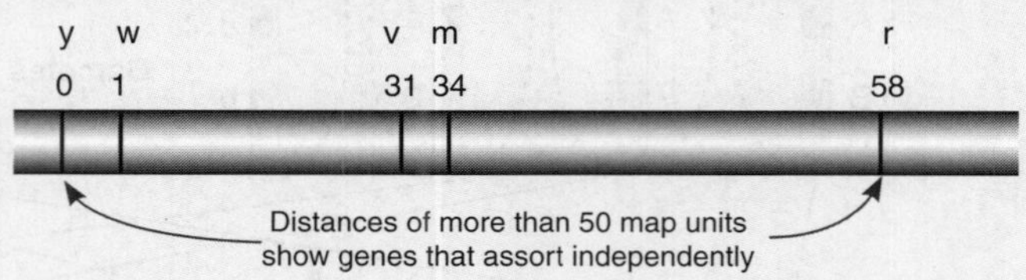

Map of the X chromosome of *Drosophila*, showing the relative distances between five different genes (in map units).

2. Complete the recombination diagram above, adding the gametes in the ovals and offspring genotypes and phenotypes in the rectangles:

3. Explain how recombination increases the amount of genetic variation in offspring: ________________

4. Explain why it is not possible to have a recombination frequency of greater than 50% (half recombinant progeny):

5. A second pair of *Drosophila* are mated. The female is Cucu YY (straight wing, grey body), while the male is Cucu yy (straight wing, yellow body). Assuming recombination, perform the cross and list the offspring genotypes and phenotypes:

ISBN:978-1-927309-20-9

93 Detecting Linkage in Dihybrid Crosses

Key Idea: Linkage between genes can be detected by observing the phenotypic ratios in the offspring.

Shortly after the rediscovery of Mendel's work early in the 20th century, it became apparent that his ratios of 9:3:3:1 for heterozygous dihybrid crosses did not always hold true. Experiments on sweet peas by William Bateson and Reginald Punnett, and on *Drosophila* by Thomas Hunt Morgan, showed that there appeared to be some kind of coupling between genes. This coupling, which we now know to be linkage, did not follow any genetic relationship known at the time.

Sweet pea cross

P: Red flowers, round pollen (ppll) X Purple flowers, long pollen (PPLL)

F1: Purple flowers, long pollen (PpLl) X Purple flowers, long pollen (PpLl)

Bateson and Punnett studied sweet peas in which purple flowers (P) are dominant to red (p), and long pollen grains (L) are dominant to round (l). If these genes were unlinked, the outcome of an cross between two heterozygous sweet peas should have been a 9:3:3:1 ratio.

Table 1: Sweet pea cross results

	Observed	Expected
Purple long (P_L_)	284	
Purple round (P_ll)	21	
Red long (ppL_)	21	
Red round (ppll)	55	
Total	381	381

Morgan performed experiments to investigate linked genes in *Drosophila*. He crossed a heterozygous red-eyed normal-winged (Prpr Vgvg) fly with a homozygous purple-eyed vestigial-winged (prpr vgvg) fly. The table (below) shows the outcome of the cross.

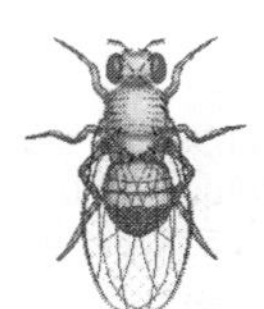

Red eyed normal winged (Prpr Vgvg) X Purple eyed vestigial winged (prpr vgvg)

Table 2: *Drosophila* cross results

Genotype	Observed	Expected	Gamete type
Prpr Vgvg	1339	710	Parental
prpr Vgvg	152		
Prpr vgvg	154		
prpr vgvg	1195		
Total	2840	2840	

1. Fill in the missing numbers in the **expected** column of **Table 1**, remembering that a 9:3:3:1 ratio is expected:

2. (a) Fill in the missing numbers in the **expected** column of **Table 2**, remembering that a 1:1:1:1 ratio is expected:

   (b) Add the gamete type (parental/recombinant) to the gamete type column in Table 2:

   (c) What type of cross did Morgan perform here?

   ___

3. (a) Use the pedigree chart below to determine if nail-patella syndrome is dominant or recessive, giving reasons for your choice:

   ___

   (b) What evidence is there that nail-patella syndrome is linked to the ABO blood group locus?

   ___

   (c) Suggest a likely reason why individual III-3 is not affected despite carrying the B allele:

   ___

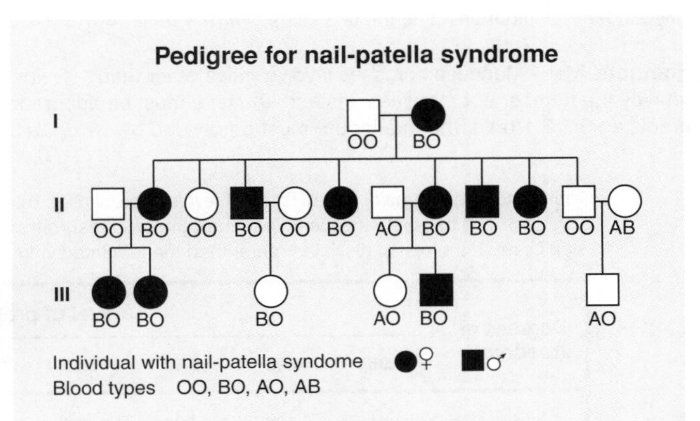

Linked genes can be detected by pedigree analysis. The diagram above shows the pedigree for the inheritance of nail-patella syndrome, which results in small, poorly developed nails and kneecaps in affected people. Other body parts such as elbows, chest, and hips can also be affected. The nail-patella syndrome gene is linked to the ABO blood group locus.

ISBN: 978-1-927309-20-9

LINK 94 KNOW

94 Chi-Squared Test for Goodness of Fit

Key Idea: The chi-squared test for goodness of fit is used to compare sets of categorical data and evaluate if differences between them are statistically significant or due to chance.

The chi-squared test (χ^2) for goodness of fit is a statistical test used to compare an experimental result with an expected theoretical outcome, e.g. an expected Mendelian ratio or a theoretical value indicating "no difference" between groups, e.g. in distribution or abundance in relation to an environmental factor. It is a simple test but data must be raw counts and sample sizes should be >20. When using χ^2 to test the outcome of a genetic cross, the null hypothesis predicts the phenotypic ratio of offspring according to the expected Mendelian ratio for the cross, assuming independent assortment (no linkage). Significant departures from the predicted Mendelian ratio may indicate linkage of the alleles in question.

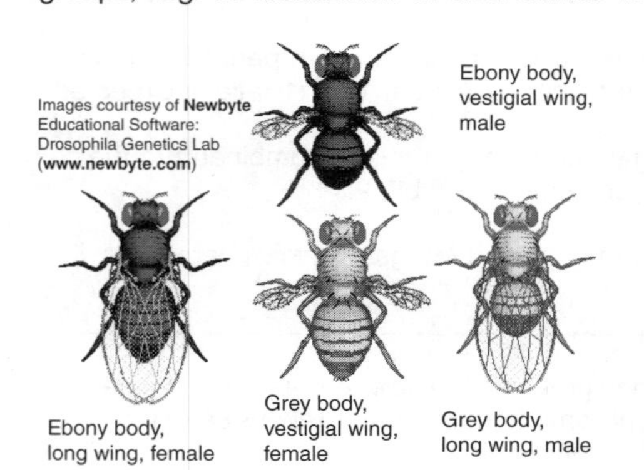

Using χ^2 in Mendelian genetics

In a genetic cross between two *Drosophila* individuals, the predicted Mendelian ratios for the offspring were 1:1:1:1 for each of the four following phenotypes: grey body-long wing, grey body-vestigial wing, ebony body-long wing, ebony body-vestigial wing. The observed results of the cross were not exactly as predicted. The following numbers for each phenotype were observed in the offspring of the cross:

Observed results of the example *Drosophila* cross

Phenotype	Number	Phenotype	Number
Grey body, long wing	**98**	Ebony body, long wing	**102**
Grey body, vestigial wing	**88**	Ebony body, vestigial wing	**112**

Using χ^2, the probability of this result being consistent with a 1:1:1:1 ratio could be tested:

Step 1: Calculate the expected value (E)

In this case, this is the sum of the observed values divided by the number of categories (see note below)

$$\frac{400}{4} = 100$$

Step 2: Calculate O – E

The difference between the observed and expected values is calculated as a measure of the deviation from a predicted result. Since some deviations are negative, they are all squared to give positive values. This step is usually performed as part of a tabulation (right, darker blue column).

Category	O	E	O – E	$(O-E)^2$	$\frac{(O-E)^2}{E}$
Grey, long wing	98	100	–2	4	0.04
Grey, vestigial wing	88	100	–12	144	1.44
Ebony, long wing	102	100	2	4	0.04
Ebony, vestigial wing	112	100	12	144	1.44
	Total = 400			χ^2	$\Sigma = 2.96$

Step 3: Calculate the value of χ^2

$$\chi^2 = \sum \frac{(O-E)^2}{E}$$

Where: O = the observed result
E = the expected result
Σ = sum of

The calculated χ^2 value is given at the bottom right of the last column in the tabulation.

Step 4: Calculating degrees of freedom

The probability that any particular χ^2 value could be exceeded by chance depends on the number of degrees of freedom. This is simply ***one less than the total number of categories*** (this is the number that could vary independently without affecting the last value). ***In this case: 4–1 = 3.***

Step 5a: Using the χ^2 table

On the χ^2 table (part reproduced in Table 1 below) with 3 degrees of freedom, the calculated value for χ^2 of 2.96 corresponds to a probability of between 0.2 and 0.5 (see arrow). *This means that by chance alone a χ^2 value of 2.96 could be expected between 20% and 50% of the time.*

Step 5b: Using the χ^2 table

The probability of between 0.2 and 0.5 is higher than the 0.05 value which is generally regarded as significant. The null hypothesis cannot be rejected and we have no reason to believe that the observed results differ significantly from the expected (at $P = 0.05$).

Footnote: Many Mendelian crosses involve ratios other than 1:1. For these, calculation of the expected values is not simply a division of the total by the number of categories. Instead, the total must be apportioned according to the ratio. For example, for a total of 400 as above, in a predicted 9:3:3:1 ratio, the total count must be divided by 16 (9+3+3+1) and the expected values will be 225: 75: 75: 25 in each category.

Table 1: Critical values of χ^2 at different levels of probability. By convention, the critical probability for rejecting the null hypothesis (H_0) is 5%. If the test statistic is less than the tabulated critical value for $P = 0.05$ we cannot reject H_0 and the result is not significant. If the test statistic is greater than the tabulated value for $P = 0.05$ we reject H_0 in favour of the alternative hypothesis.

Degrees of freedom	Level of probability (*P*) 0.98	0.95	0.80	0.50	0.20	0.10	0.05	0.02	0.01	0.001
1	0.001	0.004	0.064	0.455	1.64	2.71	3.84	5.41	6.64	10.83
2	0.040	0.103	0.466	1.386	3.22	4.61	5.99	7.82	9.21	13.82
3	0.185	0.352	1.005	2.366	4.64	6.25	7.82	9.84	11.35	16.27
4	0.429	0.711	1.649	3.357	5.99	7.78	9.49	11.67	13.28	18.47
5	0.752	0.145	2.343	4.351	7.29	9.24	11.07	13.39	15.09	20.52

← *Do not reject H_0* | *Reject H_0* →

REFER WEB 94 LINK 92 LINK 93

ISBN:978-1-927309-20-9

95 Using the Chi-Squared Test in Genetics

Key Idea: The following problems examine the use of the chi-squared (χ^2) test in genetics.

A worked example illustrating the use of the chi-squared test for a genetic cross is provided on the previous page.

1. In a tomato plant experiment, two heterozygous individuals were crossed (the details of the cross are not relevant here). The predicted Mendelian ratios for the offspring of this cross were **9:3:3:1** for each of the **four following phenotypes**: purple stem-jagged leaf edge, purple stem-smooth leaf edge, green stem-jagged leaf edge, green stem-smooth leaf edge.

   The observed results of the cross were not exactly as predicted.
   The numbers of offspring with each phenotype are provided below:

Observed results of the tomato plant cross

Phenotype	Number	Phenotype	Number
Purple stem-jagged leaf edge	**12**	Green stem-jagged leaf edge	**8**
Purple stem-smooth leaf edge	**9**	Green stem-smooth leaf edge	**0**

   (a) State your null hypothesis for this investigation (Ho): ____________________

   ____

   (b) State the alternative hypothesis (HA): ____________________

   ____

2. Use the chi-squared (χ^2) test to determine if the differences observed between the phenotypes are significant. The table of critical values of χ^2 at different *P* values is provided on the previous page.

   (a) Enter the observed values (number of individuals) and complete the table to calculate the χ^2 value:

Category	O	E	O — E	(O — E)^2	$\frac{(O - E)^2}{E}$
Purple stem, jagged leaf					
Purple stem, smooth leaf					
Green stem, jagged leaf					
Green stem, smooth leaf					
	Σ				Σ

   (b) Calculate χ^2 value using the equation:

$$\chi^2 = \sum \frac{(O - E)^2}{E}$$

   χ^2 = ____________

   (c) Calculate the degrees of freedom: ____________

   (d) Using the χ^2 table, state the *P* value corresponding to your calculated χ^2 value:

   ____

   (e) State your decision: *(circle one)*
   reject Ho / do not reject Ho

3. Students carried out a pea plant experiment, where two heterozygous individuals were crossed. The predicted Mendelian ratios for the offspring were **9:3:3:1** for each of the **four following phenotypes**: round-yellow seed, round-green seed, wrinkled-yellow seed, wrinkled-green seed.

   The observed results were as follows:

Phenotype	Number	Phenotype	Number
Round-yellow seed	**441**	Wrinkled-yellow seed	**143**
Round-green seed	**159**	Wrinkled-green seed	**57**

   Use a separate piece of paper to complete the following:

   (a) State the null and alternative hypotheses (Ho and HA).

   (b) Calculate the χ^2 value.

   (c) Calculate the degrees of freedom and state the *P* value corresponding to your calculated χ^2 value.

   (d) State whether or not you reject your null hypothesis: reject Ho / do not reject Ho (circle one)

4. Comment on the whether the χ^2 values obtained above are similar. Suggest a reason for any difference:

   ____

   ____

   ____

ISBN: 978-1-927309-20-9

96 Problems Involving Dihybrid Inheritance

Key Idea: For dihybrid crosses involving autosomal unlinked genes, the offspring appear in predictable ratios.

Test your understanding of dihybrid inheritance by solving problems involving the inheritance of two genes.

1. In cats, the following alleles are present for coat characteristics: black (**B**), brown (**b**), short (**L**), long (**l**), tabby (**T**), blotched tabby (**tb**). Use the information to complete the dihybrid crosses below:

(a) A black short haired (**BBLl**) male is crossed with a black long haired (**Bbll**) female. Determine the genotypic and phenotypic ratios of the offspring:

Genotype ratio: ______________________

Phenotype ratio: ______________________

(b) A tabby, short haired male (**TtbLl**) is crossed with a blotched tabby, short haired (**tbtbLl**) female. Determine ratios of the offspring:

Genotype ratio: ______________________

Phenotype ratio: ______________________

2. A plant with orange-striped flowers was cultivated from seeds. The plant was self-pollinated and the F_1 progeny appeared in the following ratios: 89 orange with stripes, 29 yellow with stripes, 32 orange without stripes, 9 yellow without stripes.

(a) Describe the dominance relationships of the alleles responsible for the phenotypes observed: ______________________

(b) Determine the genotype of the original plant with orange striped flowers: ______________________

3. In rabbits, spotted coat **S** is dominant to solid colour **s,** while for coat colour, black **B** is dominant to brown **b**. A brown spotted rabbit is mated with a solid black one and all the offspring are black spotted (the genes are not linked).

(a) State the genotypes:

Parent 1: ______________________

Parent 2: ______________________

Offspring: ______________________

(b) Use the Punnett square to show the outcome of a cross between the F_1 (the F_2):

(c) Using ratios, state the phenotypes of the F_2 generation: ______________________

ISBN:978-1-927309-20-9

4. In guinea pigs, rough coat **R** is dominant over smooth coat **r** and black coat **B** is dominant over white **b**. The genes are not linked.
A homozygous rough black animal was crossed with a homozygous smooth white:

(a) State the genotype of the F_1: ______________________

(b) State the phenotype of the F_1: ______________________

(c) Use the Punnett square (top right) to show the outcome of a cross between the F_1 (the F_2):

(d) Using ratios, state the phenotypes of the F_2 generation: ______________________

____

____

(e) Use the Punnett square (right) to show the outcome of a **back cross** of the F_1 to the rough, black parent:

(f) Using ratios, state the phenotype of the F_2 generation: ______________________

____

(g) A rough black guinea pig was crossed with a rough white one produced the following offspring: 28 rough black, 31 rough white, 11 smooth black, and 10 smooth white. Determine the genotypes of the parents:

____

5. The Himalayan colour-pointed, long-haired cat is a breed developed by crossing a pedigree (true-breeding), uniform-coloured, long-haired Persian with a pedigree colour-pointed (darker face, ears, paws, and tail) short-haired Siamese.

The genes controlling hair colouring and length are on separate chromosomes: uniform colour **U**, colour pointed **u**, short hair **S**, long hair **s**.

Persian | Siamese | Himalayan

(a) Using the symbols above, indicate the genotype of each breed below its photograph (above, right).

(b) State the genotype of the F_1 (Siamese X Persian): ______________________

(c) State the phenotype of the F_1: ______________________

(d) Use the Punnett square to show the outcome of a cross between the F_1 (the F_2):

(e) State the ratio of the F_2 that would be Himalayan: ______________________

(f) State whether the Himalayan would be true breeding: ______________________

(g) State the ratio of the F_2 that would be colour-point, short-haired cats: ____________

6. A *Drosophila* male with genotype **Cucu Ebeb** (straight wing, grey body) is crossed with a female with genotype **cucu ebeb** (curled wing, ebony body). The phenotypes of the F_1 were recorded and the percentage of each type calculated. The percentages were: Straight wings, grey body 45%, curled wings, ebony body 43%, straight wings, ebony body 6%, and curled wings grey body 6%.

Straight wing Cucu
Grey body, Ebeb

Curled wing cucu
Ebony body, ebeb

(a) Is there evidence of crossing over in the offspring? ______________________

(b) Explain your answer: ______________________

____

____

(c) Determine the genotypes of the offspring: ______________________

____

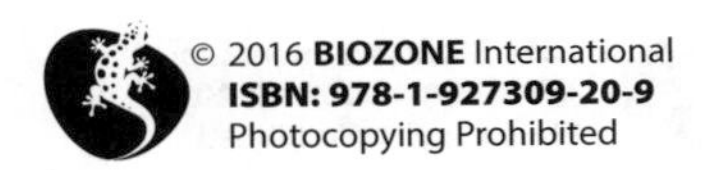

ISBN: 978-1-927309-20-9

97 Gene Interactions

Key Idea: Epistatic genes interact to control the expression of a single characteristic.

Although one gene product (e.g. an enzyme) may independently produce a single phenotypic character, genes frequently interact to produce the phenotype we see. Two or more loci may interact to produce new phenotypes, or an allele at one locus may mask or modify the effect of alleles at other loci. These epistatic gene interactions result in characteristic phenotypic ratios in the offspring that are different from those expected under independent assortment.

In the example of pea seeds the alleles R and r, and Y and y act independently on seed shape and colour respectively. If the Y or y allele is present there is no effect on the phenotype produced by the R or r allele. But consider a situation where the Y or y allele works together with the R and r alleles to produce a single phenotype? A simple example of this is the comb type of chickens.

There are four comb types in chickens (right). The four phenotypes are produced by the interactions of two alleles R and P. In a cross between two individuals heterozygous for both alleles (RrPp), the phenotypic ratio is 9:3:3:1, as would be expected in a dihybrid cross involving independent assortment. However, there are many examples of gene interactions that produce different phenotypic ratios. The ratio can be diagnostic in that it can indicate the type of interaction involved.

Single comb	Pea comb	Rose comb	Walnut comb
Genotypes:	Genotypes:	Genotypes:	Genotypes:
rrpp	**rrP_**	**R_pp**	**R_P_**
rrpp	rrPp, rrPP	Rrpp, RRpp	RRPP RrPP, RrPp, RRPp

In chickens, new phenotypes result from interaction between dominant alleles, as well as from the interaction between homozygous recessives.

How do genes interact?

Gene interaction usually occurs when the protein products or enzymes of several genes are all part of the same metabolic process. In the example below, enzyme A (produced from gene A) acts on a precursor substance to produce a colourless intermediate. Enzyme B (from gene B) acts on the intermediate to produce the final product. If either gene A or gene B produces a non-functional enzyme, the end product will be affected. The way gene A or B act on their substrates affects the appearance of the final product and produces different phenotypic ratios. The ratios can be used to identify the type of gene interaction, but they all come under the title of **epistasis**.

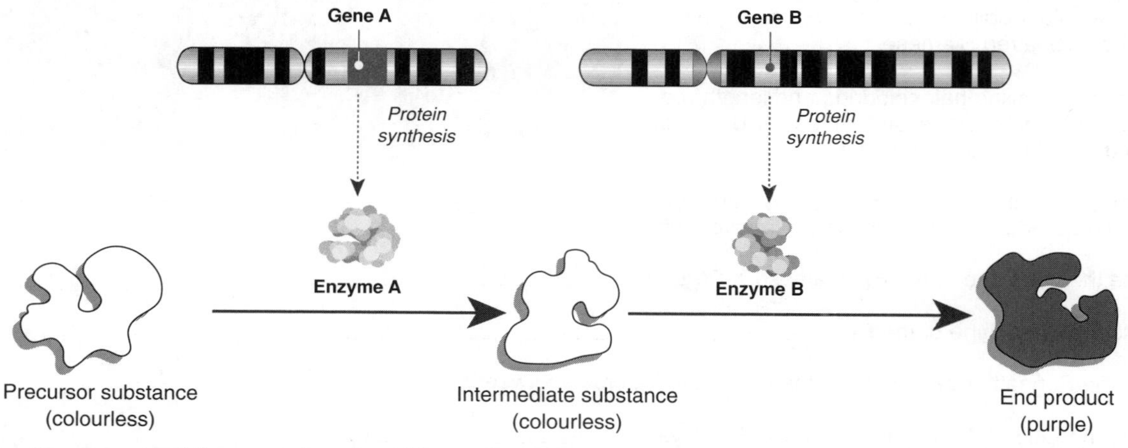

1. (a) For the example of the comb type in chickens determine the ratios of each comb type in a heterozygous cross:

   ____

   (b) What phenotype is produced by the interaction of dominant alleles? ______________

   (c) What phenotype is produced by the interaction of recessive alleles? ______________

2. In the metabolic process shown above, gene A has the alleles A and a. Allele a produces a non-functional enzyme. Gene B has the alleles B and b. Allele b produces a non-functional enzyme. What is the effect on the end product if:

   (a) Gene A has the allele a and Gene B has the allele B: ______________

   (b) Gene A has the allele A and Gene B has the allele b: ______________

   (c) Gene A has the allele A and Gene B has the allele B: ______________

   (d) Gene A has the allele a and Gene B has the allele b: ______________

   (e) On a separate sheet, use a Punnett square to determine the ratio of purple to colourless from a cross AABb x AaBb:

   ____

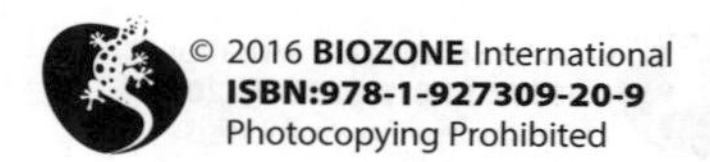

ISBN:978-1-927309-20-9

Table of gene interactions

This table shows five common dihybrid gene interactions and a dihybrid cross with no gene interaction as a comparison. The important point to note is the change to the expected dihybrid 9:3:3:1 ratio in each case. Note that there is independent assortment at the genotypic level. Epistasis is indicated by the change in the phenotypic ratio. Collaboration is usually not considered an epistatic effect because the phenotypic ratio is unchanged, but it does fit in the broad definition of epistasis being an interaction of two or more genes to control a single phenotype.

No of offspring (out of 16)

Possible genotypes from AaBb x AaBb

<table>
<tr><th rowspan="2">Type of gene interaction</th><th rowspan="2">AABB
1</th><th rowspan="2">AABb
2</th><th rowspan="2">AaBB
2</th><th rowspan="2">AaBb
4</th><th rowspan="2">AAbb
1</th><th rowspan="2">Aabb
2</th><th rowspan="2">aaBB
1</th><th rowspan="2">aaBb
2</th><th rowspan="2">aabb
1</th><th rowspan="2">F_1 dihybrid ratio</th><th colspan="2">Example</th></tr>
<tr><th>Organism</th><th>Character</th></tr>
<tr><td>No interaction</td><td colspan="4">Yellow round</td><td colspan="2">Yellow wrinkled</td><td colspan="2">Green round</td><td>Green wrinkled</td><td>9:3:3:1</td><td>Peas</td><td>seed colour/ coat colour</td></tr>
<tr><td>Collaboration</td><td colspan="4">Walnut</td><td colspan="2">Rose</td><td colspan="2">Pea</td><td>Single</td><td>9:3:3:1</td><td>Chickens</td><td>Comb shape</td></tr>
<tr><td>Recessive epistasis</td><td colspan="4">Black</td><td colspan="2">Brown</td><td colspan="3">White</td><td>9:3:4</td><td>Mice</td><td>Coat colour</td></tr>
<tr><td>Duplicate recessive epistasis</td><td colspan="4">Purple</td><td colspan="5">Colourless</td><td>9:7</td><td>Sweet pea flowers</td><td>Colour</td></tr>
<tr><td>Dominant epistasis</td><td colspan="6">White</td><td colspan="2">Yellow</td><td>Green</td><td>12:3:1</td><td>Squash</td><td>Fruit colour</td></tr>
<tr><td>Duplicate dominant epistasis</td><td colspan="8">Coloured</td><td>Colour-less</td><td>15:1</td><td>Wheat</td><td>Kernel colour</td></tr>
</table>

3. In a species of freshwater fish, the allele G produces a green pattern on the dorsal fin when in the presence of the allele Y. A fish that is homozygous recessive for the both genes produces no green pattern to the dorsal fin.

   (a) Write the genotypes for a fish with a green patterned dorsal fin: ____________

   (b) Write the genotypes for a fish with no green pattern: ____________

   (c) Carry out a cross for the heterozygous genotype (GgYy). From the phenotypic ratio, what kind of epistatic interaction is occurring here?

   ____

4. In a second species of freshwater fish, the tail can be one of three colours, red, pink, or orange. The colours are controlled by the genes B and H. The following information is known about the breeding of the fish:
   i Any orange tailed fish crossed together only produce orange tailed fish.
   ii Any red tailed fish crossed with any orange tailed fish may produce offspring with red, pink, or orange tails.
   iii Any pink tailed fish crossed with any orange tailed fish produces offspring that have either all pink tails or 50% with pink tails and 50% with orange tails.

   Use the information above and the space below to work out the type of gene interaction that is being examined here:

The gene interaction is: ____________

ISBN: 978-1-927309-20-9

98 Epistasis

Key Idea: Epistatic interactions by genes that control coat colour typically produce three coat colours.

Epistatic genes often mask the effect of other genes. In this case there are typically three possible phenotypes for a dihybrid cross involving masking interactions. One example is in the genes controlling coat colour in rodents and other mammals. Skin and hair colour is the result of melanin, a pigment which may be either black/brown (eumelanin) or reddish/yellow (phaeomelanin). Melanin itself is synthesised via several biochemical steps from the amino acid tyrosine. The control of coat colour and patterning in mammals is complex and involves at least five major interacting genes. One of these genes (gene C), controls the production of the pigment melanin, while another gene (gene B), is responsible for whether the colour is black or brown (this interaction is illustrated for mice, below). In albinism, the homozygous recessive condition, cc, blocks the expression of the other coat colours.

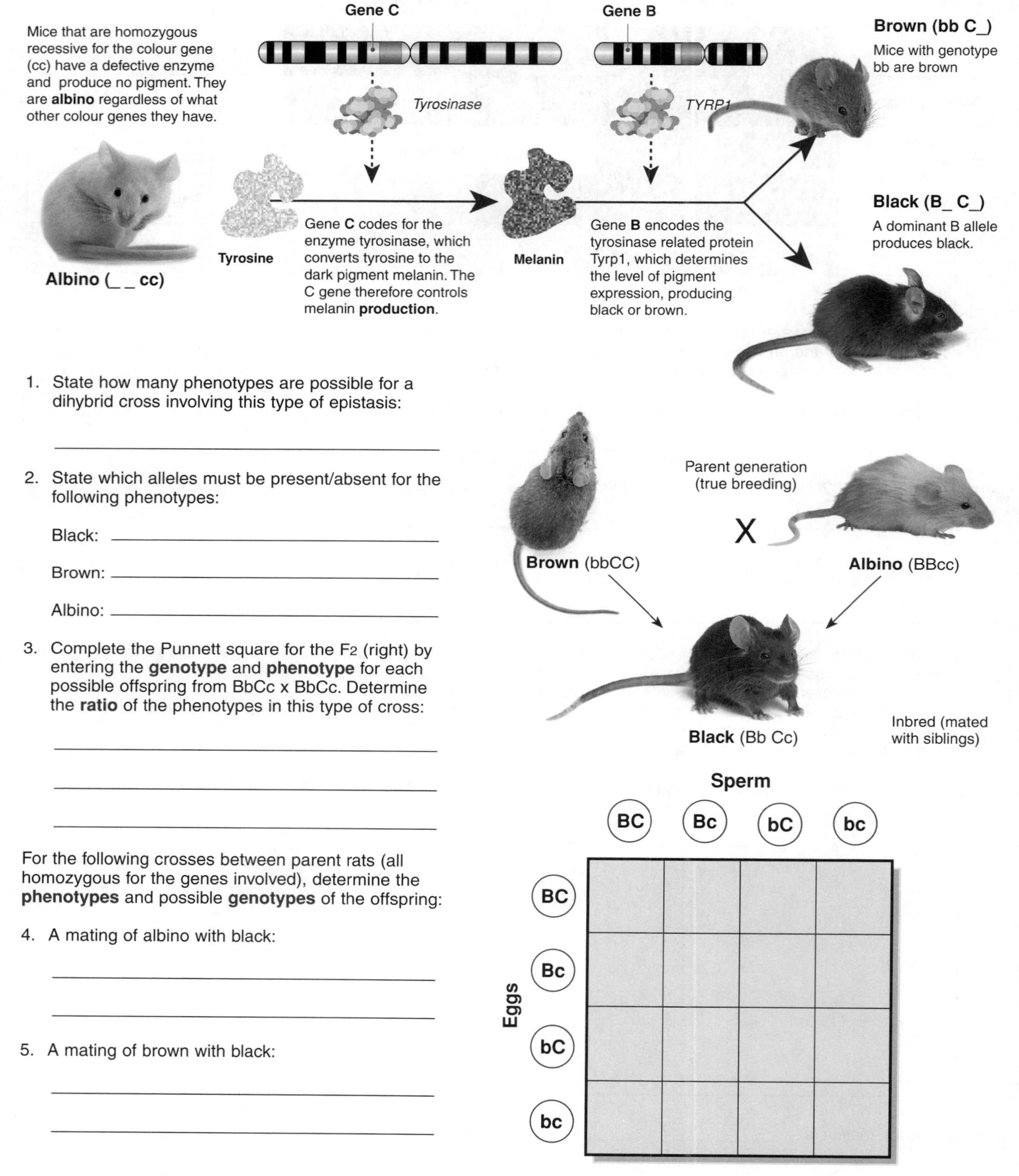

1. State how many phenotypes are possible for a dihybrid cross involving this type of epistasis:

2. State which alleles must be present/absent for the following phenotypes:

   Black:

   Brown:

   Albino:

3. Complete the Punnett square for the F_2 (right) by entering the **genotype** and **phenotype** for each possible offspring from BbCc x BbCc. Determine the **ratio** of the phenotypes in this type of cross:

For the following crosses between parent rats (all homozygous for the genes involved), determine the **phenotypes** and possible **genotypes** of the offspring:

4. A mating of albino with black:

5. A mating of brown with black:

ISBN:978-1-927309-20-9

Allele combinations and coat colour

Epistatic interactions also regulate coat colour in Labrador dogs. The basic coat colour is controlled by the interaction of two genes, each with two alleles. The epistatic E gene determines if pigment will be present in the coat, and the B gene determines the pigment density (depth of colour). Dogs with genotype ee will always be yellow. As a result, three main coat colour variations are possible in Labradors: black, chocolate, and yellow. The yellow coat colour can have two possible phenotypes, giving a total of 4 phenotypes.

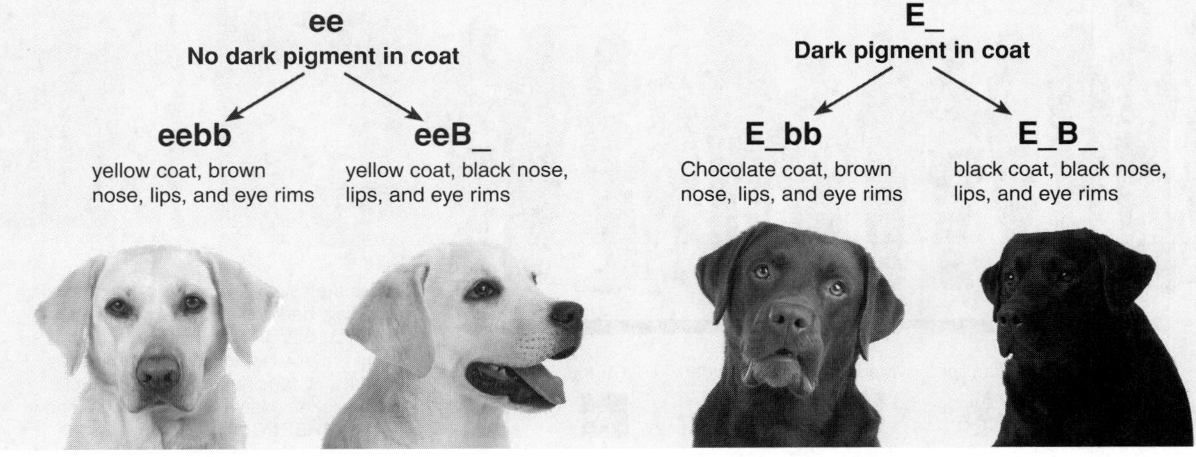

6. State how many main phenotypes are possible for a dihybrid cross involving the genes E and B: ____________________

7. State which alleles must be present and absent for the following phenotypes:

   Black: ____________________

   Brown: ____________________

   Yellow: ____________________

8. (a) State the phenotype and genotype of the F_1 in a cross between a male black and a female yellow Labrador. Both dogs are homozygous for the alleles involved:

   ____

   (b) F_1 male Labradors were crossed with F_1 female Labradors. In the working space provided right, show this cross and state the genotype and phenotype ratios of the offspring:

   ____

   ____

   (c) From the phenotypic ratio, what type of gene interaction is this?

   ____

Working space

9. Yellow Labradors can have either a black nose, lips, and eye rims, or a brown nose, lips, and eye rims.

   (a) What are the genotypes for a yellow coated dog with a black nose, lips, and eye rims:

   ____

   (b) What is the genotype for a yellow coated dog with a brown nose, lips, and eye rims:

   ____

   (c) Furthermore, yellow coated Labradors display variations of the C gene from dark (CC) to light (cc). What is the genotype of a dark yellow Labrador with a black nose, lips, and eye rims:

   ____

   ____

ISBN: 978-1-927309-20-9

99 Polygenes

Key Idea: Many phenotypes are affected by multiple genes. Some phenotypes (e.g. kernel colour in maize and skin colour in humans) are determined by more than one gene and show **continuous variation** in a population. The production of the skin pigment melanin in humans is controlled by at least three genes. The amount of melanin produced is directly proportional to the number of dominant alleles for either gene (from 0 to 6).

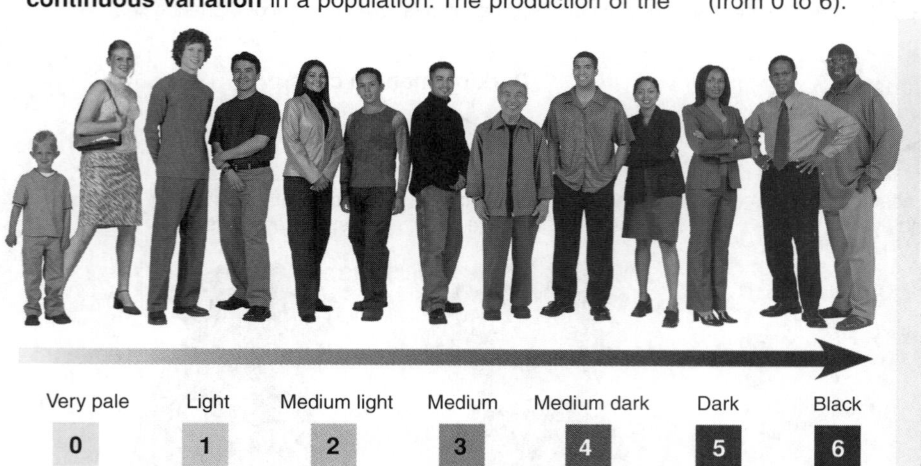

Very pale	Light	Medium light	Medium	Medium dark	Dark	Black
0	1	2	3	4	5	6

A light-skinned person

A dark-skinned person

There are seven shades skin colour ranging from very dark to very pale, with most individual being somewhat intermediate in skin colour. No dominant allele results in a lack of dark pigment (aabbcc). Full pigmentation (black) requires six dominant alleles (AABBCC).

1. Complete the Punnett square for the F_2 generation (below) by entering the genotypes and the number of dark alleles resulting from a cross between two individuals of intermediate skin colour. Colour-code the offspring appropriately for easy reference.

   (a) How many of the 64 possible offspring of this cross will have darker skin than their parents:

   ____

   (b) How many genotypes are possible for this type of gene interaction:

   ____

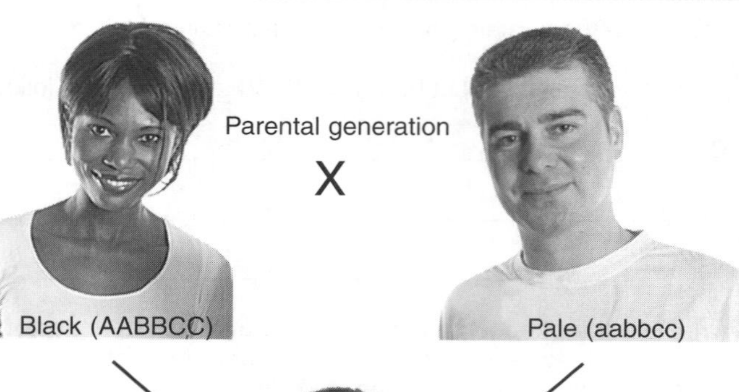

2. Explain why we see many more than seven shades of skin colour in reality:

   ____

   ____

F_2 generation (AaBbCc X AaBbCc)

GAMETES	ABC	ABc	AbC	Abc	aBC	aBc	abC	abc
ABC								
ABc								
AbC								
Abc								
aBC								
aBc								
abC								
abc								

ISBN:978-1-927309-20-9

3. Discuss the differences between **continuous** and **discontinuous variation**, giving examples to illustrate your answer:

4. From a sample of no less than 30 adults, collect data for one continuous variable (e.g. height, weight, shoe size, hand span). Record and tabulate your results in the space below, and then plot a frequency histogram on the grid below:

Raw data

Tally chart (frequency table)

Variable: ....

Frequency

(a) Calculate the following for your data and attach your working.

Mean: ______ **Mode**: ______ **Median**: ______

Standard deviation: ______

(b) Describe the pattern of distribution shown by the graph, giving a reason for your answer: ______

(c) What is the genetic basis of this distribution? ______

(d) What is the importance of a large sample size when gathering data relating to a continuous variable? ______

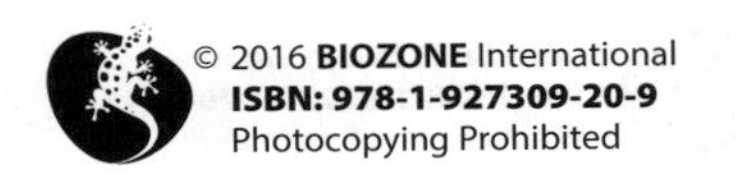

ISBN: 978-1-927309-20-9

100 Genes, Inheritance, and Natural Selection

Key Idea: Natural selection acts on phenotypic variation, produced by the combination of alleles in an individual.

Each individual in a population is the carrier of its own particular combination of genetic material. In sexually reproducing organisms, different combinations of genes arise because of the shuffling of the chromosomes during gamete formation. New allele combinations also occur as a result of mate selection and the chance meeting of different gametes from each of the two parents. Some organisms have allele combinations well suited to the prevailing environment. Those organisms will have greater reproductive success (fitness) than those with less favourable allele combinations and consequently, their genes (alleles) will be represented in a greater proportion in subsequent generations. For asexual species, offspring are essentially clones. New alleles can arise through mutation and some of these may confer a selective advantage. Of course, environments are rarely static, so new allele combinations are always being tested for success.

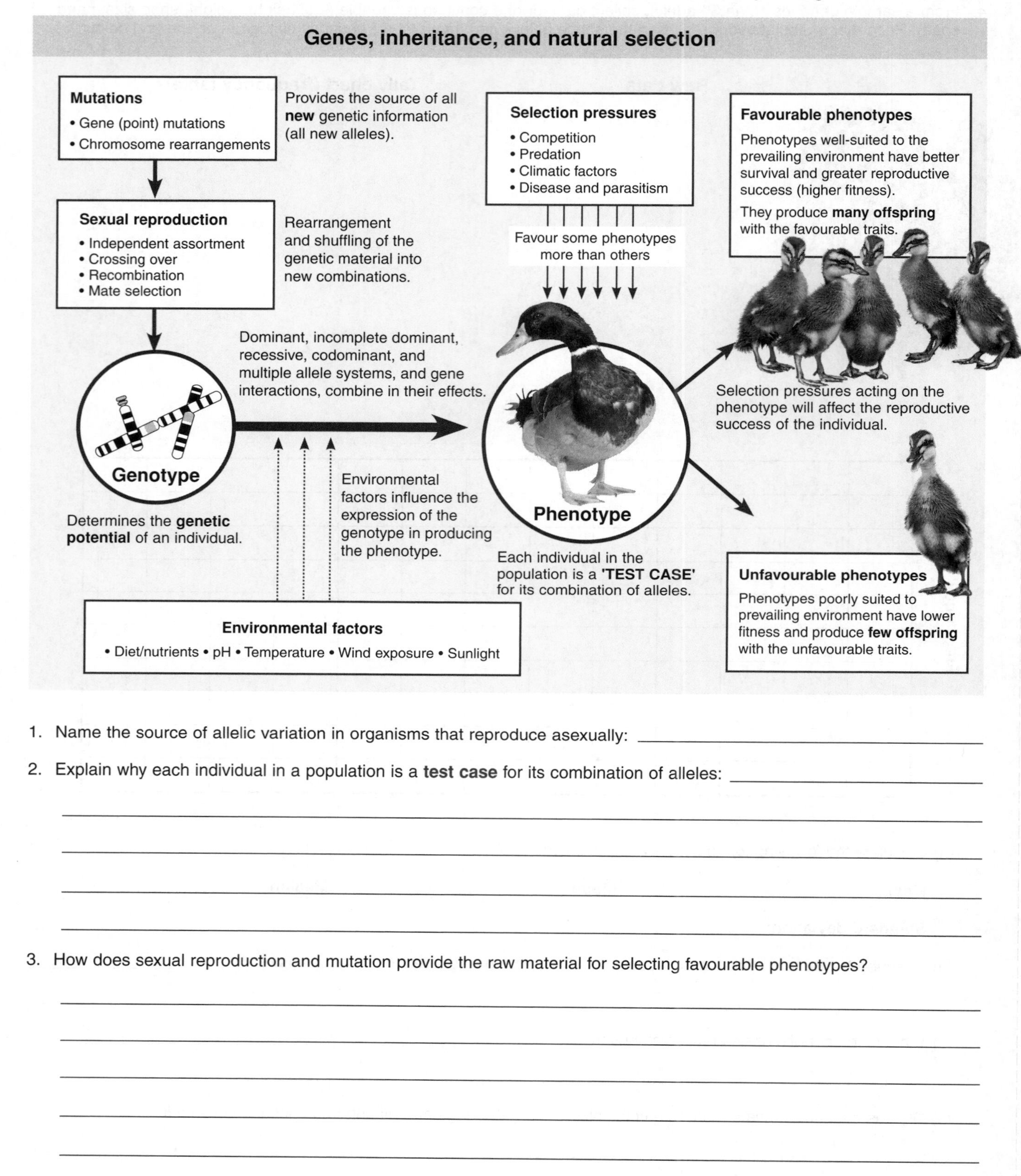

1. Name the source of allelic variation in organisms that reproduce asexually: ______________________

2. Explain why each individual in a population is a **test case** for its combination of alleles: ______________________

____

____

____

____

3. How does sexual reproduction and mutation provide the raw material for selecting favourable phenotypes?

____

____

____

____

____

KNOW LINK 101

ISBN:978-1-927309-20-9

101 Gene Pools and Evolution

Key Idea: The proportions of alleles in a gene pool can be altered by the processes that increase or decrease variation. This activity portrays two populations of a beetle species. Each beetle is a 'carrier' of genetic information, represented by the alleles (A and a) for a gene that controls colour and has a dominant/recessive inheritance pattern. There are normally two phenotypes: black and pale. Mutations may create new alleles. Some of the **microevolutionary processes** (natural selection, genetic drift, gene flow, and mutation) that affect the genetic composition (**allele frequencies**) of gene pools are shown below. Simulate the effect of these processes using the cut out, *Gene Pool Exercise*.

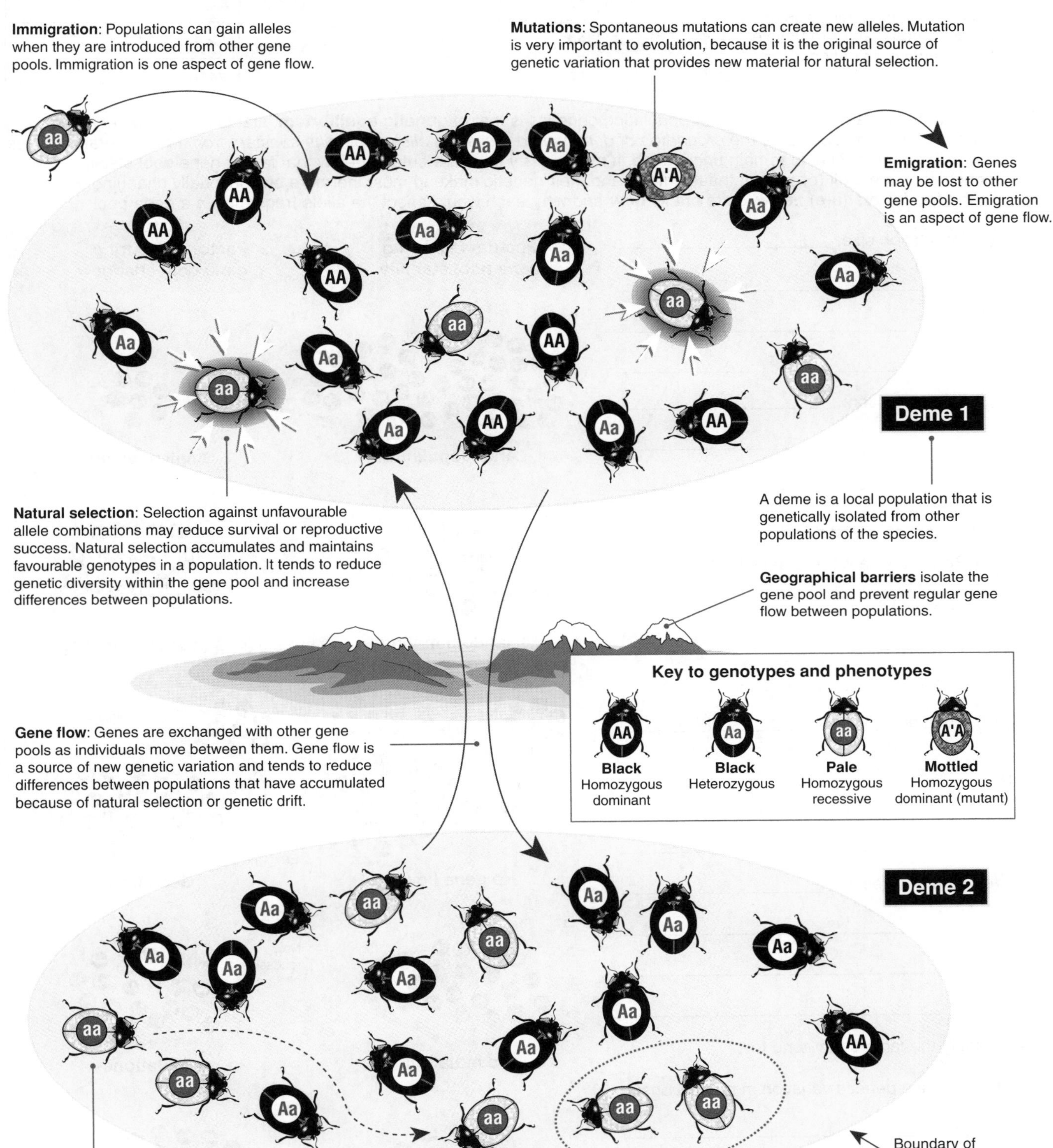

ISBN: 978-1-927309-20-9

1. For each of the two demes shown on the previous page (treating the mutant in deme 1 as a AA):

   (a) Count up the numbers of **allele types** (**A** and **a**).

   (b) Count up the numbers of **allele combinations** (**AA, Aa, aa**).

2. Calculate the frequencies as percentages (%) for the allele types and combinations:

Deme 1		Number counted	%
Allele types	A		
	a		
Allele combinations	AA		
	Aa		
	aa		

Deme 2		Number counted	%
Allele types	A		
	a		
Allele combinations	AA		
	Aa		
	aa		

3. One of the fundamental concepts for population genetics is that of **genetic equilibrium**, stated as: *"For a very large, randomly mating population, the proportion of dominant to recessive alleles remains constant from one generation to the next"*. If a gene pool is to remain unchanged, it must satisfy all of the criteria below that favour gene pool stability. Few populations meet all (or any) of these criteria and their genetic makeup must therefore by continually changing. For each of the five factors (a-e) below, state briefly **how** and **why** each would affect the allele frequency in a gene pool:

   (a) Population size: ______________________

   (b) Mate selection: ______________________

   (c) Gene flow between populations: ______________________

   (d) Mutations: ______________________

   (e) Natural selection: ______________________

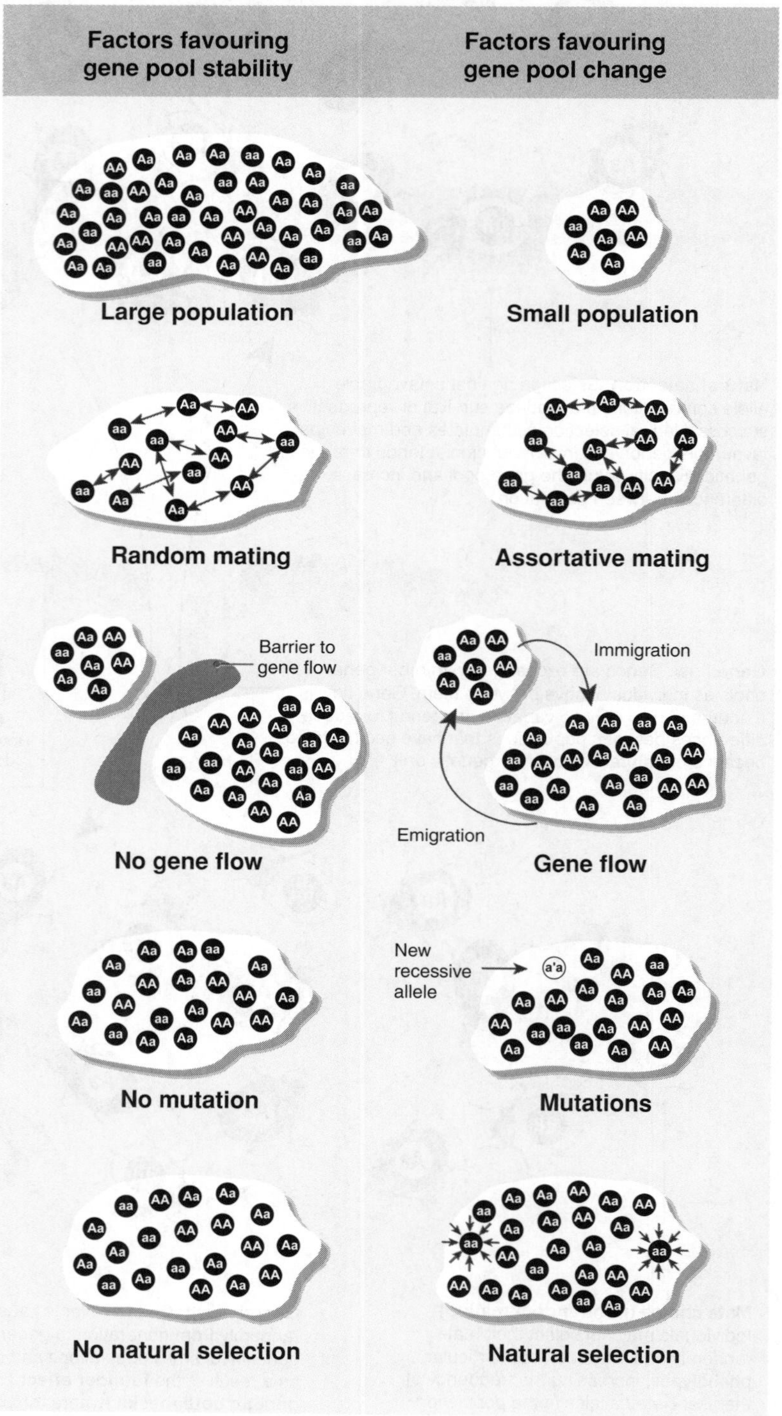

Factors favouring gene pool stability	Factors favouring gene pool change
Large population	Small population
Random mating	Assortative mating
No gene flow	Gene flow
No mutation	Mutations
No natural selection	Natural selection

4. Identify the factors that tend to:

   (a) Increase genetic variation in populations: ______________________

   (b) Decrease genetic variation in populations: ______________________

ISBN:978-1-927309-20-9

102 Gene Pool Exercise

The set of all the versions of all the genes in a population (it genetic make-up) is called the **gene pool**. Cut out the squares below and use them to model the events described in *Modeling Natural Selection*.

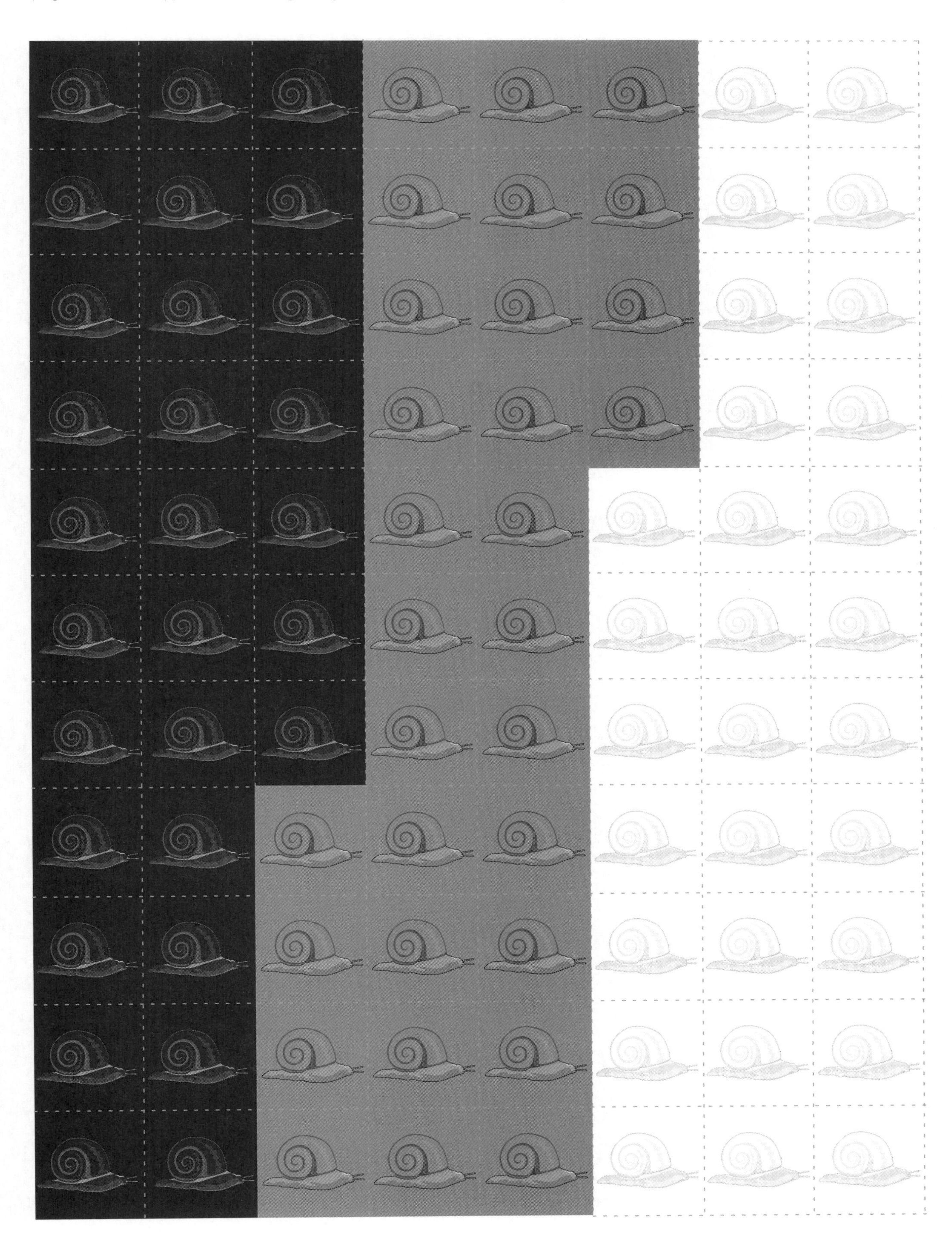

ISBN: 978-1-927309-20-9

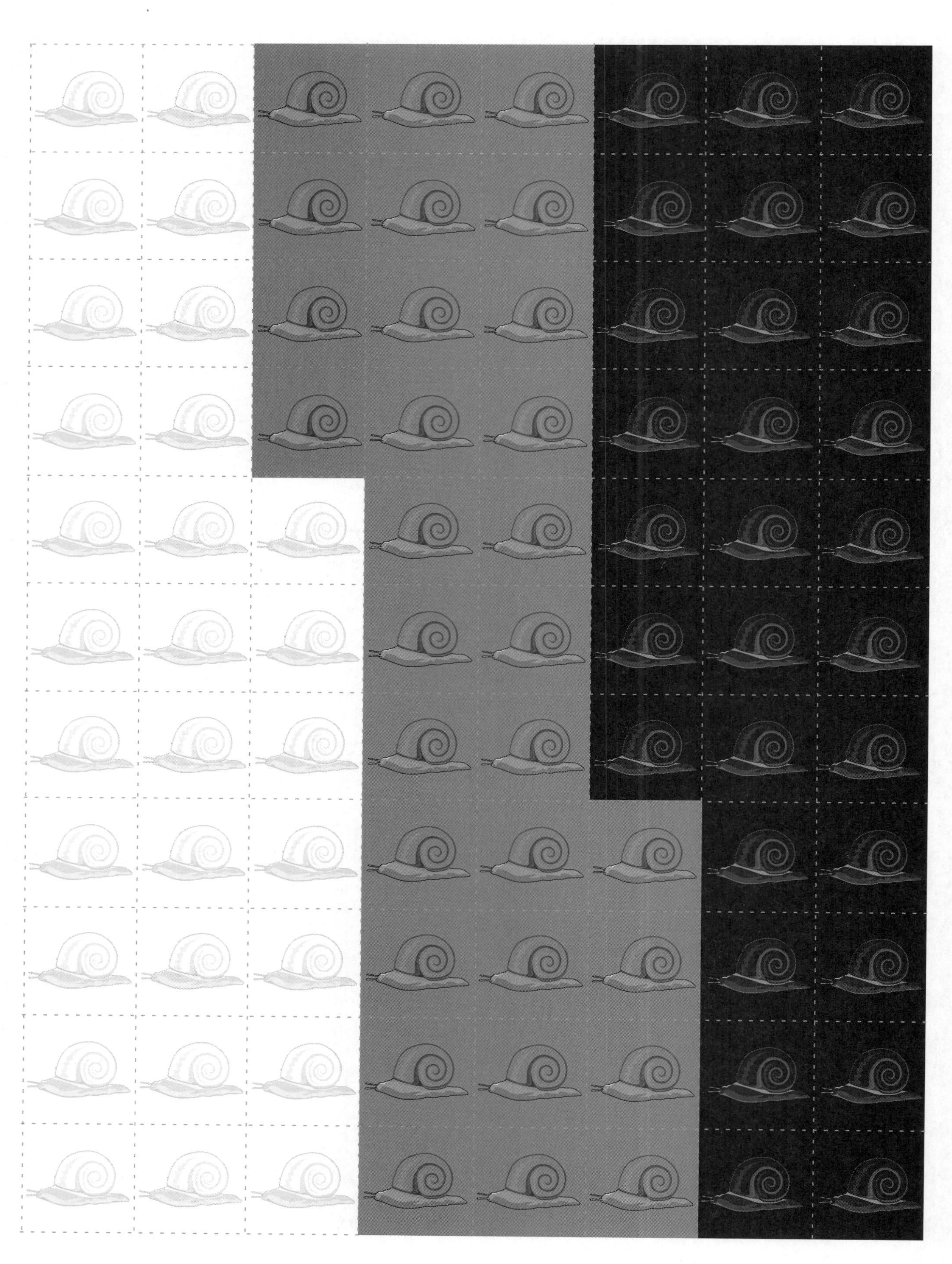

ISBN:978-1-927309-20-9

103 Modelling Natural Selection

Key Idea: The way that natural selection acts on phenotypes can be modelled for a hypothetical population in which individuals differ with respect to one phenotypic character.

Natural selection can be modelled in a simple activity based on predation. You can carry out the following activity by yourself, or work with a partner to increase the size of the population. The black, grey, and white squares on the preceding pages represent phenotypes of a population. Cut them out and follow the instructions below to model natural selection. You will also need a sheet of white paper and a sheet of black paper.

1. Cut out the squares on the preceding pages and record the number of black, grey, and white squares.
2. For the first half of the activity you will also need a black sheet of paper or material that will act as the environment (A3 is a good size). For the second half of the activity you will need a white sheet of paper.
3. Place 10 black, 10 white, and 22 grey squares in a bag and shake them up to mix them. Keep the other squares for making up population proportions later. Write the values in the numbers row of generation 1 below.
4. Work out the proportion of each phenotype in the population (e.g. 10/42 = 0.24) and place these values in the table below. This represents your starting population (you can combine populations with a partner to increase the population size for more reliable results).
5. Now take the squares out of the bag and randomly distribute them over the sheet of black paper (this works best if your partner does this while you aren't looking).
6. You will act the part of a predator on the snails. For 15 seconds, pick up the squares that stand out (are obvious) on the black paper using your thumb and forefinger. These squares represent animals in the population that have been preyed upon and killed. Place them to one side. The remaining squares represent the population that survived to reproduce.
7. Count the remaining phenotypes. In this population, black carries the alleles BB, grey the alleles Bb, and white the alleles bb. On a separate sheet, calculate the frequency of the B and b alleles in the remaining population (hint: if there are 5 black and 10 grey snails then there are 20 B alleles).
8. These frequencies are what is passed on to the next generation. To produce the next generation, the number of black, grey, and white snails must be calculated. This can be done using the original population number and Hardy - Weinberg equations ($p^2 + 2pq + q^2 = 1$ and $p + q = 1$).
9. For example. If there are 24 snails left with the numbers 5 black, 10 grey, and 9 white then the frequency of B (p) = (5 x 2 + 10) / (24 x 2) = 0.4167 and b (q) = 0.5833. The number of black snails in the next generation will therefore be p^2 x 42 = 0.4167^2 x 42 = 7.3 = 7 (you can't have 0.3 of a snail).
10. Record the number of black, grey, and white snails in the table below in generation 2, along with their phenotype frequencies.
11. Repeat steps 4 to 10 for generation 2, and 3 more generations (5 generations in total or more if you wish).
12. On separate graph paper, draw a line graph of the proportions of each colour over the five generations. Which colours have increased, which have decreased?
13. Now repeat the whole activity using a white sheet background instead of the black sheet. What do you notice about the proportions this time?

Generation		Black	Grey	White
1	Number			
	Proportion			
2	Number			
	Proportion			
3	Number			
	Proportion			
4	Number			
	Proportion			
5	Number			
	Proportion			

ISBN: 978-1-927309-20-9

104 Changes in a Gene Pool

Key Idea: Natural selection and migration can alter the allele frequencies in gene pools.

The diagram below shows an hypothetical population of beetles undergoing changes as it is subjected to two 'events'. The three phases represent a progression in time (i.e. the same gene pool, undergoing change). The beetles have two phenotypes (black and pale) determined by the amount of pigment deposited in the cuticle. The gene controlling this character is represented by two alleles **A** and **a**. Your task is to analyse the gene pool as it undergoes changes.

1. For each phase in the gene pool below fill in the tables provided as follows; (some have been done for you):

   (a) Count the number of A and a alleles separately. Enter the count into the top row of the table (left hand columns).
   (b) Count the number of each type of allele combination (AA, Aa and aa) in the gene pool. Enter the count into the top row of the table (right hand columns).
   (c) For each of the above, work out the frequencies as percentages (bottom row of table):

Allele frequency = No. counted alleles ÷ Total no. of alleles x 100

Phase 1: Initial gene pool

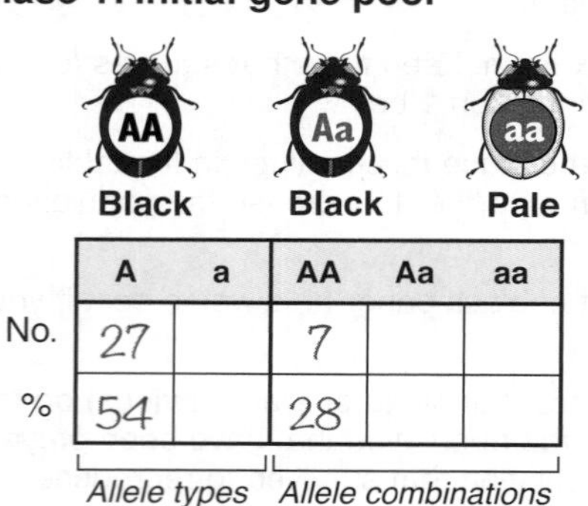

	A	a	AA	Aa	aa
No.	27		7		
%	54		28		

Allele types | *Allele combinations*

Two pale individuals died. Their alleles are removed from the gene pool.

Phase 2: Natural selection

In the same gene pool at a later time there was a change in the allele frequencies. This was due to the loss of certain allele combinations due to natural selection. Some of those with a genotype of aa were eliminated (poor fitness).

These individuals (surrounded by small white arrows) are not counted for allele frequencies; they are dead!

	A	a	AA	Aa	aa
No.					
%					

This individual is entering the population and will add its alleles to the gene pool.

This individual is leaving the population, removing its alleles from the gene pool.

Phase 3: Immigration and emigration

This particular kind of beetle exhibits wandering behaviour. The allele frequencies change again due to the introduction and departure of individual beetles, each carrying certain allele combinations.

Individuals coming into the gene pool (AA) are counted for allele frequencies, but those leaving (aa) are not.

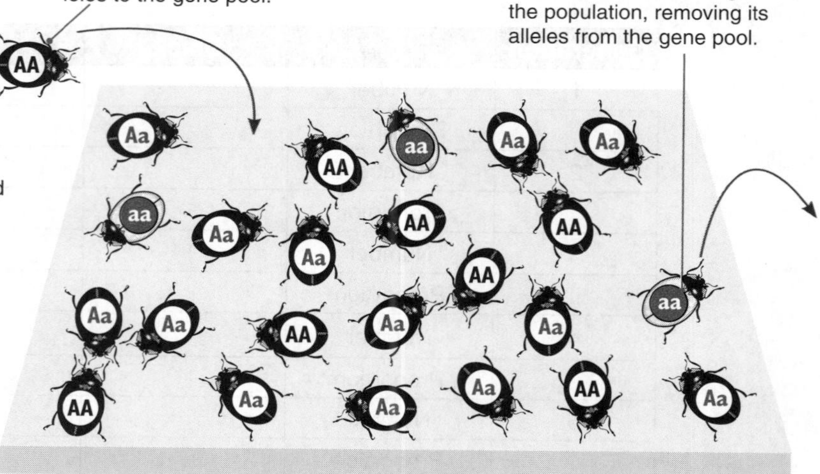

	A	a	AA	Aa	aa
No.					
%					

ISBN:978-1-927309-20-9

105 Hardy-Weinberg Calculations

Key Idea: The Hardy-Weinberg equation is a mathematical model used to calculate allele and genotype frequencies in populations.

The Hardy-Weinberg equation provides a simple mathematical model of genetic equilibrium in a gene pool, but its main application in population genetics is in calculating allele and genotype frequencies in populations, particularly as a means of studying changes and measuring their rate.

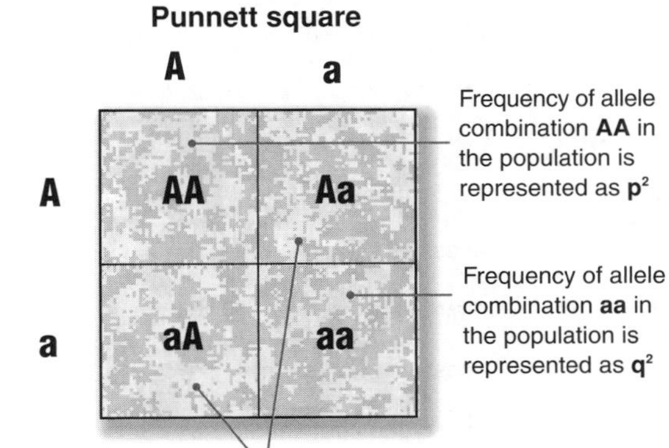

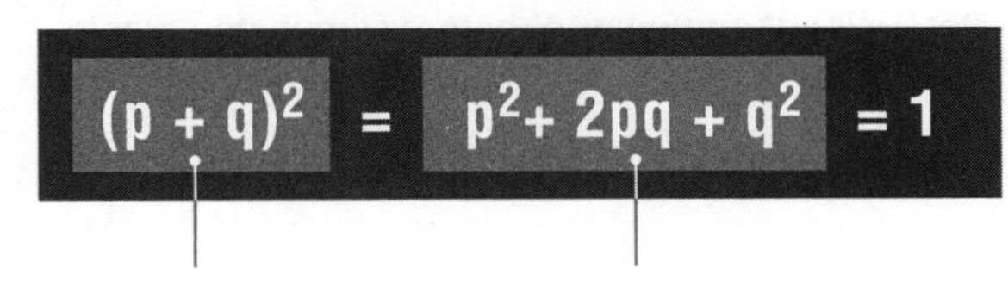

Frequency of allele types

p = Frequency of allele A
q = Frequency of allele a

Frequency of allele combinations

p^2 = Frequency of AA (homozygous dominant)
2pq = Frequency of Aa (heterozygous)
q^2 = Frequency of aa (homozygous recessive)

The Hardy-Weinberg equation is applied to populations with a simple genetic situation: dominant and recessive alleles controlling a single trait. The frequency of all of the dominant (A) and recessive alleles (a) equals the total genetic complement, and adds up to 1 or 100% of the alleles present (i.e. p + q = 1).

How to solve Hardy-Weinberg problems

In most populations, the frequency of two alleles of interest is calculated from the proportion of homozygous recessives (q^2), as this is the only genotype identifiable directly from its phenotype. If only the dominant phenotype is known, q^2 may be calculated (1 – the frequency of the dominant phenotype). The following steps outline the procedure for solving a Hardy-Weinberg problem:

***Remember that all calculations must be carried out using proportions,* NOT PERCENTAGES!**

1. Examine the question to determine what piece of information you have been given about the population. In most cases, this is the percentage or frequency of the homozygous recessive phenotype q^2, or the dominant phenotype $p^2 + 2pq$ (see note above).
2. The first objective is to find out the value of p or q, If this is achieved, then every other value in the equation can be determined by simple calculation.
3. Take the square root of q^2 to find q.
4. Determine p by subtracting q from 1 (i.e. p = 1 – q).
5. Determine p^2 by multiplying p by itself (i.e. $p^2 = p \times p$).
6. Determine 2pq by multiplying p times q times 2.
7. Check that your calculations are correct by adding up the values for $p^2 + q^2 + 2pq$ (the sum should equal 1 or 100%).

Worked example

Among white-skinned people in the USA, approximately 70% of people can taste the chemical phenylthiocarbamide (PTC) (the dominant phenotype), while 30% are non-tasters (the recessive phenotype).

Determine the frequency of:	***Answers***
(a) Homozygous recessive phenotype(**q^2**).	30% - provided
(b) The dominant allele (**p**).	45.2%
(c) Homozygous tasters (**p^2**).	20.5%
(d) Heterozygous tasters (**2pq**).	49.5%

Data: The frequency of the dominant phenotype (70% tasters) and recessive phenotype (30% non-tasters) are provided.

Working:

Recessive phenotype: **q^2** = 30%
use 0.30 for calculation

therefore: **q** = 0.5477
square root of 0.30

therefore: **p** = 0.4523
1 – q = p
1 – 0.5477 = 0.4523

Use p and q in the equation (top) to solve any unknown:

Homozygous dominant **p^2** = 0.2046
(p x p = 0.4523 x 0.4523)

Heterozygous: **2pq** = 0.4953

1. A population of hamsters has a gene consisting of 90% M alleles (black) and 10% m alleles (grey). Mating is random.
Data: Frequency of recessive allele (10% m) and dominant allele (90% M).

Determine the proportion of offspring that will be black and the proportion that will be grey (show your working):

____

____

Recessive allele:	q =	
Dominant allele:	p =	
Recessive phenotype:	q^2 =	
Homozygous dominant:	p^2 =	
Heterozygous:	2pq =	

ISBN: 978-1-927309-20-9

2. You are working with pea plants and found 36 plants out of 400 were dwarf.
Data: Frequency of recessive phenotype (36 out of 400 = 9%)

(a) Calculate the frequency of the tall gene: ____________________

(b) Determine the number of heterozygous pea plants:

____

____

Recessive allele:	q =	
Dominant allele:	p =	
Recessive phenotype:	q^2 =	
Homozygous dominant:	p^2 =	
Heterozygous:	$2pq$ =	

3. In humans, the ability to taste the chemical phenylthiocarbamide (PTC) is inherited as a simple dominant characteristic. Suppose you found out that 360 out of 1000 college students could not taste the chemical.
Data: Frequency of recessive phenotype (360 out of 1000).

(a) State the frequency of the gene for tasting PTC:

____

(b) Determine the number of heterozygous students in this population:

____

Recessive allele:	q =	
Dominant allele:	p =	
Recessive phenotype:	q^2 =	
Homozygous dominant:	p^2 =	
Heterozygous:	$2pq$ =	

4. A type of deformity appears in 4% of a large herd of cattle. Assume the deformity was caused by a recessive gene.
Data: Frequency of recessive phenotype (4% deformity).

(a) Calculate the percentage of the herd that are carriers of the gene:

____

(b) Determine the frequency of the dominant gene in this case:

____

Recessive allele:	q =	
Dominant allele:	p =	
Recessive phenotype:	q^2 =	
Homozygous dominant:	p^2 =	
Heterozygous:	$2pq$ =	

5. Assume you placed 50 pure bred black guinea pigs (dominant allele) with 50 albino guinea pigs (recessive allele) and allowed the population to attain genetic equilibrium (several generations have passed).
Data: Frequency of recessive allele (50%) and dominant allele (50%).

Determine the proportion (%) of the population that becomes white:

____

Recessive allele:	q =	
Dominant allele:	p =	
Recessive phenotype:	q^2 =	
Homozygous dominant:	p^2 =	
Heterozygous:	$2pq$ =	

6. It is known that 64% of a large population exhibit the recessive trait of a characteristic controlled by two alleles (one is dominant over the other).
Data: Frequency of recessive phenotype (64%). Determine the following:

(a) The frequency of the recessive allele: ____________________

(b) The percentage that are heterozygous for this trait: ____________________

(c) The percentage that exhibit the dominant trait: ____________________

(d) The percentage that are homozygous for the dominant trait: ____________________

(e) The percentage that has one or more recessive alleles: ____________________

7. Albinism is recessive to normal pigmentation in humans. The frequency of the albino allele was 10% in a population.
Data: Frequency of recessive allele (10% albino allele).

Determine the proportion of people that you would expect to be albino:

____

____

____

Recessive allele:	q =	
Dominant allele:	p =	
Recessive phenotype:	q^2 =	
Homozygous dominant:	p^2 =	
Heterozygous:	$2pq$ =	

ISBN:978-1-927309-20-9

106 Analysis of a Squirrel Gene Pool

Key Idea: Allele frequencies for real populations can be calculated using the Hardy-Weinberg equation. Analysis of those allele frequencies can show how the population's gene pool changes over time.

In Olney, Illinois, there is a unique population of albino (white) and grey squirrels. Between 1977 and 1990, students at Olney Central College carried out a study of this population. They recorded the frequency of grey and albino squirrels. The albinos displayed a mutant allele expressed as an albino phenotype only in the homozygous recessive condition. The data they collected are provided in the table below. Using the **Hardy-Weinberg equation**, it was possible to estimate the frequency of the normal 'wild' allele (G) providing grey fur colouring, and the frequency of the mutant albino allele (g) producing white squirrels when homozygous.

Thanks to **Dr. John Stencel**, Olney Central College, Olney, Illinois, US, for providing the data for this exercise.

Grey squirrel, usual colour form

Albino form of grey squirrel

Population of grey and white squirrels in Olney, Illinois (1977-1990)

Year	Grey	White	Total	GG	Gg	gg
1977	602	182	784	26.85	49.93	23.21
1978	511	172	683	24.82	50.00	25.18
1979	482	134	616	28.47	49.77	21.75
1980	489	133	622	28.90	49.72	21.38
1981	536	163	699	26.74	49.94	23.32
1982	618	151	769	31.01	49.35	19.64
1983	419	141	560	24.82	50.00	25.18
1984	378	106	484	28.30	49.79	21.90
1985	448	125	573	28.40	49.78	21.82
1986	536	155	691	27.71	49.86	22.43
1987	*No data collected this year*					
1988	652	122	774	36.36	47.88	15.76
1989	552	146	698	29.45	49.64	20.92
1990	603	111	714	36.69	47.76	15.55

Freq. of g	Freq. of G
48.18	51.82
50.18	49.82
46.64	53.36
46.24	53.76
48.29	51.71
44.31	55.69
50.18	49.82
46.80	53.20
46.71	53.29
47.36	52.64
39.70	60.30
45.74	54.26
39.43	60.57

1. **Graph population changes**: Use the data in the first 3 columns of the table above to plot a line graph. This will show changes in the phenotypes: numbers of grey and white (albino) squirrels, as well as changes in the total population. Plot: **grey**, **white**, and **total** for each year:

   (a) Determine by how much (as a %) total population numbers have fluctuated over the sampling period:

   (b) Describe the overall trend in total population numbers and any pattern that may exist:

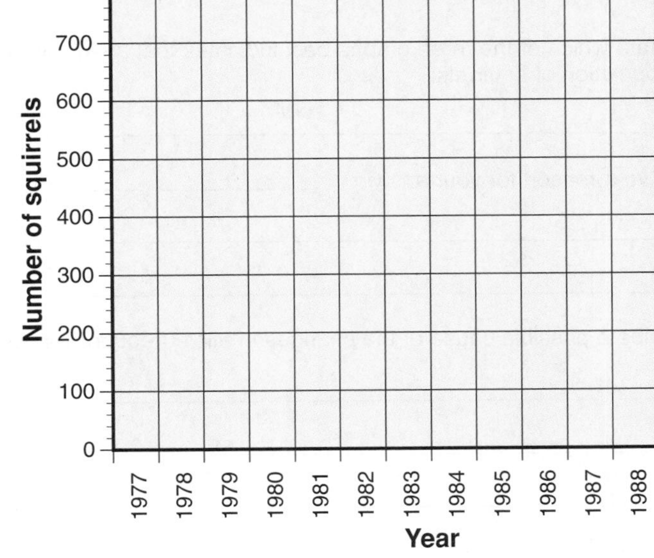

ISBN: 978-1-927309-20-9

2. Graph genotype changes: Use the data in the genotype columns of the table on the opposite page to plot a line graph. This will show changes in the allele combinations (**GG, Gg, gg**). Plot: **GG**, **Gg**, and **gg** for each year:

Describe the overall trend in the frequency of:

(a) Homozygous dominant (**GG**) genotype:

(b) Heterozygous (**Gg**) genotype:

(c) Homozygous recessive (**gg**) genotype:

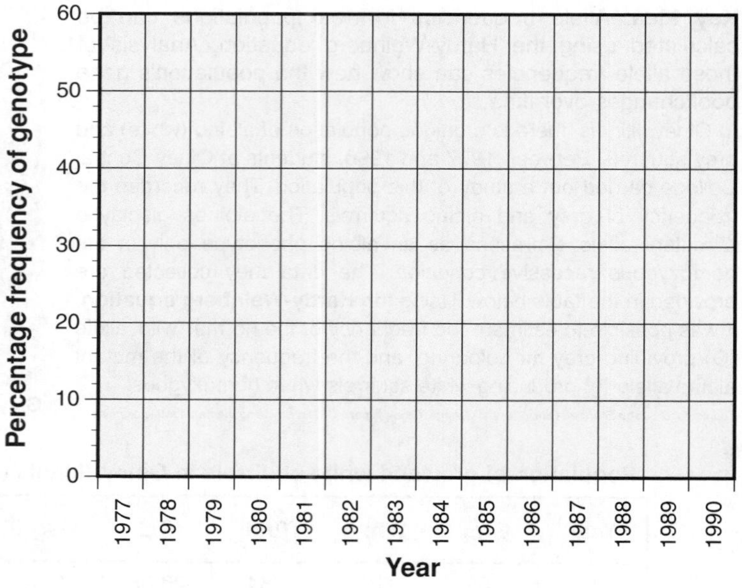

Percentage frequency of genotype

60
50
40
30
20
10
0

1977 1978 1979 1980 1981 1982 1983 1984 1985 1986 1987 1988 1989 1990

Year

3. Graph allele changes: Use the data in the last two columns of the table on the previous page to plot a line graph. This will show changes in the allele frequencies for each of the dominant (**G**) and recessive (**g**) alleles. Plot: the frequency of **G** and the frequency of **g**:

(a) Describe the overall trend in the frequency of the dominant allele (**G**):

(b) Describe the overall trend in the frequency of the recessive allele (**g**):

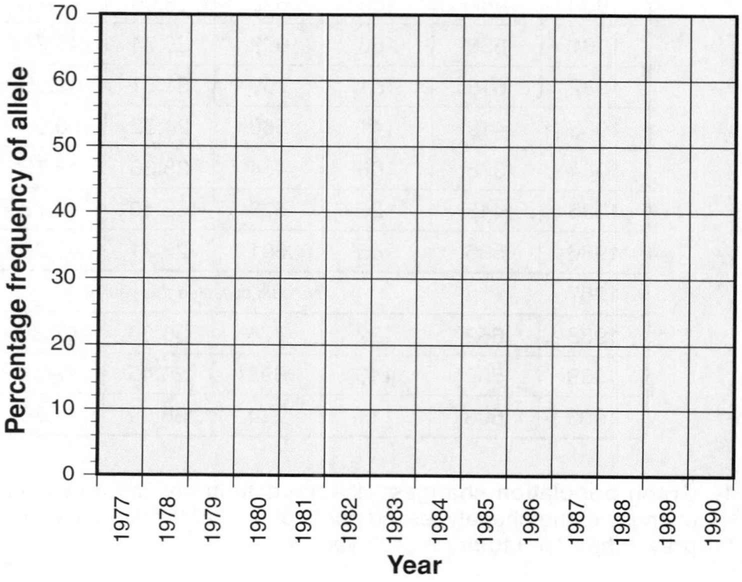

Percentage frequency of allele

70
60
50
40
30
20
10
0

1977 1978 1979 1980 1981 1982 1983 1984 1985 1986 1987 1988 1989 1990

Year

4. (a) State which of the three graphs best indicates that a significant change may be taking place in the gene pool of this population of squirrels:

(b) Give a reason for your answer:

5. Describe a possible cause of the changes in allele frequencies over the sampling period:

ISBN:978-1-927309-20-9

107 Directional Selection in Moths

Key Idea: Directional selection pressures on the peppered moth during the Industrial Revolution shifted the common phenotype from the grey form to the melanic (dark) form.

Natural selection may act on the frequencies of phenotypes (and hence genotypes) in populations in one of three different ways (through stabilising, directional, or disruptive selection).

Colour change in the **peppered moth** (*Biston betularia*) during the Industrial Revolution is often used to show **directional selection** in a polymorphic population (polymorphic means having two or more forms). Intensive coal burning during this time caused trees to become dark with soot, and the dark form (morph) of peppered moth became dominant.

The gene controlling colour in the peppered moth, is located on a single locus. The allele for the melanic (dark) form (**M**) is dominant over the allele for the grey (light) form (**m**).

The peppered moth, *Biston betularia*, has two forms: a grey mottled form, and a dark melanic form. During the Industrial Revolution, the relative abundance of the two forms changed to favour the dark form. The change was thought to be the result of selective predation by birds. It was proposed that the grey form was more visible to birds in industrial areas where the trees were dark. As a result, birds preyed upon them more often, resulting in higher numbers of the dark form surviving.

Museum collections of the peppered moth over the last 150 years show a marked change in the frequency of the melanic form (above right). Moths collected in 1850, prior to the major onset of the Industrial Revolution in England, were mostly the grey form (above left). Fifty years later the frequency of the darker melanic forms had increased.

In the 1940s and 1950s, coal burning was still at intense levels around the industrial centres of Manchester and Liverpool. During this time, the melanic form of the moth was still very dominant. In the rural areas further south and west of these industrial centres, the occurrence of the grey form increased dramatically. With the decline of coal burning factories and the introduction of the Clean Air Act in cities, air quality improved between 1960 and 1980. Sulfur dioxide and smoke levels dropped to a fraction of their previous levels. This coincided with a sharp fall in the relative numbers of melanic moths (right).

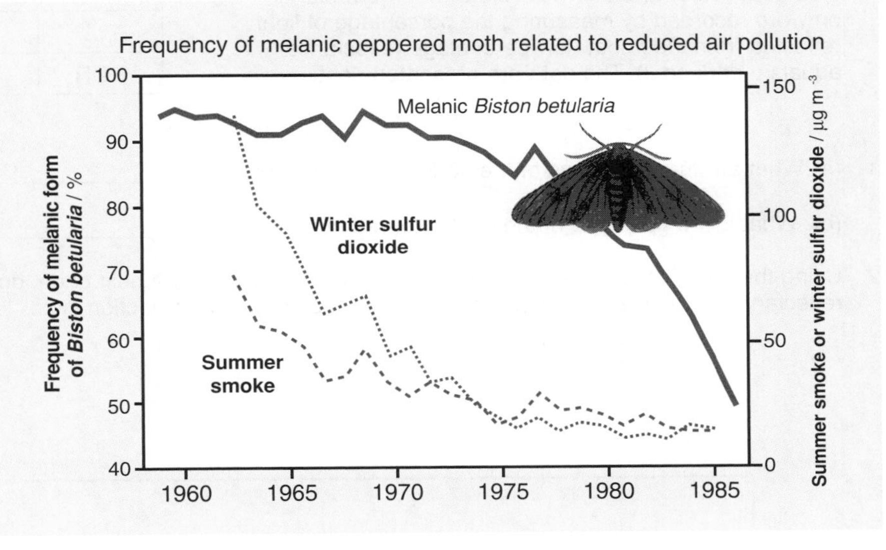

1. The populations of peppered moth in England have undergone changes in the frequency of an obvious phenotypic character over the last 150 years. What is the phenotypic character?

2. Describe how the selection pressure on the grey form has changed with change in environment over the last 150 years:

3. Describe the relationship between allele frequency and phenotype frequency:

4. The level of pollution dropped around Manchester and Liverpool between 1960 and 1985. How did the frequency of the darker melanic form change during this period?

ISBN: 978-1-927309-20-9

108 Natural Selection in Pocket Mice

Key Idea: The need to blend into their surroundings to avoid predation is an important selection pressure acting on the coat colour of rock pocket mice.

Rock pocket mice are found in the deserts of southwestern United States and northern Mexico. They are nocturnal, foraging at night for seeds, while avoiding owls (their main predator). During the day they shelter from the desert heat in their burrows. The coat colour of the mice varies from light brown to very dark brown. Throughout the desert environment in which the mice live there are outcrops of dark volcanic rock. The presence of these outcrops and the mice that live on them present an excellent study in natural selection.

- The coat colour of the Arizona rock pocket mice is controlled by the Mc1r gene (a gene that in mammals is commonly associated with the production of the dark pigment melanin).

  There are variations for the gene that controls coat colour. These variations are called alleles. Homozygous dominant (**DD**) and heterozygous mice (**Dd**) have dark coats, while homozygous recessive mice (**dd**) have light coats. The coat colour of mice in New Mexico is not related to the Mc1r gene.

- 107 rock pocket mice from 14 sites were collected and their coat colour and the rock colour they were found on were recorded by measuring the percentage of light reflected from their coat (low percentage reflectance equals a dark coat). The data are presented right:

Site	Rock type (V volcanic)	Percent reflectance / %	
		Mice coat	Rock
KNZ	V	4	10.5
ARM	V	4	9
CAR	V	4	10
MEX	V	5	10.5
TUM	V	5	27
PIN	V	5.5	11
AFT		6	30
AVR		6.5	26
WHT		8	42
BLK	V	8.5	15
FRA		9	39
TIN		9	39
TUL		9.5	25
POR		12	34.5

1. (a) What are the genotypes of the dark coloured mice? ______________________

   (b) What is the genotype of the light coloured mice? ______________________

2. Using the data in the table above and the grids below and on the next page, draw column graphs of the percent reflectance of the mice coats and the rocks at each of the 14 collection sites.

ISBN:978-1-927309-20-9

3. (a) What do you notice about the reflectance of the rock pocket mice coat colour and the reflectance of the rocks they were found on?

(b) Suggest a cause for the pattern in 3(a). How do the phenotypes of the mice affect where the mice live?

(c) What are two exceptions to the pattern you have noticed in 3(a)?

(d) How might these exceptions have occurred?

4. What type of selection appears to be operating in each of the environments (dark and light rock)? Explain:

5. The rock pocket mice populations in Arizona use a different genetic mechanism to control coat colour than the New Mexico populations. What does this tell you about the evolution of the genetic mechanism for coat colour?

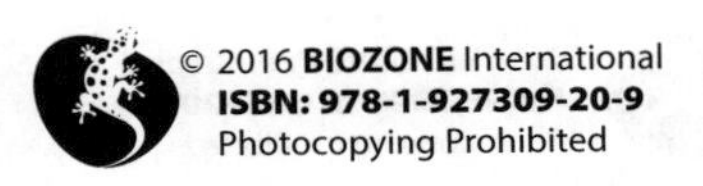

ISBN: 978-1-927309-20-9

109 Disruptive Selection in Darwin's Finches

Key Idea: Disruptive selection in the finch *Geospiza fortis* produces a bimodal distribution for beak size.

The Galápagos Islands, 970 km west of Ecuador, are home to the finch species *Geospiza fortis*. A study during a prolonged drought on Santa Cruz Island showed how **disruptive selection** can change the distribution of genotypes in a population. During the drought, large and small seeds were more abundant than the preferred intermediate seed size.

Beak sizes of *G. fortis* were measured over a three year period (2004-2006), at the start and end of each year. At the start of the year, individuals were captured, banded, and their beaks were measured.

The presence or absence of banded individuals was recorded at the end of the year when the birds were recaptured. Recaptured individuals had their beaks measured.

The proportion of banded individuals in the population at the end of the year gave a measure of fitness. Absent individuals were presumed dead (fitness = 0).

Fitness related to beak size showed a bimodal distribution (left) typical of disruptive selection.

Beak size vs fitness in *Geospiza fortis*

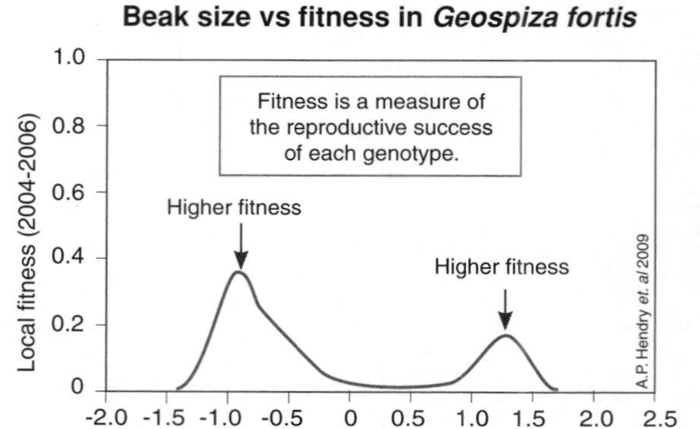

*Fitness showed a **bimodal distribution** (arrowed) being highest for smaller and larger beak sizes.*

Measurements of the beak length, width, and depth were combined into one **single measure**.

Beak size pairing in *Geospiza fortis*

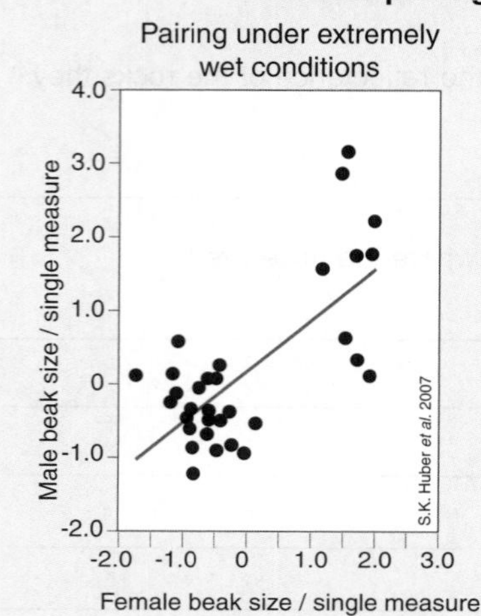

Pairing under dry conditions

Male beak size / single measure

Female beak size / single measure

S.K. Huber et al. 2007

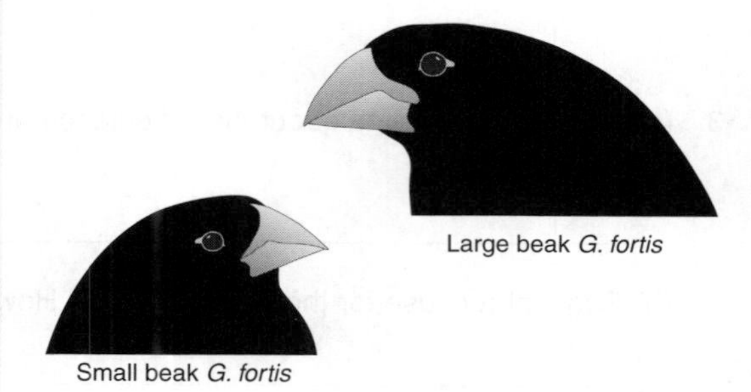

A 2007 study found that breeding pairs of birds had similar beak sizes. Male and females with small beaks tended to breed together, and males and females with large beaks tended to breed together. Mate selection maintained the biomodal distribution in the population during extremely wet conditions. If beak size wasn't a factor in mate selection, the beak size would even out.

1. (a) How did the drought affect seed size on Santa Cruz Island? ______________________

   (b) How did the change in seed size during the drought create a selection pressure for changes in beak size?

2. How does beak size relate to fitness (differential reproductive success) in *G. fortis*? ______________________

3. (a) Is mate selection in *G. fortis* random / non-random? (delete one)

   (b) Give reasons for your answer: ______________________

ISBN:978-1-927309-20-9

110 Selection for Skin Colour in Humans

Key Idea: Skin colour is an evolutionary response to the need to synthesise vitamin D which requires sunlight, and to conserve folate which breaks down in sunlight.

Pigmented skin of varying tones is a feature of humans that evolved after early humans lost the majority of their body hair. However, the distribution of skin colour globally is not random; people native to equatorial regions have darker skin tones than people from higher latitudes. For many years, biologists postulated that this was because darker skins had evolved to protect against skin cancer. The problem with this explanation was that skin cancer is not tied to evolutionary fitness because it affects post-reproductive individuals and cannot therefore provide a mechanism for selection. More complex analyses of the physiological and epidemiological evidence has shown a more complex picture in which selection pressures on skin colour are finely balanced to produce a skin tone that regulates the effects of the sun's ultraviolet radiation on the nutrients vitamin D and folate, both of which are crucial to successful human reproduction, and therefore evolutionary fitness. The selection is stabilising within each latitudinal region.

Skin colour in humans is a product of natural selection

Adapted from Jablonski & Chaplin, Sci. Am. Oct. 2002

Human skin colour is the result of two opposing selection pressures. Skin pigmentation has evolved to protect against destruction of folate from ultraviolet light, but the skin must also be light enough to receive the light required to synthesise vitamin D. Vitamin D synthesis is a process that begins in the skin and is inhibited by dark pigment. Folate is needed for healthy neural development in humans and a deficiency is associated with fatal neural tube defects. Vitamin D is required for the absorption of calcium from the diet and therefore normal skeletal development. Women also have a high requirement for calcium during pregnancy and lactation. Populations that live in the tropics receive enough ultraviolet (UV) radiation to synthesise vitamin D all year long. Those that live in northern or southern latitudes do not. In temperate zones, people lack sufficient UV light to make vitamin D for one month of the year. Those nearer the poles lack enough UV light for vitamin D synthesis most of the year (above). Their lighter skins reflect their need to maximise UV absorption (the photos show skin colour in people from different latitudes).

ISBN: 978-1-927309-20-9

The skin of people who have inhabited particular regions for millennia has adapted to allow sufficient vitamin D production while still protecting folate stores. In the photos above, some of these original inhabitants are illustrated to the left of each pair and compared with the skin tones of more recent immigrants (to the right of each pair, with the number of years since immigration). The numbered locations are on the map.

1. (a) Describe the role of folate in human physiology: __________

   (b) Describe the role of vitamin D in human physiology: __________

2. (a) Early hypotheses to explain skin colour linked pigmentation level only to the degree of protection it gave from UV-induced skin cancer. Explain why this hypothesis was inadequate in accounting for how skin colour evolved:

   (b) Explain how the new hypothesis for the evolution of skin colour overcomes these deficiencies:

3. Explain why, in any given geographical region, women tend to have lighter skins (by 3-4% on average) than men:

4. The Inuit people of Alaska and northern Canada have a diet rich in vitamin D and their skin colour is darker than predicted on the basis of UV intensity at their latitude. Explain this observation:

5. (a) What health problems might be expected for people of African origin now living in northern UK?

   (b) How could these people avoid these problems in their new higher latitude environment?

ISBN:978-1-927309-20-9

111 Genetic Drift

Key Idea: Genetic drift is the term for the random changes in allele frequency that occur in all populations. It has a more pronounced effect in small populations.

Not all individuals, for various reasons, will be able to contribute their genes to the next generation. In a small population, the effect of a few individuals not contributing their alleles to the next generation can have a great effect on allele frequencies. Alleles may even become **lost** from the gene pool altogether (frequency becomes 0%) or **fixed** as the only allele for the gene present (frequency becomes 100%). The random change in allele frequencies is called **genetic drift**.

The genetic makeup (allele frequencies) of the population changes randomly over a period of time

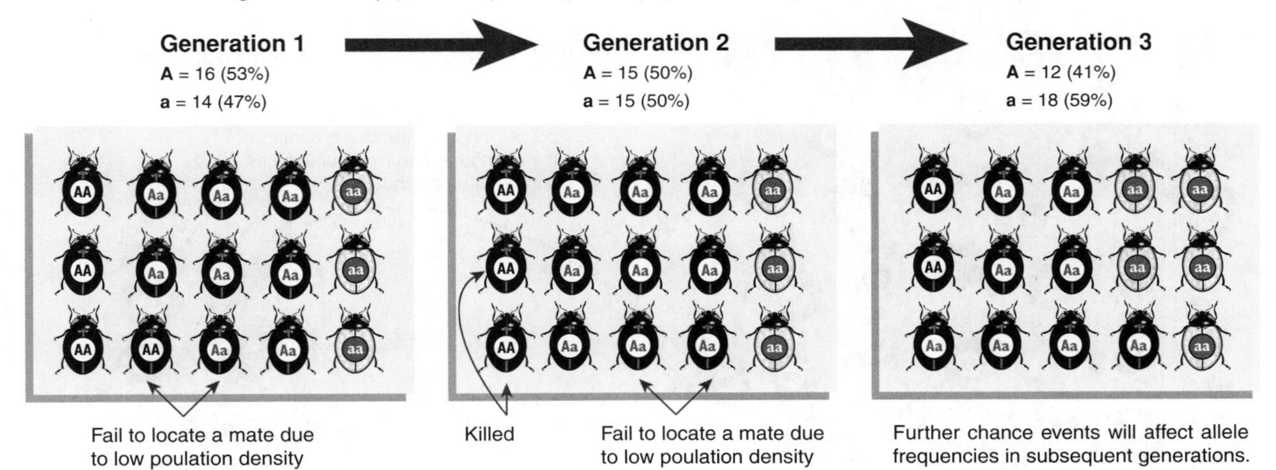

This diagram shows the gene pool of a hypothetical small population over three generations. For various reasons, not all individuals contribute alleles to the next generation. With the random loss of the alleles carried by these individuals, the allele frequency changes from one generation to the next. The change in frequency is directionless as there is no selecting force. The allele combinations for each successive generation are determined by how many alleles of each type are passed on from the preceding one.

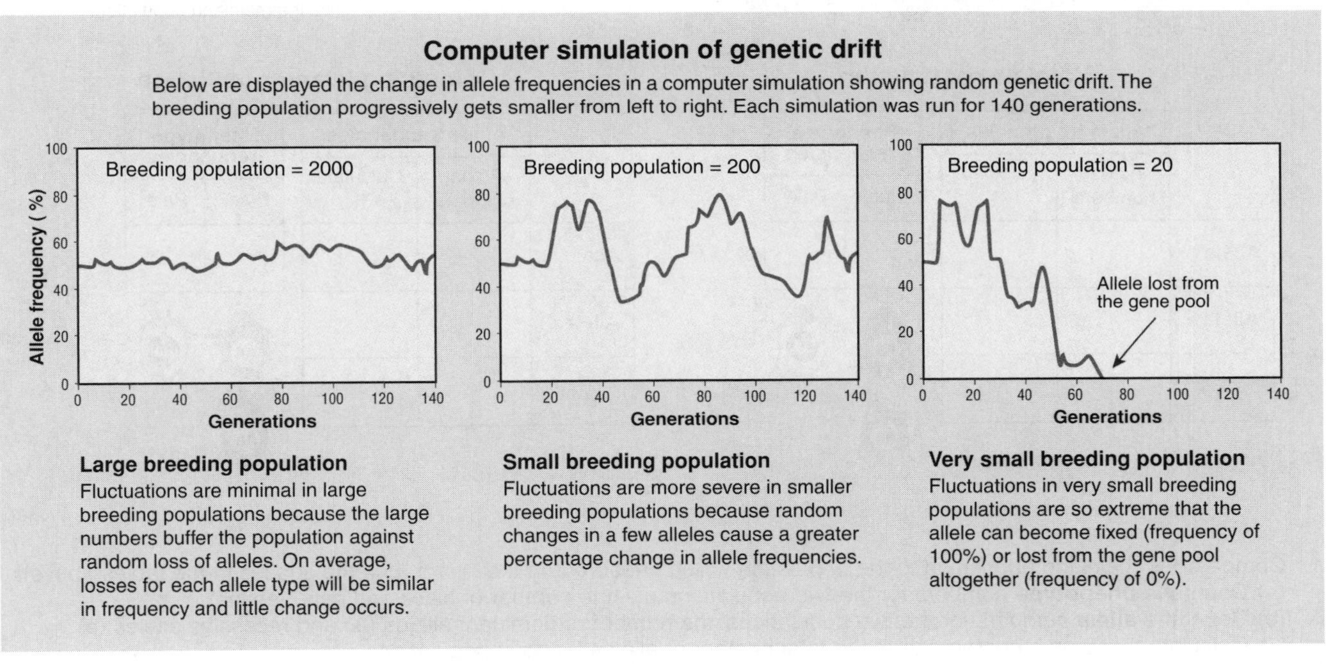

Computer simulation of genetic drift

Below are displayed the change in allele frequencies in a computer simulation showing random genetic drift. The breeding population progressively gets smaller from left to right. Each simulation was run for 140 generations.

Large breeding population
Fluctuations are minimal in large breeding populations because the large numbers buffer the population against random loss of alleles. On average, losses for each allele type will be similar in frequency and little change occurs.

Small breeding population
Fluctuations are more severe in smaller breeding populations because random changes in a few alleles cause a greater percentage change in allele frequencies.

Very small breeding population
Fluctuations in very small breeding populations are so extreme that the allele can become fixed (frequency of 100%) or lost from the gene pool altogether (frequency of 0%).

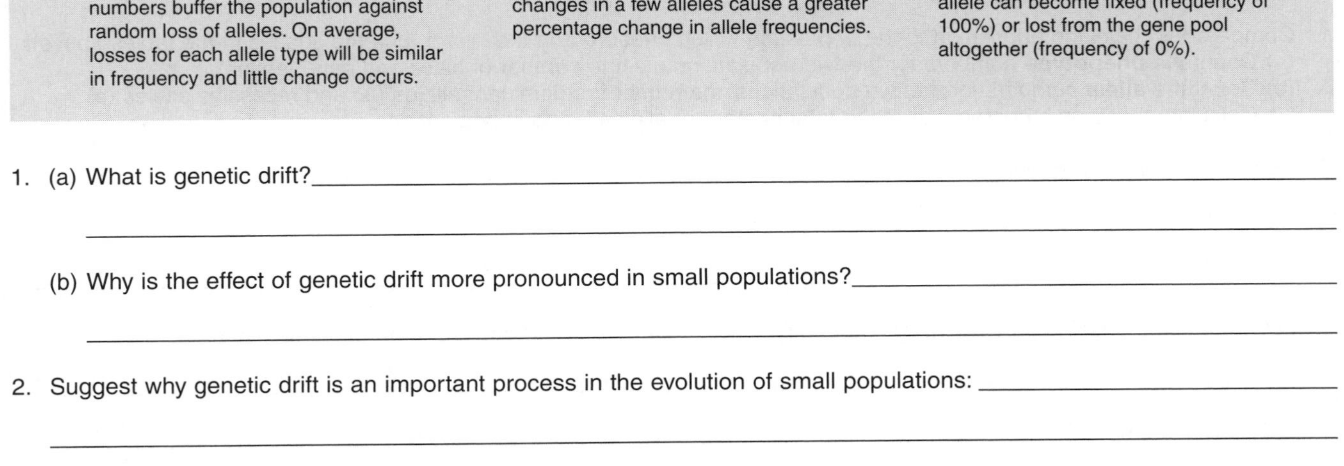

1. (a) What is genetic drift? ______________________________

   (b) Why is the effect of genetic drift more pronounced in small populations? ______________________________

2. Suggest why genetic drift is an important process in the evolution of small populations: ______________________________

ISBN: 978-1-927309-20-9

112 The Founder Effect

Key Idea: The founder effect can result in differences in allele frequencies between a parent and founder populations.

If a small number of individuals from a large population becomes isolated from their original parent population, their sample of alleles is unlikely to represent the allele proportions of the parent population. This phenomenon is called the **founder effect** and it can result in the colonising (founder) population evolving in a different direction to the parent population. This is particularly the case if the founder population is subjected to different selection pressures in a new environment and if the population is missing alleles that are present in the parent population.

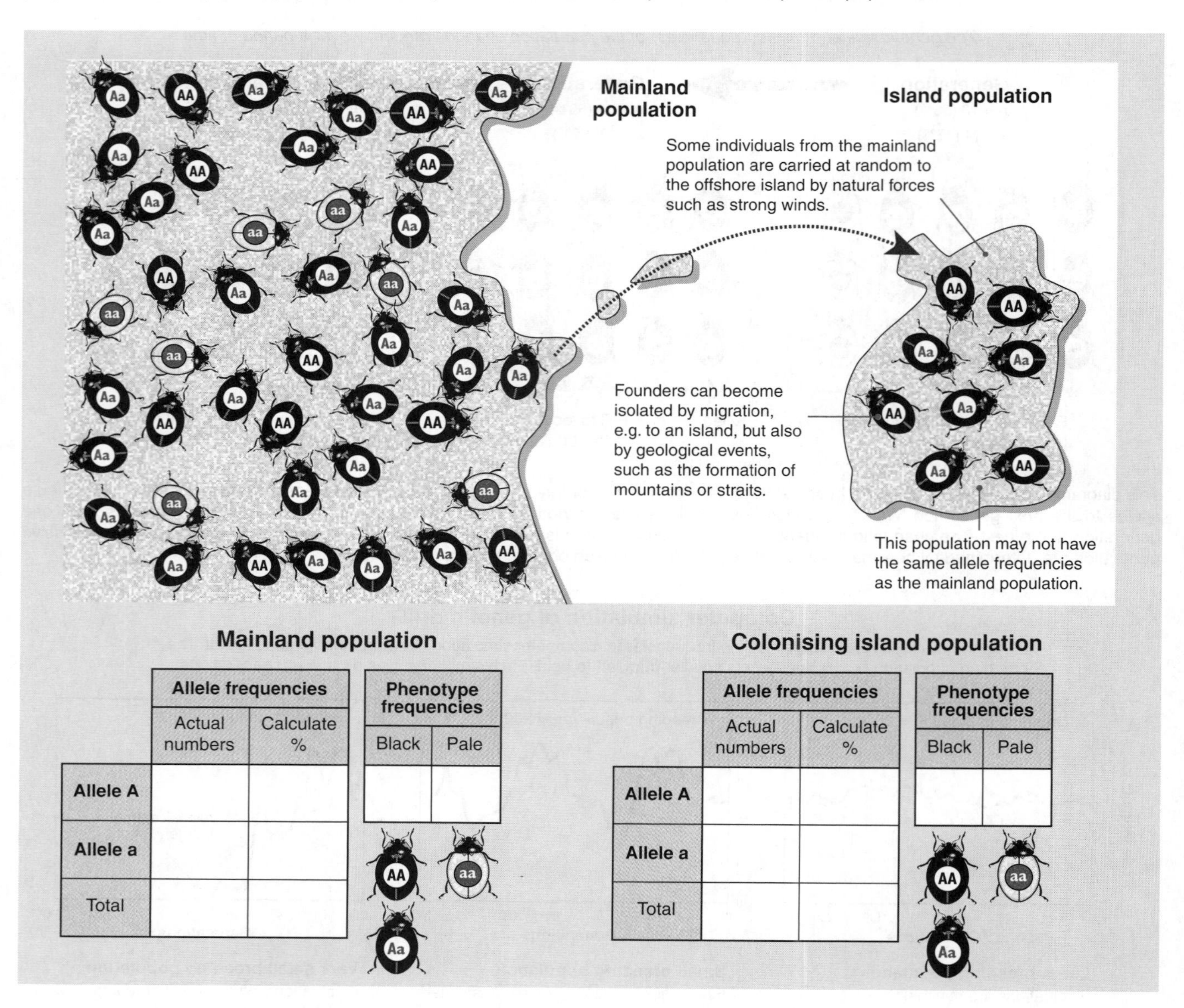

Mainland population

	Allele frequencies	
	Actual numbers	Calculate %
Allele A		
Allele a		
Total		

Phenotype frequencies	
Black	Pale

Colonising island population

	Allele frequencies	
	Actual numbers	Calculate %
Allele A		
Allele a		
Total		

Phenotype frequencies	
Black	Pale

1. Compare the mainland population to the population which ended up on the island (use the spaces in the tables above):
   (a) Count the **phenotype** numbers for the two populations (i.e. the number of black and pale beetles).
   (b) Count the **allele** numbers for the two populations: the number of dominant alleles (A) and recessive alleles (a). Calculate these as a percentage of the total number of alleles for each population.

2. How are the allele frequencies of the two populations different? ______

3. Describe some possible ways in which various types of organism can be **carried** to an offshore island:
   (a) Plants: ______
   (b) Land animals: ______
   (c) Non-marine birds: ______

ISBN:978-1-927309-20-9

Microgeographic isolation in garden snails

The European garden snail (*Cornu aspersum*, formerly *Helix aspersa*) is widely distributed throughout the world, both naturally and by human introduction. However because of its relatively slow locomotion and need for moist environments it can be limited in its habitat and this can lead to regional variation. The study below illustrates an investigation carried out on two snail populations in the city of Bryan, Texas. The snail populations covered two adjacent city blocks surrounded by tarmac roads.

The snails were found in several colonies in each block. Allele frequencies for the gene *MDH-1* (alleles A and a) were obtained and compared. Statistical analysis of the allele frequencies of the two populations showed them to be significantly different ($P << 0.05$). Note: A Mann-Whitney *U* test was used in this instance. It is similar to a Student's *t* test, but does not assume a normal distribution of data (it is non-parametric).

Block A

Block B

Road (not to sclae)

Source: Evolution, Vol 29, No. 3, 1975

Snail colony (circle size is proportional to colony size).

Building

	Colony	1	2	3	4	5	6	7	8	9	10	11	12	13	14	15
Block A	*MDH-1* A %	39	39	36	42	39	47	32	42	44	42	44	50	50	58	75
Block A	*MDH-1* a %															
Block B	*MDH-1* A %	81	61	75	68	70	61	70	60	58	61	54	54	47		
Block B	*MDH-1* a %															

4. Complete the table above by filling in the frequencies of the *MDH-1* a allele:

5. Suggest why these snail populations are effectively geographically isolated: ______________________

6. Both the *MDH-1* alleles produce fully operative enzymes. Suggest why the frequencies of the alleles have become significantly different.

7. Identify the colony in block A that appears to be isolated from the rest of the block itself: ______________________

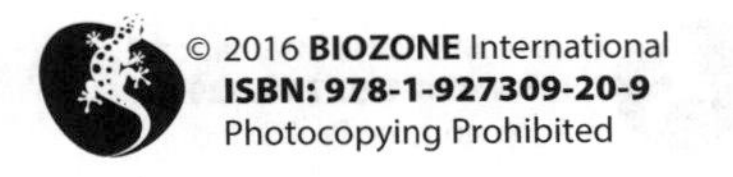

ISBN: 978-1-927309-20-9

113 Genetic Bottlenecks

Key Idea: Genetic bottlenecks occur when population numbers and diversity fall dramatically. Although a population's numbers may recover, its genetic diversity often does not.

Populations may sometimes be reduced to low numbers by predation, disease, or periods of climatic change. These large scale reductions are called genetic (or population) bottlenecks. The sudden population decline is not necessarily selective and it may affect all phenotypes equally. Large scale catastrophic events, such as fire or volcanic eruptions, are examples of such non-selective events. Affected populations may later recover, having squeezed through a 'bottleneck' of low numbers. The diagram below illustrates how population numbers may be reduced as a result of a catastrophic event. Following such an event, the gene pool of the surviving remnant population may be markedly different to that of the original gene pool. Genetic drift may cause further changes to allele frequencies. The small population may return to previous levels but with a reduced genetic diversity.

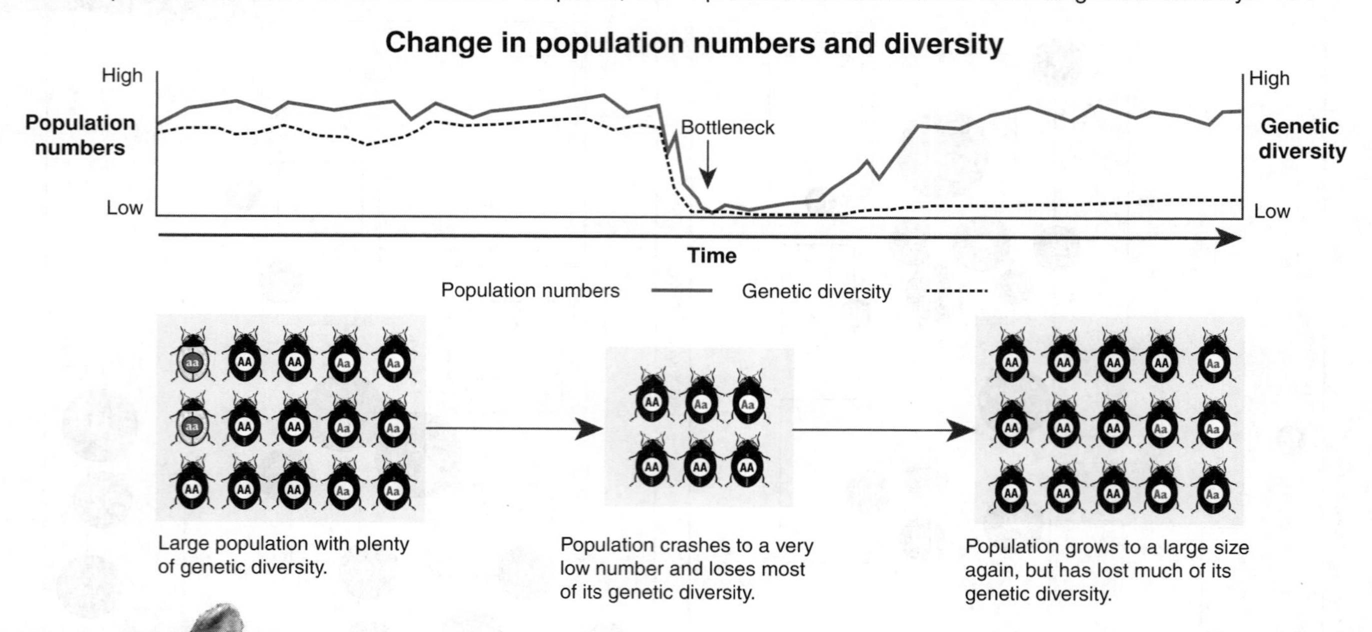

Large population with plenty of genetic diversity.

Population crashes to a very low number and loses most of its genetic diversity.

Population grows to a large size again, but has lost much of its genetic diversity.

Modern examples of genetic bottlenecks

Cheetahs: The world population of cheetahs currently stands at fewer than 20 000. Recent genetic analysis has found that the entire population exhibits very little genetic diversity. It appears that cheetahs may have narrowly escaped extinction at the end of the last ice age, about 10-20 000 years ago. If all modern cheetahs arose from a very limited genetic stock, this would explain their present lack of genetic diversity. The lack of genetic variation has resulted in a number of problems that threaten cheetah survival, including sperm abnormalities, decreased fecundity, high cub mortality, and sensitivity to disease.

Illinois prairie chicken: When Europeans first arrived in North America, there were millions of prairie chickens. As a result of hunting and habitat loss, the Illinois population of prairie chickens fell from about 100 million in 1900 to fewer than 50 in the 1990s. A comparison of the DNA from birds collected in the mid-twentieth century and DNA from the surviving population indicated that most of the genetic diversity has been lost.

Photo: Dept. of Natural Resources, Illinois

1. Endangered species are often subjected to genetic bottlenecks. Explain how genetic bottlenecks affect the ability of a population of an endangered species to recover from its plight:

2. Why has the lack of genetic diversity in cheetahs increased their sensitivity to disease?

3. Describe the effect of a genetic bottleneck on the potential of a species to adapt to changes (i.e. its ability to evolve):

ISBN:978-1-927309-20-9

114 Isolation and Species Formation

Key Idea: Ecological and geographical isolation are important in separating populations prior to reproductive isolation.

Isolating mechanisms are barriers to successful interbreeding between species. **Reproductive isolation** is fundamental to the biological species concept, which defines a species by its inability to breed with other species to produce fertile offspring. **Geographical barriers** are not regarded as reproductive isolating mechanisms because they are not part of the species' biology, although they are often a necessary precursor to reproductive isolation in sexually reproducing populations. Ecological isolating mechanisms are those that isolate gene pools on the basis of ecological preferences, e.g. habitat selection. Although ecological and geographical isolation are sometimes confused, they are quite distinct, as ecological isolation involves a component of the species biology.

Geographical isolation

Geographical isolation describes the isolation of a species population (gene pool) by some kind of physical barrier, for example, mountain range, water body, isthmus, desert, or ice sheet. Geographical isolation is a frequent first step in the subsequent reproductive isolation of a species. For example, geological changes to the lake basins has been instrumental in the subsequent proliferation of cichlid fish species in the rift lakes of East Africa (right). Similarly, many Galapagos Island species (e.g. iguanas, finches) are now quite distinct from the Central and South American species from which they arose after isolation from the mainland.

Ecological (habitat) isolation

Ecological isolation describes the existence of a **prezygotic reproductive barrier** between two species (or sub-species) as a result of them occupying or breeding in different habitats within the same general geographical area. Ecological isolation includes small scale differences (e.g. ground or tree dwelling) and broad differences (e.g. desert vs grasslands). The red-browed and brown **treecreepers** (*Climacteris* spp.) are sympatric in south-eastern Australia and both species feed largely on ants. However the brown spends most of its time foraging on the ground or on fallen logs while the red-browed forages almost entirely in the trees. Ecological isolation often follows geographical isolation, but in many cases the geographical barriers may remain in part. For example, five species of **antelope squirrels** occupy different habitat ranges throughout the southwestern United States and northern Mexico, a region divided in part by the Grand Canyon. The white tailed antelope squirrel is widely distributed in desert areas to the north and south of the canyon, while the smaller, more specialised Harris' antelope squirrel has a much more limited range only to the south in southern Arizona. The Grand Canyon still functions as a barrier to dispersal but the species are now ecologically isolated as well.

Geographical and ecological isolation of species

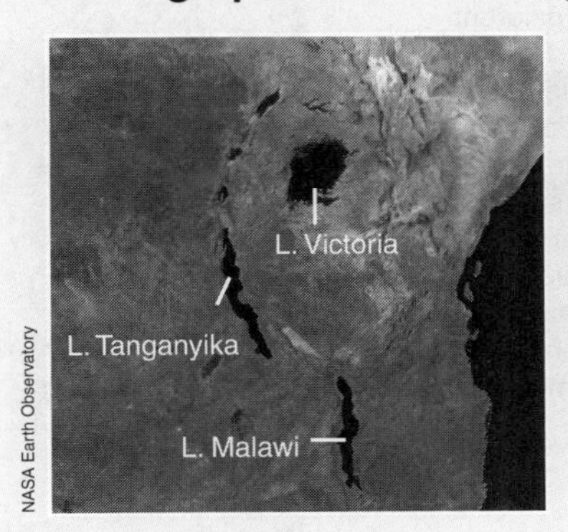

Red-browed treecreeper

Brown treecreeper

White-tailed antelope squirrel

The Grand Canyon - a massive rift in the Colorado Plateau

Harris' antelope squirrel

1. Describe the role of isolating mechanisms in maintaining the integrity of a species: ______

2. (a) Why is geographical isolation not regarded as a reproductive isolating mechanism? ______

   (b) Explain why, despite this, it often precedes reproductive isolation: ______

3. Distinguish between geographical and ecological isolation: ______

ISBN: 978-1-927309-20-9

115 Reproductive Isolation

Key Idea: Reproductive isolating mechanisms acting before and after fertilisation, prevent interbreeding between species. Reproductive isolation is a defining feature of biological species. Any mechanism that prevents two species from producing viable, fertile hybrids contributes to reproductive isolation. Single barriers to gene flow (such as geographical barriers) are usually insufficient to isolate a gene pool, so most species commonly have more than one type of barrier. Most reproductive isolating mechanisms (RIMs) are prezygotic and operate before fertilisation. Postzyotic RIMs, which act after fertilisation, are important in maintaining the integrity of closely related species.

Prezygotic isolating mechanisms

Temporal isolation

Individuals from different species do not mate because they are active during different times of the day or in different seasons. Plants flower at different times of the year or at different times of the day to avoid hybridisation (e.g. species of the orchid genus *Dendrobium* occupy the same location but flower on different days). Closely related animal species may have different breeding seasons or periods of emergence. Species of **periodical cicadas** (*Magicicada*) in a particular region are developmentally synchronised, despite very long life cycles. Once their underground period of development (13 or 17 years depending on the species) is over, the entire population emerges at much the same time to breed.

Temporal isolation: periodical cicadas

Cicada emergence

Gamete isolation

The gametes from different species are often incompatible, so even if they meet they do not survive. Where fertilisation is internal, the sperm may not survive in the reproductive tract of another species. If the sperm does survive and reach the ovum, chemical differences in the gametes prevent fertilisation. Gamete isolation is particularly important in aquatic environments where the gametes are released into the water and fertilised externally, such as in reproduction in frogs. Chemical recognition is also used by flowering plants to recognise pollen from the same species.

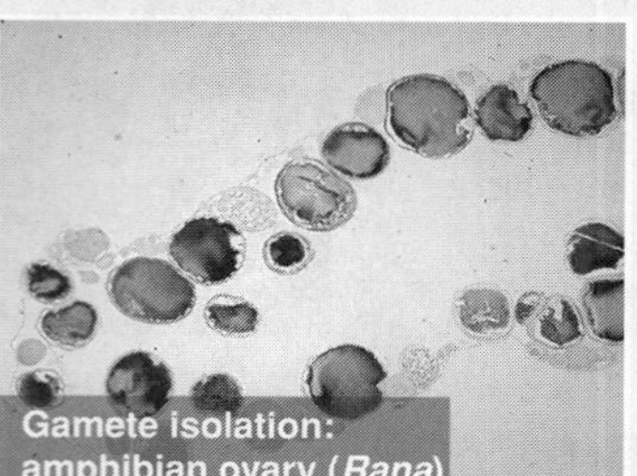
Gamete isolation: amphibian ovary (*Rana*)

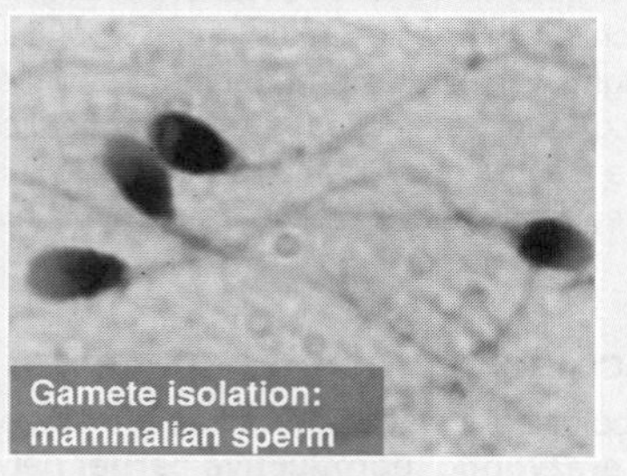
Gamete isolation: mammalian sperm

Behavioural isolation

Behavioural isolation operates through differences in species courtship behaviours. Courtship is a necessary prelude to mating in many species and courtship behaviours are species specific. Mates of the same species are attracted with distinctive dances, vocalisations, and body language. Courtship behaviours are not easily misinterpreted and will be unrecognised and ignored by individuals of another species. Birds exhibit a remarkable range of courtship displays. The use of song is widespread but ritualised movements, including nest building, are also common. For example, the elaborate courtship bowers of bowerbirds are well known, and Galápagos frigatebirds have an elaborate display in which they inflate a bright red gular pouch (right). Amongst insects, empid flies have some of the most elaborate of courtship displays. They are aggressive hunters so ritualised behaviour involving presentation of a prey item facilitates mating. The sexual organs of the flies are also like a lock-and-key, providing mechanical reproductive isolation as well (see below).

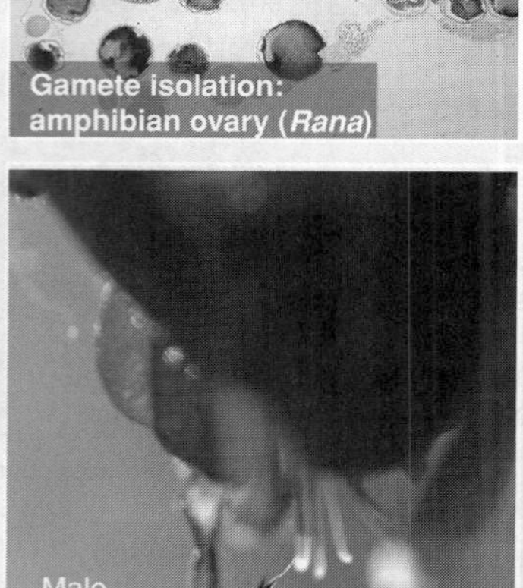

Behaviour and morphology: empid flies mating

Behaviour: male frigatebird display

Behaviour: male tree frog calling

Behaviour: wing beating in male sage grouse

Mechanical (morphological) isolation

Structural differences (incompatibility) in the anatomy of reproductive organs prevents sperm transfer between individuals of different species. This is an important isolating mechanism preventing breeding between closely related species of arthropods. Many flowering plants have coevolved with their animal pollinators and have flowers structures to allow only that insect access. Structural differences in the flowers and pollen of different plant species prevents cross breeding because pollen transfer is restricted to specific pollinators and the pollen itself must be species compatible.

Mechanical: Damselflies mating

Mechanical: flower shape in orchids

KNOW LINK 114 LINK 116 LINK 117 LINK 119

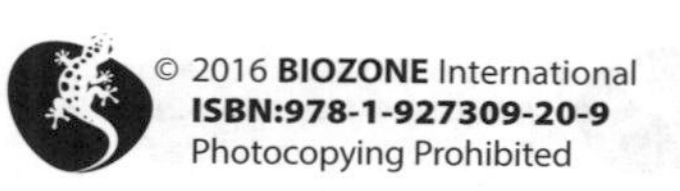

ISBN:978-1-927309-20-9

Postzygotic isolating mechanisms

Hybrid sterility

Even if two species mate and produce hybrid offspring that are vigorous, the species are still reproductively isolated if the hybrids are sterile (genes cannot flow from one species' gene pool to the other). Such cases are common among the horse family (such as the zebra and donkey shown on the right). One cause of this sterility is the failure of meiosis to produce normal gametes in the hybrid. This can occur if the chromosomes of the two parents are different in number or structure (see the "**zebronkey**" karyotype on the right). The **mule**, a cross between a donkey stallion and a horse mare, is also an example of **hybrid vigour** (they are robust) as well as **hybrid sterility**. Female mules sometimes produce viable eggs but males are infertile.

Hybrid inviability

Mating between individuals of two species may produce a zygote, but genetic incompatibility may stop development of the zygote. Fertilised eggs often fail to divide because of mis-matched chromosome numbers from each gamete. Very occasionally, the hybrid zygote will complete embryonic development but will not survive for long. For example, although sheep and goats seem similar and can be mated together, they belong to different genera. Any offspring of a sheep-goat pairing is generally stillborn.

Hybrid breakdown

Hybrid breakdown is common feature of some plant hybrids. The first generation (F_1) may be fertile, but the second generation (F_2) are infertile or inviable. Examples include hybrids between cotton species (near right), species within the genus *Populus*, and strains of the cultivated rice *Oryza* (far right)

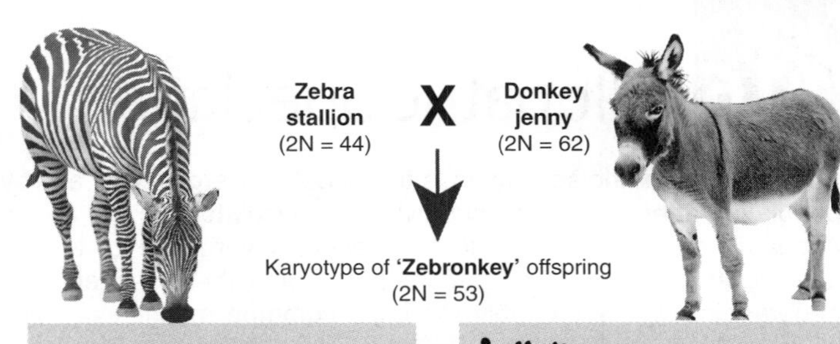

Sheep (*Ovis*) 54 chromosomes

Goat (*Capra*) 60 chromosomes

1. In the following examples, classify the reproductive isolating mechanism as either **prezygotic** or **postzygotic** and describe the mechanisms by which the isolation is achieved (e.g. structrual isolation, hybrid sterility etc.):

   (a) Some different cotton species can produce fertile hybrids, but breakdown of the hybrid occurs in the next generation when the offspring of the hybrid die in their seeds or grow into defective plants:

   Prezygotic / postzygotic (delete one) Mechanism of isolation: ____________

   (b) Many plants have unique arrangements of their floral parts that stops transfer of pollen between plants:

   Prezygotic / postzygotic (delete one) Mechanism of isolation: ____________

   (c) Two skunk species do not mate despite having habitats that overlap because they mate at different times of the year:

   Prezygotic / postzygotic (delete one) Mechanism of isolation: ____________

   (d) Several species of the frog genus *Rana*, live in the same regions and habitats, where they may occasionally hybridise. The hybrids generally do not complete development, and those that do are weak and do not survive long:

   Prezygotic / postzygotic (delete one) Mechanism of isolation: ____________

2. Postzygotic isolating mechanisms are said to reinforce prezygotic ones. Explain why this is the case:

ISBN: 978-1-927309-20-9

116 Allopatric Speciation

Key Idea: Allopatric speciation is the genetic divergence of a population after it becomes subdivided and isolated.

Allopatric speciation refers to the genetic divergence of a species after a population becomes split and then isolated geographically. It is probably the most common mechanism by which new species arise and has certainly been important in regions where there have been cycles of geographical fragmentation, e.g. as a result of ice expansion and retreat (and accompanying sea level changes) during glacial and interglacial periods.

Stage 1: Moving into new environments

There are times when the range of a species expands for a variety of different reasons. A single population in a relatively homogeneous environment will move into new regions of their environment when they are subjected to intense competition (whether it is interspecific or intraspecific). The most severe form of competition is between members of the same species since they are competing for identical resources in the habitat. In the diagram on the right there is a 'parent population' of a single species with a common gene pool with regular 'gene flow' (theoretically any individual has access to all members of the opposite sex for mating purposes).

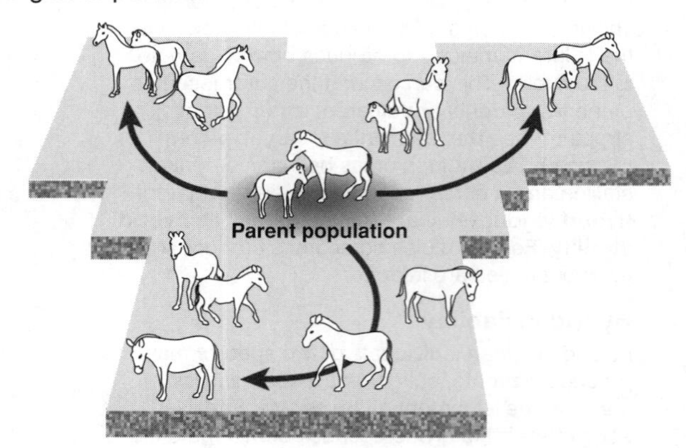

Stage 2: Geographical isolation

Isolation of parts of the population may occur due to the formation of **physical barriers**, such as mountains, deserts, or stretches of water. These barriers may cut off those parts of the population that are at the extremes of the range and gene flow is prevented or rare. The rise and fall of the sea level has been particularly important in functioning as an isolating mechanism. Climatic change can leave 'islands' of habitat separated by large inhospitable zones that the species cannot traverse.

Example: In mountainous regions, alpine species can populate extensive areas of habitat during cool climatic periods. During warmer periods, they may become isolated because their habitat is reduced to 'islands' of high ground surrounded by inhospitable lowland habitat.

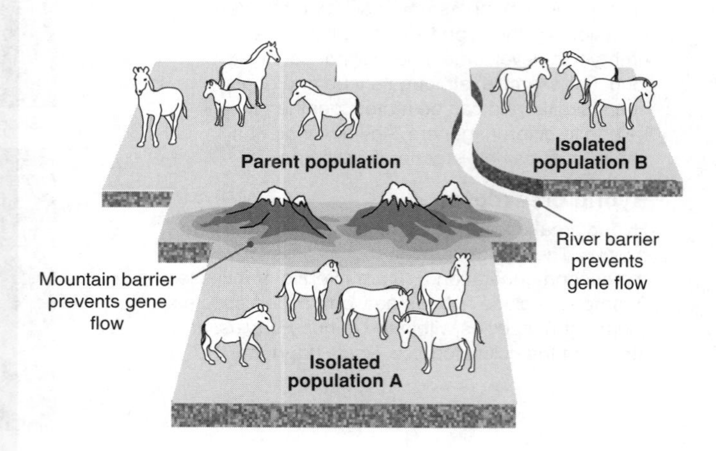

Stage 3: Different selection pressures

The isolated populations (A and B) may be subjected to quite different selection pressures. These will favour individuals with traits that suit each particular environment. For example, population A will be subjected to selection pressures that relate to drier conditions. This will favour those individuals with phenotypes (and therefore genotypes) that are better suited to dry conditions. They may for instance have a better ability to conserve water. This would result in improved health, allowing better disease resistance and greater reproductive performance (i.e. more of their offspring survive). Finally, as allele frequencies for certain genes change, the population takes on the status of a subspecies. Reproductive isolation is not yet established but the **subspecies** are significantly different genetically from other related populations.

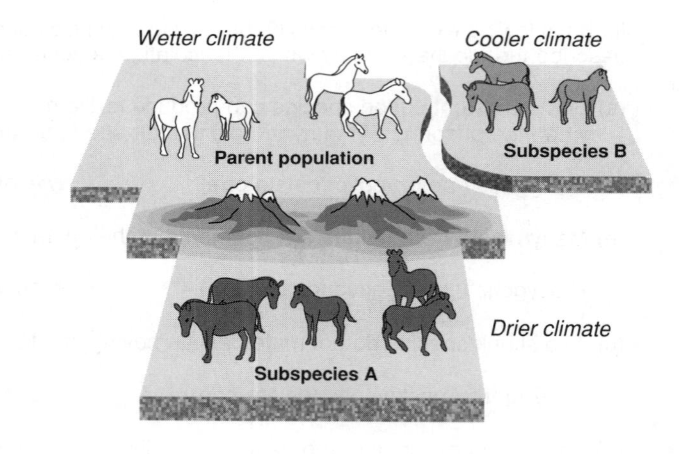

Stage 4: Reproductive isolation

The separated populations (isolated subspecies) undergo genetic and behavioural changes. These ensure that the gene pool of each population remains isolated and 'undiluted' by genes from other populations, even if the two populations should be able to remix (due to the removal of the geographical barrier). Gene flow does not occur. The arrows (diagram, right) indicate the zone of overlap between two species after Species B has moved back into the range inhabited by the parent population. Closely-related species whose distribution overlaps are said to be **sympatric species**. Those that remain geographically isolated are called **allopatric species**.

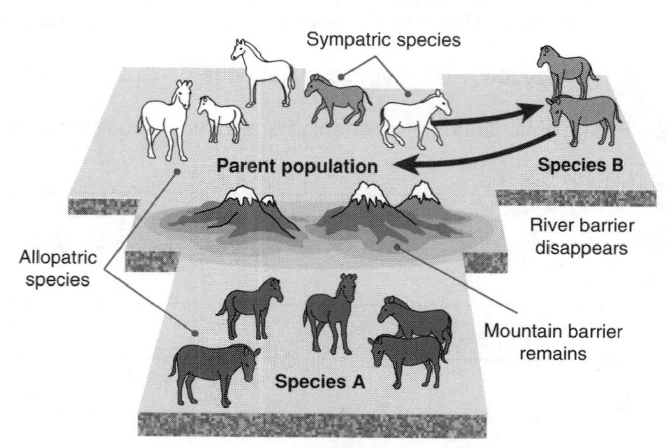

ISBN:978-1-927309-20-9

1. Why do some animals, given the opportunity, move into new environments? ______

2. Plants are unable to move. How might plants disperse to new environments? ______

3. Describe the amount of **gene flow** within a parent population prior to and during the expansion of a species' range:

4. Explain how cycles of climate change can cause large changes in **sea level** (up to 200 m):

5. (a) What kinds of **physical barriers** could isolate different parts of the same population? ______

   (b) How might emigration achieve the same effect as geographical isolation? ______

6. (a) How might **selection pressures** differ for a population that becomes isolated from the parent population?

   (b) Describe the general effect of the change in selection pressures on the **allele frequencies** of the isolated gene pool:

7. Explain how reproductive isolation could develop in geographically separated populations (see previous pages):

8. What is the difference between an allopatric and sympatric species? ______

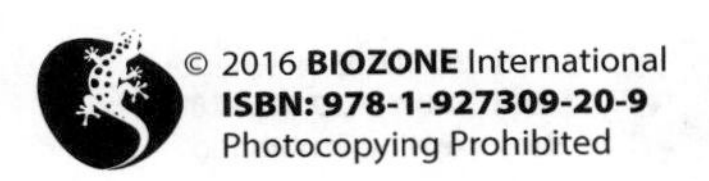

ISBN: 978-1-927309-20-9

117 Stages in Species Formation

Key Idea: Speciation may occur in stages marked by increasing isolation of diverging gene pools. Physical separation is followed by increasing reproductive isolation.

The diagram below shows a possible sequence of events in the origin of two new species from an ancestral population. Over time, the genetic differences between two populations increase and the populations become increasingly isolated from each other. The isolation of the two gene pools may begin with a geographical barrier. This may be followed by progressively greater reduction in gene flow between the populations until the two gene pools are isolated and they each attain species status.

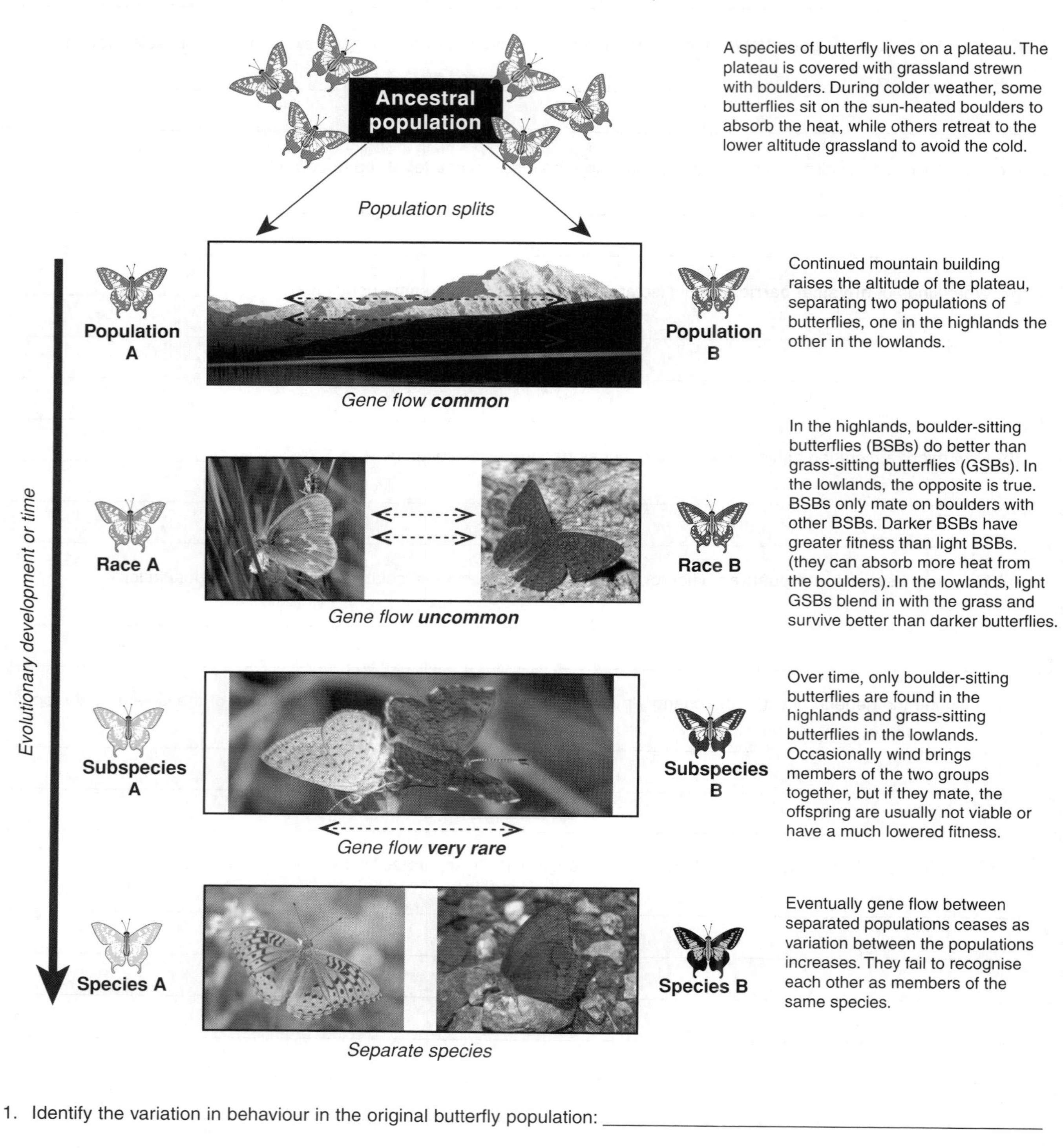

A species of butterfly lives on a plateau. The plateau is covered with grassland strewn with boulders. During colder weather, some butterflies sit on the sun-heated boulders to absorb the heat, while others retreat to the lower altitude grassland to avoid the cold.

Continued mountain building raises the altitude of the plateau, separating two populations of butterflies, one in the highlands the other in the lowlands.

In the highlands, boulder-sitting butterflies (BSBs) do better than grass-sitting butterflies (GSBs). In the lowlands, the opposite is true. BSBs only mate on boulders with other BSBs. Darker BSBs have greater fitness than light BSBs. (they can absorb more heat from the boulders). In the lowlands, light GSBs blend in with the grass and survive better than darker butterflies.

Over time, only boulder-sitting butterflies are found in the highlands and grass-sitting butterflies in the lowlands. Occasionally wind brings members of the two groups together, but if they mate, the offspring are usually not viable or have a much lowered fitness.

Eventually gene flow between separated populations ceases as variation between the populations increases. They fail to recognise each other as members of the same species.

1. Identify the variation in behaviour in the original butterfly population: ______________________

____

2. What were the selection pressures acting on BSBs in the highlands and GSBs in the lowlands respectively?

____

____

____

ISBN:978-1-927309-20-9

118 Small Flies and Giant Buttercups

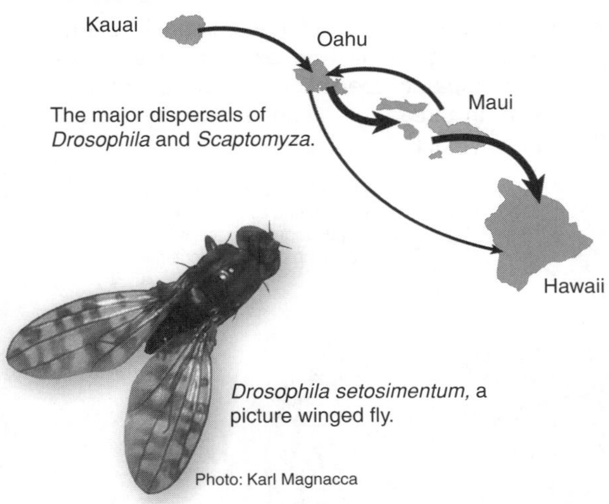

Drosophilidae (commonly known as fruit flies) are a group of small flies found almost everywhere in the world. Two genera, *Drosophila* and *Scaptomyza* are found in the Hawaiian islands and between them there are more than 800 species present on a land area of just 16 500 km^2; it is one of the densest concentrations of related species found anywhere. The flies range from 1.5 mm to 20 mm in length and display a startling range of wing forms and patterns, body shapes and colours, and head and leg shapes. This diverse array of species and characteristics has made these flies the subject of much evolutionary and genetics research. Genetic analyses show that they are all related to a single species that may have arrived on the islands around 8 million years ago and diversified to exploit a range of unoccupied niches. Older species appear on the older islands and more recent species appear as one moves from the oldest to the newest islands. Such evidence points to numerous colonisation events as new islands emerged from the sea. The volcanic nature of the islands means that newly isolated environments are a frequent occurrence. For example, forested areas may become divided by lava flows, so that flies in one region diverge rapidly from flies in another just tens of metres away. One such species is *D. silvestris.* Males have a series of hairs on their forelegs, which they brush against females during courtship. Males in the northeastern part of the island have many more of these hairs than the males on the southwestern side of the island. While still the same species, the two demes are already displaying structural and behavioural isolation. Behavioural isolation is clearly an important phenomenon in drosophilid speciation. A second species, *D. heteroneura*, is closely related to *D. silvestris* and the two species live sympatrically. Although hybrid offspring are fully viable, hybridisation rarely occurs because male courtship displays are very different.

New Zealand alpine buttercups (*Ranunculus*) are some of the largest in the world and are also the product of repeated speciation events. There are 14 species of *Ranunculus* in New Zealand; more than in the whole of North and South America combined. They occupy five distinct habitats ranging from snowfields and scree slopes to bogs. Genetic studies have shown that this diversity is the result of numerous isolation events following the growth and recession of glaciers. As the glaciers retreat, alpine habitat becomes restricted and populations are isolated at the tops of mountains. This restricts gene flow and provides the environment for species divergence. When the glaciers expand again, the extent of the alpine habitat increases, allowing isolated populations to come in contact and closely related species to hybridise.

1. Explain why so many drosophilidae are present in Hawaii: ______

2. Explain why these flies are of interest: ______

3. Describe the relationship between the age of the islands and the age of the fly species: ______

4. Explain why New Zealand has so many alpine buttercups: ______

ISBN: 978-1-927309-20-9

119 Sympatric Speciation

Key Idea: Sympatric speciation is speciation occurring in the absence of physical barriers between gene pools.

Sympatric speciation refers to the formation of new species within the same place (sympatry). Sympatric speciation is rarer than allopatric speciation because it is difficult to prevent gene flow. However, it is not uncommon in plants that form **polyploids** (organisms with extra complete sets of chromosomes). Sympatric speciation can occur through niche differentiation in areas of sympatry, or by instant speciation through polyploidy.

Speciation through niche differentiation

Niche isolation

In a heterogeneous environment (one that is not the same everywhere), a population exists within a diverse collection of **microhabitats**. Some organisms prefer to occupy one particular type of 'microhabitat' most of the time, only rarely coming in contact with fellow organisms that prefer other microhabitats. Some organisms become so dependent on the resources offered by their particular microhabitat that they never meet up with their counterparts in different microhabitats.

Reproductive isolation

Finally, the individual groups have remained genetically isolated for so long because of their microhabitat preferences, that they have become reproductively isolated. They have become new species that have developed subtle differences in behaviour, structure, or physiology. Gene flow (via sexual reproduction) is limited to organisms that share a similar microhabitat preference (as shown in the diagram on the right).

Example: Some beetles prefer to find plants identical to the species they grew up on, when it is time for them to lay eggs. Individual beetles of the same species have different preferences.

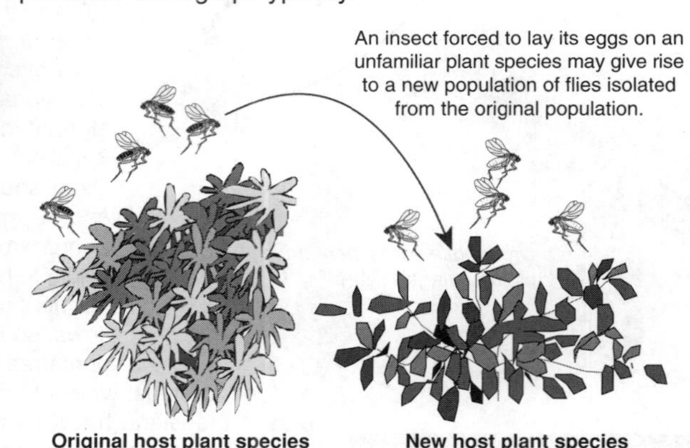

Original host plant species

New host plant species

Gene flow

No gene flow

Instant speciation by polyploidy

Polyploidy (duplication of chromosome sets) may result in the formation of a new species without physical isolation from the parent species. Polyploidy produces sudden reproductive isolation for the new group. Polyploids in animals are rarely viable. Many plants, on the other hand, are able to reproduce vegetatively, or carry out self pollination. This ability to reproduce on their own enables such polyploid plants to produce a breeding population.

Polyploidy in a hybrid between two different species can often make the hybrid fertile. This occurred in modern wheat. Swedes are also a polyploid species formed from a hybrid between a type of cabbage and a type of turnip.

Origin of **polyploid event**

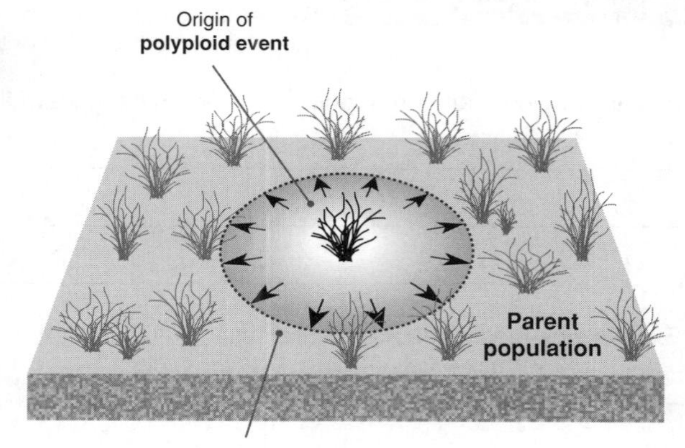

New polyploid plant species spreads outwards through the existing parent population

1. Explain what is meant by **sympatric speciation** (do not confuse this with sympatric species):

2. Explain how **polyploidy** can result in the formation of a new species:

3. Identify an example of a species that has been formed by polyploidy:

4. Explain how **niche differentiation** can result in the formation of a new species:

ISBN:978-1-927309-20-9

120 Components of an Ecosystem

Key Idea: An ecosystem consists of all the organisms living in a particular area and their physical environment.

An **ecosystem** is a community of living organisms and the physical (non-living) components of their environment. The community (living component of the ecosystem) is in turn made up of a number of **populations**, these being organisms of the same species living in the same geographical area. The structure and function of an ecosystem is determined by the physical (abiotic) and the living (biotic) factors, which determine species distribution and survival.

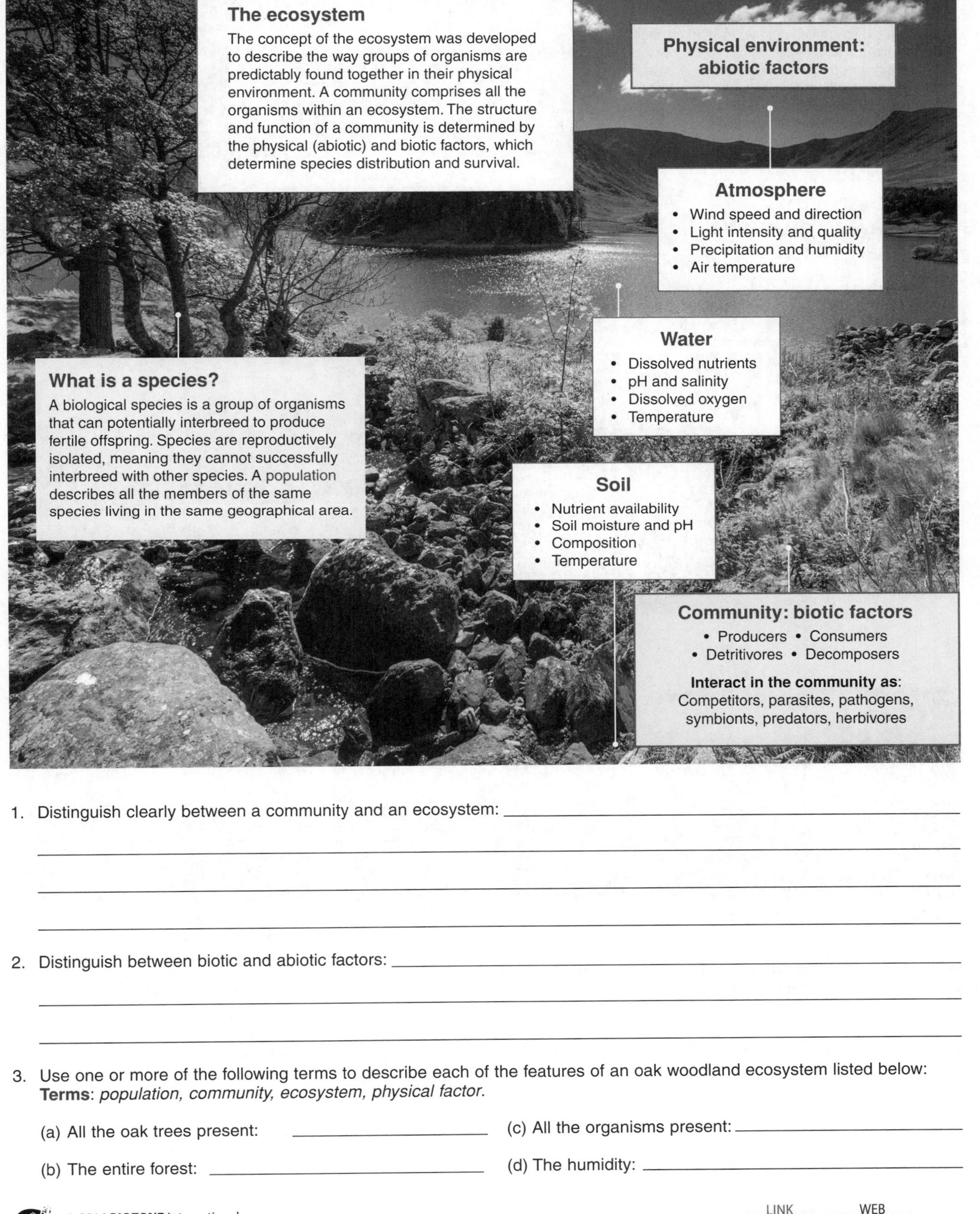

1. Distinguish clearly between a community and an ecosystem: ______________________

____

____

____

2. Distinguish between biotic and abiotic factors: ______________________

____

____

3. Use one or more of the following terms to describe each of the features of an oak woodland ecosystem listed below:
Terms: *population, community, ecosystem, physical factor.*

(a) All the oak trees present: ______________

(b) The entire forest: ______________

(c) All the organisms present: ______________

(d) The humidity: ______________

ISBN: 978-1-927309-20-9

121 Types of Ecosystems

Key Idea: Ecosystems have no fixed boundaries and so can vary in size.

Ecosystems can be of any size. The only limit is the size determined by the human observer. For example, a tree can be thought of as an ecosystem, if we ignore the individual comings and goings of animals and look at the system as a whole. But the tree may be part of a larger ecosystem, a forest, which again is part of a larger biome, and so on until we encompass the entire biosphere, that narrow belt around the Earth containing all the Earth's living organisms.

Ecosystems can be on vastly different scales. Yosemite National Park in northern California covers 3000 km^2. Large parts of it are covered in mixed coniferous forests. The forest ecosystem comprises various tree species (e.g. Douglas fir, giant sequoia, and black oak). There are over 250 species of vertebrates including deer, bear, mountain lion, and a variety of bird life.

The ecosystem of a tree can be quite varied. The tree provides energy and materials for insects and other invertebrates that live on or in it. Bacteria and fungi decompose leaves and dead material on the tree or in the soil. The tree provides roosts for birds and fruit or seeds as a food source.

Within the forested areas there are clearings that consist of grasses and scrub with the occasional isolated tree. These areas provide good grazing for deer and open hunting areas for owls.

Tidal rock pools are micro-ecosystems. Each one will be slightly different to the next, with different species assemblages and abiotic factors. The ocean in the background is an ecosystem on a vastly larger scale.

The border of a garden or back yard can be used to define an ecosystem. Gardens can provide quite different ecosystems, ranging from tropical to dry depending on the type of plants and watering system.

Animals can be ecosystems in the same way as trees. All animals carry populations of microbes in their gut or on their bodies. Invertebrates, such as lice, may live in the fur and spend their entire life cycle there.

1. Describe the borders that would define each of the three Yosemite ecosystems described above:

(a) ______

(b) ______

(c) ______

ISBN:978-1-927309-20-9

122 Physical Factors and Gradients

Key Idea: Physical factors influence the habitat and distribution of organisms.

Gradients in abiotic factors influence habitats and create microclimates, and so are important in determining patterns of distribution in communities. This activity examines the physical gradients and microclimates that typically might occur in three very different environments: a rocky shore, a stratified forest, and a small lake. The gradients in physical factors create a more heterogeneous environment that may allow species with different physical tolerances to co exist.

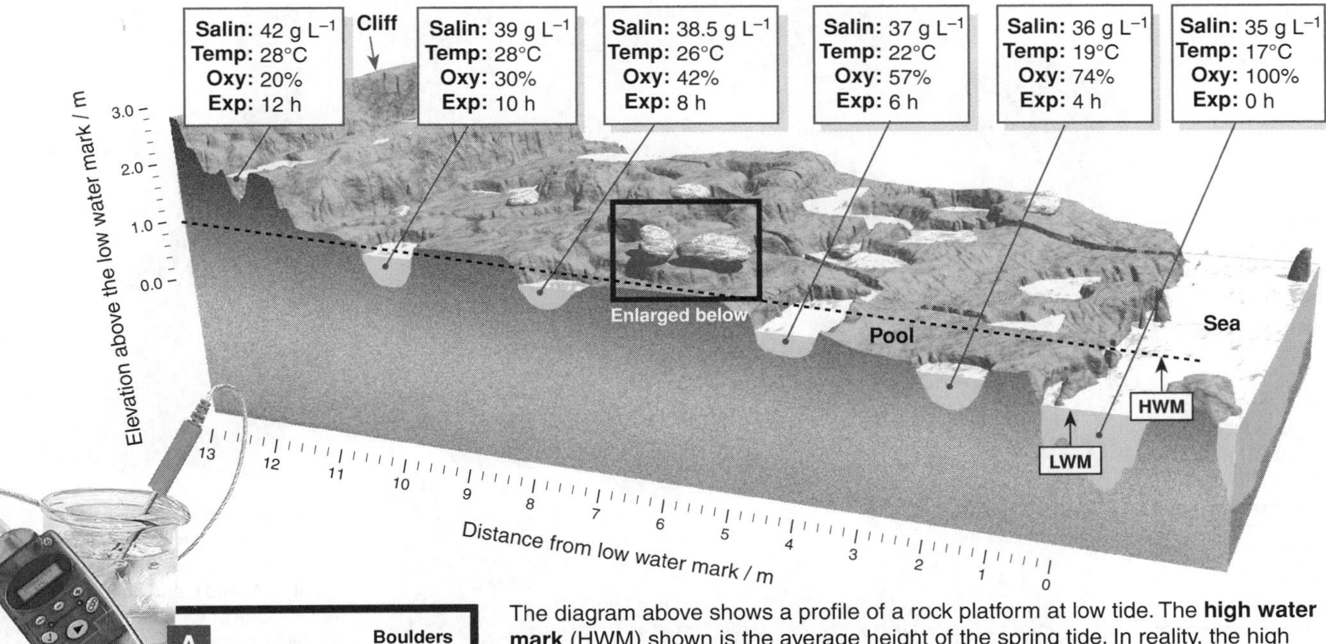

The diagram above shows a profile of a rock platform at low tide. The **high water mark** (HWM) shown is the average height of the spring tide. In reality, the high tide level varies with the phases of the moon. The low water mark (LWM) is an average level subject to the same variations. The rock pools vary in size, depth, and position on the platform. They are isolated at different elevations, trapping water from the ocean for time periods that may be relatively brief or up to 10-12 hours duration. Pools near the HWM are exposed for longer periods of time than those near the LWM. The difference in exposure times results in some of the physical factors exhibiting a gradient, i.e. the factor's value gradually changes over a horizontal and/or vertical distance. Physical factors sampled in the pools include salinity, or the amount of dissolved salts (g) per litre (Salin), temperature (Temp), dissolved oxygen compared to that of open ocean water (Oxy), and exposure, or the amount of time isolated from the ocean water (Exp).

1. Describe the environmental gradient (general trend) from the low water mark (LWM) to the high water mark (HWM) for:

   (a) Salinity: ______________________ (c) Dissolved oxygen: ______________________

   (b) Temperature: ______________________ (d) Exposure: ______________________

2. (a) The inset diagram (above, right) is an enlarged view of two boulders on the rock platform. Describe how the physical factors listed below might differ at each of the labelled points A, B, and C:

   Mechanical force of wave action: ______________________

   ____

   Surface temperature when exposed: ______________________

   ____

   (b) State the term given to these localised variations in physical conditions: ______________________

3. Rock pools above the normal high water mark (HWM), such as the uppermost pool in the diagram above, can have wide extremes of salinity. Explain the conditions under which these pools might have either:

   (a) Very low salinity: ______________________

   (b) Very high salinity: ______________________

ISBN: 978-1-927309-20-9

We have seen how an environmental gradient can occur with horizontal distance along a shore, but environmental gradients can also occur as a result of vertical distance from the ground. In a forest, the light quantity and quality, humidity, wind speed, and temperature change gradually from the canopy to the forest floor. These changes give rise to a layered or stratified community in which different species occupy different vertical positions in the forest according to their particular tolerances.

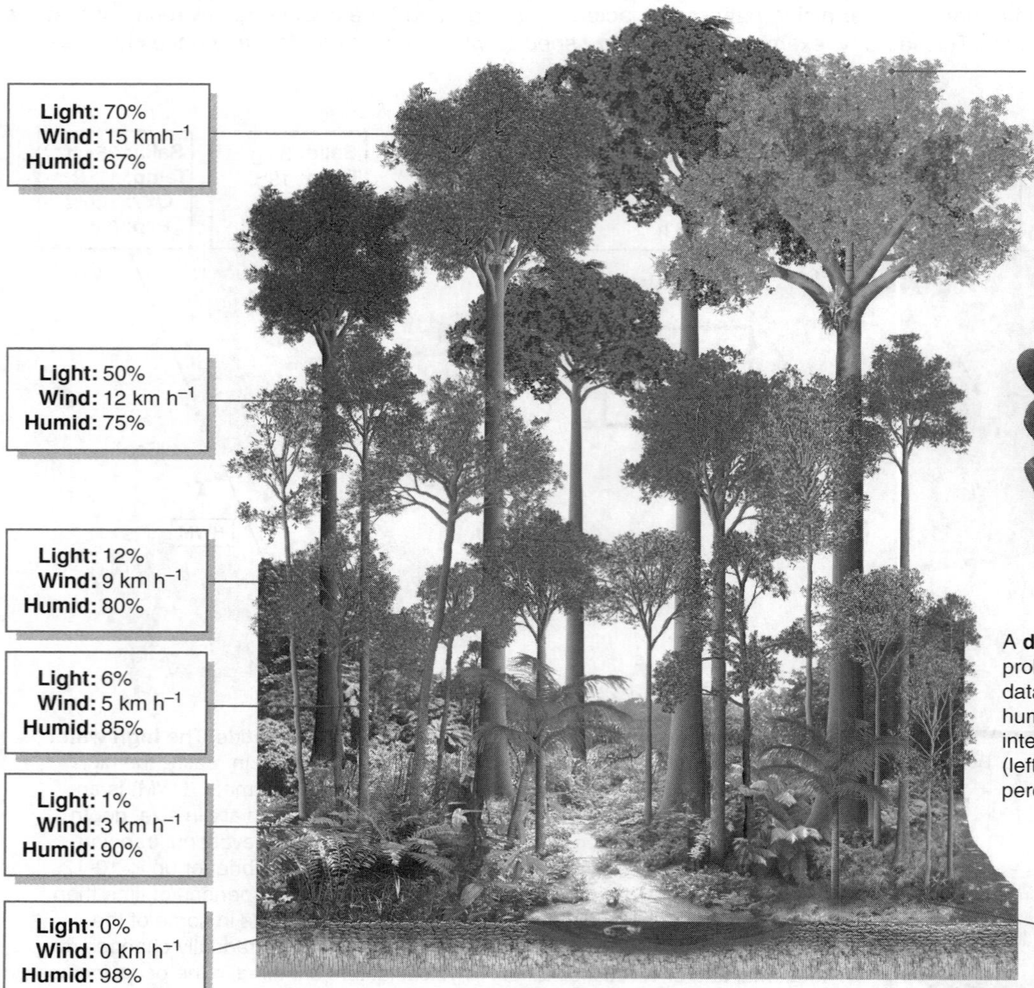

Canopy

Light: 70%
Wind: 15 kmh^{-1}
Humid: 67%

Light: 50%
Wind: 12 km h^{-1}
Humid: 75%

Light: 12%
Wind: 9 km h^{-1}
Humid: 80%

Light: 6%
Wind: 5 km h^{-1}
Humid: 85%

Light: 1%
Wind: 3 km h^{-1}
Humid: 90%

Light: 0%
Wind: 0 km h^{-1}
Humid: 98%

Leaf litter

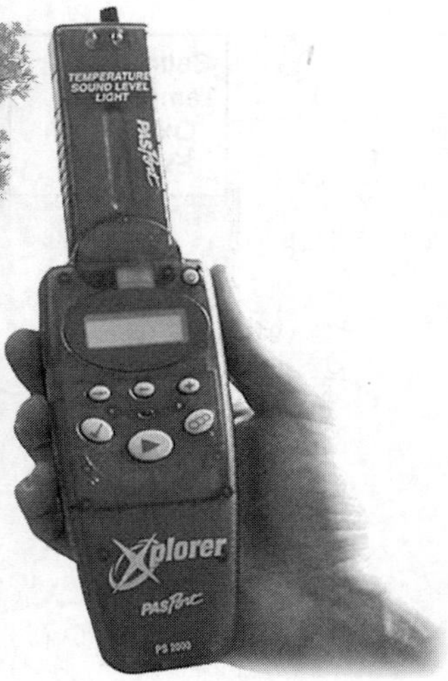

A **datalogger** fitted with suitable probes was used to gather data on wind speed (**Wind**), humidity (**Humid**), and light intensity (**Light**) for each layer (left). Light intensity is given as a percentage of full sunlight.

Tropical rainforests are complex communities with a vertical structure which divides the vegetation into layers. This pattern of vertical layering is called **stratification.**

4. With respect to the diagram above, describe the general trend from canopy to leaf litter for:

   (a) Light intensity: ____________ (b) Wind speed: ____________ (c) Humidity: ____________

5. Explain why each of these factors changes as the distance from the canopy increases:

   (a) Light intensity: ____________

   (b) Wind speed: ____________

   (c) Humidity: ____________

6. Apart from light intensity, describe what other feature of the light will change with distance from the canopy:

   ____

7. Plants growing on the forest floor have some advantages and disadvantages with respect to the physical factors.

   (a) Describe one advantage: ____________

   (b) Describe one disadvantage: ____________

ISBN:978-1-927309-20-9

Oxbow lakes are formed from old river meanders that have been cut off and isolated when a river changes course. They are shallow (about 2-9 m deep) but often deep enough to develop temporary, but relatively stable, temperature gradients from top to bottom (below). Oxbows are commonly very productive and this can influence values for abiotic factors such as dissolved oxygen and light penetration, which can vary widely both with depth and proximity to the shore. Typical values for water temperature (Temp), dissolved oxygen (Oxygen), and light penetration as a percentage of the light striking the surface (Light) are indicated below.

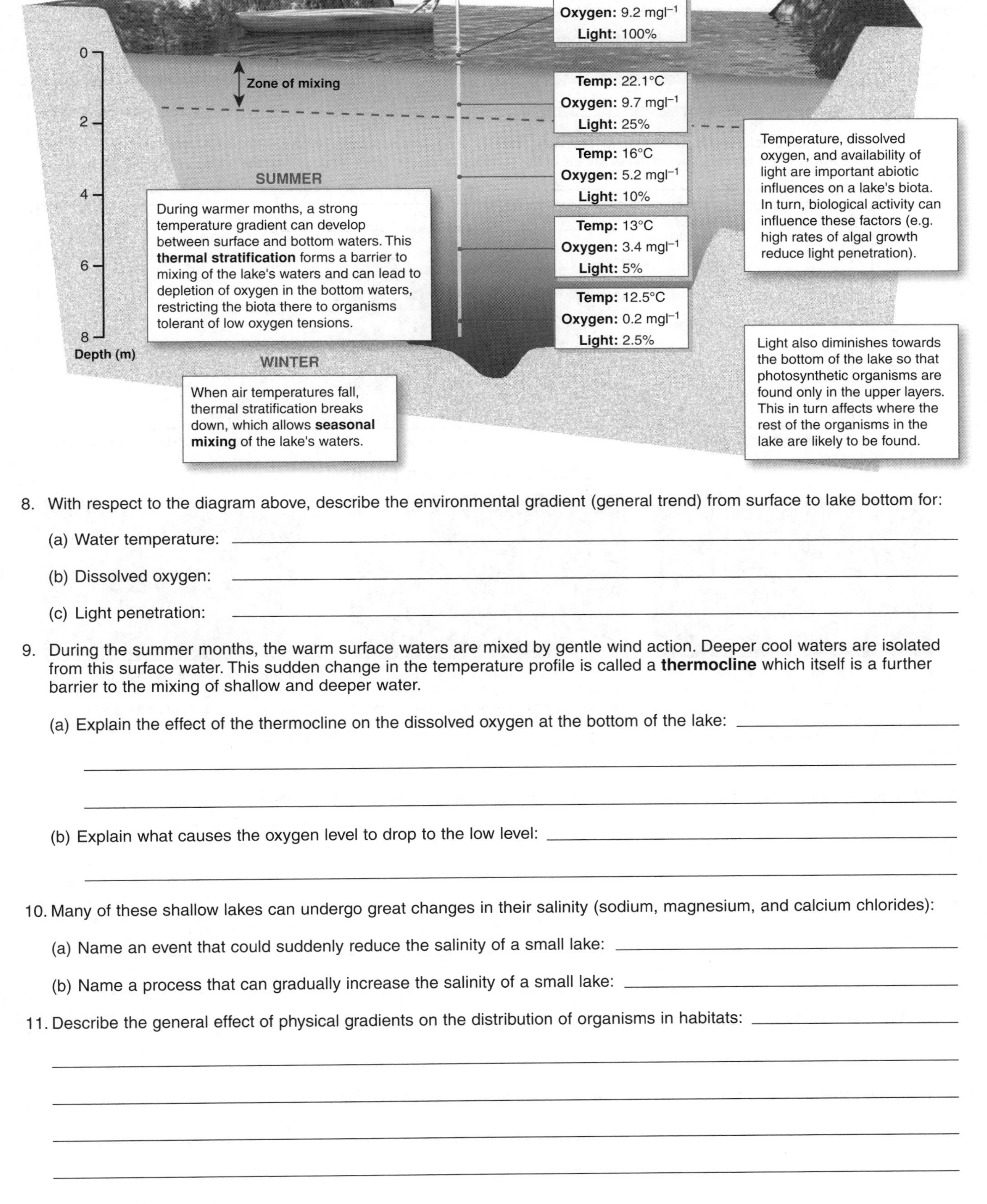

8. With respect to the diagram above, describe the environmental gradient (general trend) from surface to lake bottom for:

   (a) Water temperature: ______

   (b) Dissolved oxygen: ______

   (c) Light penetration: ______

9. During the summer months, the warm surface waters are mixed by gentle wind action. Deeper cool waters are isolated from this surface water. This sudden change in the temperature profile is called a **thermocline** which itself is a further barrier to the mixing of shallow and deeper water.

   (a) Explain the effect of the thermocline on the dissolved oxygen at the bottom of the lake: ______

   (b) Explain what causes the oxygen level to drop to the low level: ______

10. Many of these shallow lakes can undergo great changes in their salinity (sodium, magnesium, and calcium chlorides):

    (a) Name an event that could suddenly reduce the salinity of a small lake: ______

    (b) Name a process that can gradually increase the salinity of a small lake: ______

11. Describe the general effect of physical gradients on the distribution of organisms in habitats: ______

ISBN: 978-1-927309-20-9

123 Habitat

Key Idea: The environment in which an organism lives is its habitat. The habitat may not be homogeneous in its quality.

The environment in which an organism (or species) lives (including all the physical and biotic factors) is its **habitat**. Within any habitat, each species has a range of tolerance to variations in its environment. Within the population, individuals will have slightly different tolerance ranges based on small differences in genetic make-up, age, and health. The wider an organism's tolerance range for any one factor (e.g. temperature) the more likely it is that the organism will survive variations in that factor. For the same reasons, species with a wider tolerance range are likely to be more widely distributed. Organisms have a narrower **optimum range** within which they function best. This may vary seasonally or during development. Organisms are usually most abundant where the abiotic factors are closest to the optimum range.

Habitat occupation and tolerance range

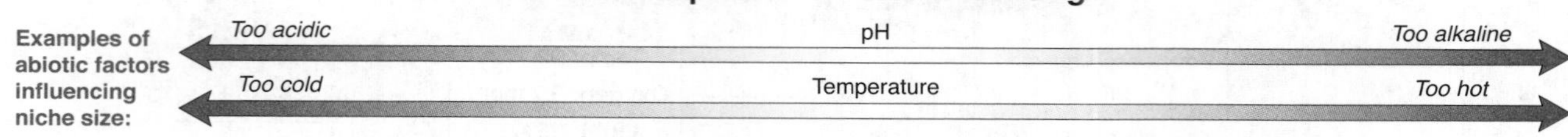

The law of tolerances states that *for each abiotic factor, a species population (or organism) has a tolerance range within which it can survive. Toward the extremes of this range, that abiotic factor tends to limit the organism's ability to survive.*

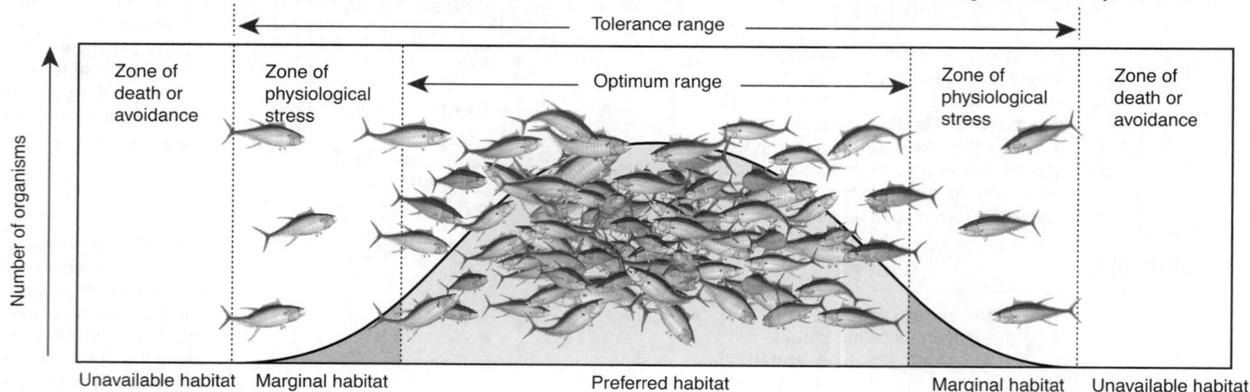

The scale of available habitats

A habitat may be vast and relatively homogeneous for the most part, as is the open ocean. Barracuda (above) occur around reefs and in the open ocean where they are aggressive predators.

For non-motile organisms, such as the fungus pictured above, a suitable habitat may be defined by the particular environment in a relatively small area, such as on this decaying log.

For microbial organisms, such as the bacteria and protozoans of the ruminant gut, the habitat is defined by the chemical environment within the rumen (R) of the host animal, in this case, a cow.

1. What is the relationship between an organism's tolerance range and the habitat it occupies? ______________________________

2. (a) In the diagram above, in which range is most of the population found? Explain your answer: ______________________________

(b) What are the greatest constraints on an organism's growth and reproduction within this range? ______________________________

3. Describe some probable stresses on an organism forced into a marginal habitat: ______________________________

KNOW LINK 124 LINK 129

ISBN:978-1-927309-20-9

124 Ecological Niche

Key Idea: An organism's niche describes its functional role within its environment.

The **ecological niche** describes the functional position of a species in its ecosystem. It includes how the species responds to the distribution of resources and how it alters those resources for other species. The full range of environmental conditions under which an organism can exist describes its fundamental niche. As a result of interactions with other organisms, species usually occupy a realised niche that is narrower than this. Central to the niche concept is the idea that two species with exactly the same niche cannot coexist, because they would compete for the same resources and one would exclude the other. This is known as **Gause's competitive exclusion principle**. More often, species compete for some of the same resources. Competition will be intense where their resource use curves overlap.

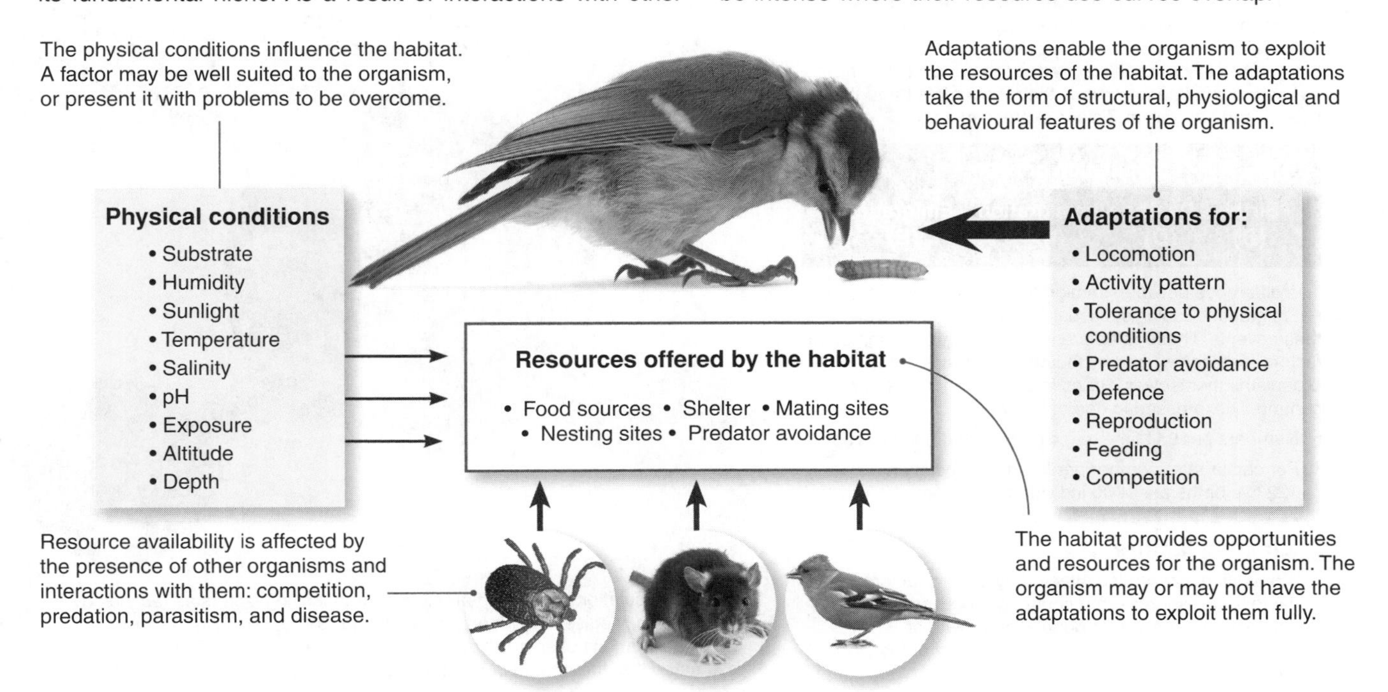

The realised niche

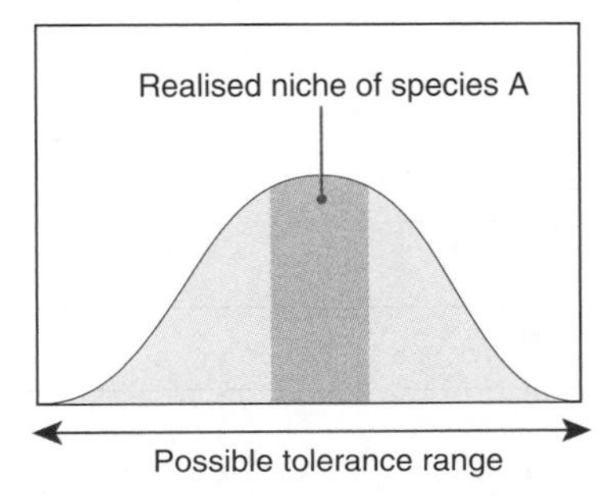

The tolerance range represents the **fundamental niche** a species could exploit. The actual or **realised niche** of a species is narrower than this because of competition with other species.

Intraspecific competition

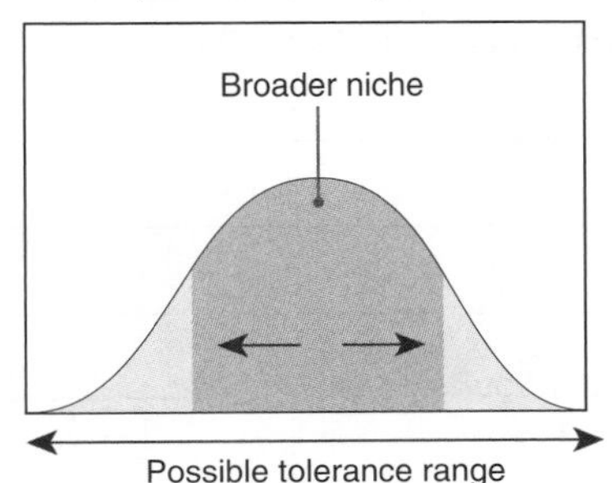

Competition is intense between individuals of the same species because they exploit the same resources. Individuals must exploit resources at the extremes of their tolerance range and the realised niche expands.

Interspecific competition

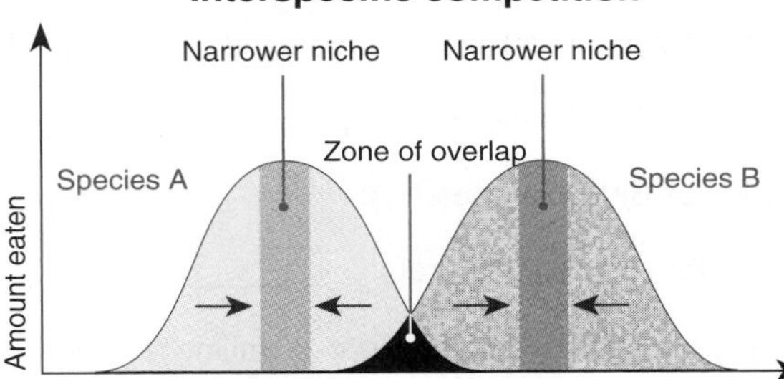

If two (or more) species compete for some of the same resources, their resource use curves will overlap. Within the zone of overlap, resource competition will be intense and selection will favour niche specialisation so that one or both species occupy a narrower niche.

1. (a) In what way could the realised niche be regarded as flexible? ___

(b) What factors might further constrain the extent of the realised niche? ___

2. Contrast the effects of interspecific competition and intraspecific competition on niche breadth: ___

ISBN: 978-1-927309-20-9

LINK 134 KNOW

125 Factors Determining Population Growth

Key Idea: Population size increases through births or immigration and decreases through deaths and emigration.

Populations are dynamic and the number of individuals in a population may fluctuate considerably over time. Populations gain individuals through births or immigration, and lose individuals through deaths and emigration. For a population in equilibrium, these factors balance out and there is no net change in the population abundance. When losses exceed gains, the population declines. When gains exceed losses, the population increases.

Births, deaths, immigrations (movements into the population) and emigrations (movements out of the population) are events that determine the numbers of individuals in a population. Population growth depends on the number of individuals added to the population from births and immigration, minus the number lost through deaths and emigration. This is expressed as:

Population growth =

Births – Deaths + Immigration – Emigration
(B) (D) (I) (E)

The difference between immigration and emigration gives net migration. Ecologists usually measure the **rate** of these events. These rates are influenced by environmental factors (see below) and by the characteristics of the organisms themselves. Rates in population studies are commonly expressed in one of two ways:

- Numbers per unit time, e.g. 20,150 live births per year.
- Per capita rate (number per head of population), e.g. 122 live births per 1000 individuals per year (12.2%).

Limiting factors

Population size is also affected by limiting factors; factors or resources that control a process such as organism growth, or population growth or distribution. Examples include availability of food, predation pressure, or available habitat.

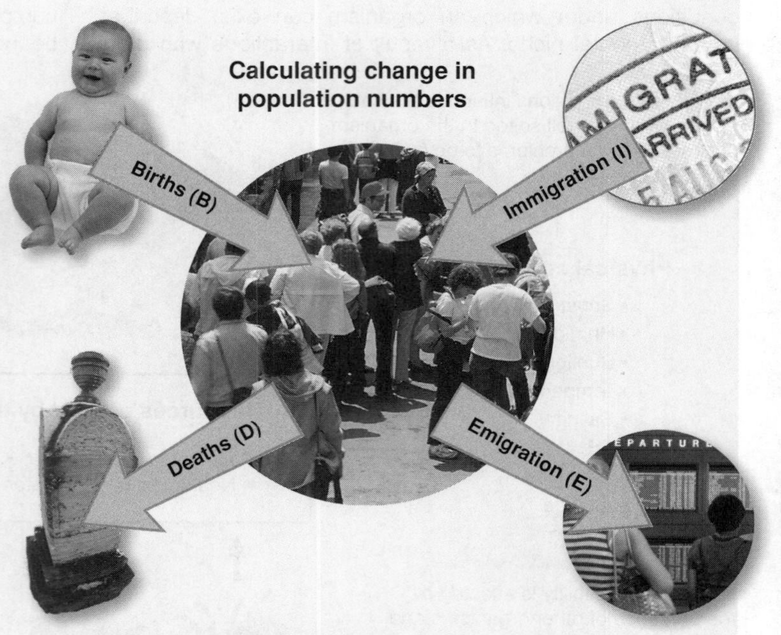

Human populations often appear exempt from limiting factors as technology and efficiency solve many food and shelter problems. However, as the last arable land is used and agriculture reaches its limits of efficiency, it is estimated that the human population may peak at around 10 billion by 2050.

1. Define the following terms used to describe changes in population numbers:

   (a) Death rate (mortality): ______________________

   (b) Birth rate (natality): ______________________

   (c) Net migration rate: ______________________

2. Explain how the concept of limiting factors applies to population biology: ______________________

   ____

3. Using the terms, B, D, I, and E (above), construct equations to express the following (the first is completed for you):

   (a) A population in equilibrium: B + I = D + E

   (b) A declining population: ______________________

   (c) An increasing population: ______________________

4. A population started with a total number of 100 individuals. Over the following year, population data were collected. Calculate birth rates, death rates, net migration rate, and rate of population change for the data below (as percentages):

   (a) Births = 14: Birth rate = ____________ (b) Net migration = +2: Net migration rate = ____________

   (c) Deaths = 20: Death rate = ____________ (d) Rate of population change = ____________

   (e) State whether the population is increasing or declining: ______________________

5. The human population is just over 7 billion. Describe and explain two limiting factors for population growth in humans:

   ____

   ____

DATA LINK 126 LINK 129

ISBN:978-1-927309-20-9

126 Population Growth Curves

Key Idea: Populations typically show either exponential or logistic growth. The maximum sustainable population size is limited by the environment's carrying capacity.

Population growth is the change in a population's numbers over time (dN/dt). It is regulated by the **carrying capacity** (K), which is the maximum number the environment can sustain. Population growth falls into two main types: exponential or logistic (sigmoidal). Both can be defined mathematically (below). In these mathematical models, the per capita rate of increase is denoted by a lower case *r*, which is given by the births minus deaths, divided by N (B-D/N) **Exponential growth** occurs when resources are essentially unlimited. **Logistic growth** begins exponentially, but slows as the population approaches the carrying capacity.

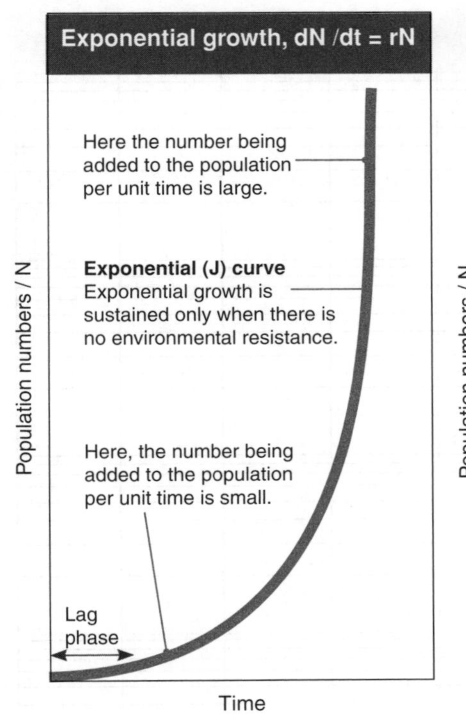

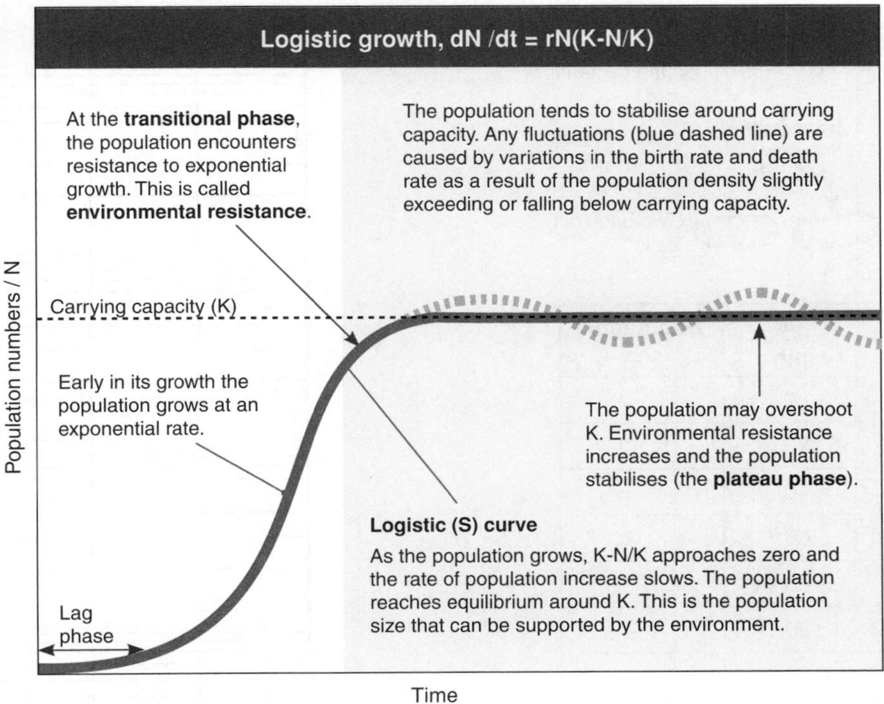

1. Why don't populations continue to increase exponentially in an environment? ________________________

2. What is meant by environmental resistance? ________________________

3. (a) What is meant by **carrying capacity**? ________________________

   (b) Explain the importance of carrying capacity to the growth and maintenance of population numbers: ________________________

4. (a) Describe and explain the phases of the logistic growth curve: ________________________

   (b) Explain why a population might overshoot carrying capacity before stabilising around carrying capacity:

5. Use a spreadsheet to demonstrate how the equations for exponential and logistic growth produce their respective curves. Watch the weblink animations provided or use the spreadsheet on the Teacher's Digital Edition if you need help.

ISBN: 978-1-927309-20-9

127 Plotting Bacterial Growth

Key Idea: Microbial growth can be plotted on a graph and used to predict microbial cell numbers at a set time.

Bacteria normally reproduce by a **binary fission**, a simple cell division that is preceded by cell elongation and involves one cell dividing in two. The time required for a cell to divide is the **generation time** and it varies between species and with environmental conditions such as temperature. When actively growing bacteria are inoculated into a liquid growth medium and the population is counted at intervals, a line can be plotted to show the growth of the cell population over time. You can simulate this for a hypothetical bacterial population with a generation time of 20 minutes.

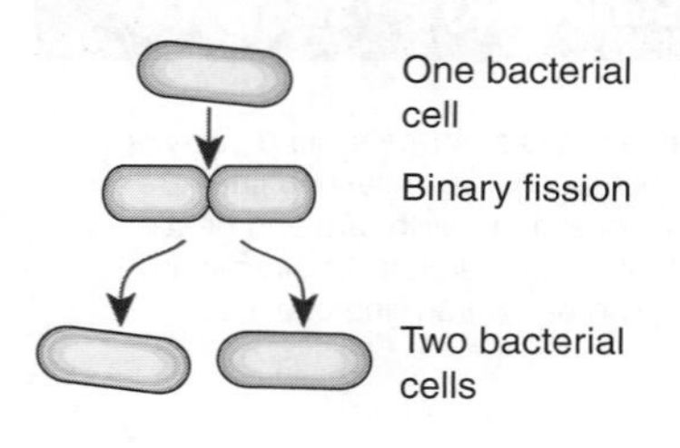

Time / min	Population size
0	1
20	2
40	4
60	8
80	
100	
120	
140	
160	
180	
200	
220	
240	
260	
280	
300	
320	
340	
360	

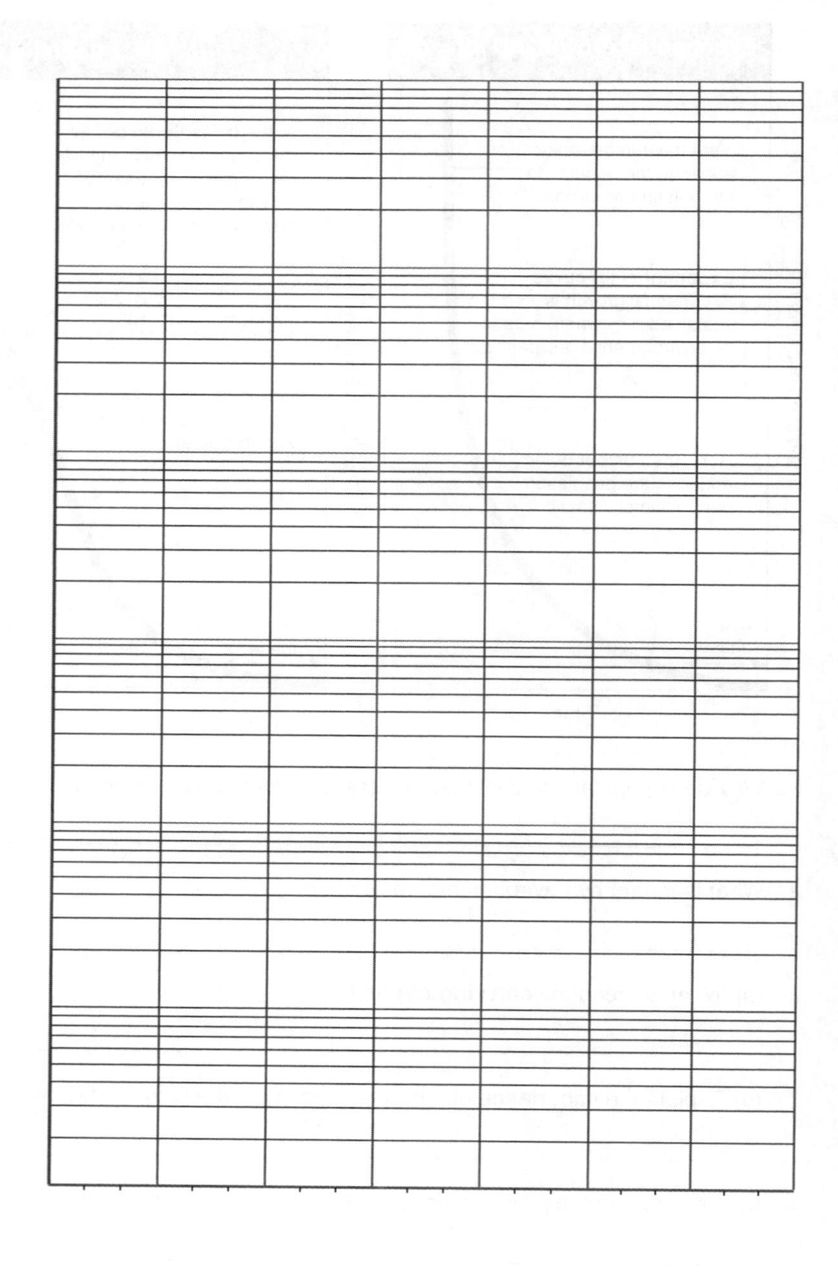

1. Complete the table above by doubling the number of bacteria for every 20 minute interval.
2. State how many bacteria were present after: 1 hour: ____________ 3 hours: ____________ 6 hours: ____________
3. Graph the results on the grid above. Make sure that you choose suitable scales and labels for each axis.
4. (a) Predict the number of cells present after 380 minutes: ____________

   (b) Plot this value on the graph above.
5. Why is a semi-log graph used to plot microbial growth? ____________

ISBN:978-1-927309-20-9

128 Investigating Bacterial Growth

Key Idea: Bacterial growth can be measured over time using a spectrophotometer.

Bacteria divide and increase their cell numbers by a process called binary fission. The rate of cell division varies between species; some divide rapidly (every 20 minutes) while others can take days to divide. The increase in cell numbers can be measured in the laboratory with a spectrophotometer as an increase in culture turbidity. The experiment below describes how students investigated bacterial growth in *E.coli* on two different growth media.

The aim

To investigate the growth rate of *E.coli* in two different liquid cultures, a minimal growth medium and a nutrient enriched complex growth medium.

The method

Using aseptic technique, the students added 0.2 mL of a pre-prepared *E.coli* culture to two test tubes. One test tube contained 5.0 mL of a minimal growth medium and the second contained 5.0 mL of a complex medium. Both samples were immediately mixed, and 0.2 mL samples removed from each and added to a cuvette. The absorbance of the sample was measured using a spectrophotometer at 660 nm. This was the 'time zero' reading. The test tubes were covered with parafilm, and placed in a 37°C water bath. Every 30 minutes, the test tubes were lightly shaken and 0.2 mL samples were taken from each so the absorbance could be measured. The results are presented in the table (right).

A spectrophotometer (left) is an instrument used to measure transmittance of a solution and so can be used to quantify bacterial growth where an increase in cell numbers results in an increase in turbidity.

In this experiment, students measured the absorbance of the solution. Absorbance measures the amount of light absorbed by the sample. Often, transmission (the amount of light that passes through a sample) is used to measure cell growth.

All bacteria should be treated as pathogenic and strict hygiene practices should be followed. These include wearing gloves, using aseptic techniques, not consuming food or drink in the laboratory, washing all surfaces with disinfectant afterwards, and hand washing. These precautions prevent the accidental introduction of the bacteria into the environment, and prevent accidental infection.

Results

Incubation time / min	Absorbance at 660 nm Minimal medium	Complex medium
0	0.021	0.014
30	0.022	0.015
60	0.025	0.019
90	0.034	0.033
120	0.051	0.065
150	0.078	0.124
180	0.118	0.238
210	0.179	0.460
240	0.273	0.698
270	0.420	0.910
300	0.598	1.070

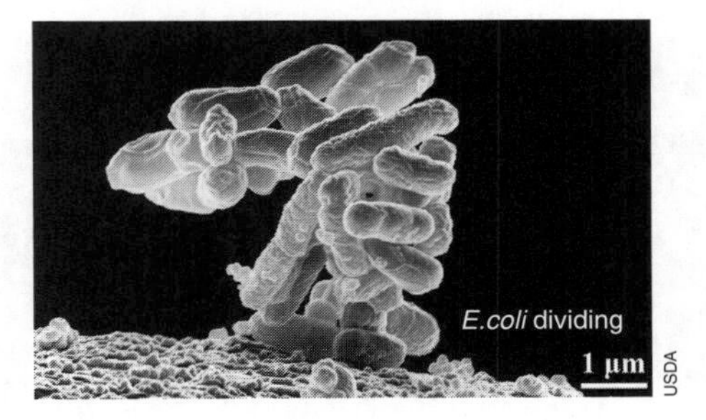

E.coli dividing

USDA

1. Why is it important to follow strict hygiene precautions when working with bacteria?

2. (a) On the grid (right) plot the results for *E.coli* growth on the two media:

   (b) What is the absorbance measuring?

   (c) Describe the effect of the complex medium on *E.coli* growth:

3. Another group of students wanted to calculate the number of cells in each sample. Explain how this could be achieved:

ISBN: 978-1-927309-20-9

129 Population Size and Carrying Capacity

Key Idea: Carrying capacity is the maximum number of organisms a particular environment can support.

An ecosystem's carrying capacity, i.e. the maximum number of individuals of a given species that the resources can sustain indefinitely, is limited by the ecosystem's resources. Factors affecting carrying capacity of an ecosystem can be biotic (e.g. food supply) or abiotic (e.g. water, climate, and available space).The carrying capacity of an ecosystem is determined by the most limiting factor and can change over time (e.g. as a result of seasonal changes). Below carrying capacity, population size increases because resources are not limiting. As the population approaches carrying capacity (or exceeds it) resources become limiting and environmental resistance increases, decreasing population growth.

Limiting factors

The effect of limiting factors and the type of factor that is limiting may change over time. The graph, right, shows how the carrying capacity of a forest-dwelling species changes based on changes to the limiting factors:

1. A population moves into the forest and rapidly increases in numbers due to abundant resources.
2. The population overshoots the carrying capacity.
3. The environment is damaged due to large numbers and food becomes more limited, lowering the original carrying capacity.
4. The population becomes stable at the new carrying capacity.
5. The forest experiences a drought and the carrying capacity is reduced as a result.
6. The drought breaks and the carrying capacity rises but is less than before because of habitat damage during the drought.

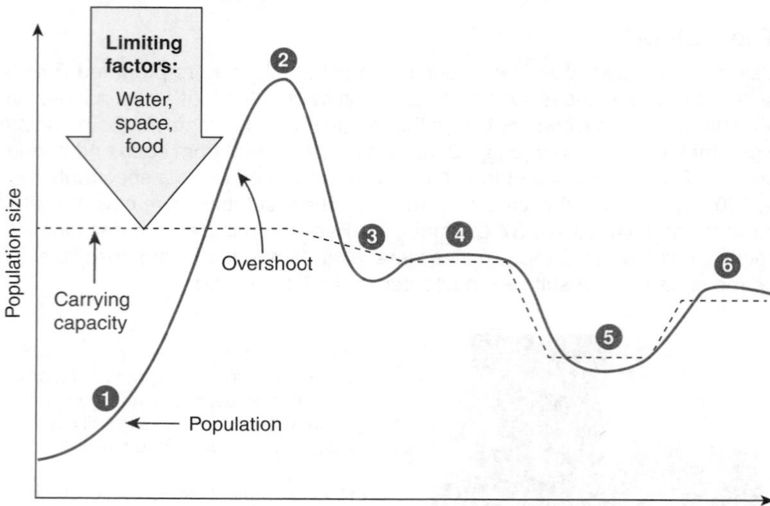

Factors affecting population size

Density dependent factors
The effect of these on population size is influenced by population density.

They include:

- Competition
- Predation
- Disease

Density dependent factors tend to be biotic and are less important when population density is low.

They regulate population size by decreasing birth rates and increasing death rates.

Density independent factors
The effect of these on population size does not depend on population density.

They include catastrophic events such as:

- Volcanic eruptions, fire
- Drought, flood, tsunamis
- Earthquakes

Density independent factors tend to be abiotic.

They regulate population size by increasing death rates.

1. What is carrying capacity? ______________________________

2. How does carrying capacity limit population numbers? ______________________________

3. What limiting factors have changed at points 3, 5, and 6 in the graph above, and how have they changed?

(a) 3: ______________________________

(b) 5: ______________________________

(c) 6: ______________________________

ISBN:978-1-927309-20-9

130 A Case Study in Carrying Capacity

Key Idea: Environmental factors influence predator-prey interactions so the outcomes are not always predictable. Environmental carrying capacity can be studied when an organism is introduced to a new environment. One such study involves the introduction of wolves to Coronation Island in Alaska in an attempt to control deer numbers.

When wolves were introduced to Coronation Island

Coronation Island is a small, 116 km^2 island off the coast of Alaska. In 1960, the Alaska Department of Fish and Game released two breeding pairs of wolves to the island. Their aim was to control the black-tailed deer that had been overgrazing the land. The results (below) were not what they expected. Introduction of the wolves initially appeared to have the desired effect. The wolves fed off the deer and successfully bred, and deer numbers fell. However, within a few years the deer numbers crashed. The wolves ran out of food (deer) and began eating each other, causing a drop in wolf numbers. Within eight years, only one wolf inhabited the island, and the deer were abundant. By 1983, there were no wolves on the island, and the deer numbers were high.

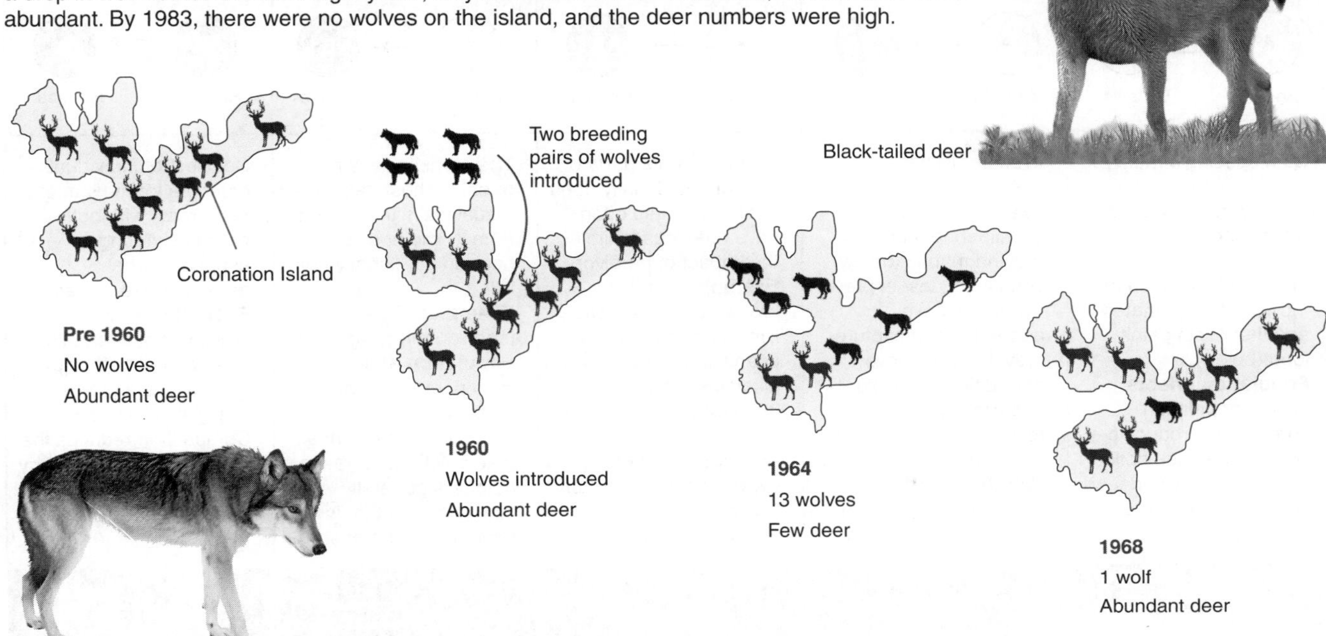

What went wrong?

- The study showed Coronation Island was too small to sustain both the wolf and the deer populations.
- The deer could not easily find refuge from the wolves, so their numbers were quickly reduced.
- Reproductive rates in the deer may have been low because of poor quality forage following years of over-grazing. When wolves were introduced, predation and low reproductive rates caused deer numbers to fall.
- The deer were the only food source for the wolves. When deer became scarce the wolves ate each other because there was no other prey available.

1. Why were wolves introduced to Coronation Island? ______________________

____

____

2. (a) What were some of the factors that caused the unexpected result? ______________________

____

____

____

(b) What do these results tell you about the carrying capacity of Coronation Island? ______________________

____

____

____

ISBN: 978-1-927309-20-9

131 Species Interactions

Key Idea: All species interact with other species. These interactions frequently regulate population growth.

Every organism interacts with others of its own and other species. These interactions are the result of coevolution, in which there is a reciprocal evolution of adaptations. A symbiosis describes a very close association between two or more parties, as occurs in mutualism and parasitism. A mutually beneficial relationship is called mutualism. If one party benefits but the other is unaffected, the relationship is commensal. If one party benefits at the expense of another, the relationship is an exploitation. In competitive interactions, resources are usually limited, so both parties are detrimentally affected. Interactions within and between species ultimately regulate population numbers.

Type of interaction between species

Mutualism	Exploitation			Competition
	Predation	Herbivory	Parasitism	
A ⇄ B Benefits / Benefits	A → B Benefits / Harmed	A → B Benefits / Harmed	A → B Benefits / Harmed	A ⇄ B Harmed / Harmed
Both species benefit from the association. **Examples**: Flowering plants and their insect pollinators have a mututalistic relationship. Flowers are pollinated and the insect gains food (below). **Population effects**: Flower population spreads by producing seeds. Bees use pollen to make honey and feed larvae, ensuring the hive's survival.	Predator kills the prey outright and eats it. **Examples**: Lion preying on wildebeest or praying mantis (below) consuming insect prey. Predators have adaptations to capture prey and prey have adaptations to avoid capture. These relationships are often the result of coevolution. **Population effects**: Predator numbers lag behind prey numbers.	Herbivore eats parts of a plant and usually does not kill it. Plants often have defences to limit the impact of herbivory. **Example**: Giraffes browsing acacia trees. Browsing stimulates the acacia to produce toxic alkaloids, which cause the giraffe to move on to another plant. **Population effects**: Browser damage is self limiting, so the plant is able to recover.	The parasite lives in or on the host, taking (usually all) its nutrition from it. The host is harmed but usually not killed. **Examples**: Pork tapeworm in a pig's gut. **Population effects**: High parasite loads make the host susceptible to diseases that may kill it. Parasite numbers generally stay at a level that is tolerated by the host.	Species, or individuals, compete for the same resources, with both parties suffering, especially when resources are limited. **Examples**: Plants growing close to each other compete for light and soil nutrients. **Population effects**: Competition reduces the maximum number of any one species in an area as resources are limited.
Honeybee and flower	Luc Viatour www.Lucnix.be *Mantid eats cricket*	*Giraffe browses acacia*	*Pork tapeworm*	*Forest plants*

1. For the purposes of this exercise, assume that species A in the diagram represents humans. Briefly describe an example of our interaction with another species (B in the diagram above) that matches each of the following interaction types:

   (a) Mutualism: ______________________________

   (b) Exploitation: ______________________________

   (c) Competition: ______________________________

2. Plants are not defenceless against herbivores. They have evolved physical and chemical defences to deter herbivores. In some cases (as in grasses) grazing stimulates growth in the plant.

   (a) What is the acacia's response to giraffe browsing? ______________________________

   (b) How might this response prevent over-browsing? ______________________________

ISBN:978-1-927309-20-9

Examples of interactions between different species are illustrated below. For each example, identify the type of interaction, and explain how each species in the relationship is affected.

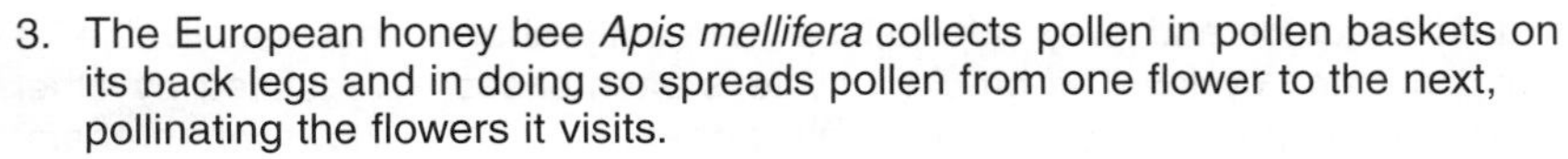

3. The European honey bee *Apis mellifera* collects pollen in pollen baskets on its back legs and in doing so spreads pollen from one flower to the next, pollinating the flowers it visits.

   (a) Identify this type of interaction: ____________________

   (b) Describe how each species is affected (benefits/harmed/no effect):

   ____

   ____

4. The squat anemone shrimp (or sexy shrimp), lives among the tentacles of sea anemones, where it gains protection and scavenges scraps of food from the anemone. The anemone is apparently neither harmed nor benefitted by the shrimp's presence although there is some evidence that ammonium released by the shrimp may benefit the anemone indirectly by supplying nutrients to the mutualistic photosynthetic algae that reside in the anemone's tissues.

   (a) Identify this type of interaction: ____________________

   (b) Describe how each species is affected (benefits/harmed/no effect):

   ____

   ____

   ____

5. Hyaenas will kill and scavenge a range of species. They form large groups and attack and kill large animals, such as wildebeest, but will also scavenge carrion or drive other animals off their kills.

   (a) Identify this type of interaction: ____________________

   (b) Describe how each species is affected (benefits/harmed/no effect):

   ____

   ____

6. Ticks are obligate haemtaophages and must obtain blood to pass from one life stage to the next. Ticks attach to the outside of hosts where they suck blood and fluids and cause irritation.

   (a) Identify this type of interaction: ____________________

   (b) Describe how each species is affected (benefits/harmed/no effect):

   ____

   ____

7. Large herbivores expose insects in the vegetation as they graze. The cattle egret, which is widespread in tropical and subtropical regions, follows the herbivores as they graze, feeding on the insects disturbed by the herbivore.

   (a) Identify this type of interaction: ____________________

   (b) Describe how each species is affected (benefits/harmed/no effect):

   ____

   ____

8. Explain the similarities and differences between a predator and a parasite: ____________________

____

____

ISBN: 978-1-927309-20-9

132 Interpreting Predator-Prey Relationships

Key Idea: Predator and prey populations frequently show regular population cycles. The predator cycle is often based on the intrinsic population cycle of the prey species.

It was once thought that predators regulated the population numbers of their prey. However, we now know that this is not usually the case. Prey species are more likely to be regulated by other factors such as the availability of food. However, predator population cycles are often regulated by the availability of prey, especially when there is little opportunity for switching to alternative prey species.

A case study in predator-prey numbers

In some areas of Northeast India, a number of woolly aphid species colonise and feed off bamboo plants. The aphids can damage the bamboo so much that it is no longer able to be used by the local people for construction and the production of textiles.

Giant ladybird beetles (*Anisolemnia dilatata*) feed exclusively on the woolly aphids of bamboo plants. There is some interest in using them as biological control agents to reduce woolly aphid numbers, and limit the damage woolly aphids do to bamboo plants.

The graph below shows the relationship between the giant lady bird beetle and the woolly aphid when grown in controlled laboratory conditions.

Bamboo plants are home to many insect species, including ladybirds and aphids.

Aphids feed off the bamboo sap, and the ladybirds are predators of the aphids (below).

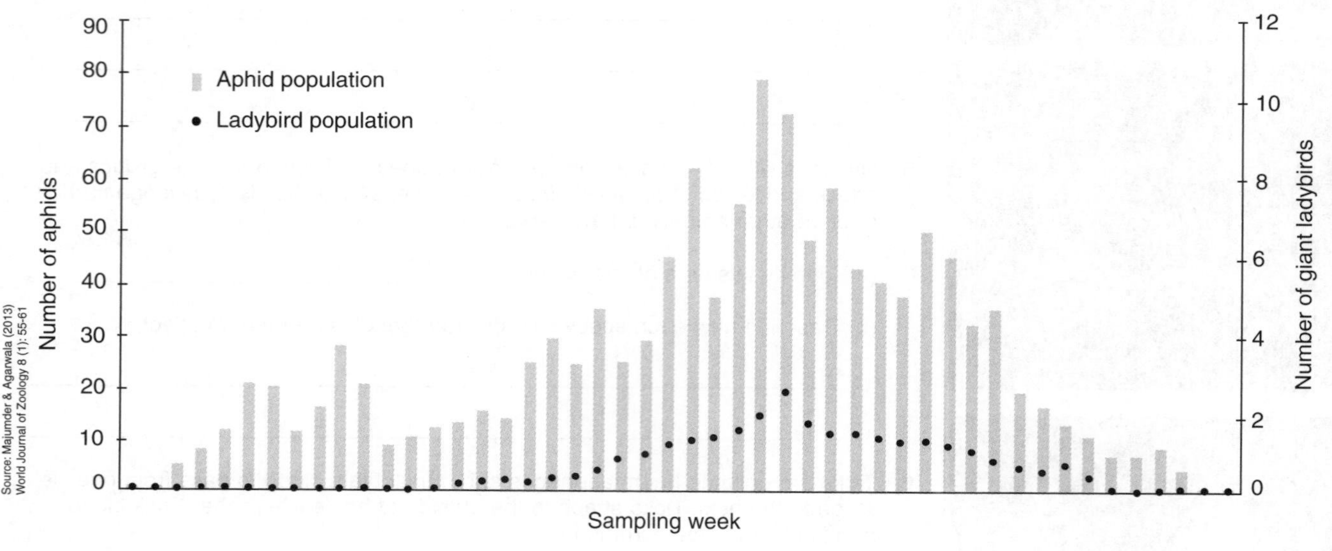

1. (a) On the graph above, mark (using different coloured pens) where the peak numbers of woolly aphids and giant ladybirds occurs:

   (b) Do the peak numbers for both species occur at the same time? ______

   (c) Why do you think this is? ______

2. (a) Is the trend between the giant ladybirds woolly aphids positive or negative (circle one).

   (b) Explain your answer: ______

ISBN:978-1-927309-20-9

133 Interspecific Competition

Key Idea: Interspecific competition occurs between individuals of different species for resources. It can affect the size and distribution of populations sharing the same environment.

Interspecific competition (competition between different species) is usually less intense than intraspecific (same species) competition because coexisting species have evolved slight differences in their realised niches. However, when two species with very similar niche requirements are brought into direct competition through the introduction of a foreign species, one usually benefits at the expense of the other, which is excluded (the **competitive exclusion principle**). The introduction of alien species is implicated in the competitive displacement and decline of many native species. Displacement of native species by introduced ones is more likely if the introduced competitor is adaptable and hardy, with high fertility. In Britain, introduction of the larger, more aggressive, grey squirrel in 1876 has contributed to a contraction in range of the native red squirrel (below), and on the Scottish coast, this phenomenon has been well documented in barnacle species (see next page)

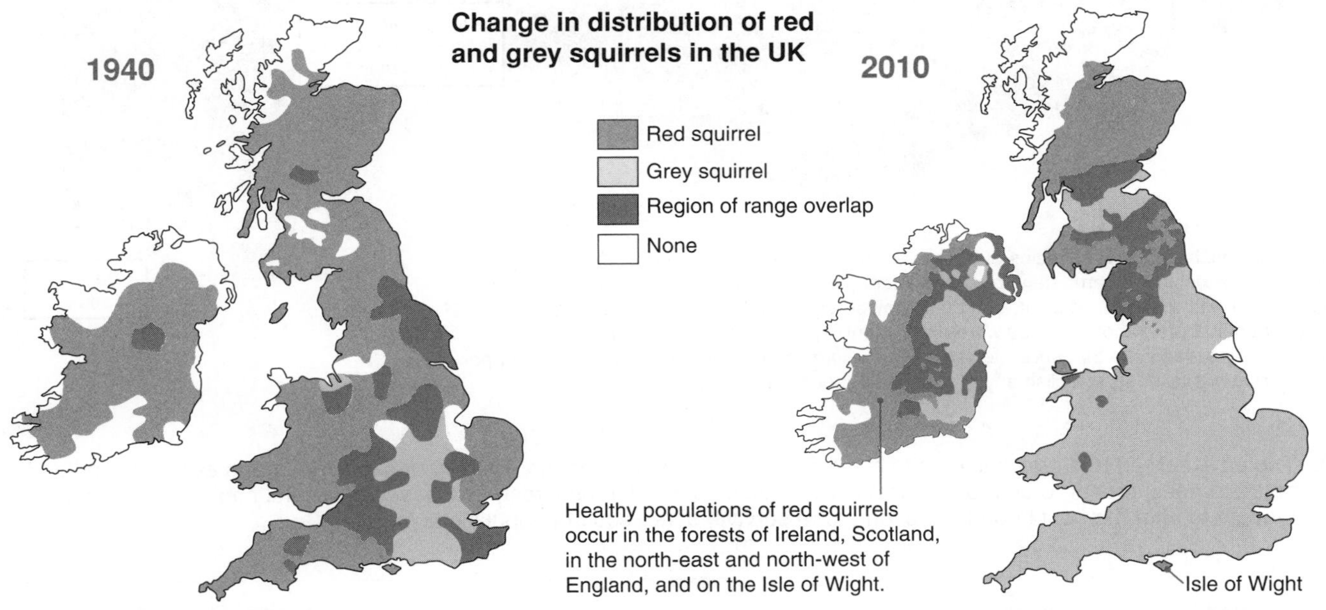

Paul Whippey cc 3.0

Red squirrel, *Sciurus vulgaris*

The European red squirrel was the only squirrel species in Britain until the introduction of the American gray squirrel in 1876. Regular distribution surveys (above) have recorded the range contraction of the reds, with the larger, more aggressive grey squirrel displacing populations of reds over much of England. Grey squirrels can exploit tannin-rich foods, which are unpalatable to reds. In mixed woodland and in competition with greys, reds may not gain enough food to survive the winter and breed. Reds are also very susceptible to several viral diseases, including squirrelpox, which is transmitted by greys.

Whereas red squirrels once occupied a range of forest types, they are now almost solely restricted to coniferous forest. The data suggest that the grey squirrel is probably responsible for the red squirrel decline, but other factors, such as habitat loss, are also likely to be important.

BirdPhotos.com cc 3.0

Gray squirrel, *Sciurus carolinesis*

1. Outline the evidence to support the view that the red-grey squirrel distributions in Britain are an example of the competitive exclusion principle:

2. Some biologists believe that competition with grey squirrels is only one of the factors contributing to the decline in the red squirrels in Britain. Explain the evidence from the 2010 distribution map that might support this view:

ISBN: 978-1-927309-20-9

LINK 134 LINK 131 WEB 133 KNOW

Competitive exclusion in barnacles

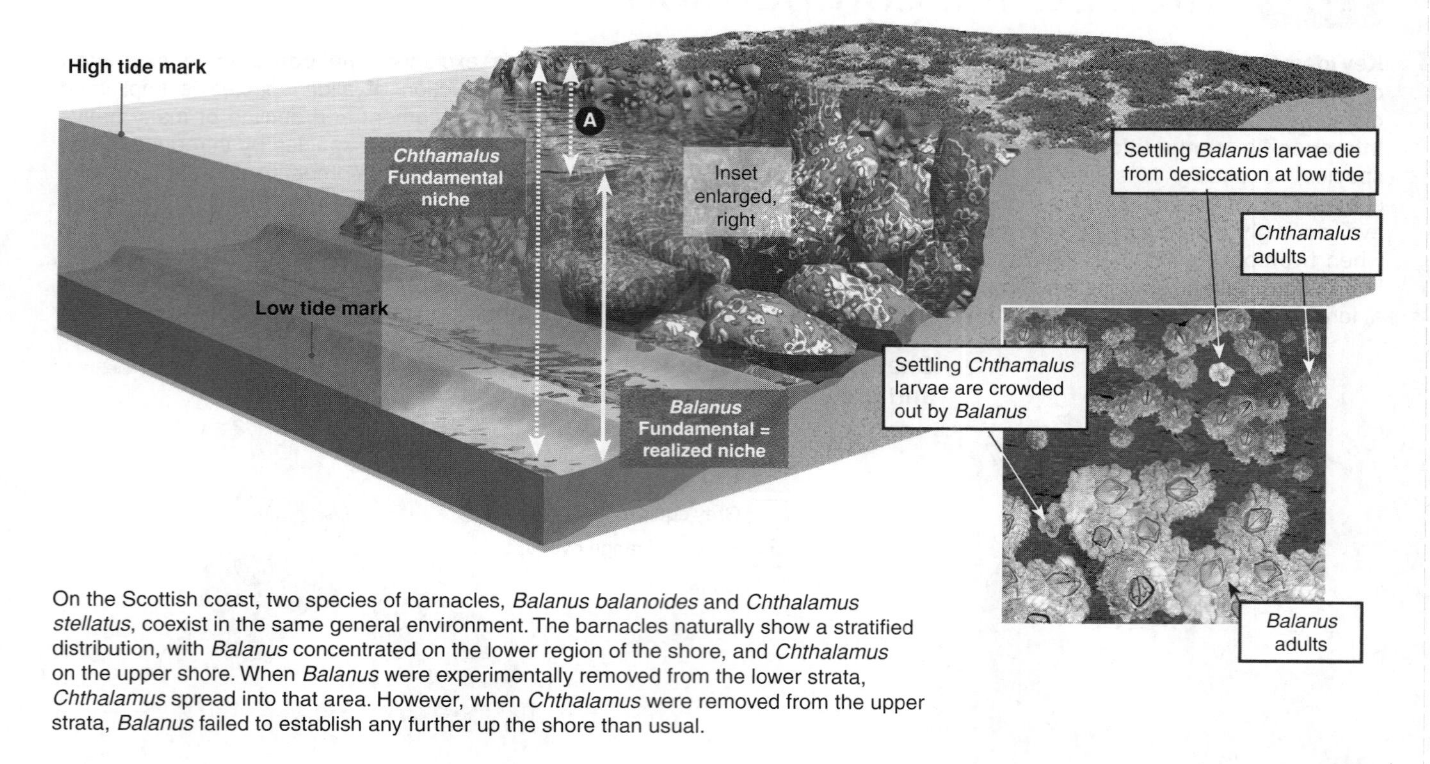

On the Scottish coast, two species of barnacles, *Balanus balanoides* and *Chthalamus stellatus*, coexist in the same general environment. The barnacles naturally show a stratified distribution, with *Balanus* concentrated on the lower region of the shore, and *Chthalamus* on the upper shore. When *Balanus* were experimentally removed from the lower strata, *Chthalamus* spread into that area. However, when *Chthalamus* were removed from the upper strata, *Balanus* failed to establish any further up the shore than usual.

3. The ability of red and grey squirrels to coexist appears to depend on the diversity of habitat type and availability of food sources (reds appear to be more successful in regions of coniferous forest). Suggest why careful habitat management is thought to offer the best hope for the long term survival of red squirrel populations in Britain:

4. Suggest other conservation methods that could possibly aid the survival of viability of red squirrel populations:

5. (a) In the example of the barnacles (above), describe what is represented by the zone labeled with the arrow A:

(b) Outline the evidence for the barnacle distribution being the result of competitive exclusion:

ISBN:978-1-927309-20-9

134 Niche Differentiation

Key Idea: Competition between species for similar resources can be reduced if competing species have slightly different niches and exploit available resources in different ways.

Although Interspecific competition is usually less intense than competition between members of the same species, many species exploit at least some of the same resources. Different species with similar ecological requirements may reduce direct competition by exploiting the resources within different microhabitats or by exploiting the same resources at different times of the day or year. This is called niche differentiation.

Foraging behaviour of tits

Tits are omnivorous passerine (perching) birds, all belonging to same family. They feed on insects, fruits, nuts, and berries and are widespread throughout the UK, Europe, Africa, and North America. In the UK, tits are common in both coniferous forest and broadleaf forests. The feeding behaviours of a large number of tit species have been intensively studied. During most of the year there is plenty of food for coexisting species and they tend to forage in much the same way and in much the same parts of the forest. However during the winter when food is less abundant feeding behaviours change and species diverge in their foraging patterns, as illustrated for four tit species below.

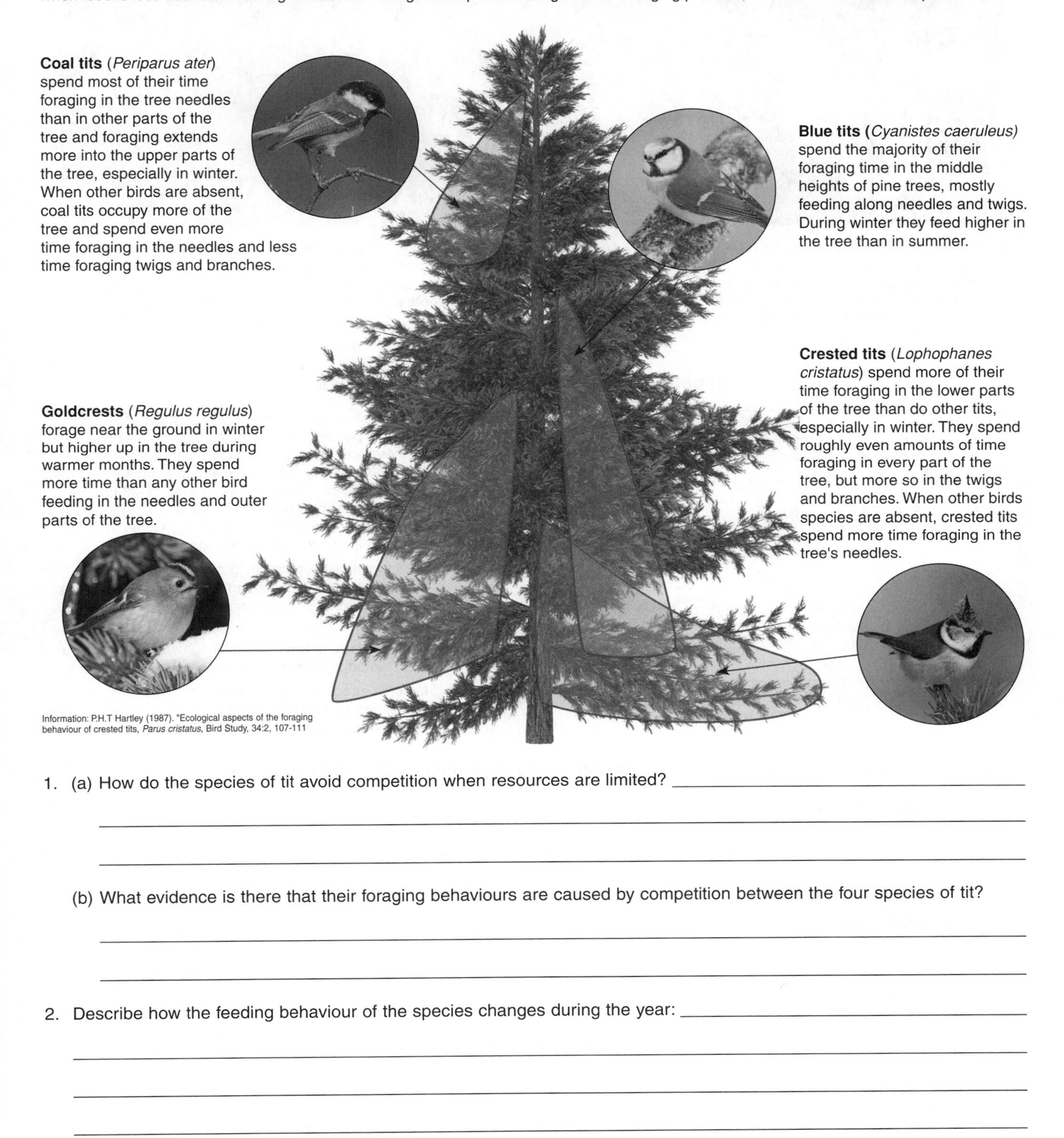

Coal tits (*Periparus ater*) spend most of their time foraging in the tree needles than in other parts of the tree and foraging extends more into the upper parts of the tree, especially in winter. When other birds are absent, coal tits occupy more of the tree and spend even more time foraging in the needles and less time foraging twigs and branches.

Blue tits (*Cyanistes caeruleus*) spend the majority of their foraging time in the middle heights of pine trees, mostly feeding along needles and twigs. During winter they feed higher in the tree than in summer.

Goldcrests (*Regulus regulus*) forage near the ground in winter but higher up in the tree during warmer months. They spend more time than any other bird feeding in the needles and outer parts of the tree.

Crested tits (*Lophophanes cristatus*) spend more of their time foraging in the lower parts of the tree than do other tits, especially in winter. They spend roughly even amounts of time foraging in every part of the tree, but more so in the twigs and branches. When other birds species are absent, crested tits spend more time foraging in the tree's needles.

Information: P.H.T Hartley (1987). "Ecological aspects of the foraging behaviour of crested tits, *Parus cristatus*, Bird Study, 34:2, 107-111

1. (a) How do the species of tit avoid competition when resources are limited? ______

   (b) What evidence is there that their foraging behaviours are caused by competition between the four species of tit?

2. Describe how the feeding behaviour of the species changes during the year: ______

ISBN: 978-1-927309-20-9

135 Intraspecific Competition

Key Idea: Individuals of the same species exploit the same resources, so competition between them is usually intense and will act to limit population growth.

As a population grows, the resources available to each individual become fewer and **intraspecific competition** (competition between members of the same species) increases. When the demand for a resource (e.g. food or light) exceeds supply, that resource becomes a limiting factor to the number of individuals the environment can support (the **carrying capacity**). Populations respond to resource limitation by reducing growth rate (e.g. lower birth rates or higher mortality). The response of individuals to limited resources varies. In many invertebrates and some vertebrates, individuals reduce their growth rate and mature at a smaller size. In many vertebrates, territories space individuals apart according to resource availability and only those individuals able to secure a territory will have sufficient resources to breed.

Scramble competition in caterpillars

Direct competition for available food between members of the same species is called **scramble competition.** In some situations where scramble competition is intense, none of the competitors gets enough food to survive.

Contest competition in wolves

In some cases, competition is limited by hierarchies existing within a social group. Dominant individuals receive adequate food, but individuals low in the hierarchy must **contest** the remaining resources and may miss out.

Display of a male anole

Intraspecific competition may be for mates or breeding sites, or for food. In anole lizards (above), males have a bright red throat pouch and use much of their energy displaying to compete with other males for available mates.

Competition between tadpoles of *Rana tigrina*

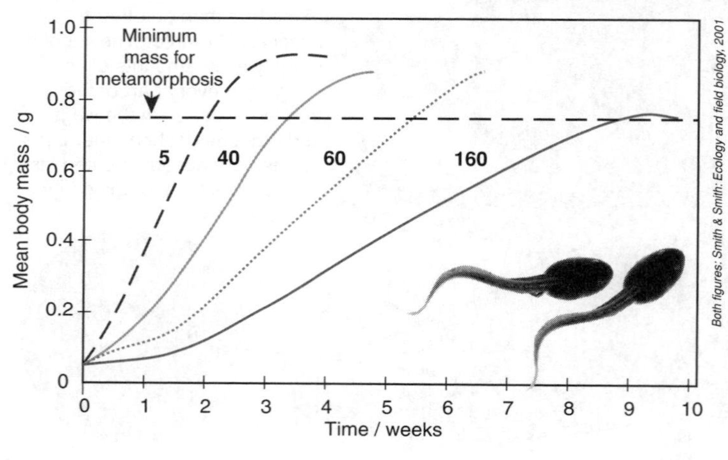

Food shortage reduces both individual growth rate and survival, and population growth. In some organisms, where there is a metamorphosis or a series of moults before adulthood (e.g. frogs, crustacean zooplankton, and butterflies), individuals may die before they mature.

The graph (left) shows how the growth rate of tadpoles (*Rana tigrina*) declines as the density increases from 5 to 160 individuals (in the same sized space).

- At high densities, tadpoles grow more slowly, take longer to reach the minimum size for metamorphosis (0.75 g), and have less chance of metamorphosing into frogs.
- Tadpoles held at lower densities grow faster to a larger size, metamorphosing at an average size of 0.889 g.
- In some species, such as frogs and butterflies, the adults and juveniles reduce the intensity of intraspecific competition by exploiting different food resources.

1. Using an example, predict the likely effects of **intraspecific competition** on each of the following:

   (a) Individual growth rate: ______________________

   (b) Population growth rate: ______________________

   (c) Final population size: ______________________

ISBN:978-1-927309-20-9

Golden eagle breeding territories in Northern Scotland, 1967

Single site
Group of sites belonging to one pair
Marginal site, not regularly occupied
Breeding, year of survey 1967
Low ground unsuitable for breeding eagles.

Territoriality in birds and other animals is usually a result of intraspecific competition. It frequently produces a pattern of uniform distribution over an area of suitable habitat, although this depends somewhat on the distribution of resources. The diagram above shows the territories of golden eagles (*Aquila chrysaetos*) in Scotland. Note the relatively uniform distribution of the breeding sites.

Territoriality in great tits (*Parus major*)

Six breeding pairs of great tits were removed from an oak woodland (below). Within three days, four new pairs had moved into the unoccupied areas (below, right) and some residents had expanded their territories. The new birds moved in from territories in hedgerows, considered to be suboptimal habitat. This type of territorial behaviour limits the density of breeding animals in areas of optimal habitat.

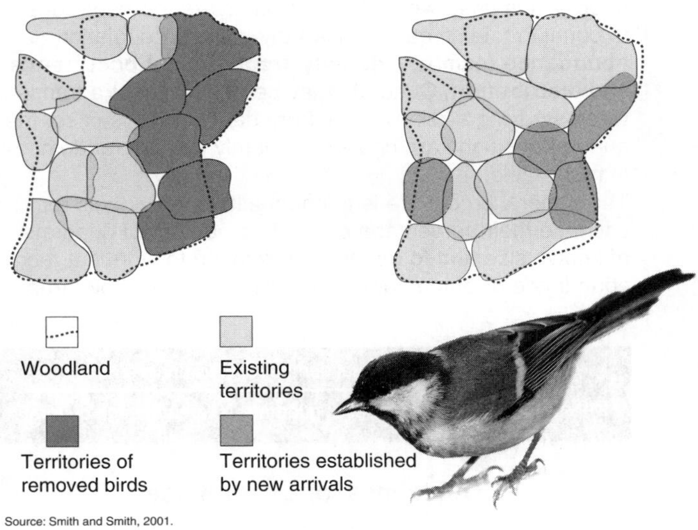

Source: Smith and Smith, 2001.

2. In the tank experiment with *Rana* (see previous page), the tadpoles were contained in a fixed volume with a set amount of food:

   (a) Describe how *Rana* tadpoles respond to resource limitation: ______

   (b) Categorise the effect on the tadpoles as density-dependent / density-independent (delete one).

   (c) Comment on how much the results of this experiment are likely to represent what happens in a natural population:

3. Identify two ways in which animals can reduce the intensity of intraspecific competition:

   (a) ______

   (b) ______

4. (a) Suggest why carrying capacity of an ecosystem might decline: ______

   (b) Predict how a decline in carrying capacity might affect final population size: ______

5. Using appropriate examples, discuss the role of territoriality in reducing intraspecific competition:

ISBN: 978-1-927309-20-9

136 Quadrat Sampling

Key Idea: Quadrat sampling involves a series of random placements of a frame of known size over an area of habitat to assess the abundance or diversity of organisms.

Quadrat sampling is a method by which organisms in a certain proportion (sample) of the habitat are counted directly. It is used when the organisms are too numerous to count in total. It can be used to estimate population **abundance** (number), **density, frequency of occurrence**, and **distribution**. Quadrats may be used without a transect when studying a relatively uniform habitat. In this case, the quadrat positions are chosen randomly using a random number table.

The general procedure is to count all the individuals (or estimate their percentage cover) in a number of quadrats of known size and to use this information to work out the abundance or percentage cover value for the whole area.

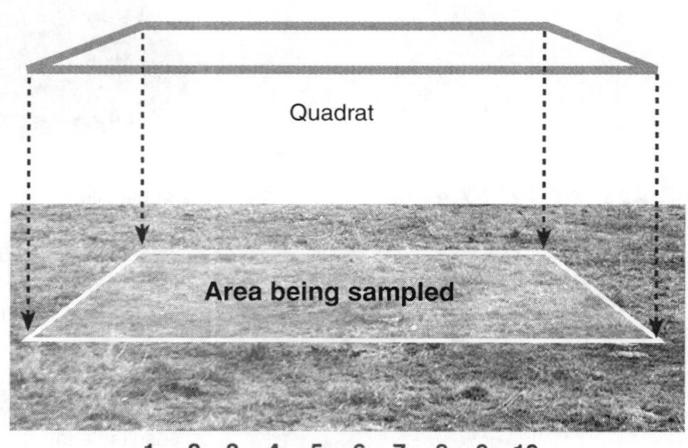

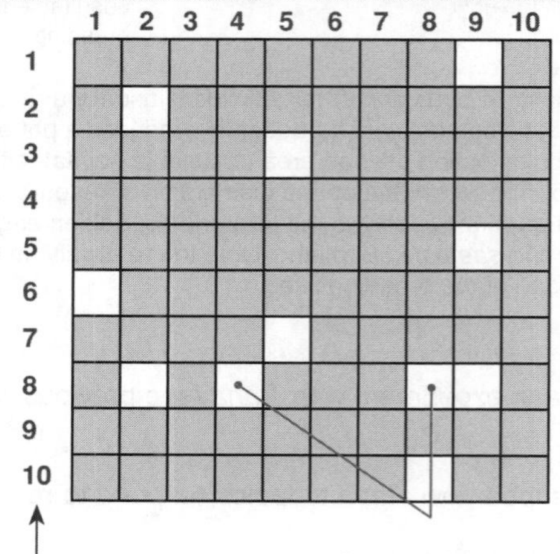

The area to be sampled is divided up into a grid pattern with indexed coordinates

Quadrats are applied to the predetermined grid on a random basis. This can be achieved by using a random number table.

$$\text{Estimated average density} = \frac{\text{Total number of individuals counted}}{\text{Number of quadrats} \times \text{area of each quadrat}}$$

Guidelines for quadrat use:

1. The **area of each quadrat** must be known. Quadrats should be the same shape, but not necessarily square.
2. **Enough quadrat samples** must be taken to provide results that are representative of the total population.
3. The **population of each quadrat** must be known. Species must be distinguishable from each other, even if they have to be identified at a later date. It has to be decided beforehand what the count procedure will be and how organisms over the quadrat boundary will be counted.
4. The size of the quadrat should be appropriate to the organisms and habitat, e.g. a large size quadrat for trees.
5. The quadrats must be **representative of the whole area.** This is usually achieved by **random sampling** (right).

Sampling a centipede population

A researcher by the name of Lloyd (1967) sampled centipedes in Wytham Woods, near Oxford in England. A total of 37 hexagon–shaped quadrats were used, each with a diameter of 30 cm (see diagram on right). These were arranged in a pattern so that they were all touching each other. Use the data in the diagram to answer the following questions.

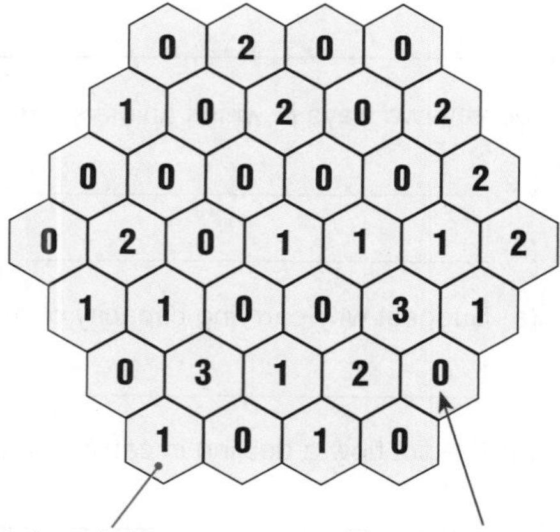

Each quadrat was a hexagon with a diameter of 30 cm and an area of 0.08 square meters.

The number in each hexagon indicates how many centipedes were caught in that quadrat.

1. Determine the average number of centipedes captured per quadrat:

2. Calculate the estimated average density of centipedes per square metre (remember that each quadrat is 0.08 square metres in area):

3. Looking at the data for individual quadrats, describe in general terms the distribution of the centipedes in the sample area:

4. Describe one factor that might account for the distribution pattern:

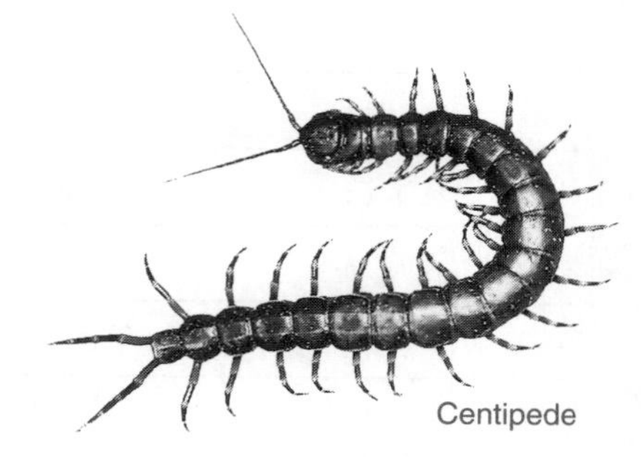

Centipede

ISBN:978-1-927309-20-9

137 Quadrat-Based Estimates

Key Idea: The size and number of quadrats used to sample a community must be sufficient to be representative of that community without taking an excessively long time to use. The simplest description of a community is a list of the species present. This does not provide information about the relative abundance of the species, although this can be estimated using abundance scales (e.g. ACFOR). Quadrats can provide quantitative information about a community. The size of the quadrat and the number of samples taken must represent the community as fairly as possible.

What size quadrat?

Quadrats are usually square, and cover 0.25 m^2 (0.5 m x 0.5 m) or 1 m^2, but they can be of any size or shape, even a single point. The quadrats used to sample plant communities are often 0.25 m^2. This size is ideal for low-growing vegetation, but quadrat size needs to be adjusted to habitat type. The quadrat must be large enough to be representative of the community, but not so large as to take a very long time to use.

A quadrat covering an area of 0.25 m^2 is suitable for most low growing plant communities, such as this alpine meadow, fields, and grasslands.

Larger quadrats (e.g.1m^2) are needed for communities with shrubs and trees. Quadrats as large as 4 m x 4 m may be needed in woodlands.

Small quadrats (0.01 m^2 or 100 mm x 100 mm) are appropriate for lichens and mosses on rock faces and tree trunks.

How many quadrats?

As well as deciding on a suitable quadrat size, the other consideration is how many quadrats to take (the sample size). In species-poor or very homogeneous habitats, a small number of quadrats will be sufficient. In species-rich or heterogeneous habitats, more quadrats will be needed to ensure that all species are represented adequately.

Determining the number of quadrats needed

- Plot the cumulative number of species recorded (on the y axis) against the number of quadrats already taken (on the x axis).
- The point at which the curve levels off indicates the suitable number of quadrats required.

Fewer quadrats are needed in species-poor or very uniform habitats, such as this bluebell woodland.

Describing vegetation

Density (number of individuals per unit area) is a useful measure of abundance for animal populations, but can be problematic in plant communities where it can be difficult to determine where one plant ends and another begins. For this reason, plant abundance is often assessed using **percentage cover**. Here, the percentage of each quadrat covered by each species is recorded, either as a numerical value or using an abundance scale such as the ACFOR scale.

The ACFOR Abundance Scale

A = Abundant (30% +)
C = Common (20-29%)
F = Frequent (10-19%)
O = Occasional (5-9%)
R = Rare (1-4%)

The ACFOR scale could be used to assess the abundance of species in this wildflower meadow. Abundance scales are subjective, but it is not difficult to determine which abundance category each species falls into.

1. Describe one difference between the methods used to assess species abundance in plant and in animal communities:

____

2. What is the main consideration when determining appropriate quadrat size? _______________

3. What is the main consideration when determining number of quadrats? _______________

4. Explain two main disadvantages of using the ACFOR abundance scale to record information about a plant community:

(a) _______________

(b) _______________

ISBN: 978-1-927309-20-9

138 Sampling a Rocky Shore Community

Key Idea: The estimates of a population gained from using quadrat sampling may vary depending on where the quadrats are placed. Larger samples can account for variation.

The diagram (next page) represents an area of seashore with its resident organisms. The distribution of coralline algae and four animal species are shown. This exercise is designed to prepare you for planning and carrying out a similar procedure to practically investigate a natural community.

1. **Decide on the sampling method**
For the purpose of this exercise, it has been decided that the populations to be investigated are too large to be counted directly and a quadrat sampling method is to be used to estimate the average density of the four animal species as well as that of the algae.

2. **Mark out a grid pattern**
Use a ruler to mark out 3 cm intervals along each side of the sampling area (area of quadrat = 0.03 x 0.03 m). **Draw lines** between these marks to create a 6 x 6 grid pattern (total area = 0.18 x 0.18 m). This will provide a total of 36 quadrats that can be investigated.

3. **Number the axes of the grid**
Only a small proportion of the possible quadrat positions will be sampled. It is necessary to select the quadrats in a random manner. It is not sufficient to simply guess or choose your own on a 'gut feeling'. The best way to choose the quadrats randomly is to create a numbering system for the grid pattern and then select the quadrats from a random number table. Starting at the *top left hand corner*, **number the columns** and **rows** from 1 to 6 on each axis.

4. **Choose quadrats randomly**
To select the required number of quadrats randomly, use random numbers from a random number table. The random numbers are used as an index to the grid coordinates. Choose 6 quadrats from the total of 36 using table of random numbers provided for you at the bottom of the next page. Make a note of which column of random numbers you choose. Each member of your group should choose a different set of random numbers (i.e. different column: A–D) so that you can compare the effectiveness of the sampling method.

Column of random numbers chosen: ______

NOTE: Highlight the boundary of each selected quadrat with coloured pen/highlighter.

5. **Decide on the counting criteria**
Before the counting of the individuals for each species is carried out, the criteria for counting need to be established.

There may be some problems here. You must decide before sampling begins as to what to do about individuals that are only partly inside the quadrat. Possible answers include:

(a) Only counting individuals that are completely inside the quadrat.
(b) Only counting individuals with a clearly defined part of their body inside the quadrat (such as the head).
(c) Allowing for 'half individuals' (e.g. 3.5 barnacles).
(d) Counting an individual that is inside the quadrat by half or more as one complete individual.

Discuss the merits and problems of the suggestions above with other members of the class (or group). You may even have counting criteria of your own. Think about other factors that could cause problems with your counting.

6. **Carry out the sampling**
Carefully examine each selected quadrat and **count the number of individuals** of each species present. Record your data in the spaces provided on the next page.

7. **Calculate the population density**
Use the combined data TOTALS for the sampled quadrats to estimate the average density for each species by using the formula:

$$\text{Density} = \frac{\text{Total number in all quadrats sampled}}{\text{Number of quadrats sampled} \times \text{area of a quadrat}}$$

Remember that a total of 6 quadrats are sampled and each has an area of 0.0009 m^2. The density should be expressed as the number of individuals *per square metre (no. m^{-2})*.

Plicate barnacle: ______ Snakeskin chiton: ______

Oyster borer: ______ Coralline algae: ______

Limpet: ______

8. (a) In this example the animals are not moving. Describe the problems associated with sampling moving organisms. Explain how you would cope with sampling these same animals if they were really alive and very active:

______

______

______

(b) Carry out a direct count of all 4 animal species and the algae for the whole sample area (all 36 quadrats). Apply the data from your direct count to the equation given in (7) above to calculate the actual population density (remember that the number of quadrats in this case = 36):

Barnacle: ______ Oyster borer: ______ Chiton: ______ Limpet: ______ Algae: ______

Compare your estimated population density to the actual population density for each species:

______

______

______

ISBN:978-1-927309-20-9

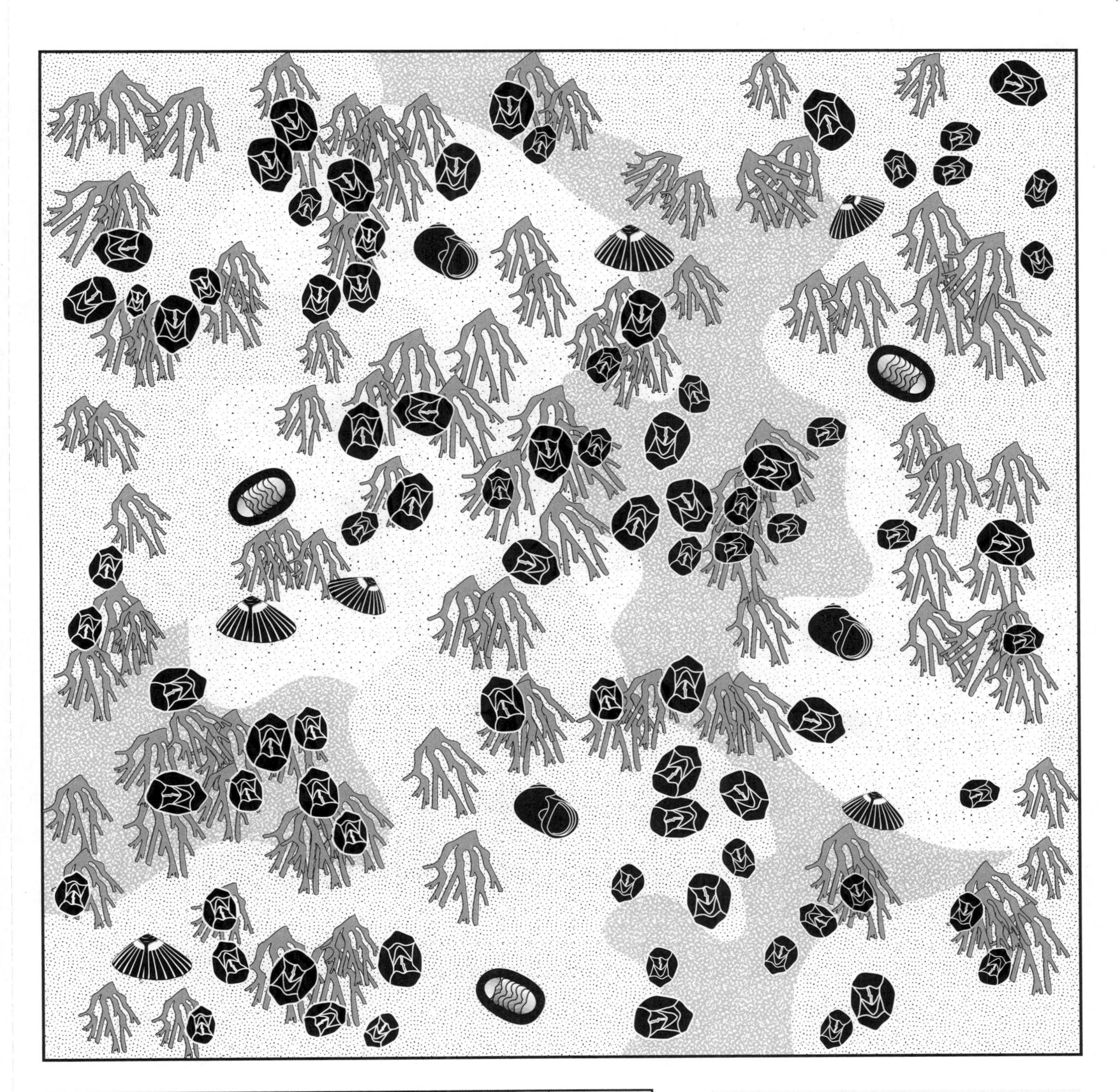

Coordinates for each quadrat	Plicate barnacle	Oyster borer	Snakeskin chiton	Limpet	Coralline algae
1:					
2:					
3:					
4:					
5:					
6:					
TOTAL					

Table of random numbers

A	B	C	D
22	31	62	22
32	15	63	43
31	56	36	64
46	36	13	45
43	42	45	35
56	14	31	14

The table above has been adapted from a table of random numbers from a statistics book. Use this table to select quadrats randomly from the grid above. Choose one of the columns (A to D) and use the numbers in that column as an index to the grid. The first digit refers to the row number and the second digit refers to the column number. To locate each of the 6 quadrats, find where the row and column intersect, as shown below:

Example: 52 refers to the 5th row and the 2nd column

ISBN: 978-1-927309-20-9

139 Transect Sampling

Key Idea: Transect sampling is useful for providing information on species distribution along an environmental gradient.

A **transect** is a line placed across a community of organisms. Transects provide information on the distribution of species in the community. They are particularly valuable when the transect records community composition along an **environmental gradient** (e.g. up a mountain or across a seashore). The usual practice for small transects is to stretch a string between two markers. The string is marked off in measured distance intervals and the species at each marked point are noted. The sampling points along the transect may also be used for the siting of quadrats, so that changes in density and community composition can be recorded. Belt transects are essentially a form of continuous quadrat sampling. They provide more information on community composition but can be difficult to carry out. Some transects provide information on the vertical, as well as horizontal, distribution of species (e.g. tree canopies in a forest).

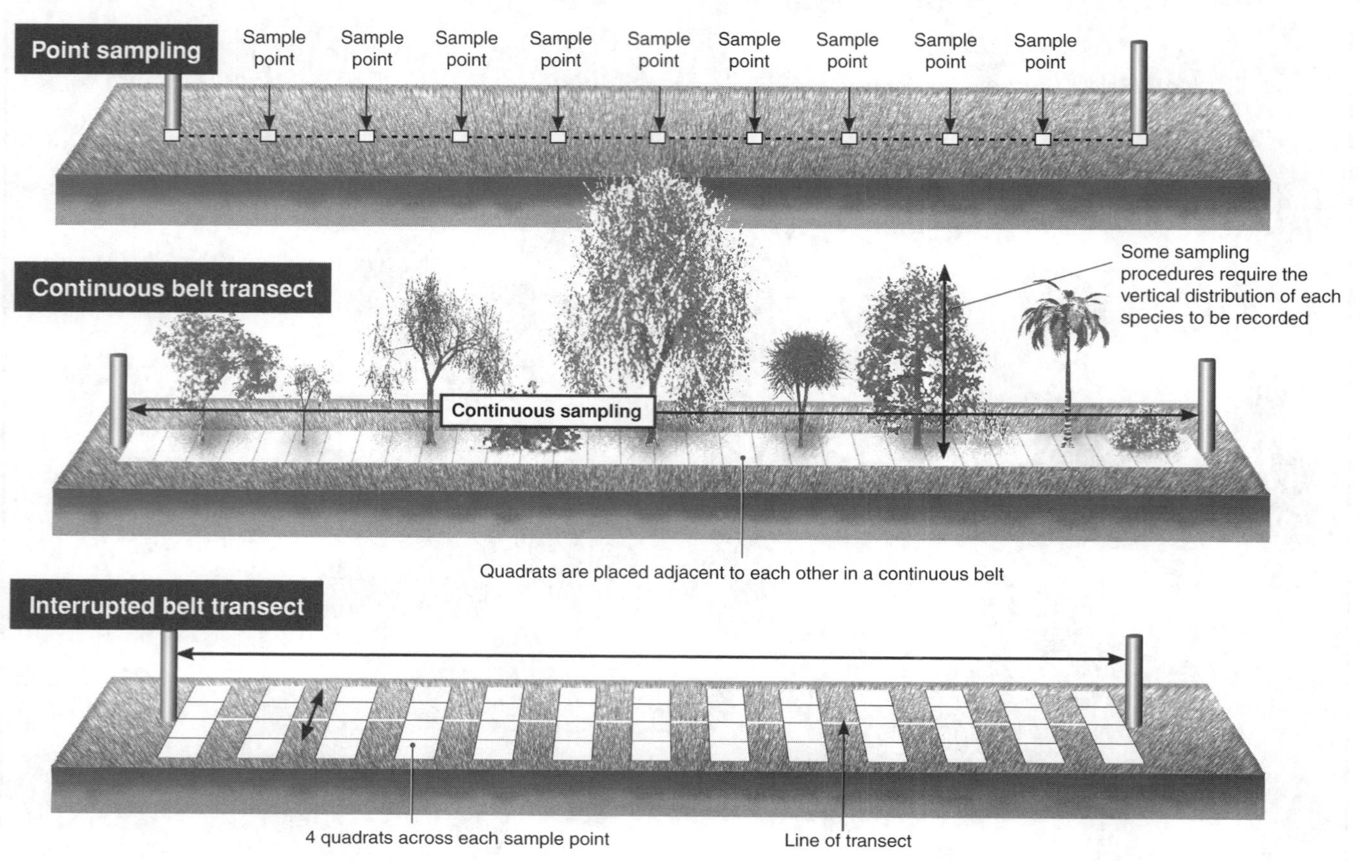

1. Belt transect sampling uses quadrats placed along a line at marked intervals. In contrast, point sampling transects record only the species that are touched or covered by the line at the marked points.

   (a) Describe one disadvantage of belt transects: ______________________________

   ____

   (b) Why might line transects give an unrealistic sample of the community in question? ______________________________

   ____

   ____

   (c) How do belt transects overcome this problem? ______________________________

   ____

   (d) When would it not be appropriate to use transects to sample a community? ______________________________

   ____

2. How could you test whether or not a transect sampling interval was sufficient to accurately sample a community?

   ____

   ____

   ____

ISBN:978-1-927309-20-9

A **kite graph** is a good way to show the distribution of organisms sampled using a belt transect. Data may be expressed as abundance or percentage cover along an environmental gradient. Several species can be shown together on the same plot so that the distributions can be easily compared.

3. The data on the right were collected from a rocky shore field trip. Four common species of barnacle were sampled in a continuous belt transect from the low water mark, to a height of 10 m above that level. The number of each of the four species in a 1 m^2 quadrat was recorded.

Plot a **kite graph** of the data for all four species on the grid below. Be sure to choose a scale that takes account of the maximum number found at any one point and allows you to include all the species on the one plot. Include the scale on the diagram so that the number at each point on the kite can be calculated.

An example of a kite graph

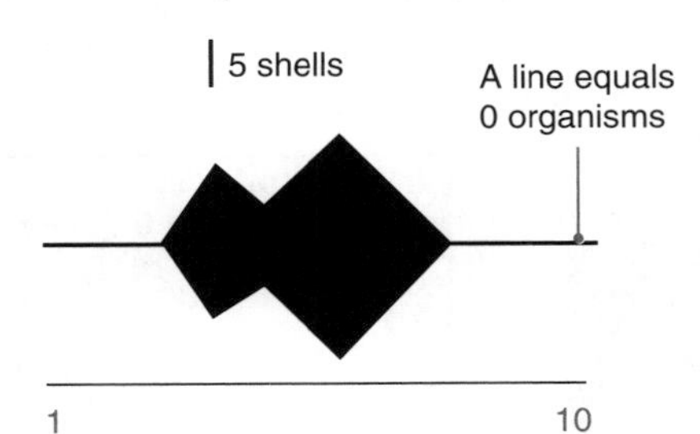

Field data notebook

Numbers of barnacles (4 common species) showing distribution on a rocky shore

Height above low water / m	Barnacle species: Plicate barnacle	Columnar barnacle	Brown barnacle	Sheet barnacle
0	0	0	0	65
1	10	0	0	12
2	32	0	0	0
3	55	0	0	0
4	100	18	0	0
5	50	124	0	0
6	30	69	2	0
7	0	40	11	0
8	0	0	47	0
9	0	0	59	0
10	0	0	65	0

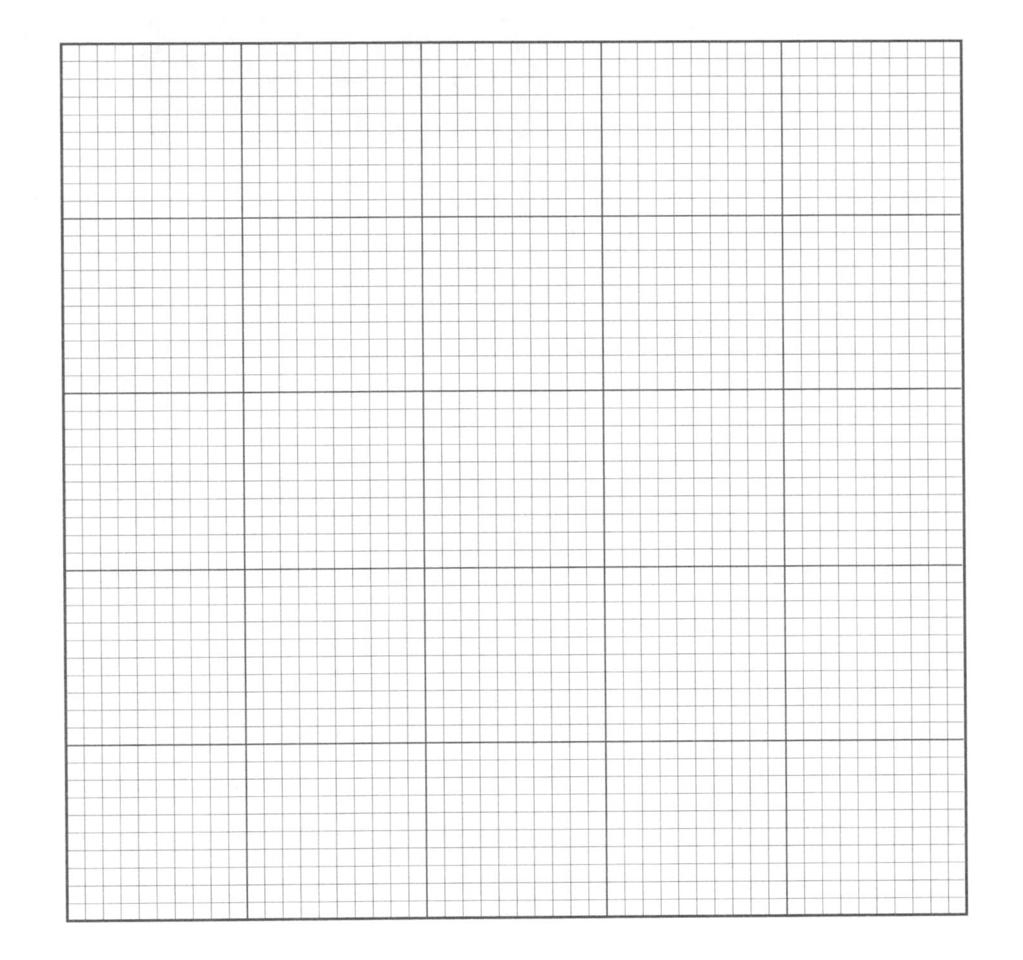

ISBN: 978-1-927309-20-9

140 Qualitative Practical Work: Seaweed Zonation

Key Idea: Qualitative and quantitative data can be used to explain patterns of zonation in seashore communities.

Three species of brown algae (genus *Fucus*), together with the brown alga *Ascophyllum nodosum*, form the dominant seaweeds on rocky shores in Britain, where they form distinct zones along the shore. Zonation is a characteristic feature of many seashore communities where species' distribution is governed by tolerances to particular physical conditions (e.g. time of exposure to air). When collecting data on the distribution and abundance of *Fucus* species, it is useful to also make qualitative observations about the size, vigour, and degree of desiccation of specimens at different points on the shore. These observations provide biological information which can help to explain the observed patterns.

Spiral wrack (*Fucus spiralis*)

Andreas Trepte

Fucus is a genus of marine brown algae, commonly called wracks, which are found in the midlittoral zone of rocky seashores (i.e. the zone between the low and high levels). A group of students made a study of a rocky shore dominated by three species of *Fucus*: spiral wrack, bladder wrack, and serrated wrack. Their aim was to investigate the distribution of three *Fucus* species in the midlittoral zone and relate this to the size and vigour (V) of the seaweeds and the degree of desiccation (D) evident.

Bladder wrack (*F. vesiculosus*)

Thalli

Stemonitis

Serrated wrack (*F. serratus*)

Stemonitis

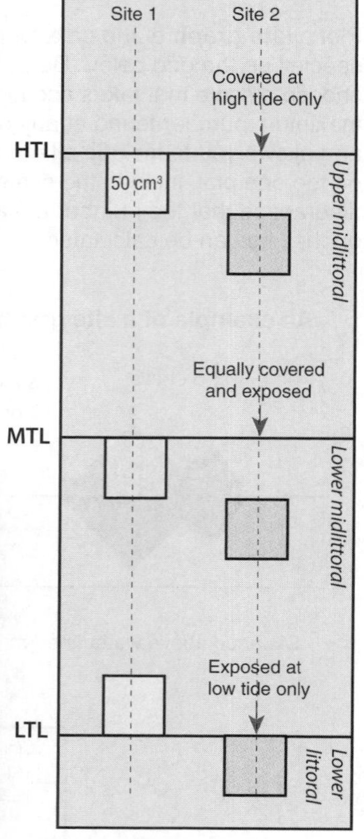

Procedure

Three 50 cm^3 quadrats were positioned from the LTL to the HTL at two sites on the shore as shown in the diagram (far right). An estimate of **percentage cover** (**C**) of each species of *Fucus* was made for each sample. Information on vigour and degree of desiccation was collected at the same time.

Qualitative data were collected as simple scores:

+ = vigorous with large thalli no evidence of dessication
0 = less vigorous with smaller thalli some evidence of dessication
– = small, poorly grown thalli obvious signs of desiccation

1. (a) Describe the quantitative component of this study:

(b) Describe the qualitative component of this study:

	SITE 1									SITE 2								
	HTL			MTL			LTL			HTL			MTL			LTL		
Species	C	D	V	C	D	V	C	D	V	C	D	V	C	D	V	C	D	V
Spiral wrack	50	0	+	0	na	na	0	na	na	30	+	0	0	na	na	0	na	na
Bladder wrack	15	–	–	80	+	+	20	+	0	50	0	–	70	+	+	0	na	na
Serrated wrack	0	na	na	0	na	na	75	+	+	0	na	na	10	–	–	80	+	+

2. The results of the quadrat survey are tabulated above. On a separate sheet, plot a column graph of the percentage coverage of each species at each position on the shore and at sites 1 and 2. Staple it to this page.

3. Relate the distribution pattern to the changes in degree of desiccation and in size and vigour of the seaweed thalli:

4. Suggest why the position of the quadrats was staggered for the two sites and describe a disadvantage of this design:

ISBN:978-1-927309-20-9

141 Mark and Recapture Sampling

Key Idea: Mark and recapture sampling allows the population size of highly mobile organisms to be estimated.

The mark and recapture method of estimating population size is used in the study of animal populations in which the individuals are highly mobile. It is of no value where animals do not move or move very little. The number of animals caught in each sample must be large enough to be valid. The technique is outlined in the diagram below.

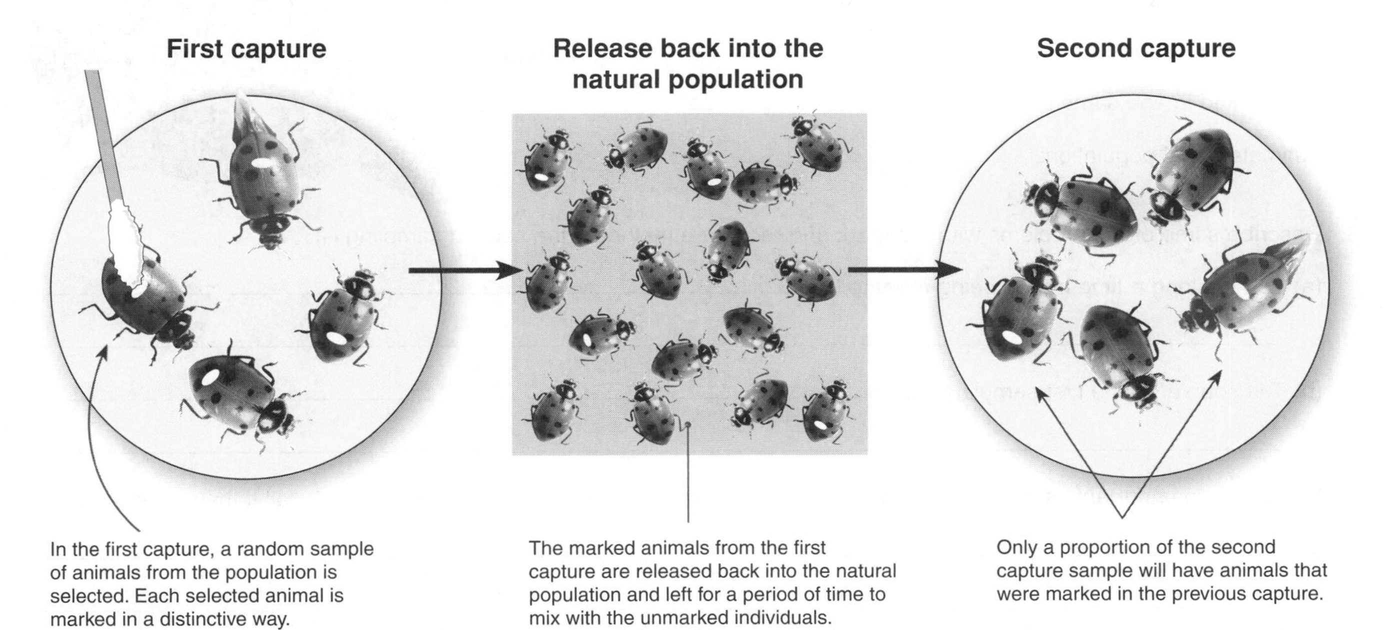

The Lincoln Index

$$\text{Total population} = \frac{\text{No. of animals in 1st sample (all marked)} \times \text{Total no. of animals in 2nd sample}}{\text{Number of marked animals in the second sample (recaptured)}}$$

The mark and recapture technique comprises a number of simple steps:

1. The population is sampled by capturing as many of the individuals as possible and practical.
2. Each animal is marked in a way to distinguish it from unmarked animals (unique mark for each individual not required).
3. Return the animals to their habitat and leave them for a long enough period for complete mixing with the rest of the population to take place
4. Take another sample of the population (this does not need to be the same sample size as the first sample, but it does have to be large enough to be valid).
5. Determine the numbers of marked to unmarked animals in this second sample. Use the equation above to estimate the size of the overall population.

1. For this exercise you will need several boxes of matches and a pen. Work in a group of 2-3 students to 'sample' the population of matches in the full box by using the mark and recapture method. Each match will represent one animal.

   (a) Take out 10 matches from the box and mark them on 4 sides with a pen so that you will be able to recognise them from the other unmarked matches later.
   (b) Return the marked matches to the box and shake the box to mix the matches.
   (c) Take a sample of 20 matches from the same box and record the number of marked matches and unmarked matches.
   (d) Determine the total population size by using the equation above.
   (e) Repeat the sampling 4 more times (steps b–d above) and record your results:

	Sample 1	Sample 2	Sample 3	Sample 4	Sample 5
Estimated population					

   (f) Count the actual number of matches in the matchbox : ______

   (g) Compare the actual number to your estimates and state by how much it differs: ______

ISBN: 978-1-927309-20-9

2. In 1919 a researcher by the name of Dahl wanted to estimate the number of trout in a Norwegian lake. The trout were subject to fishing so it was important to know how big the population was in order to manage the fish stock. He captured and marked 109 trout in his first sample. A few days later, he caught 177 trout in his second sample, of which 57 were marked. Use the **Lincoln index** (on the previous page) to estimate the total population size:

   Size of 1st sample: ______________________

   Size of 2nd sample: ______________________

   No. marked in 2nd sample: ______________________

   Estimated total population: ______________________

3. Describe some of the problems with the mark and recapture method if the second sampling is:

   (a) Left too long a time before being repeated: ______________________

   (b) Too soon after the first sampling: ______________________

4. Describe two important assumptions in this method of sampling that would cause the method to fail if they were not true:

   (a) ______________________

   (b) ______________________

5. Some types of animal would be unsuitable for this method of population estimation (i.e. would not work).

   (a) Name an animal for which this method of sampling would not be effective: ______________________

   (b) Explain your answer above: ______________________

6. Describe three methods for marking animals for mark and recapture sampling. Take into account the possibility of animals shedding their skin, or being difficult to get close to again:

   (a) ______________________

   (b) ______________________

   (c) ______________________

7. Scientists in the UK and Canada have, at various times since the 1950s, been involved in computerised tagging programs for Northern cod (a species once abundant in Northern Hemisphere waters but now severely depleted). Describe the type of information that could be obtained through such tagging programs:

   ____

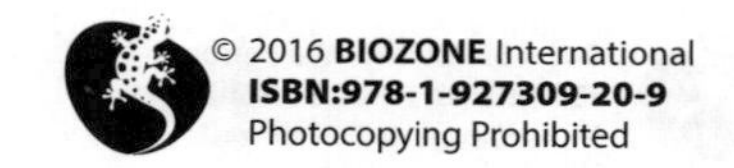

ISBN:978-1-927309-20-9

142 Field Study of a Rocky Shore

Key Idea: Field studies collect physical and biological data that measure aspects of community structure or function.

Many biological investigations require the collection of data from natural communities. Biotic data may include the density or distribution of organisms at a site. Recording physical (abiotic) data of the site allows the site to be compared with others. The investigation below looks at the populations of animals found on an exposed and a sheltered rocky shore.

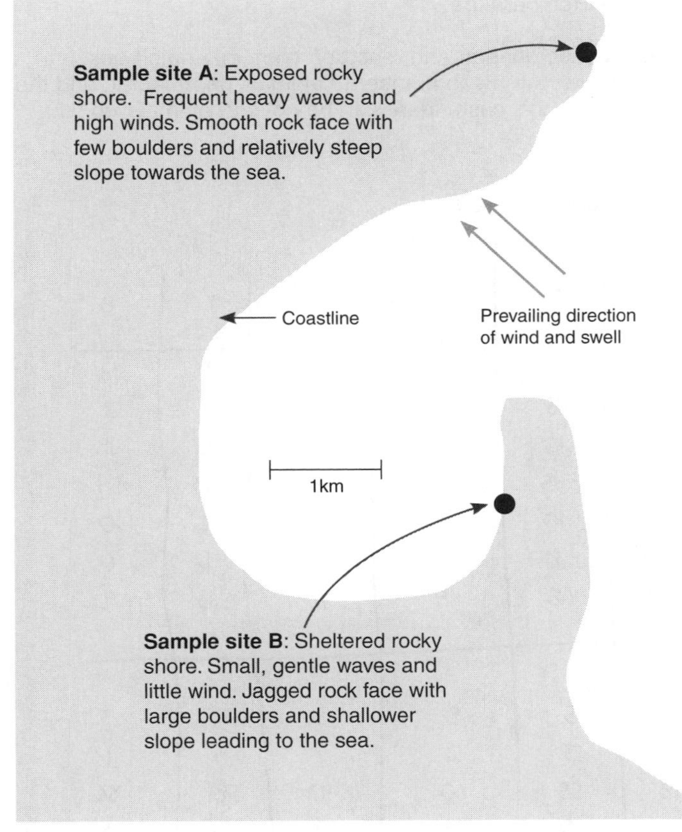

The aim

To investigate the differences in the abundance of intertidal animals on an exposed rocky shore and a sheltered rocky shore.

Background

The composition of rocky shore communities is strongly influenced by the shore's physical environment. Animals that cling to rocks must keep their hold on the substrate while being subjected to intense wave action and currents. However, the constant wave action brings high levels of nutrients and oxygen. Communities on sheltered rocky shores, although encountering less physical stress, may face lower nutrient and oxygen levels.

To investigate differences in the abundance of intertidal animals, students laid out 1 m^2 quadrats at regular intervals along one tidal zone at two separate but nearby sites: a rocky shore exposed to wind and heavy wave action and a rocky shore with very little heavy wave action. The animals were counted and their numbers in each quadrat recorded.

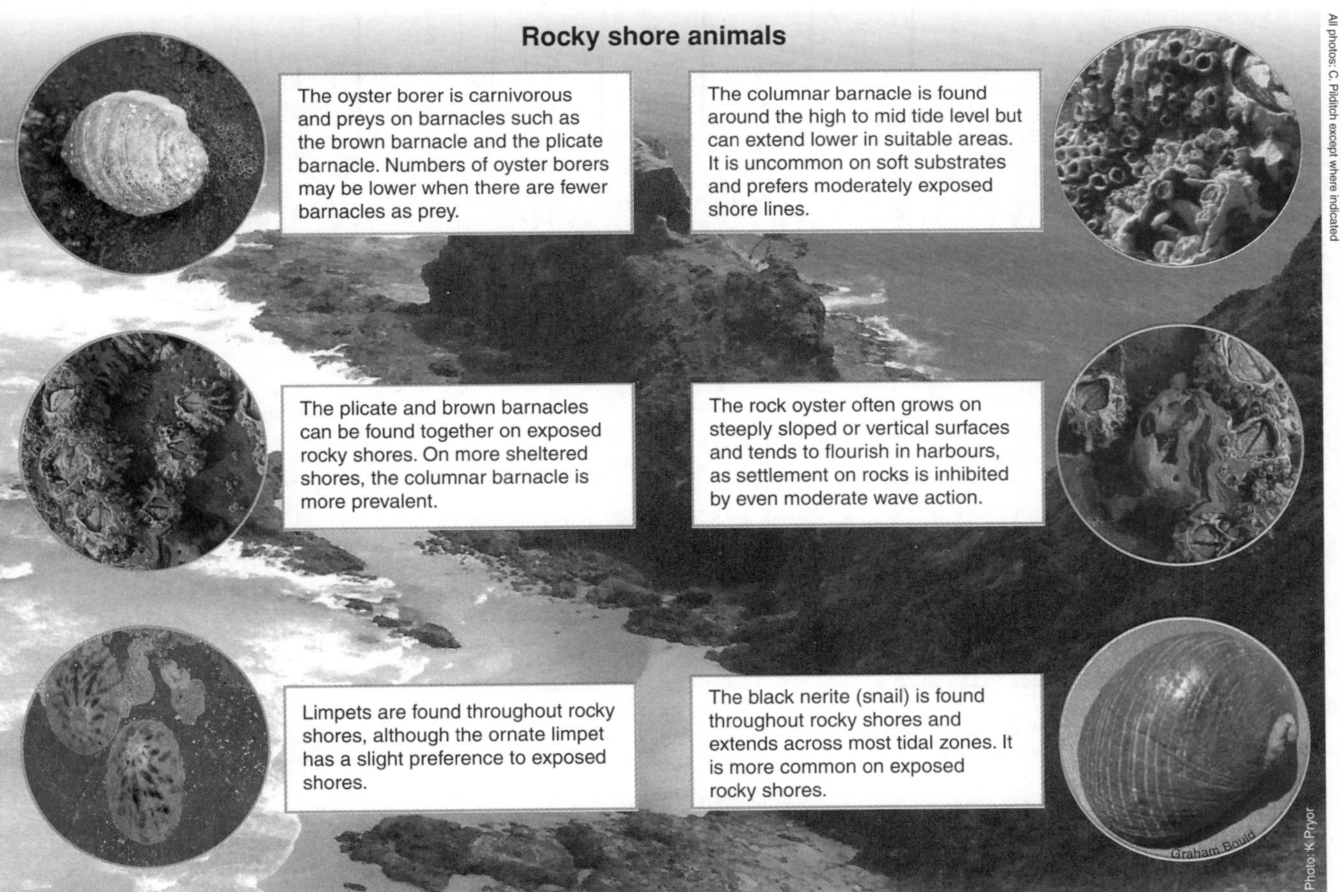

All photos: C. Pilditch except where indicated

ISBN: 978-1-927309-20-9

1. Underline an appropriate hypothesis for this field study from the four possible hypotheses below:

   (a) Rocky shore communities differ because of differences in wave action.

   (b) Rocky shore communities differ because of the topography of the coastline.

   (c) The physical conditions of exposed rocky shores and sheltered rocky shores are very different and so the intertidal communities will also be different.

   (d) Rocky shore communities differ because of differences in water temperature.

2. During the field study, students counted the number of animals in each quadrat and recorded them in a note book. Complete the table with the total number of each species at each site, the mean number of animals per quadrat, and the median and mode for each set of samples per species. Remember, in this case, there can be no 'part animals' so you will need to round your values to the nearest whole number:

Field data notebook

Count per quadrat. Quadrats 1 m^2

Site A	1	2	3	4	5	6	7	8
Brown barnacle	39	38	37	21	40	56	36	41
Oyster borer	6	7	4	3	7	8	9	2
Columnar barnacle	6	8	14	10	9	12	8	11
Plicate barnacle	50	52	46	45	56	15	68	54
Ornate limpet	9	7	8	10	6	7	6	10
Radiate limpet	5	6	4	8	6	7	5	6
Black nerite	7	7	6	8	4	6	8	9
Site B								
Brown barnacle	7	6	7	5	8	5	7	7
Oyster borer	2	3	1	3	2	2	1	1
Columnar barnacle	56	57	58	55	60	47	58	36
Plicate barnacle	11	11	13	10	14	9	9	8
Rock oyster	7	8	8	6	2	4	8	6
Ornate limpet	7	8	5	6	5	7	9	3
Radiate limpet	13	14	11	10	14	12	9	13
Black nerite	6	5	3	1	4	5	2	3

		Brown barnacle	Oyster borer	Columnar barnacle	Plicate barnacle	Rock oyster	Ornate limpet	Radiate limpet	Black nerite
Site A	Total number of animals								
	Mean number of animals per m^2								
	Median value								
	Modal value								
Site B	Total number of animals								
	Mean number of animals per m^2								
	Median value								
	Modal value								

ISBN:978-1-927309-20-9

3. Use the grid below to draw a column graph of the mean number of species per 1 m^2 at each sample site. Remember to include a title, correctly labelled axes, and a key.

4. (a) Compare the mean, median, and modal values obtained for the samples at each site:

   (b) What does this tell you about the distribution of the data:

5. (a) Which species was entirely absent from site A?

   (b) Suggest why this might be the case:

6. (a) Explain why more brown barnacles and plicate barnacles were found at site A:

   (b) Explain why more oyster borers were found at site A:

7. (a) Comment on the numbers of limpets at each site:

   (b) What does this suggest to you about their biology:

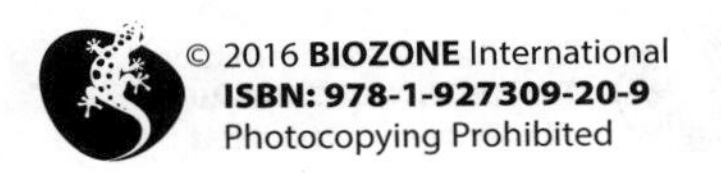

ISBN: 978-1-927309-20-9

143 Investigating Distribution and Abundance

Key Idea: Sampling populations *in-situ* can reveal patterns of distribution, which can be attributed to habitat preference. These investigations are common in ecological studies.

Use this activity to practise analysing data from a field study in which the aim was to identify and describe an existing pattern of species distribution.

The aim

To investigate the effect of fallen tree logs on the distribution of pill millipedes in a forest.

Background

Millipedes consume decaying vegetation and live in the moist conditions beneath logs and in the leaf litter of forest floors. The moist environment protects them from drying out as their cuticle is not a barrier to water loss.

Experimental method

The distribution of millipede populations in relation to fallen tree logs was investigated in a small forest reserve. Six logs of similar size were chosen from similar but separate regions of the forest. Logs with the same or similar surrounding environment (e.g. leaf litter depth, moisture levels) were selected.

For each log, eight samples of leaf litter at varying distances from the fallen tree log were taken using 30 cm^2 quadrats. Samples were taken from two transects, one each side of the log. The sample distances were: directly below the log (0 m), 1.5 m, 2.5 m, and 3.5 m from the log. It was assumed that the conditions on each side of the log would be essentially the same. The leaf litter was placed in Tullgren funnels and the invertebrates extracted. The number of millipedes in each sample was counted. The raw data are shown below.

Pill millipede
Glomeris marginata

Experimental setup

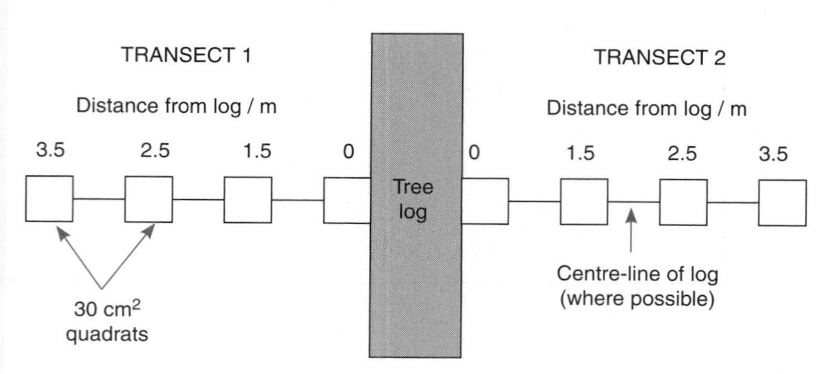

Environmental conditions for each transect position either side of the log were assumed to be equal.

Raw data for tree log and millipede investigation

Tree log	Transect	Distance from log / m: 0	1.5	2.5	3.5
1	1	12	11	3	2
	2	10	12	2	1
2	1	8	3	4	4
	2	9	5	2	1
3	1	14	6	3	3
	2	3	8	7	2

Tree log	Transect	Distance from log / m: 0	1.5	2.5	3.5
4	1	2	4	1	6
	2	4	5	2	2
5	1	12	10	16	10
	2	6	3	2	5
6	1	10	9	7	2
	2	11	11	8	1

LINK
REFER 142

ISBN:978-1-927309-20-9

1. Plot column graphs on the grids below to show the distribution of millipedes at each log. Plot transect 1 and 2 data separately but side by side for each graph:

Log 1

Log 2

Log 3

Log 4

Log 5

Log 6

2. (a) Is there a relationship between distance from the tree log and the number of millipedes found? ____________________

____

(b) What physical factors might account for this? ____________________

____

____

(c) How could you improve the design of the study to obtain more information about the environment? ____________

____

____

3. During this study, an assumption was made that the environmental factors were the same on each side of the log.

(a) Do you think this assumption is valid? ____________________

(b) Explain you reasoning: ____________________

____

____

4. (a) Using pooled data at each distance, how could you use the data to test if the differences in millipede numbers with distance from the log were significant? Explain your choice:

____

____

(b) Complete your test using a spreadsheet or on a separate sheet of paper and attach it to this page.

ISBN: 978-1-927309-20-9

144 Ecosystems are Dynamic

Key Idea: Natural ecosystems are dynamic systems, responding to short-term and cyclical changes, but remaining relatively stable in the long term.

Ecosystems experience constant changes, from the daily light-dark cycle and seasonal changes, to the loss and gain of organisms. However, over the long term, a mature (or climax) ecosystem remains much the same, a situation known as a **dynamic equilibrium**.

The dynamic ecosystem

- Ecosystems are dynamic in that they are constantly changing. Temperature changes over the day, water enters as rain and leaves as water vapour, animals enter and leave. Many ecosystem changes are cyclical. Some cycles may be short term e.g. the seasons, others long term, e.g climatic cycles such as El Niño.
- Although ecosystems may change constantly over the short term, they may be relatively static over the middle to long term. For example, some tropical areas have wet and dry seasons, but over hundreds of years the ecosystem as a whole remains unchanged.
- However, over the long to very long term ecosystems change as the position of the continents and tilt of the Earth change, and as animals and plants evolve.

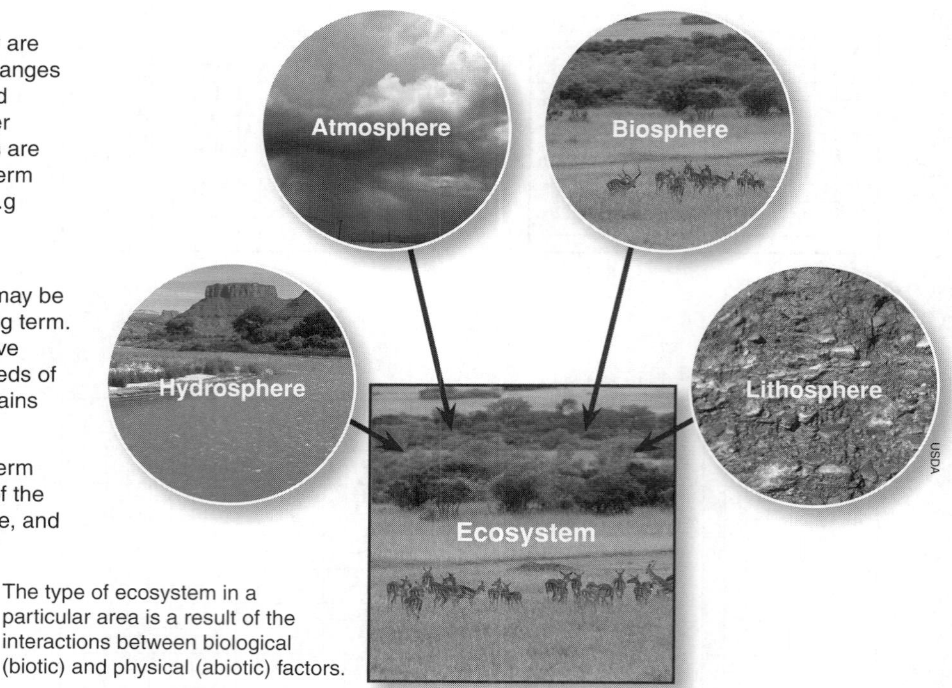

The type of ecosystem in a particular area is a result of the interactions between biological (biotic) and physical (abiotic) factors.

An ecosystem may remain stable for many hundreds or thousands of years provided that the components interacting within it remain stable.

Small scale changes usually have little effect on an ecosystem. Fire or flood may destroy some parts, but enough is left for the ecosystem to return to is original state relatively quickly.

Large scale disturbances such as volcanic eruptions, sea level rise, or large scale open cast mining remove all components of the ecosystem, changing it forever.

1. What is meant by the term dynamic ecosystem? ______________________

____

2. (a) Describe two small scale events that an ecosystem may recover from: ______________________

____

(b) Describe two large scale events that an ecosystem may not recover from: ______________________

____

3. "Climax communities are ones that have reached an equilibrium." Explain what this means: ______________________

____

____

____

ISBN:978-1-927309-20-9

145 Ecosystems are Resilient

Key Idea: An ecosystem's resilience depends on its health, biodiversity, and the frequency with which it is disturbed.

Resilience is the ability of the ecosystem to resist damage and recover after disturbance. Resilience is influenced by three main factors: the ecosystem's biodiversity, its health or intactness, and the frequency of disturbance. Some ecosystems, because of their particular species assemblages, are naturally more resilient than others,

Factors affecting ecosystem resilience

Ecosystem biodiversity

- The greater the diversity of an ecosystem the greater the chance that all the roles (niches) in an ecosystem will be occupied, making it harder for invasive species to establish and easier for the ecosystem to recover after a disturbance.

Ecosystem health

- Intact ecosystems are more likely to be resilient than ecosystems suffering from species loss or disease.

Disturbance frequency

- Single disturbances to an ecosystem can be survived, but frequent disturbances make it more difficult for an ecosystem to recover. Some ecosystems depend on frequent natural disturbances for their maintenance, e.g. grasslands rely on natural fires to prevent shrubs and trees from establishing. The keystone grass species have evolved to survive frequent fires.
- A study of coral and algae cover at two locations in Australia's Great Barrier Reef (right) showed how ecosystems recover after a disturbance. At Low Isles, frequent disturbances (e.g. from cyclones) made it difficult for corals to reestablish, while at Middle Reef, infrequent disturbances made it possible to coral to reestablish its dominant position in the ecosystem.

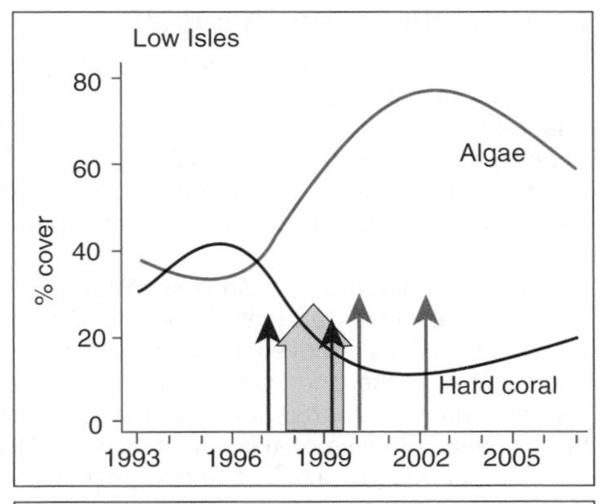

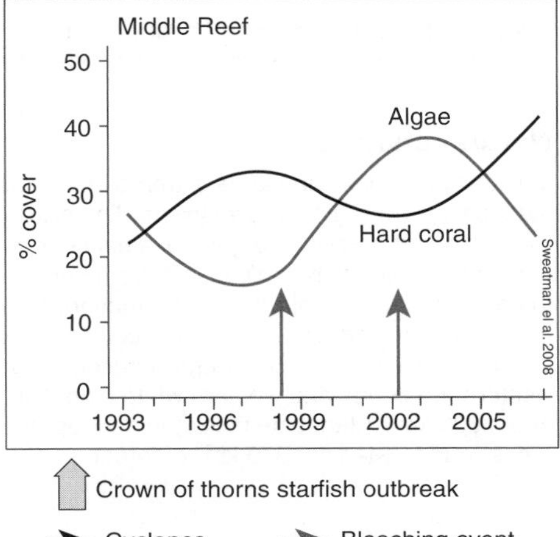

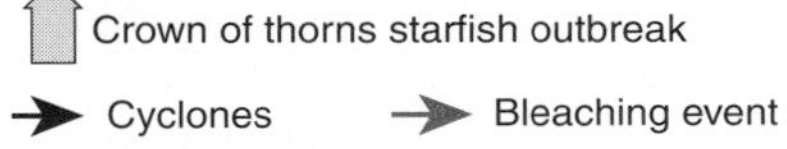

Low Island is one of two islands that form the Low Isles. It is the site of a working lighthouse and weather station, but is uninhabited.

Stability

The stability of an ecosystem can be illustrated by a ball in a tilted bowl. The system is resilient under a certain set of conditions. If the ball is given a slight push (the disturbance is small) it will eventually return to its original state (**line A**). However if the ball is pushed too hard (the disturbance is too large) then the ball will roll out of the bowl and the original state with never be restored (**line B**).

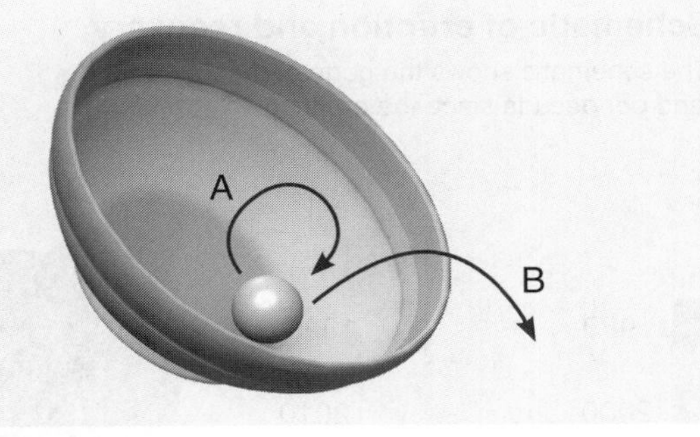

1. Define ecosystem resilience: ______________________________

____

2. Why did the coral at Middle Reef remain abundant from 1993 to 2005 while the coral at Low Isles did not?

____

____

____

3. A monoculture (single species system) of cabbages is an unstable ecosystem. What is required for a cabbage monoculture to remain unchanged?

____

____

ISBN: 978-1-927309-20-9

146 Ecosystem Changes

Key Idea: Sometimes disturbances to an ecosystem are so extreme that the ecosystem never returns to its original state. Ecosystems are dynamic, constantly fluctuating between different states, as occurs during seasonal changes. However, large scale changes to the characteristics of an ecosystem can occur as a result of climate change, volcanic eruptions, or large scale fires. These drivers for change can alter an ecosystem so much that it never returns to its previous state.

Human influenced changes

Peat bogs take thousands of years to develop. They form in wet areas where plant material is prevented from decay by acidic and anaerobic conditions. When dry, peat is used as a fuel source. Mining of peat on a large scale began in Europe in the first millennium. The mining of peat destroys the peat ecosystem. Peat forms very slowly, so it may take hundreds or thousands of years for a mined peat bog to recover (if ever).

Advances in peat mining during the 1500s allowed peat bogs to be mined far below the water table. The mined areas filled with water to form lakes. This removed any chance of the peat bog recovering and also meant the land was no longer available for agriculture. In the Netherlands, 115-230 hectares of land a year was being lost to the formation of these peat lakes. In total, the Netherlands has lost 60 000 hectares (600 km^2) of land to peat mining.

Photograph, right: Abandoned peat workings. The mining of peat below the water table has resulted in flooded man-made lakes.

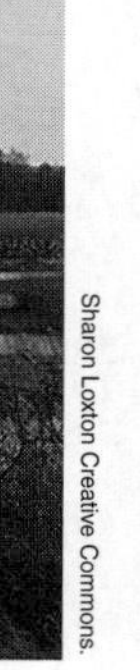

Natural changes

Volcanic eruptions can cause extreme and sudden changes to the local (or even global) ecosystem. The eruption of Mount St. Helens in 1980 provides a good example of how the natural event of a volcanic eruption can cause extreme and long lasting changes to an ecosystem. Before the 1980 eruption, Mount St. Helens had an almost perfect and classical conical structure. The forests surrounding it were predominantly conifer, including Douglas-fir, western red cedar, and western white pine. The eruption covered about 600 km^2 (dark blue below) in ash (up to 180 m deep in some areas) and blasted flat 370 km^2 of forest.

Mount St. Helens, one day before the eruption.

Schematic of eruption and recovery

The schematic shows the general area of bare land per decade since the eruption

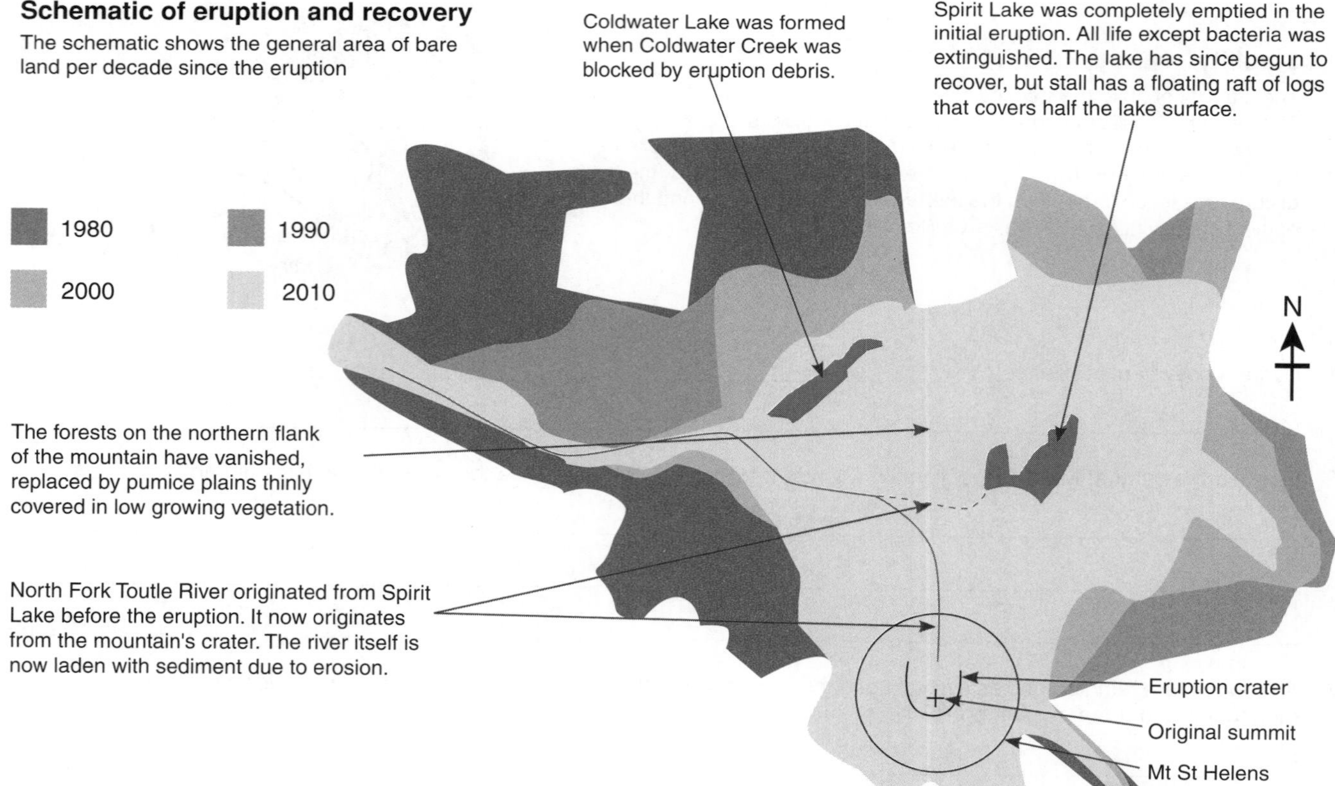

ISBN:978-1-927309-20-9

1. (a) Explain why peat is cut from peat bogs:

   (b) Why does large scale cutting of peat from peat bogs destroy the bog ecosystem?

   (c) Explain why mining peat from below the water table causes very large scale ecosystem change:

2. Describe the major change in the ecosystem on Mount St Helens' northern flank after the eruption:

3. Study the eruption schematic on the previous page. Why has the recovery of the ecosystem area shown in light blue (2010) been so much slower than elsewhere?

4. Describe the large scale change that occurred at Coldwater Creek after the eruption. What effect would this have had on the local ecosystem?

5. Describe the large scale change that occurred at Spirit Lake. Explain why this would cause an almost complete change in the lake ecosystem:

6. (a) What two major changes have occurred in the North Fork Toutle River?

   (b) How might this affect this riverine ecosystem?

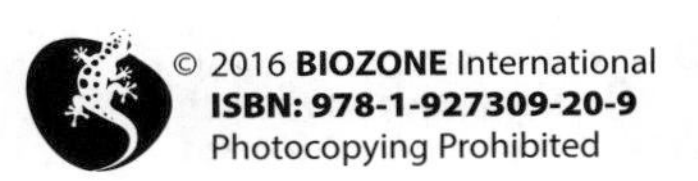

ISBN: 978-1-927309-20-9

147 Primary Succession

Key Idea: Primary succession is a type of ecological succession occurring in a region where there is no pre-existing vegetation or soil.

Ecological succession is the process by which communities change over time. Succession occurs as a result of the interactions between biotic and abiotic factors. Earlier communities modify the physical environment, making it more favourable for species that make up the later communities. Over time, a succession results in a stable climax community. **Primary succession** is a type of ecological succession describing the colonisation of a region where there is no pre-existing vegetation or soil.

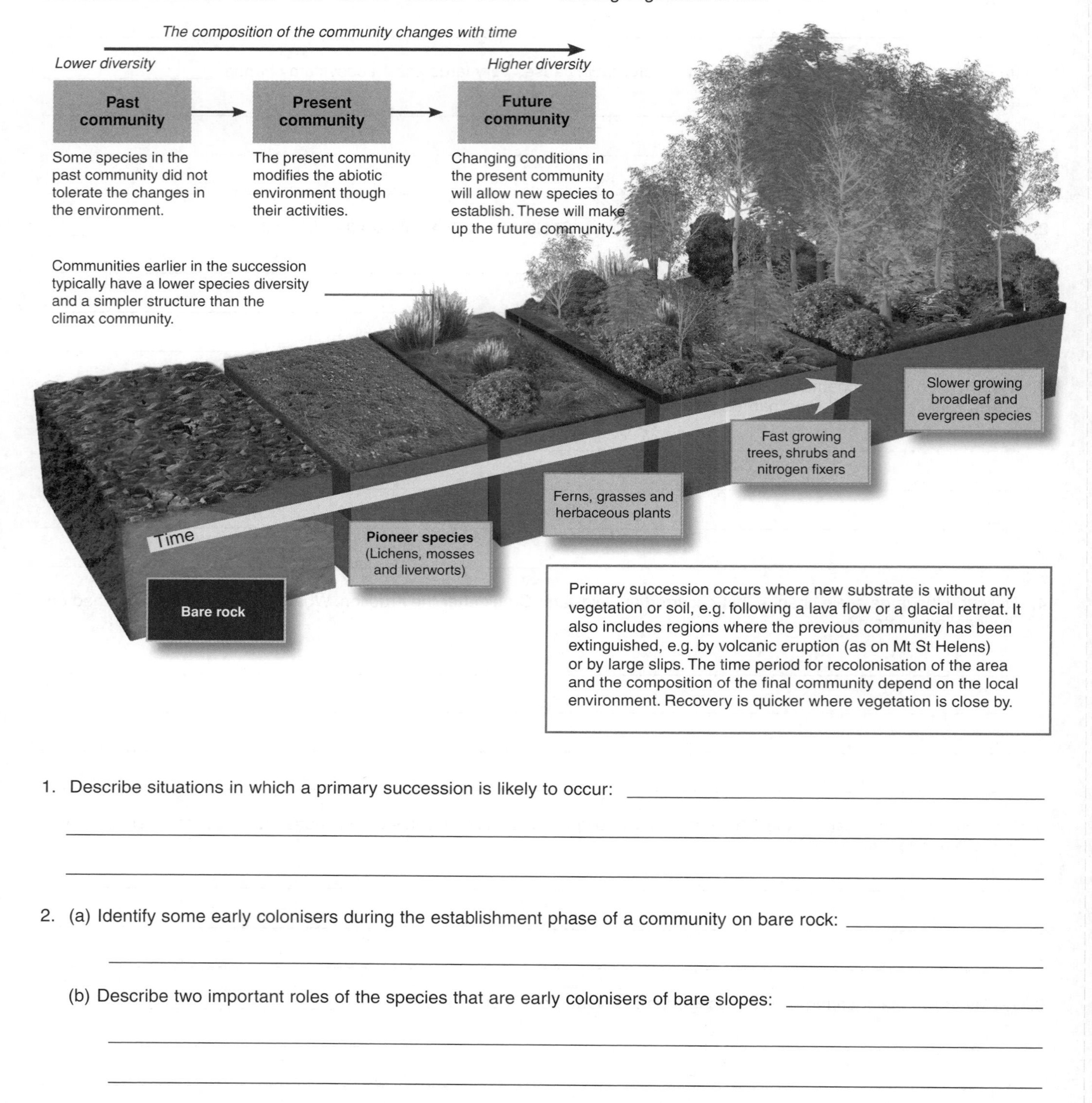

Primary succession occurs where new substrate is without any vegetation or soil, e.g. following a lava flow or a glacial retreat. It also includes regions where the previous community has been extinguished, e.g. by volcanic eruption (as on Mt St Helens) or by large slips. The time period for recolonisation of the area and the composition of the final community depend on the local environment. Recovery is quicker where vegetation is close by.

1. Describe situations in which a primary succession is likely to occur: ______________________________

2. (a) Identify some early colonisers during the establishment phase of a community on bare rock: ______________________________

   (b) Describe two important roles of the species that are early colonisers of bare slopes: ______________________________

3. Explain why climax communities are more stable and resistant to disturbance than early successional communities: ______________________________

ISBN:978-1-927309-20-9

148 Succession on Surtsey Island

Key Idea: The successional events occurring on the island of Surtsey confirm that primary succession occurs in stages.

Surtsey Island is a volcanic island, 33 km off Iceland. The island was formed over four years from 1963 to 1967 when a submerged volcano built up an island. The island is 150 m above sea level and 1.4 km^2. As an entirely new island, Surtsey provided researchers with an ideal environment to study primary succession. Its colonisation by plants and animals has been recorded since its formation. The first vascular plant was discovered in 1965, two years before the eruptions ended. Since then, 69 plant species have colonised the island and there are several established seabird colonies.

The first stage of colonisation of Surtsey (1965-1974) was dominated by shore plants colonising the northern shores of the island. The most successful coloniser was *Honckenya peploides* which established on tephra sand and gravel flats. It first set seed in 1971 and then spread across the island. Carbon and nitrogen levels in the soil were very low during this time. This initial colonisation by shore plants was followed by a lag phase (from 1975-1984). There was further establishment of shore plants but few new colonisers, which slowed the rate of succession.

After the establishment of a gull colony on the southern end of the island a number of new plant species arrived (1985 -1994). Populations of plants inside or near the gull colony expanded rapidly covering about 3 ha, while populations outside the colony remained low but stable. Grasses such as *Poa annua* formed extensive patches of vegetation. After this rapid increase in plant species, the arrival of new colonisers again slowed (1995-2008). A second wave of colonisers began to establish following this slower phase and soil organic matter increased markedly. The first bushy plants established in 1998, with the arrival of the willow *Salix phylicifolia*. The area of vegetation cover near the gull colony expanded to about 10 ha.

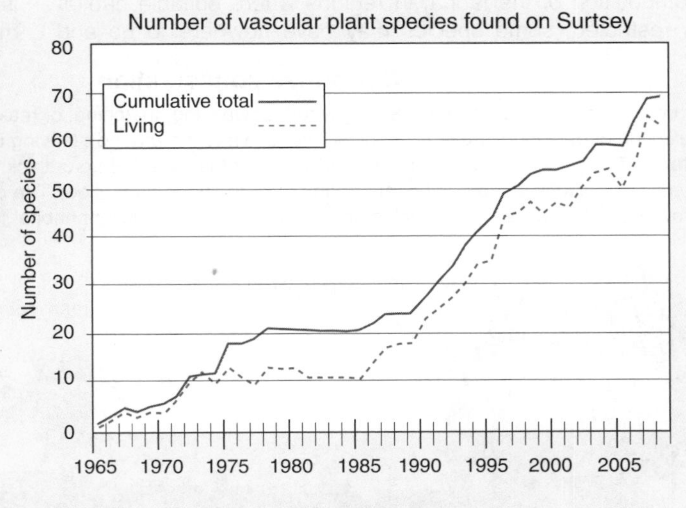

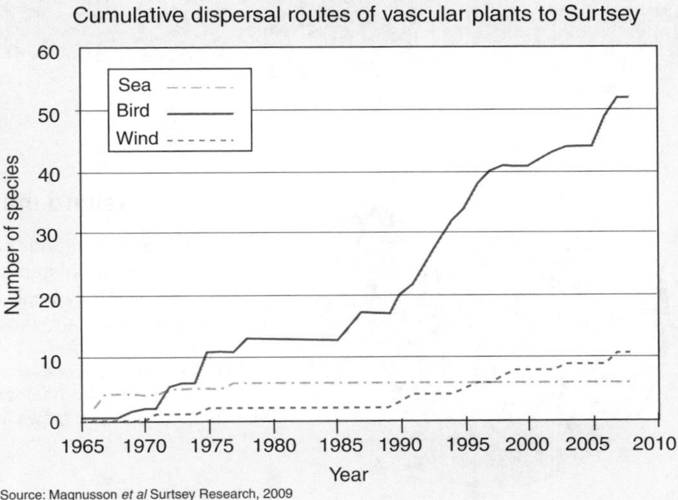

Source: Magnusson *et al* Surtsey Research, 2009

1. Explain why Surtsey provided ideal conditions for studying primary succession: __________

2. Explain why the first colonising plants established in the north of the island, while later colonisers established in the south.

3. Use the graphs to identify the following:

   (a) The year the gull colony established: __________

   (b) The most common method for new plant species to arrive on the island: __________

   (c) The year of the arrival of the second wave of plant colonisers. Suggest a reason for this second wave of colonisers:

ISBN: 978-1-927309-20-9

LINK 147 KNOW

149 Conservation and Succession

Key Idea: Some habitats need to be actively managed in order for them to be maintained over time.

All ecosystems change over time as a result of normal successional processes. In some systems, these changes may be regular and seasonal, as occurs with seasonal flooding or fires. In most environments, successional changes markedly change the habitat and result in a shift in the composition of the fauna. In regions where suitable habitat is restricted, some species may have nowhere to go and may then face local or total extinction. In these situations, conservation of particular species and their environment may be achieved through managing the habitat to delay or stop successional changes. Many British wild areas, which have existed long enough to have their own unique flora and fauna, are maintained in this way. The natural succession is **deflected** to produce a **plagioclimax** community that is maintained through management tools such as grazing, mowing, and coppicing.

Dormouse conservation

Coppicing is the practice of harvesting wood for weaving, thatching, or making charcoal. A selection of deciduous trees is coppiced (cut down to the ground), leaving the stumps known as **stools**. This forces a change in the growth form of the plant from one thick trunk to many thin stems and results in more light penetrating through to the forest floor. The extra light allows the growth of wildflowers and brambles, the latter of which is excellent habitat for dormice.

Dormice feed on nuts and berries, both of which they can find in coppiced forest lands such as hazel and oak. Numbers of dormice in the UK are in decline as a result of loss of habitat, which can be attributed partly to the increasingly uncommon practice of coppicing. Coppicing helps to maintain a stable habitat for dormice by halting the natural succession of scrubland and bramble to mature forest.

Grassland management

Grasslands and moorlands comprise low vegetation and grasses. Both areas provide food and shelter for a number of game animals including pheasants and grouse. As the plants age, their growth slows and they begin to be out-competed or over-topped by taller plants. In order to maintain grasslands and moor, the grasses and plants are grazed by sheep or periodically burned.

In moorlands, burning is normally carried out around every 12-15 years (under strict controls) because, after this period, the vegetation's growth rate slows significantly. Burning the plants maintains their vigour while at the same time removing unwanted vegetation. Where burning is not possible, mechanical cutting may be used. However, the cut vegetation must be removed to prevent the new growth being smothered.

1. Explain why many conservation projects must necessarily slow or stop succession: ______________________________

____

____

2. Describe the effect on biodiversity these kinds of conservation measures ultimately have: ______________________________

____

____

ISBN:978-1-927309-20-9

150 Conservation and Sustainability

Key Idea: Conservation and sustainability encompass the idea that resources should be managed so that they are replenished and available for future use.

Conservation is a term describing the management of a resource so that it is maintained into the future. It encompasses resources of all kinds, from plant and animal populations to mineral resources. Resource conservation has become an important theme in the twenty-first century as the rate of resource use by the expanding human population increases markedly. **Sustainability** refers to management so that the system or resource is replenished at least at same rate at which it is used. Sustainability is based on the idea of using resources within the capacity of the environment. As such, it allows for managed development and resource use.

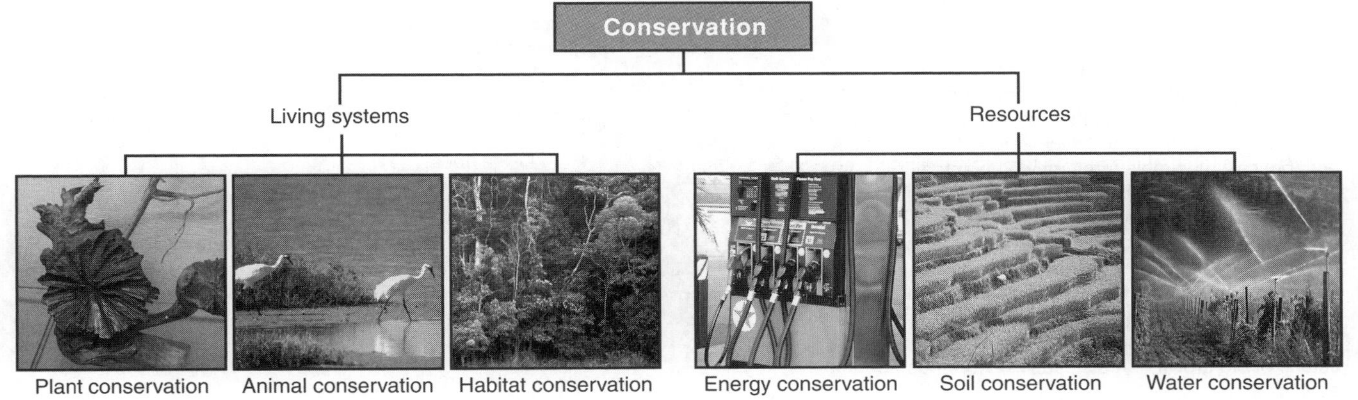

Plant conservation | Animal conservation | Habitat conservation | Energy conservation | Soil conservation | Water conservation

The conservation of living systems focuses heavily on the management of species so that their population numbers remain stable or increase over time. Many living systems have no directly measurable economic value but are important for global biodiversity. Many people also support the moral view that humans do not have the right to exterminate other organisms.

Conservation of resources focuses on the efficient use of resources so that remaining stocks are not wasted. Many of these resources are scarce or economically important so require prudent use. Others are damaging to the environment and it is better to use less of them. In recent decades, there has been a growing acknowledgement that humans cannot afford to continue to waste natural resources.

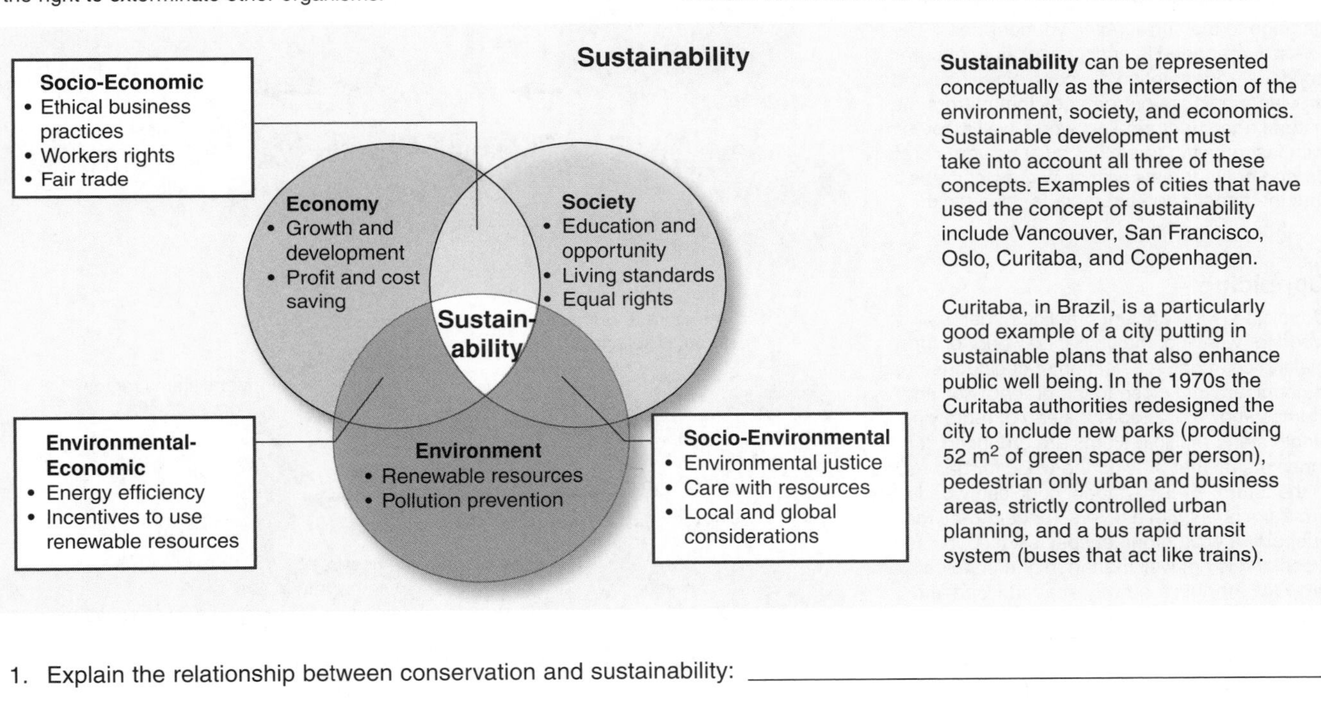

Sustainability can be represented conceptually as the intersection of the environment, society, and economics. Sustainable development must take into account all three of these concepts. Examples of cities that have used the concept of sustainability include Vancouver, San Francisco, Oslo, Curitaba, and Copenhagen.

Curitaba, in Brazil, is a particularly good example of a city putting in sustainable plans that also enhance public well being. In the 1970s the Curitaba authorities redesigned the city to include new parks (producing 52 m^2 of green space per person), pedestrian only urban and business areas, strictly controlled urban planning, and a bus rapid transit system (buses that act like trains).

1. Explain the relationship between conservation and sustainability: ______________________

____

____

2. Explain the concept of resource conservation: ______________________

____

____

3. What is the importance of society and economy in the conservation of living systems and resources?

____

____

ISBN: 978-1-927309-20-9

LINK 152 LINK 151 KNOW

151 Sustainable Forestry

Key Idea: Sustainable forestry employs a variety of removal and replacement methods that provide timber while reducing the impact of timber removal on the environment.

For forestry to be sustainable, demand for timber must be balanced with the regrowth of seedlings. Sustainable forestry allows timber demands to be met without over-exploiting the timber-producing trees. Different methods for logging are used depending on the type of forest being logged. In the UK, afforestation programmes have increased the area of forest to 12%, totalling about 23 000 square kilometres. Careful management of these forests has allowed the forested area to double since 1947. Constant management is needed to ensure this resource continues to be used sustainably without damaging the ecosystems in each case.

Clear cutting

A section of a mature forest is selected (based on tree height, girth, or species), and all the trees are removed. During this process the understorey is destroyed. A new forest of economically desirable trees may be planted. In plantation forests, the trees are generally of a single species and may even be clones. Clear cutting is a very productive and economical method of managing a forest, however it is also the most damaging to the natural environment. In plantation forests, this may not be of concern and may not affect sustainability, but clear cutting of old growth forests causes enormous ecological damage.

Selection logging

A mature forest is examined, and trees are selected for removal based on height, girth, or species. These trees are felled individually and directed to fall in such a way as to minimise the damage to the surrounding younger trees. The forest is managed in such a way as to ensure continual regeneration of young seedlings and provide a balance of tree ages that mirrors the natural age structure. This works well in forests with fast growing trees, but must be very carefully carried out in forests with slow growing trees so that the forest's age structure is not affected.

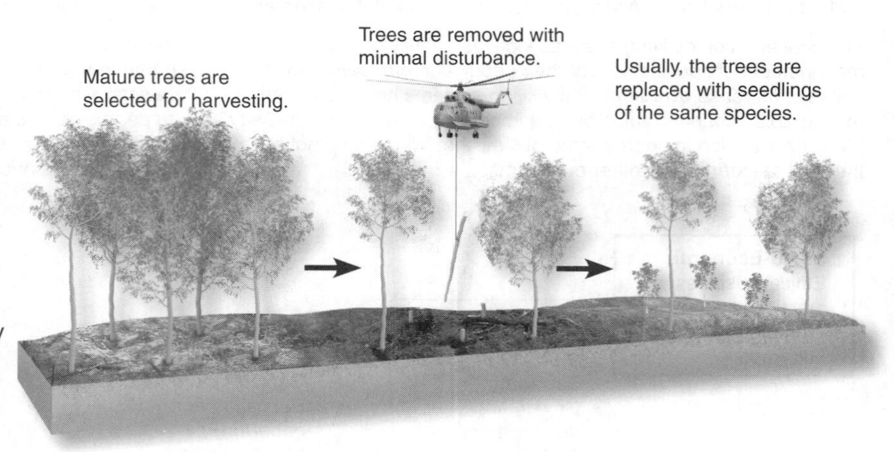

Coppicing

Coppicing is the ancient practice of harvesting wood for weaving, thatching, firewood, or for making charcoal. A selection of deciduous trees is coppiced (cut close to the ground), leaving stumps known as stools. Instead of regrowing a single stem, multiple stems are produced. It is these stems that provide the wood for harvesting in the future. Well managed, coppiced woodlands are quite open with a diverse understorey that supports a wide range of species. If coppiced woodland is not well managed, the coppice stems grow tall forming a heavily shaded woodland with little ground vegetation. These **overstood** coppices are relatively low diversity because they no longer support open woodland species, but also lack the characteristics of old growth high forest (e.g. trees of different ages).

A mature deciduous forest is selected.

Trees are coppiced, or cut close to the ground.

Many stems regrow from each stool.

Strip cutting

Strip cutting is a variation of clear cutting. Trees are clear cut out of a forest in strips. The strip is narrow enough that the forest on either side is able to reclaim the cleared land. As the cleared forest reestablishes (3-5 years) the next strip is cut. This allows the forest to be logged with minimal effort and damage to forest on either side of the cutting zone, while at the same time allowing the natural reestablishment of the original forest. Each strip is not cut again for around 30 years, depending on regeneration time.

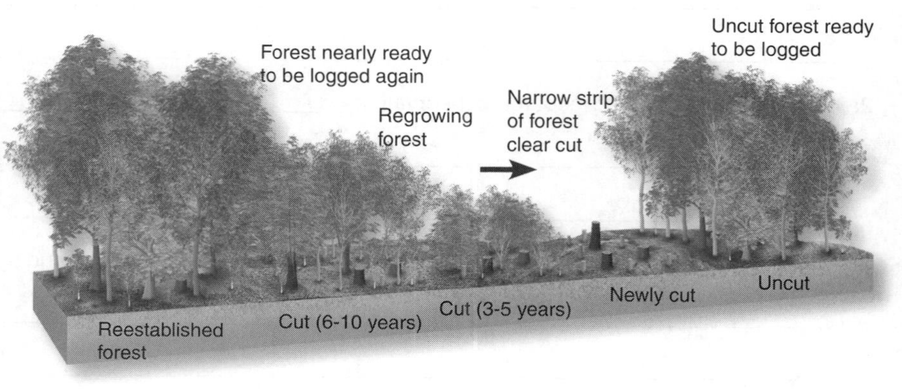

ISBN:978-1-927309-20-9

Old growth forests are climax communities. They have remained undisturbed by natural events and human interference for many hundreds of years. Old growth forests are ecologically significant because of their high biodiversity, and they are often home to endangered or endemic species. Larger forests also play a part in climate modification.

Second growth forests result from secondary ecological succession after a major forest disturbance such as fire or logging. At first, these forests may have quite different characteristics from the original community, especially if particular tree species were removed completely by logging. As the forest develops, the trees are often of the same age so that a single canopy develops.

Commercial plantations (tree farms) are specifically planted and grown for the production of timber or timber based products. These forests are virtual monocultures containing a specific timber tree, such as *Pinus radiata* (Monterey or radiata pine). These trees have often been selectively bred to produce straight-trunked, uniform trees that grow quickly and can be easily harvested and milled.

1. Describe the advantages and disadvantages of each of the following methods of logging:

   (a) Coppicing: ______

   (b) Strip cutting: ______

   (c) Clear cutting: ______

   (d) Selective logging: ______

2. With respect to sustainability, which of the four methods described on the previous page is best suited to:

   (a) Commercial plantations: ______

   (b) Traditional woodland of hardwood species such as hazel, ash, and oak: ______

   (c) Second growth forest: ______

3. Explain which logging method you think best suits the ideals of sustainability and explain your answer:

4. Discuss the role of coppicing as a method of sustainable forestry that also provides a means to conserve biodiversity:

ISBN: 978-1-927309-20-9

152 Sustainable Fishing

Key Idea: Fisheries globally have a history of unsustainable management of stocks. The depletion of fish stocks has necessitated the need for careful management strategies.
Stocks of commercially fished species must be managed carefully to ensure that the catch (take) does not undermine the long term sustainability of the fishery. This requires close attention to stock indicators, such as catch per unit of fishing effort, stock recruitment rates, population age structure, and spawning biomass. Many of the world's major fisheries, including North Sea cod, bluefin tuna, Atlantic halibut, and orange roughy have been severely over exploited and are in danger of complete collapse.

The sustainable harvesting of any food source requires that its rate of harvest is no more than its replacement rate. If the harvest rate is higher than the replacement rate then it follows that the food source will continually reduce at ever increasing percentages (assuming a constant harvest rate) and thus eventually be lost.

Sustainable yield (SY) refers to the number or weight of fish that can be removed by fishing without reducing the stock biomass from year to year. It assumes that the environmental conditions remain the same and do not contribute to fluctuations in biomass levels. The **maximum sustainable yield** (MSY) is the maximum amount of fish that can be taken without affecting the stock biomass and replacement rate. Calculating an MSY relies on obtaining precise data about a population's age structure, size, and growth rates. If the MSY is incorrectly established, unsustainable quotas may be set, and the fish stock may become depleted.

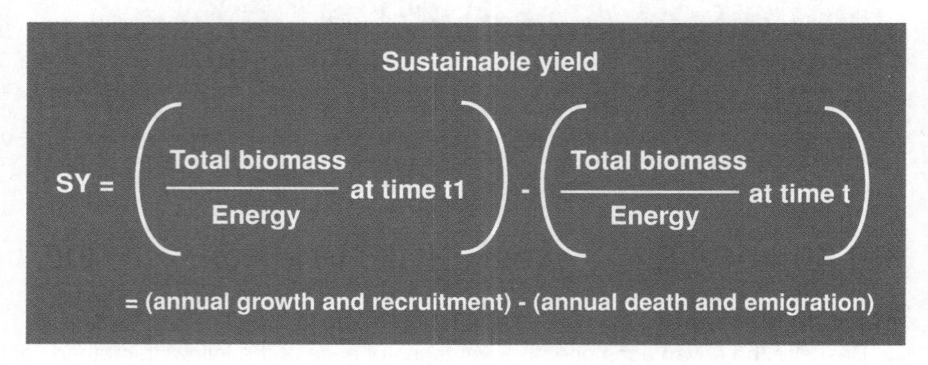

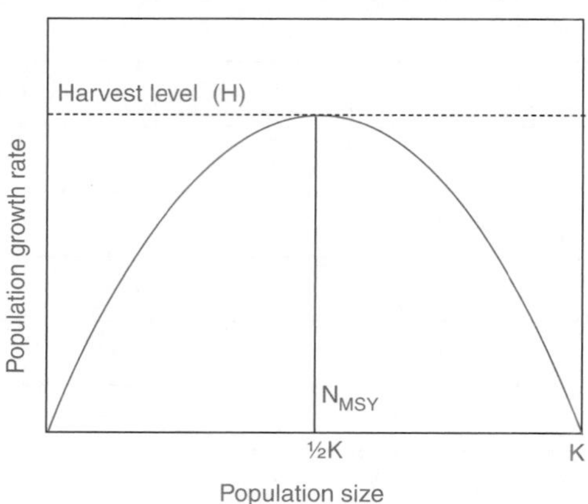

The theoretical maximum sustainable yield (N_{MSY}) occurs when a population is at half the carrying capacity (½K). At this point, the population growth rate will also be at its maximum. Under ideal conditions, harvesting at this rate (H) should be able to continue indefinitely. However, the growth rate of a population is likely to fluctuate from year to year. If a population has below-average growth for several years while the take remains the same, there is a high risk of population collapse because an ever-increasing proportion of the population will be taken with each harvest.

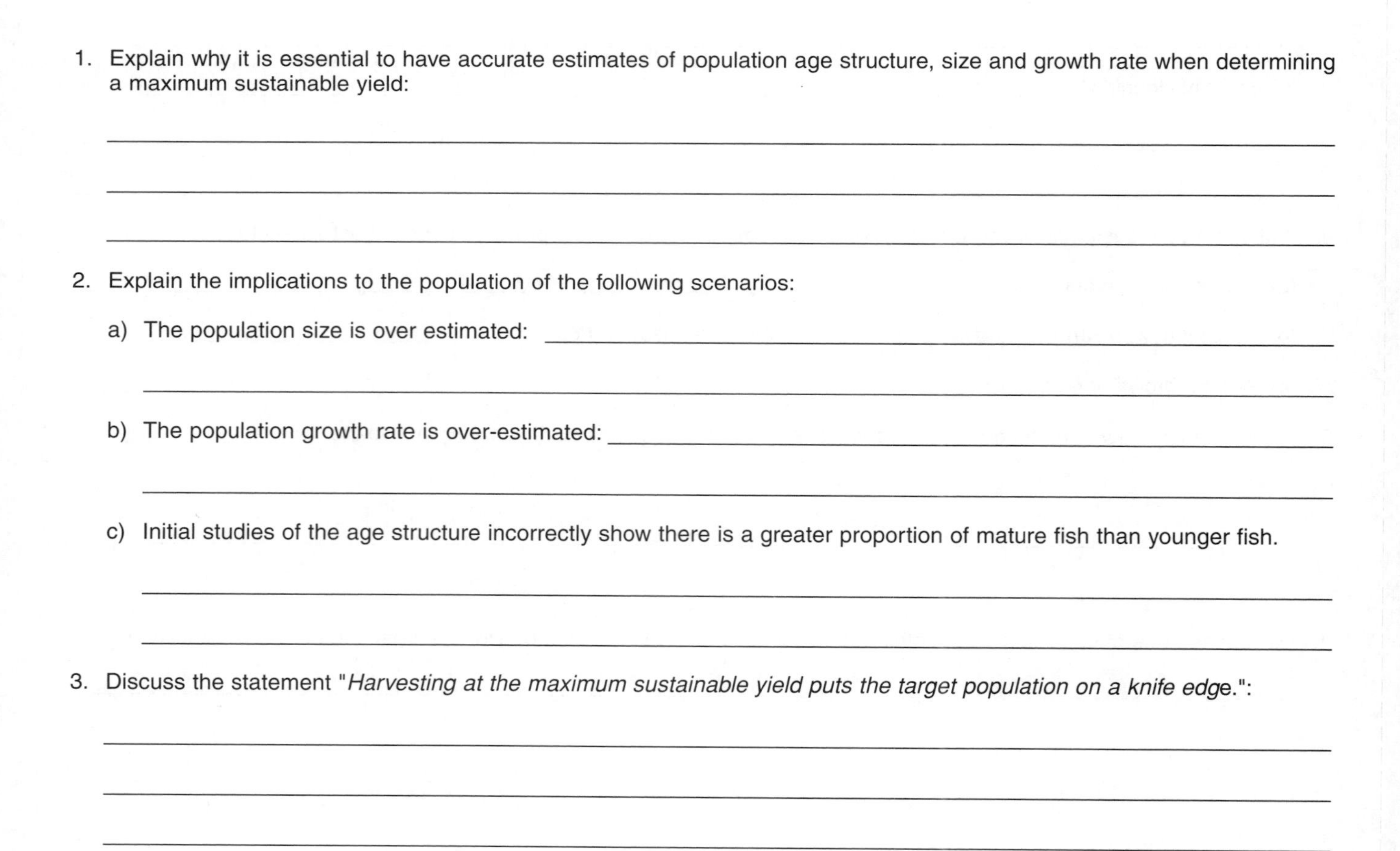

1. Explain why it is essential to have accurate estimates of population age structure, size and growth rate when determining a maximum sustainable yield:

2. Explain the implications to the population of the following scenarios:

   a) The population size is over estimated:

   b) The population growth rate is over-estimated:

   c) Initial studies of the age structure incorrectly show there is a greater proportion of mature fish than younger fish.

3. Discuss the statement "*Harvesting at the maximum sustainable yield puts the target population on a knife edge.*":

ISBN:978-1-927309-20-9

North Sea cod

The stock of North Sea cod (*Gadus morhua*) is one of the world's six large populations of this economically important species. As one of the most intensively studied, monitored, and exploited fish stocks in the North Sea, it is considered a highly relevant indicator of how well sustainable fisheries policies are operating. Currently juvenile catch rates are much higher than adult catch rates. Recent figures show approximately 54 thousand tonnes are caught annually, vastly less than the 350 000 tonnes caught in the early 1970s.

The state of the fishery

- Fishing mortality (the chance of a fish being caught) reached its maximum in 2000 and has reduced sharply since then. However it has been above the theoretical maximum mortality rate due to fishing (F_{MSY}) since at least the 1960s (right middle).
- Recruitment has been generally poor since 1987 (right bottom).
- The number of spawning adults has fallen to levels below those required to recruit new individuals into the stock (bottom right).
- ICES (the International Council for the Exploration of the Sea) advised that the spawning stock biomass (an indicator of the number of breeding adults) reached a new historic low in 2006, and that the risk of stock collapse is high.

What has been done?

- With the cod fishery at imminent collapse, measures were taken to try to save the fishery. Depending on the fishing gear being used, fishing vessels were restricted to between 9 and 25 days at sea per month.
- If stock levels do not rise on track with recovery plans, quotas and fishing days are automatically cut by 20% each year.
- ICES has recommended a zero catch limit until the stock reaches at least 70 000 tonnes.
- Net mesh sizes were increased to reduce juvenile mortality.

There has been staunch opposition by fishing companies to these measures. They argue that livelihoods will be affected and that stock has been recovering.

Throughout its range, cod has been heavily exploited. In Canada's Grand Banks, where *Gadus morhua* is known as the Atlantic cod, the fishery was closed in 1992 after the stock biomass fell to 1% of its historic levels. Nearly 22 000 jobs were lost.

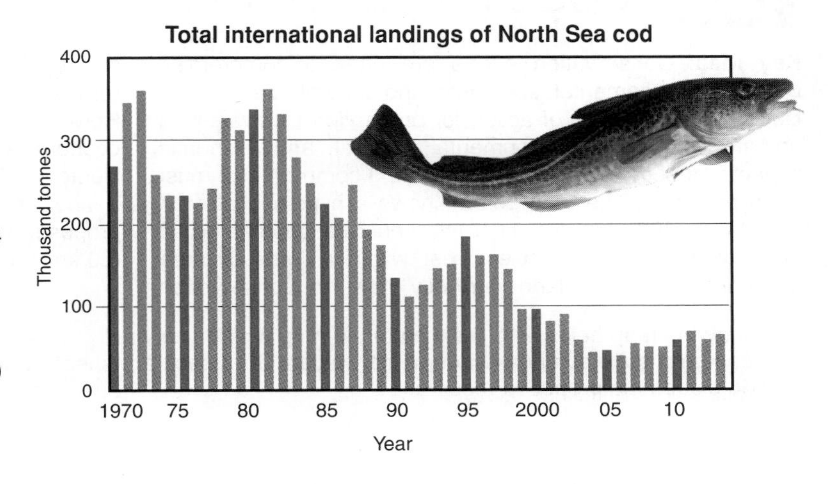

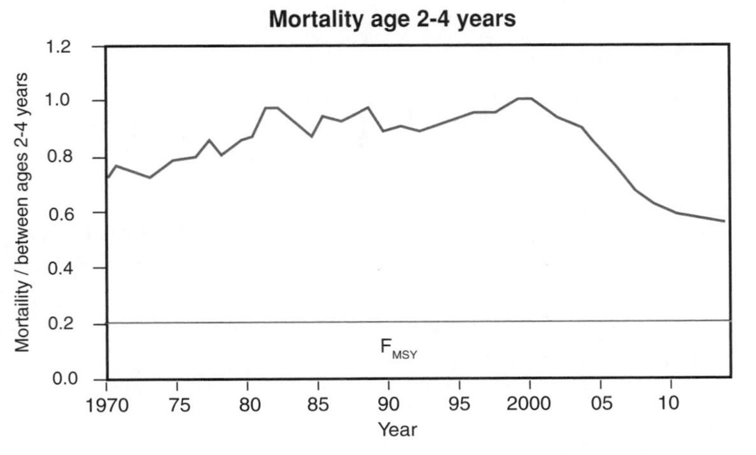

Recruitment and spawning stock biomass of North Sea cod

Recruitment at age 1 / millions: 0, 500, 1000, 1500, 2000, 2500

Spawning stock biomass / 100 tonnes: 0, 50, 100, 150, 200, 250, 300

Recruitment at age 1

Spawning stock biomass

Year: 1970, 75, 80, 85, 90, 95, 2000, 05, 10

4. (a) Describe the trends in the North Sea cod fishery since 1970: __________

(b) The current recommended fisheries target for cod mortality is 0.4. What effect do you expect this will have on stock recovery?

ISBN: 978-1-927309-20-9

153 Modelling a Solution

Key Idea: Conservation efforts are often a compromise between environmental, economic, and cultural needs.

Deciding on a course of action for preserving biodiversity is not always simple. Environmental, cultural, and economic impacts must be taken into account, and compromises must often be made. The map below shows a hypothetical area of 9300 ha (93 km^2) in which two separate populations of an endangered bird species exist within a forested area of public land. The land currently has no conservation status. A proposal to turn part of the area into an wildlife reserve has been put forward by local conservation groups. However, the area is known to have large deposits of economically viable minerals and is frequented by trampers. Hunters also spend time in the area because part of it has an established population of game animals. The proposal would allow a single area of up to 2000 ha (20 km^2) to be reserved exclusively for conservation efforts.

1. Study the map below and draw on to the map where you would place the proposed reserve, taking into account economic, cultural, and environmental values. On a separate sheet, write a report justifying your decision as to where you placed the proposed reserve.

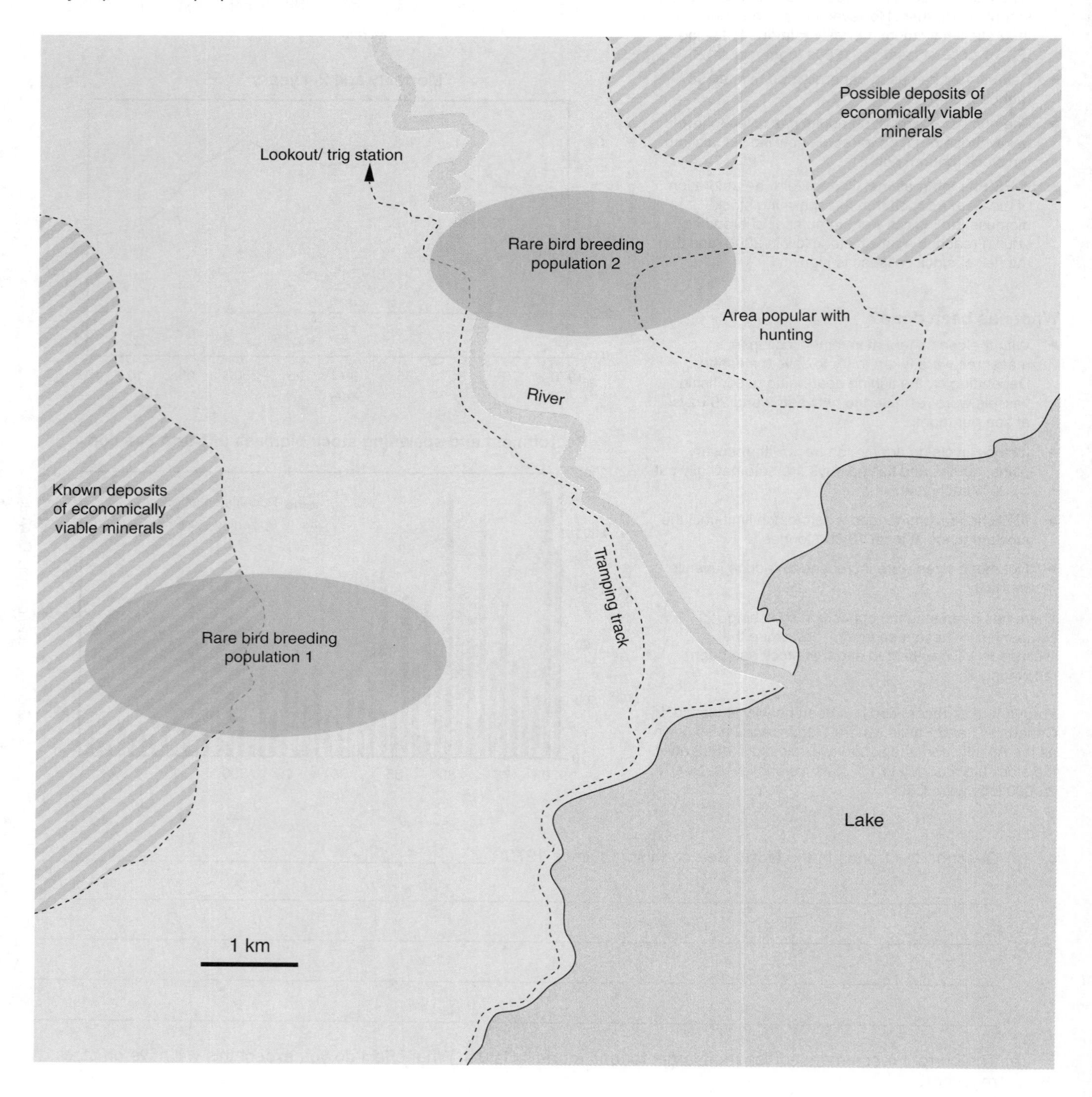

ISBN:978-1-927309-20-9

154 Chapter Review

Summarise what you know about this topic under the headings and sub-headings provided. You can draw diagrams or mind maps, or write short notes to organise your thoughts. Use the images and hints to help you and refer back to the introduction to check the points covered:

Inheritance

HINT: Alleles, monohybrid and dihybrid crosses, and gene interactions.

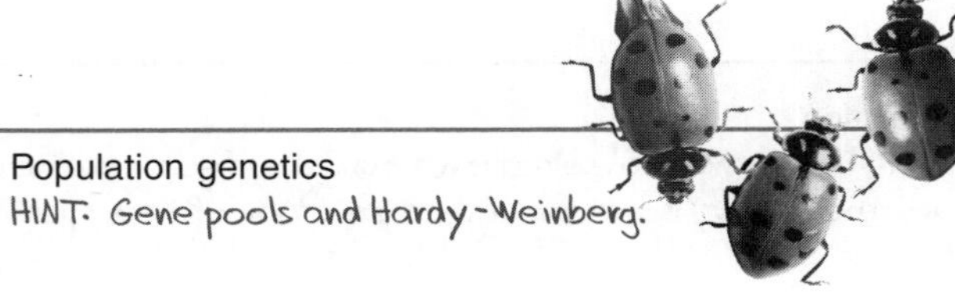

Population genetics

HINT: Gene pools and Hardy-Weinberg.

ISBN: 978-1-927309-20-9

Evolution

HINT: Explain natural selection, founder effect, genetic drift, and the role of isolation in the formation of new species.

Images: Olaf Leillinger cc 2.5

Ecology

HINT: Populations and carrying capacity, ecosystem stability and change. Sustainability.

ISBN:978-1-927309-20-9

155 KEY TERMS AND IDEAS: Did You Get It?

1. Test your vocabulary by matching each term to its definition, as identified by its preceding letter code.

alleles	A	Having two homologous copies of each chromosome (2N), usually one from the mother and one from the father.
autosome	B	Allele that will only express its trait in the absence of the dominant allele.
diploid	C	Genetic cross between two individuals that differ in one trait of particular interest.
dominant	D	Sequences of DNA occupying the same gene locus (position) on different, but homologous, chromosomes, i.e. gene variants.
genotype	E	Observable characteristics in an organism.
monohybrid cross	F	The allele combination of an organism.
phenotype	G	A non-sex chromosome.
Punnett square	H	A graphical way of illustrating the outcome of a cross.
recessive	I	Allele that expresses its trait irrespective of the other allele.

2. The following dihybrid cross shows the inheritance of colour and shape in pea seeds. Yellow (**Y**) is dominant over green (**y**) and a round shape (**R**) is dominant over the wrinkled (**r**) form.

(a) Describe the appearance (phenotype) of pea seeds with the genotype YyRr: ______________________

____

(b) Complete the Punnett square below when two seeds with the YyRr genotype are crossed. Indicate the number of each phenotype in the boxes on the right.

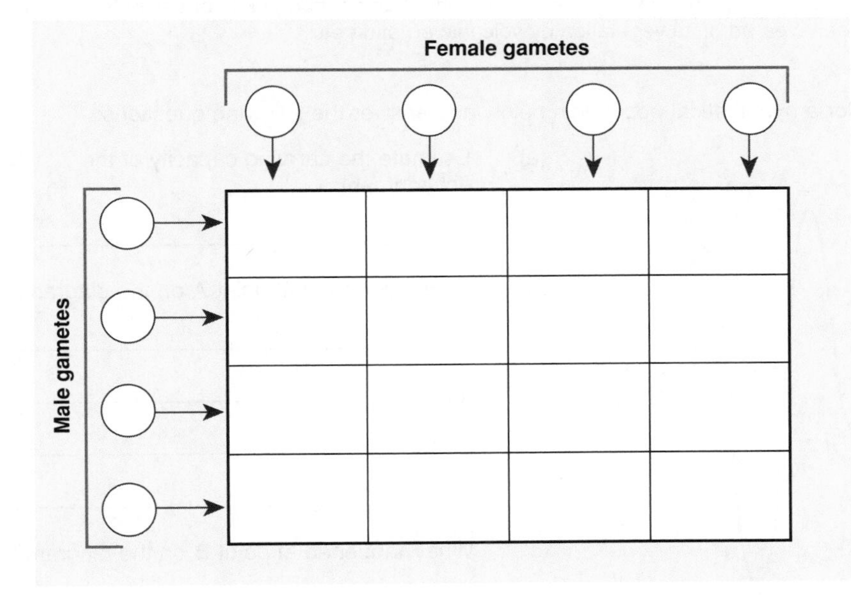

Phenotype	Number
Yellow-round	
Green-round	
Yellow-wrinkled	
Green-wrinkled	

3. Using examples, contrast the characteristics of directional and stabilising selection and their effects:

____

____

____

____

____

ISBN: 978-1-927309-20-9

4. Match each term to its definition, as identified by its preceding letter code.

allopatric speciation	**A**	The process by which heritable traits become more or less common in a population through differential survival and reproduction.
conservation	**B**	An evolutionary event in which a significant proportion of a population's alleles are lost.
founder effect	**C**	The sum total of all alleles of all breeding individuals in a population at any one time.
gene pool	**D**	Competitive interactions that occur between different species.
genetic bottleneck	**E**	The division of one species, as a result of evolutionary processes, into two or more separate species.
genetic drift	**F**	Speciation as a result of reproductive isolation without any physical separation of the populations, i.e. populations remain within the same range.
Hardy-Weinberg principle	**G**	Competitive interactions that occur between members of the same species.
interspecific competition	**H**	Speciation in which the populations are physically separated.
intraspecific competition	**I**	The loss of genetic variation when a new colony is formed by a very small number of individuals from a larger population.
mark and recapture	**J**	The principle of genetic equilibrium that describes the constancy of population allele frequencies in the absence of evolutionary influences.
natural selection	**K**	The situation in which members of a group of organisms breed with each other but not with members of other groups
primary succession	**L**	The change in allele frequency in a population as a result of random sampling. The effect is proportionally larger in small populations.
quadrat	**M**	Sampling method used to determine the size of a population in which individuals from a population are marked and released and then recapture after a set period of time.
reproductive isolation	**N**	The act of protecting a resource so that it will be available in the future.
speciation	**O**	A measured and marked region used to isolate a sample area for study.
sympatric speciation	**P**	A succession sequence that occurs on land that has not had plants or soil in the past or has been cleared of its vegetation by volcanic eruption etc.

5. Study the graph of population growth for a hypothetical population below and answer the following questions:

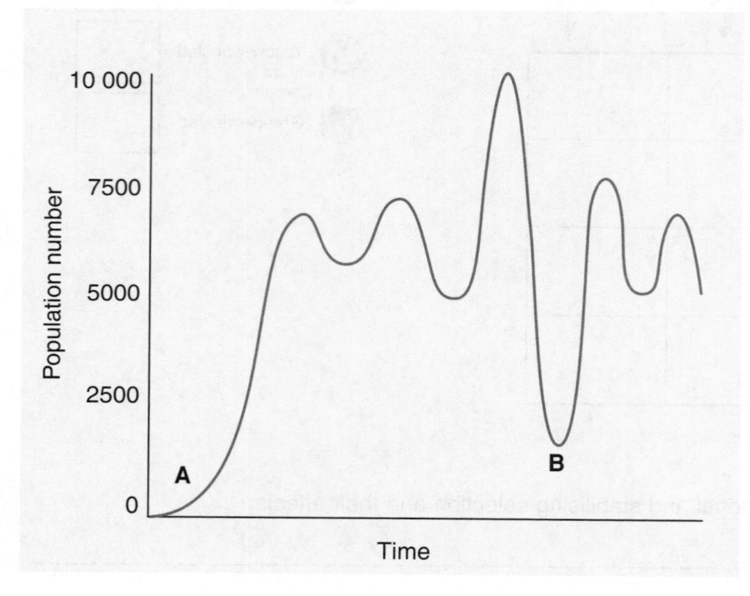

(a) Estimate the carrying capacity of the environment:

(b) What happened at point **A** on the diagram?

(c) What happened at point **B** on the diagram?

(d) What factors might have caused this?

ISBN:978-1-927309-20-9

Topic 8

The control of gene expression

Key terms

chromosome mutation
deletion (of mutation)
DNA hybridisation
DNA ligase
DNA methylation
DNA probe
epigenetics
frame shift
gel electrophoresis
gene mutation
gene therapy
genetic (gene) screening
genetic code
genetic fingerprinting
genome
genome sequencing project
GMO
insertion (of mutation)
multipotent
oncogenes
PCR
pluripotent
proteome
recombinant DNA technology
restriction enzyme
reverse transcriptase
RNA interference
silent mutation
stem cell
substitution
tissue culture
totipotent
transcription
transcription factors
transgenic organisms
translation
tumour
tumour-suppressor genes
VNTRs

8.1 Mutation

Learning outcomes — Activity number

	Learning outcome	Activity number
☐	1 Explain how mutations (changes to the DNA base sequence) can occur as a result of errors during DNA replication. Understand that the mutation rate is increased by mutagenic agents such as ionising radiation and organic solvents.	156
☐	2 Explain what is meant by a gene mutation and describe examples. Recognise base deletion, substitution, and insertion mutations and describe their effect on the encoded polypeptide.	156 157 159 160
☐	3 Describe the consequences of frame shifts downstream of the mutation.	157
☐	4 Recall the degeneracy in the genetic code and explain the consequences of this to the likely effect of mutations affecting the third base in a DNA triplet.	156
☐	5 Recognise mutations involving inversion, duplication, and translocation of larger segments of DNA involving multiple genes (chromosome mutations).	158

Stem Cell Scientist

8.2 Control of gene expression

Learning outcomes — Activity number

	Learning outcome	Activity number
☐	6 Explain what is meant by a totipotent cell and explain how regulation of gene expression in totipotent cells during development results in cell specialisation.	161
☐	7 Using examples, distinguish between pluripotent, multipotent, and unipotent stem cells in terms of their potency and their medical applications.	162
☐	8 Explain how induced pluripotent stem cells (iPS) can be generated from unipotent cells using appropriate protein transcription factors.	161
☐	9 AT Understand the principles of tissue culture in plants. Produce tissue cultures of explants of a brassica such as cauliflower.	163
☐	10 Describe the transcriptional control of gene expression in eukaryotes, including the role of oestrogen in initiating transcription.	164-166 168 169
☐	11 Describe the epigenetic control of gene expression in eukaryotes by DNA methylation of histone modification.	165 166
☐	12 Evaluate data relating to the relative influences of genetic and environmental factors and environment on phenotype.	167
☐	13 Explain the relevance of epigenetics to the development and treatment of disease.	165 166
☐	14 Describe how translation of mRNA from target genes can be inhibited by RNA interference (RNAi).	170
☐	15 Describe the main characteristics of benign and malignant tumours.	171
☐	16 Outline the role of each of the following in the development of tumours: tumour suppressor genes and oncogenes, abnormal methylation of tumour suppressor genes and oncogenes, and increased oestrogen concentrations.	171 172
☐	17 Evaluate evidence showing correlations between genetic and environmental factors and various forms of cancer.	172
☐	18 Understand how an understanding of tumour suppressor genes and oncogenes could be used to prevent, treat, or cure cancer.	171 172

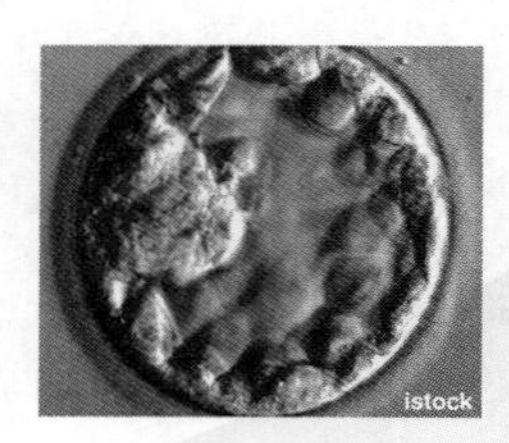

8.3 Genome projects

Learning outcomes

Activity number

- ☐ 19 Using examples, describe the scope of genome sequencing projects. Understand the role that automation has had in the feasibility of large scale genome projects. **173 175**
- ☐ 20 Explain that the purpose of genome analysis extends beyond just determining the DNA sequence. Explain how an understanding of the proteome of simple organisms may have applications in the development of new vaccines. **174 175**
- ☐ 21 Understand that much of the genome of complex organisms includes many regulatory and non-protein coding sequences. Explain the implications of this to determining the proteome for such organisms. **175**

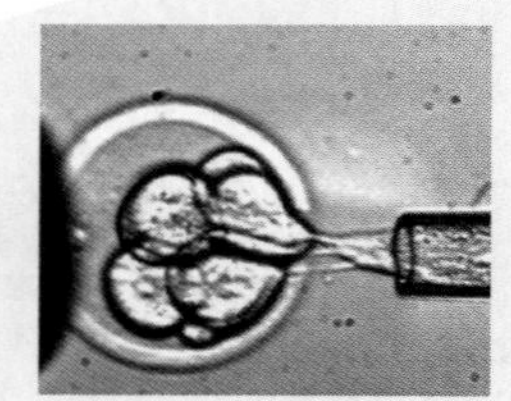

8.4 Applications of gene technologies

Learning outcomes

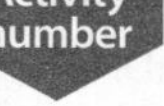
Activity number

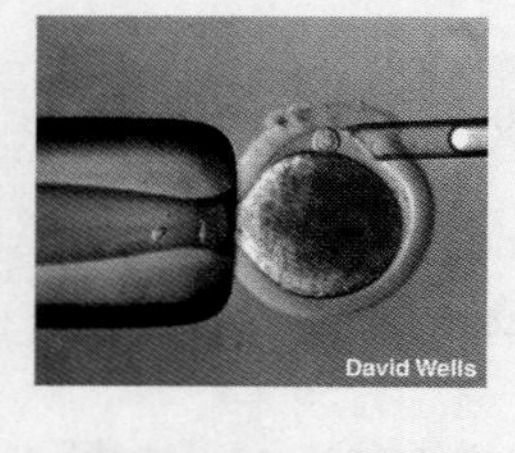

- ☐ 22 Explain what is meant by recombinant DNA technology. Explain how the universality of the genetic code and the mechanisms for transcribing and translating that code enable DNA to be transferred between different species. **178**
- ☐ 23 Describe methods for generating fragments of DNA for transfer:
  - ☐ i Conversion of mRNA to cDNA using reverse transcriptase **177**
  - ☐ ii Using restriction enzymes (endonucleases) to cut a fragment containing a desired gene or DNA sequence from DNA. **178**
  - ☐ iii Creating a synthetic gene in a gene machine. **176**
- ☐ 24 Describe and explain PCR as a *in vitro* method to amplify DNA fragments. **179**
- ☐ 25 Describe and explain the culture of transformed cells as an *in vivo* method to amplify DNA fragments. Include reference to the addition of promoter and terminator sequences to DNA fragments, the use of restriction enzymes and ligases to insert DNA fragments into vectors, the transformation of host cells, and the use of marker genes to detect transformed cells or organisms. **180**

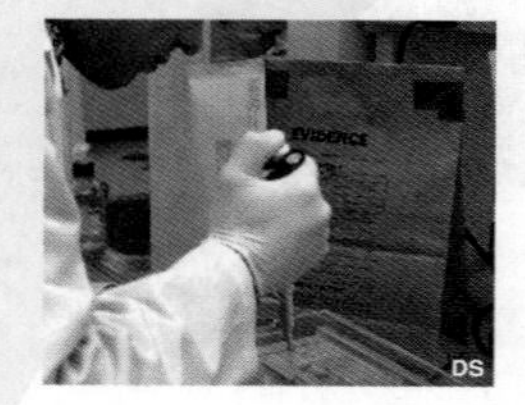

- ☐ 26 Interpret information relating to the use of recombinant DNA technology, e.g. in the production of food and therapeutic proteins. **181 182 183 187**
- ☐ 27 Evaluate the ethical, economic, and social issues associated with the use and ownership of recombinant DNA technology in agriculture, industry, and medicine. **186**
- ☐ 28 Relate recombinant DNA technology to gene therapy, including the role of vectors and transformation of cells. **184 185**
- ☐ 29 Explain how labelled DNA probes and DNA hybridisation are used to locate specific alleles of genes. Describe how gene probes can be used to screen patients for heritable conditions, drug responses, or heath risks. **190**
- ☐ 30 Describe how the information from genetic screening can be used in genetic counselling and personalised medicine. **191-193**
- ☐ 31 Describe how VNTRs in an organism's genome can be used to create a DNA profile or genetic (DNA) fingerprint. **194**
- ☐ 32 Describe how genetic fingerprinting is used to analyse DNA fragments generated by PCR. Describe the applications of DNA fingerprinting in determining genetic relationships and in quantifying the genetic variability in a population. **188 189 194**
- ☐ 33 Describe the use of genetic fingerprinting in forensic science, medical diagnosis, and plant and animal breeding. **195 196**

156 What is a Gene Mutation?

Key Idea: Gene mutations are localised changes to the DNA base sequence.

Gene mutations are small, localised changes in the DNA base sequence caused by a mutagen or an error during DNA replication. The changes may involve a single nucleotide (a **point mutation**) or a triplet. Point mutations can occur by substitution, insertion, or deletion of bases and alter the mRNA transcribed. A point mutation may not alter the amino acid sequence because the degeneracy of the genetic code means that more than one codon can code for the same amino acid. Such mutations are called silent because the change is not recorded in the amino acid sequence. Mutations that do result in a change in the amino acid sequence will most often be harmful because they alter protein functionality.

Silent mutations

Silent mutations do not change the amino acid sequence nor the final protein. In the genetic code, several codons may code for the same amino acid (this so-called degeneracy creates redundancy in the genetic code).

Silent mutations are also neutral if they do not alter fitness, although it is now known that so-called silent changes may still affect mRNA stability and transcription, even though they do not change codon information. In these cases, the changes will not be neutral.

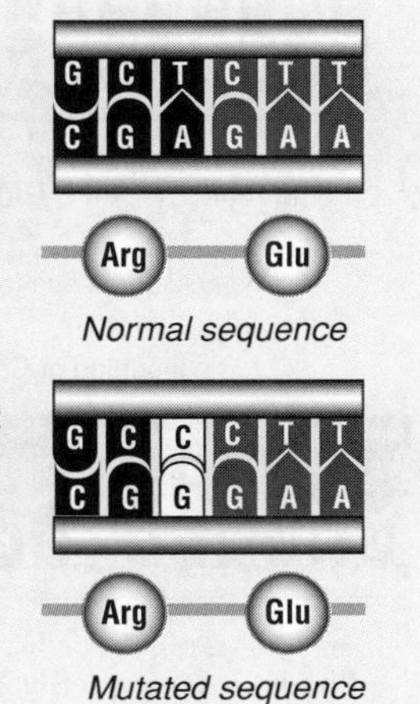

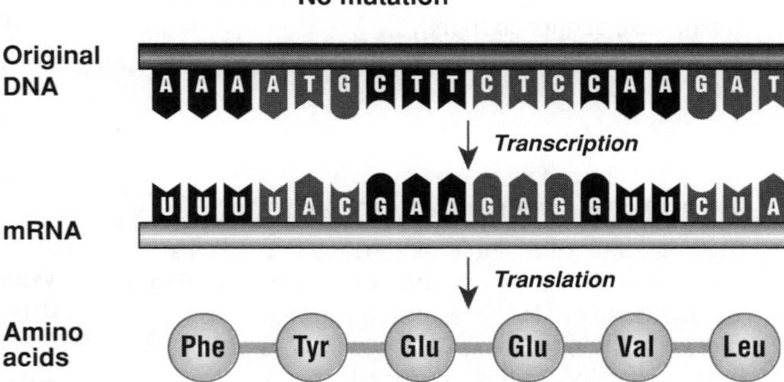

Amino acid sequence forms a normal polypeptide chain

Missense substitution

A single base is substituted for another base. Some substitutions may still code for the same amino acid (because of redundancy in the genetic code), but they may also result in a codon that codes for a different amino acid. In the example (centre right), placing a T where a C should have been in the normal sequence (top right), results in the amino acid **lysine** appearing where **glutamic acid** should be. This could affect this protein's function. If a missense substitution results in an amino acid with similar chemical properties, or the change affects a non-critical part of the protein, the change may be effectively neutral.

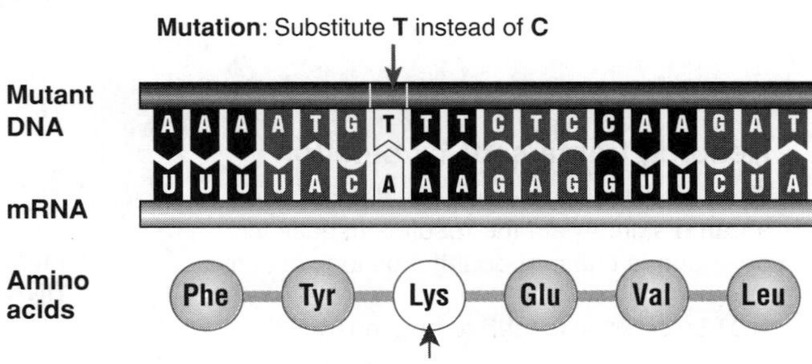

Polypeptide chain with wrong amino acid

Nonsense substitution

Some codon changes are more likely than others to cause a change in the amino acid encoded by the sequence (amino acids encoded by 4 or 6 different codons are less likely to be affected by substitutions). In the example illustrated, a single base substitution in the first nucleotide of the third codon has a dramatic effect on the nature of the encoded polypeptide chain. The codon no longer codes for an amino acid, but instead terminates the translation process of protein synthesis. This results in a very short polypeptide chain that is likely to have little or no function because the **STOP** codon is introduced near the **START** codon.

Mutation: Substitute **A** instead of **C**

Mutant DNA: A A A A T G A T T C T C C A A G A T

mRNA: U U U U A C U A A G A G G U U C U A

Amino acids: Phe – Tyr

Mutated DNA creates a STOP codon which prematurely ends synthesis of the polypeptide chain

1. Why are mutations that alter the amino acid sequence usually harmful? ___

2. Why are nonsense substitutions more damaging than missense substitutions? ___

3. (a) What is a silent mutation? ___

   (b) When is a silent mutation considered to be neutral? ___

ISBN: 978-1-927309-20-9

157 Reading Frame Shifts

Key Idea: Point mutations may cause a reading frame shift. These usually produce a non-functional protein.

A **frame shift mutation** occurs when bases are inserted or deleted from a DNA sequence and the total nucleotide sequence is no longer a multiple of three. All of the nucleotides after the frame shift mutation are displaced, creating a new reading frame sequence of codons. In most cases, there will be a change in the amino acids produced, and the protein produced will be non-functional or incorrect. Many diseases are the result of frame shift mutations.

Normal sequence

All examples of mutations below are given with reference to the normal sequence and its encoded amino acid sequence (right).

Reading frame shift by insertion

The insertion of a single extra base into a DNA sequence displaces the bases after the insertion by one position. In the example (right) instead of the next triplet being CTT it is CCT, and a different amino acid is produced.

NOTE: could also lead to nonsense

Reading frame shift by deletion

The deletion of an nucleotide also causes a frame shift. Again the result is usually a polypeptide chain of doubtful biological activity.

NOTE: could also lead to nonsense

Mutations that cause a shift in the reading frame almost always lead to a non-functional protein, but the effect is reduced the further the mutation is from the START codon.

Partial reading frame shift

Both an insertion and a deletion of bases within a gene can cause a frame shift effect in which each codon no longer has the correct triplet of three bases. In this example, three codons have been affected, along with the amino acids they code for. The error is limited to the codons between the insertion and deletion. There is no biological activity if the amino acids altered are important to the functioning of the resulting protein.

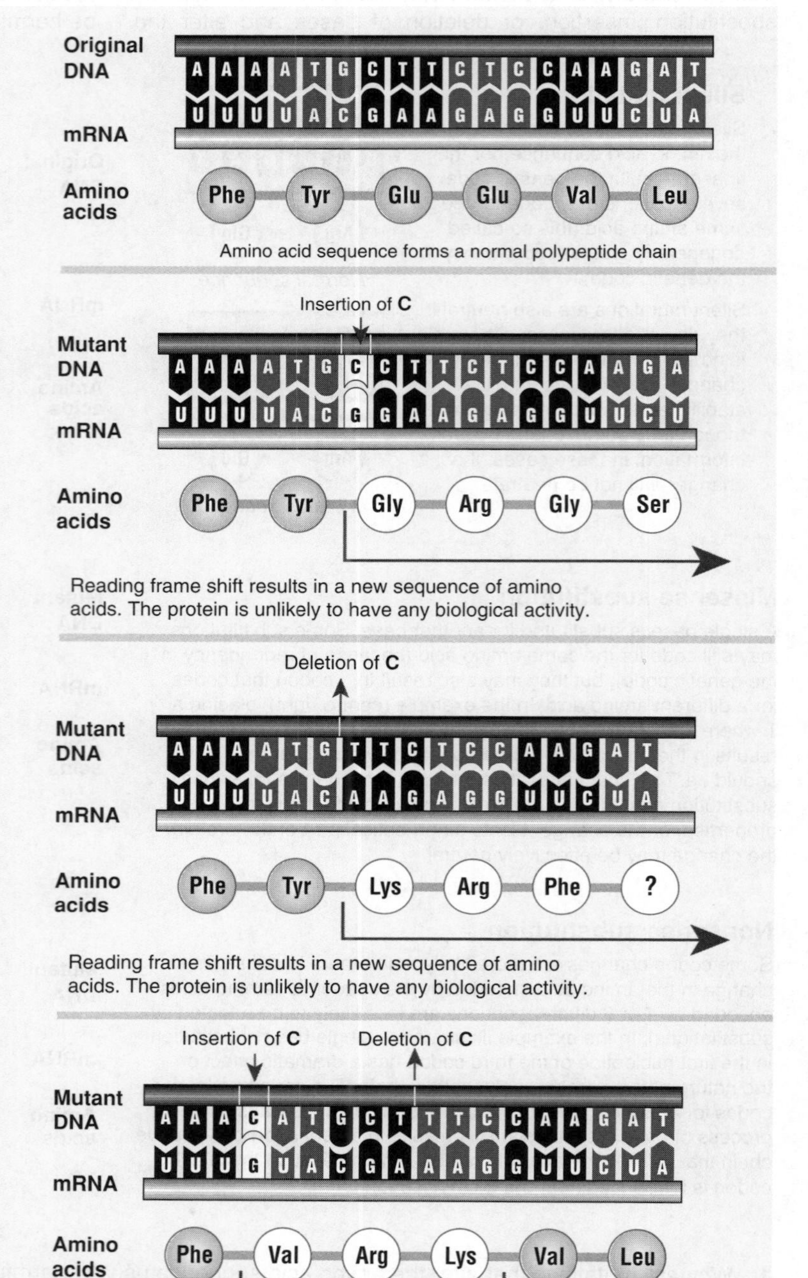

1. What is a reading frame shift? ______________________________

2. Why is a frame shift near the start codon likely to have a greater impact than one near the stop codon? ______________________________

3. What is the effect of a partial reading shift mutation on the final protein produced: ______________________________

ISBN:978-1-927309-20-9

158 Chromosome Mutations

Key Idea: Large scale mutations occurring during meiosis can fundamentally change chromosome structure.

Chromosome mutations (also called block mutations) involve the rearrangement of whole blocks of genes (involving many bases), rather than individual bases within a gene. They commonly occur during meiosis and they alter the number or sequence of whole sets of genes on the chromosome (represented by letters below). Translocations may sometimes involve the fusion of whole chromosomes, thereby reducing the chromosome number of an organism. This is thought to be an important evolutionary mechanism by which instant speciation can occur.

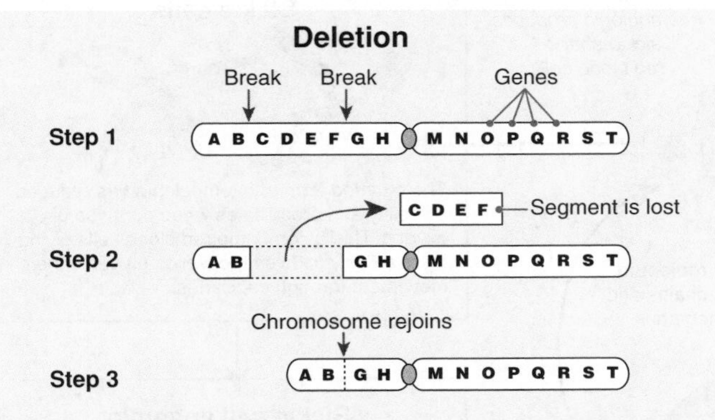

A break may occur at two points on the chromosome and the middle piece of the chromosome falls out. The two ends then rejoin to form a chromosome deficient in some genes. Alternatively, the end of a chromosome may break off and is lost.

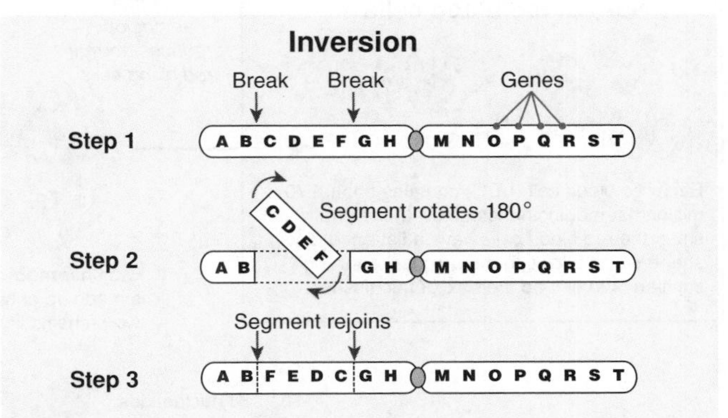

The middle piece of the chromosome falls out and rotates through 180° and then rejoins. There is no loss of genetic material. The genes will be in a reverse order for this segment of the chromosome.

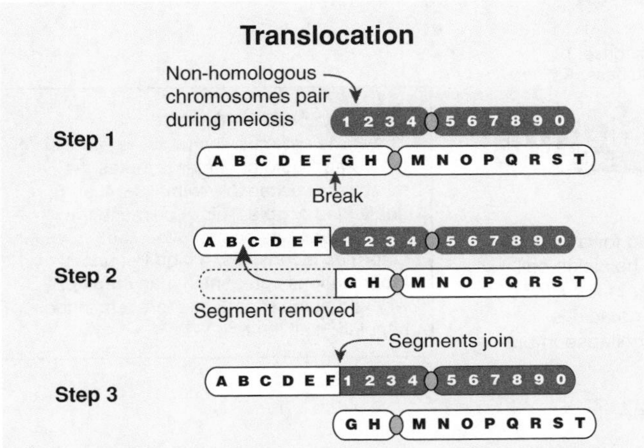

In translocation mutations, a group of genes moves between different chromosomes. The large chromosome (white) and the small chromosome (blue) are not homologous. A piece of one chromosome breaks off and joins to the other. When the chromosomes are passed to gametes, some gametes will receive extra genes, while some will be deficient.

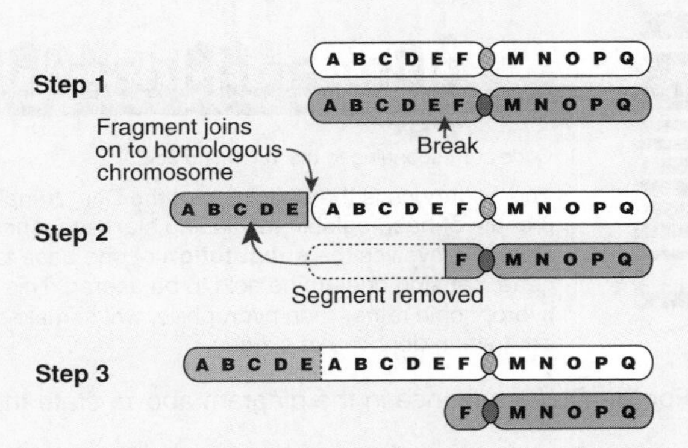

A segment is lost from one chromosome and is added to its homologue. In this diagram, the darker chromosome on the bottom is the 'donor' of the duplicated piece of chromosome. The chromosome with the segment removed is deficient in genes. Some gametes will receive double the genes while others will have no genes for the affected segment.

1. For each of the chromosome (block) mutations illustrated above, write the original gene sequence and the new gene sequence after the mutation has occurred (the first one has been done for you):

	Original sequence(s)	Mutated sequence(s)
(a) Deletion:	A B C D E F G H M N O P Q R S T	A B G H M N O P Q R S T
(b) Inversion:		
(c) Translocation:		
(d) Duplication:		

2. Which type of block mutation is likely to be the least damaging to the organism? Explain your answer:

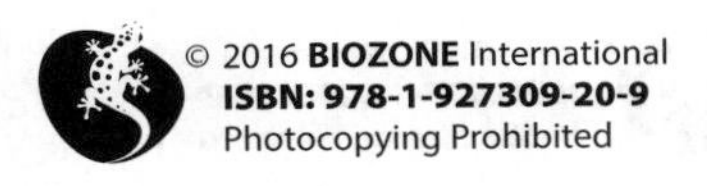

ISBN: 978-1-927309-20-9

159 Sickle Cell Mutation

Key Idea: The substitution of one nucleotide from T to A results in sickle cell disease. The mutation is codominant.

Sickle cell disease is an inherited blood disorder caused by a gene mutation (Hb^S), which produces a faulty beta (β) chain haemoglobin (Hb) protein. This in turn produces red blood cells with a deformed sickle appearance and a reduced capacity to carry oxygen. Many aspects of metabolism are also affected. The mutation is codominant (both alleles equally expressed), and people heterozygous for the mutation (carriers) have enough functional haemoglobin and suffer only minor effects.

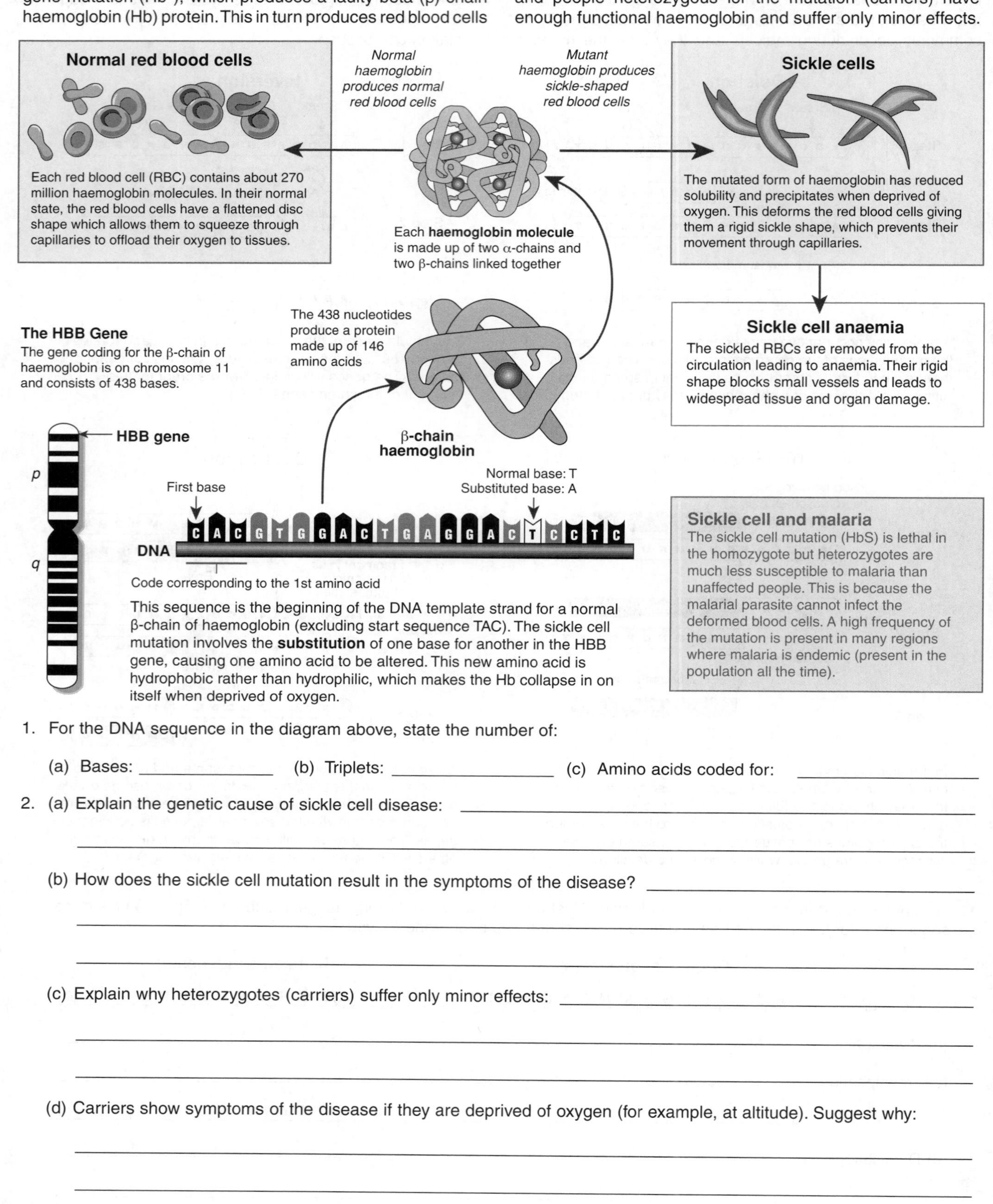

1. For the DNA sequence in the diagram above, state the number of:

   (a) Bases: ______________ (b) Triplets: ______________ (c) Amino acids coded for: ______________

2. (a) Explain the genetic cause of sickle cell disease: ______________

   (b) How does the sickle cell mutation result in the symptoms of the disease? ______________

   (c) Explain why heterozygotes (carriers) suffer only minor effects: ______________

   (d) Carriers show symptoms of the disease if they are deprived of oxygen (for example, at altitude). Suggest why:

3. Briefly explain why there is a high frequency of the sickle cell mutation in populations where malaria is endemic:

ISBN:978-1-927309-20-9

160 Cystic Fibrosis Mutation

Key Idea: Cystic fibrosis most often results from a triplet deletion in the CFTR gene, producing a protein that is unable to regulate chloride transport.

Cystic fibrosis (CF) is an inherited disorder caused by a mutation of the CFTR gene. It is one of the most common lethal autosomal recessive conditions affecting people of European descent (4% are carriers). The CFTR gene's protein product is a membrane-based protein that regulates chloride transport in cells. Over 500 mutations of the CFTR gene are known, causing disease symptoms of varying severity. The Δ(delta)F508 mutation accounts for more than 70% of all defective CFTR genes. This mutation leads to an abnormal CFTR, which cannot take its proper position in the membrane (below) nor perform its transport function.

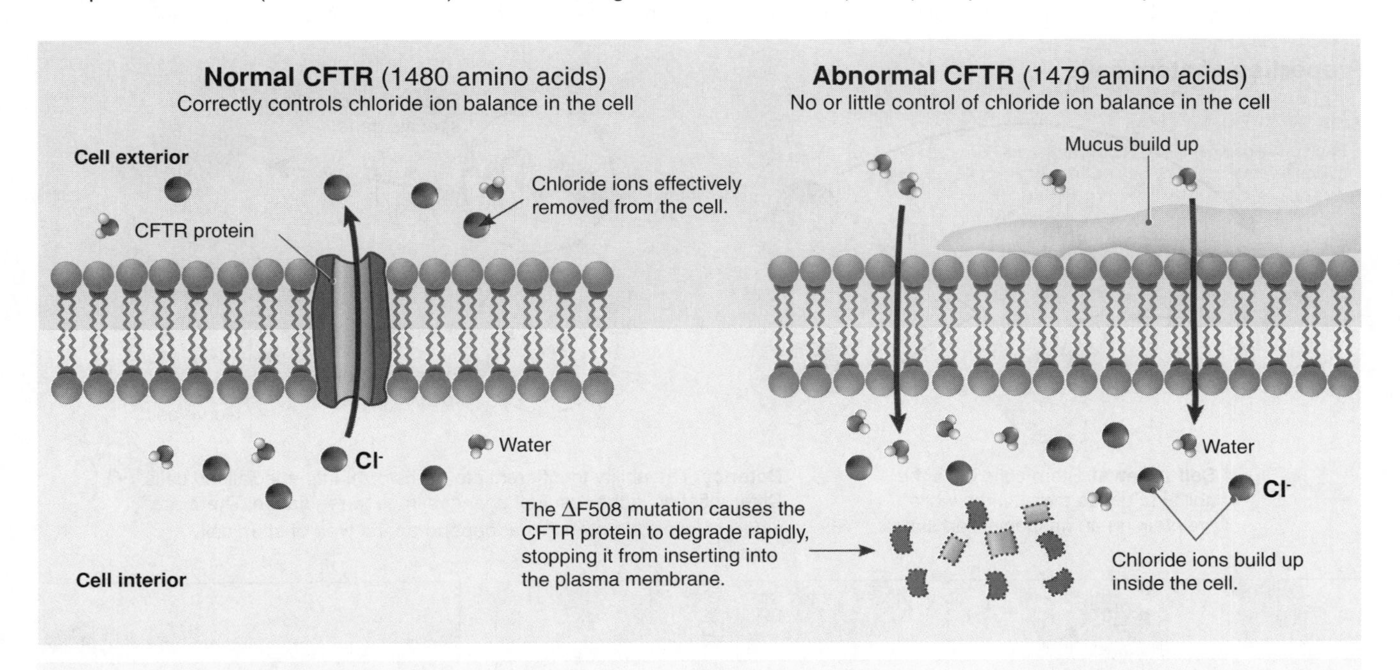

The CF gene on chromosome 7

The CF gene is located on chromosome 7. The ΔF508 mutation of the CF gene describes a deletion of the 508th triplet, which in turn causes the loss of a single **amino acid** from the gene's protein product, the cystic fibrosis transmembrane conductance regulator (CFTR). This protein normally regulates the chloride channels in cell membranes, but the mutant form fails to do this. The DNA region containing the mutation site is shown below:

p

q

CFTR gene

The CFTR protein consists of 1480 amino acids

CFTR protein

The ΔF508 mutant form of CFTR fails to take up its position in the membrane. Its absence results in defective chloride transport and leads to a net increase in water absorption by the cell. This accounts for the symptoms of cystic fibrosis, where mucus-secreting glands, particularly in the lungs and pancreas, become fibrous and produce abnormally thick mucus. The widespread presence of CFTR throughout the body also explains why CF is a multisystem condition affecting many organs.

Base 1630

DNA template strand: C C G T G G T A A T T T C T T T T A T A G T A G A A A C C A C C A

This triplet codes for the 500th amino acid

The 508th triplet is absent in the form with the ΔF508 mutation

1. (a) Write the mRNA sequence for the transcribing DNA strand above: ______________________________

   (b) Rewrite the mRNA sequence for the mutant DNA strand: ______________________________

   (c) What kind of mutation is ΔF508? ______________________________

2. (a) Explain why the abnormal CFTR fails to transport Cl^- correctly: ______________________________

   (b) What effect does this have on water movement in and out of the cell? ______________________________

ISBN: 978-1-927309-20-9

LINK 185 LINK 156 WEB 160 KNOW

161 What are Stem Cells?

Key Idea: Stem cells are undifferentiated cells found in multicellular organisms. They are characterised by the properties of self renewal and potency.

A zygote can differentiate into all the cell types of the body because its early divisions produce stem cells. Stem cells are unspecialised cells that can divide repeatedly while remaining unspecialised (**self renewal**). They give rise to the many cell types that make up the tissues of multicellular organisms. For example, bone marrow stem cells differentiate to produce all the cell types that make up blood. These multipotent (or adult) stem cells are found in most organs, where they replace old or damaged cells and replenish cells throughout life. Different types of stem cell have different abilities to differentiate (called **potency**). Fully differentiated cells cannot normally revert to an undifferentiated state, although in some conditions they can be induced to do so (opposite).

Properties of stem cells

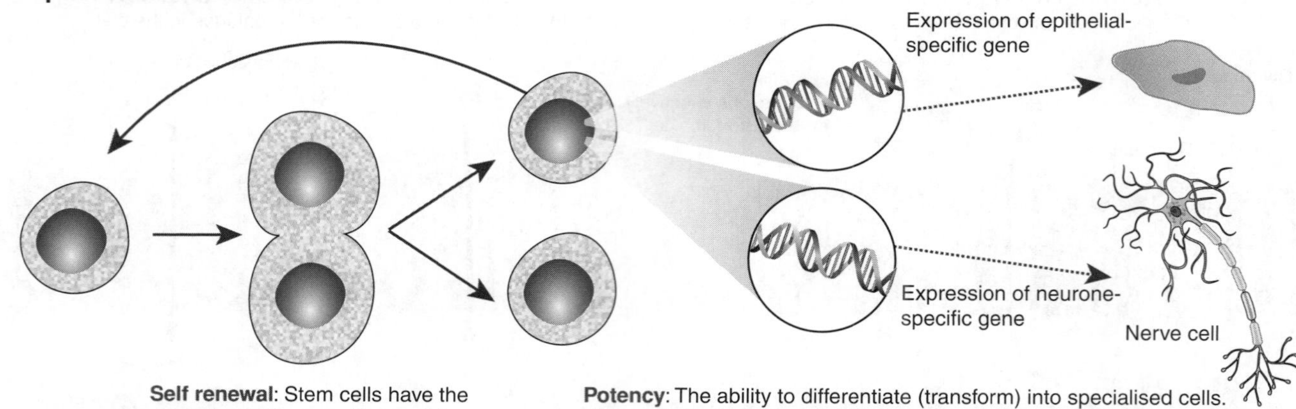

Self renewal: Stem cells have the ability to divide many times while maintaining an unspecialised state.

Potency: The ability to differentiate (transform) into specialised cells. Differentiation is the result of changes in gene regulation. There are different levels of potency that depend on the type of stem cell.

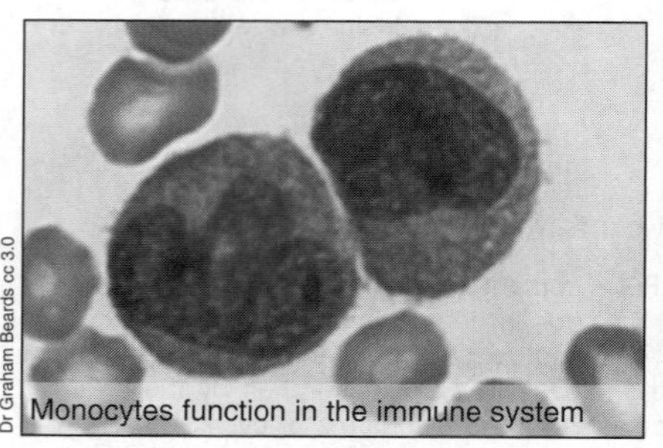

Dr Graham Beards cc 3.0

Monocytes function in the immune system

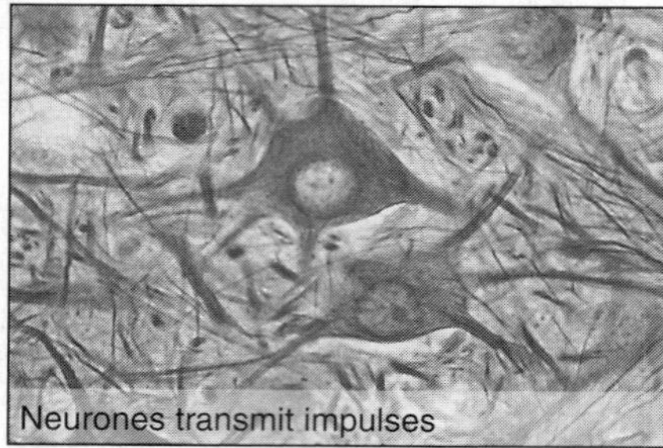

Neurones transmit impulses

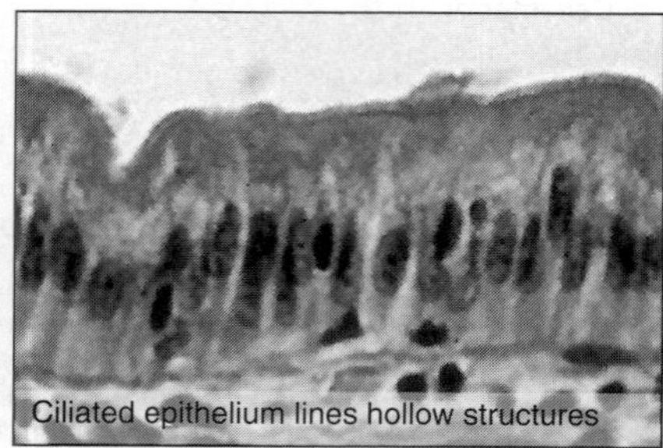

Ciliated epithelium lines hollow structures

Each specialised cell type in an organism expresses a subset of all the genes making up the organism's genome. The differentiation of cells into specialised types is the result of specific patterns of gene expression and is controlled by transcription factors. In response to cues during development (e.g. hormones), transcription factors bind to DNA and switch genes on or off, so determining the final structure and function of the cell. As a cell becomes increasingly specialised, its fate becomes fixed and it can no longer return to an undifferentiated state unless it is reprogrammed. This process, called dedifferentiation, happens naturally in some organisms but can be induced in human cells (opposite).

1. Describe the two defining features of stem cells:

   (a) ______________________________

   (b) ______________________________

2. How is the differentiation of stem cells controlled? ______________________________

3. Explain the role of stem cells in the development and maintenance of specialised tissues in multicellular organisms:

ISBN:978-1-927309-20-9

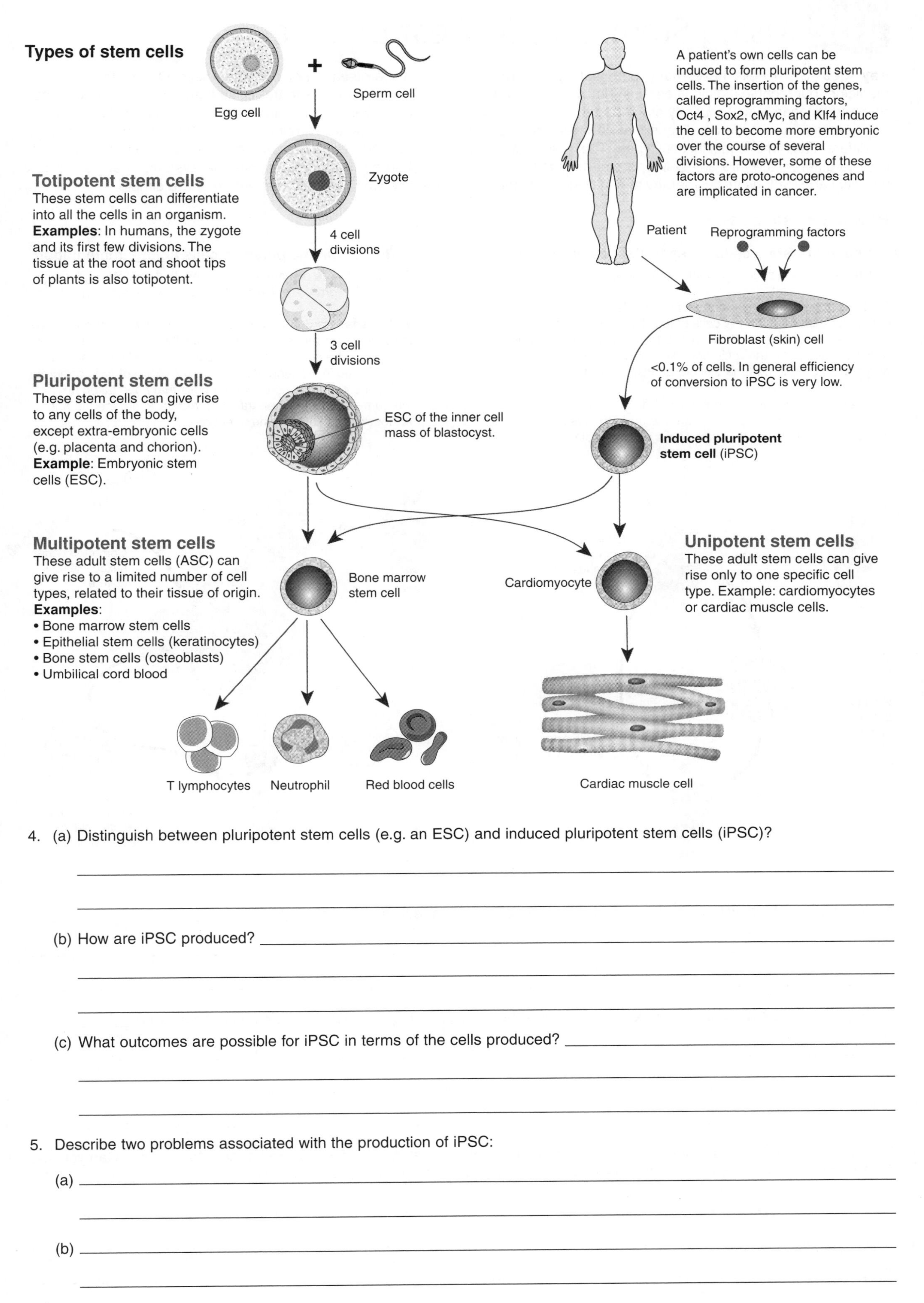

4. (a) Distinguish between pluripotent stem cells (e.g. an ESC) and induced pluripotent stem cells (iPSC)?

(b) How are iPSC produced?

(c) What outcomes are possible for iPSC in terms of the cells produced?

5. Describe two problems associated with the production of iPSC:

(a)

(b)

ISBN: 978-1-927309-20-9

162 Using Stem Cells To Treat Disease

Key Idea: Stem cells have many potential applications in medicine, but technical difficulties must be overcome first.

Stem cell research is at an early stage and there is still much to be learned about the conditions that cells require in order to differentiate into specific cell types. The ability of stem cells to differentiate into any cell type means that they have potential applications in cell therapy and in tissue engineering to replace diseased or damaged cells. One of the problems with using stem cells to treat diseased tissues is the response of the recipient's immune system. The immune system has evolved to destroy foreign objects. When stem cells cultured from another person are introduced to a recipient, they are attacked by that person's immune system. There are a few possible ways around this, but none are simple.

How to use stem cells

1. Use donor stem cells to repair tissues or organs

Problem: immune system will attack the donor's cells.

Firstly, a donor with a tissue match is selected (the types of cell surface proteins on the cells of the donor and recipient are the same or very similar). This reduces the risk of immune rejection. Secondly, the recipient takes immunosuppressant drugs to stop their immune system attacking the donated cells.

An alternative approach is to encase the donated cells in a protective shell, isolating them from the body. This is being investigated with respect to pancreatic cells and diabetes.

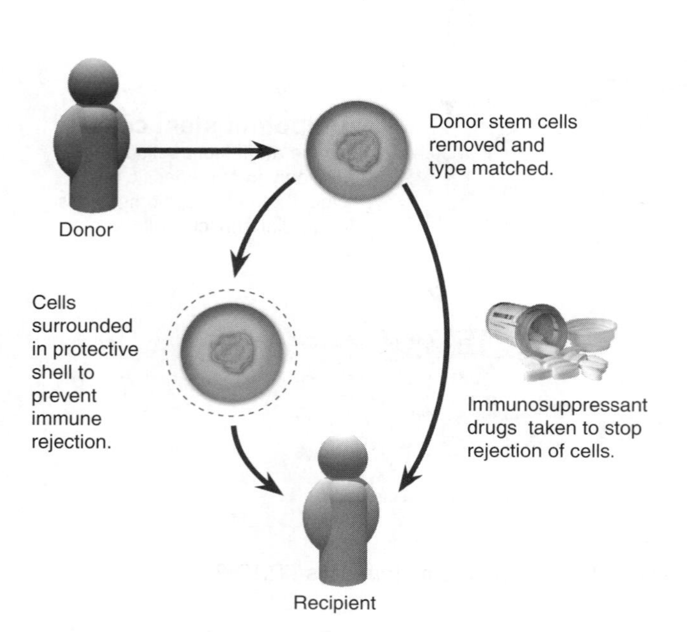

2. Reprogramme patient's cells before implantation

Problem: Some diseases are the result of defective genes. Stem cells from the patient will carry these defective genes.

If the disease is due to a genetic fault (e.g. cystic fibrosis) then the stem cells will need to be genetically corrected before use (otherwise the disease may reoccur). Stem cells are isolated and cultured in the laboratory in the presence of the corrected gene. The culture is screened for cells that have taken up the gene. These are then transplanted back into the patient, without the risk of immune rejection.

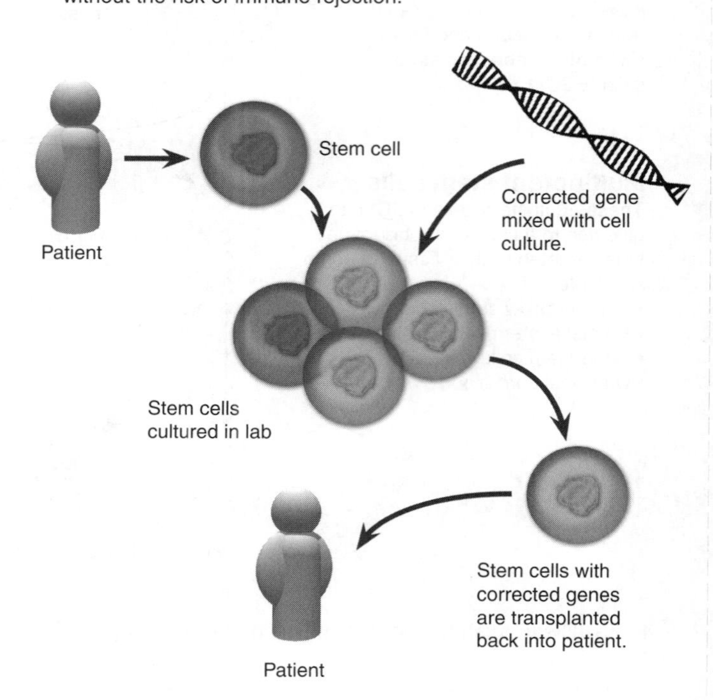

1. Identify a problem with using stem cells from a donor to treat a recipient patient: ______________________

____

____

2. Explain why stem cells with a defective gene must be corrected before reimplanting them into the patient: ______________________

____

____

____

3. Umbilical cord blood is promoted as a rich source of multipotent stem cells for autologous (self) transplants. Can you see a problem with the use of a baby's cord blood to treat a disease in that child at a later date?

____

____

____

ISBN:978-1-927309-20-9

Stem cells for Stargardt's disease

Stargardt's disease is an inherited form of juvenile macular degeneration (a loss of the central visual field of the eye). The disease is associated with a number of mutations and results in dysfunction of the retinal pigment epithelium (RPE) cells, which nourish the retinal photoreceptor cells and protect the retina from excess light. Dysfunction of the RPE causes deterioration of the photoreceptor cells in the central portion of the retina and progressive loss of central vision. This often begins between ages 6 and 12 and continues until a person is legally blind. Trials using stem cells have obtained promising results in treating the disease.

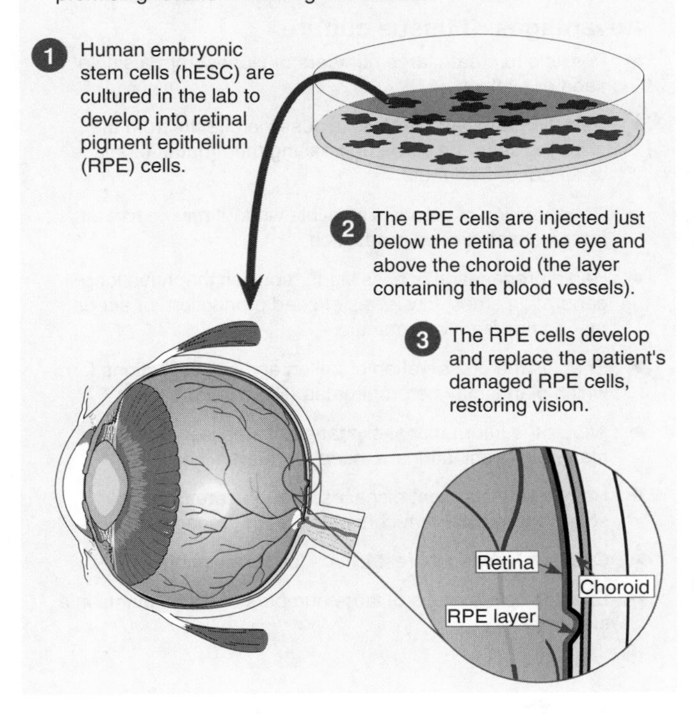

Stem cells for type 1 diabetes

Type 1 diabetes results from the body's own immune system attacking and destroying the insulin-producing beta cells of the pancreas. In theory, new beta cells could be produced using stem cells. Research is focused on how to obtain the stem cells and deliver them effectively to the patient. Many different techniques are currently being investigated. Most techniques use stem cells from non-diabetics, requiring recipients to use immunosuppressant drugs so the cells are not rejected.

A study published in 2014 described a method for treating type 1 diabetes in mice using fibroblast cells taken from the skin of mice.

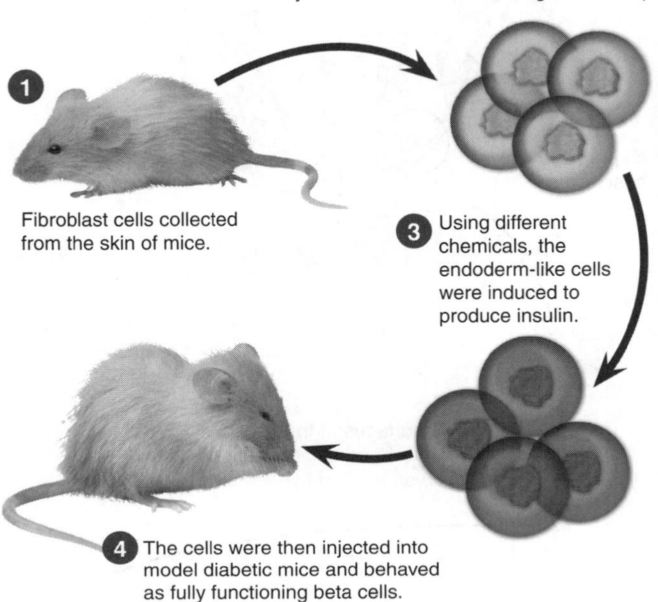

4. (a) Explain the basis for correcting Stargardt's disease using stem cell technology: ______________________

(b) Suggest why researchers derived the RPE cells from embryos rather than by reprogramming a patient's own cells:

____

(c) What advantage is there in reprogramming a patient's own cells and when would this be a preferable option?

____

5. (a) What causes type 1 diabetes? ______________________

(b) How can this be treated with stem cells? ______________________

ISBN: 978-1-927309-20-9

163 Using Totipotent Cells for Tissue Culture

Key Idea: Cells from the meristems of plants are totipotent and can be used to grow many new plant clones.

Micropropagation (or plant tissue culture) is a method used to clone plants. It is possible because plant meristematic tissue is totipotent (can develop into any tissue type) and differentiation into a complete plant can be induced by culturing the tissue in an appropriate growth environment. Micropropagation is used widely for the rapid multiplication of commercially important plant species, as well as in recovery programmes for endangered plants. However, continued culture of a limited number of cloned varieties reduces genetic diversity and plants may become susceptible to disease or environmental change. New genetic stock may be introduced into cloned lines to prevent this. Micropropagation has considerable advantages over traditional methods of plant propagation, but it is very labour intensive. Its success is affected by a variety of factors including selection of explant material, plant hormone levels, lighting, and temperature.

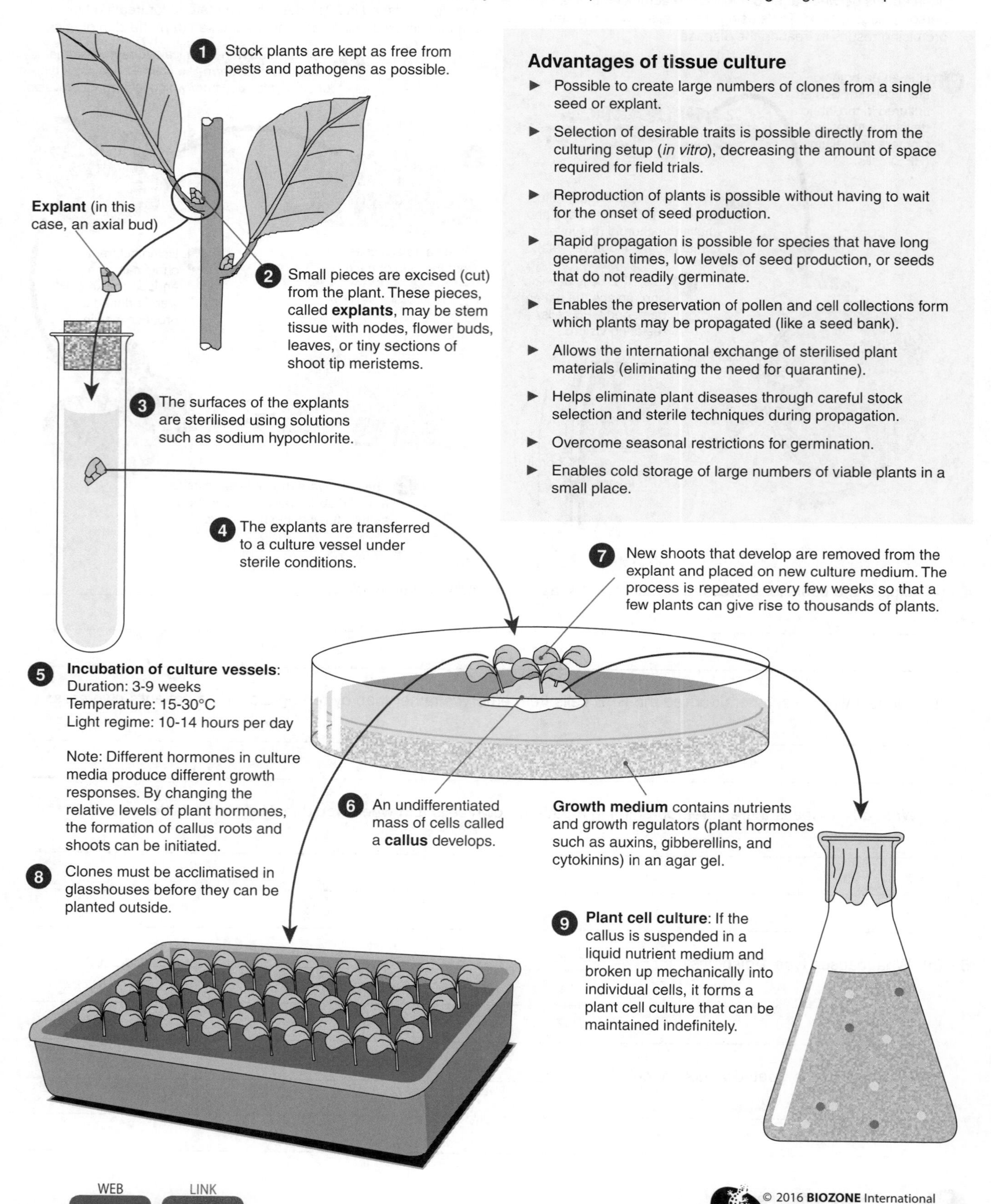

Advantages of tissue culture

- Possible to create large numbers of clones from a single seed or explant.
- Selection of desirable traits is possible directly from the culturing setup (*in vitro*), decreasing the amount of space required for field trials.
- Reproduction of plants is possible without having to wait for the onset of seed production.
- Rapid propagation is possible for species that have long generation times, low levels of seed production, or seeds that do not readily germinate.
- Enables the preservation of pollen and cell collections form which plants may be propagated (like a seed bank).
- Allows the international exchange of sterilised plant materials (eliminating the need for quarantine).
- Helps eliminate plant diseases through careful stock selection and sterile techniques during propagation.
- Overcome seasonal restrictions for germination.
- Enables cold storage of large numbers of viable plants in a small place.

ISBN:978-1-927309-20-9

Cloning cauliflower

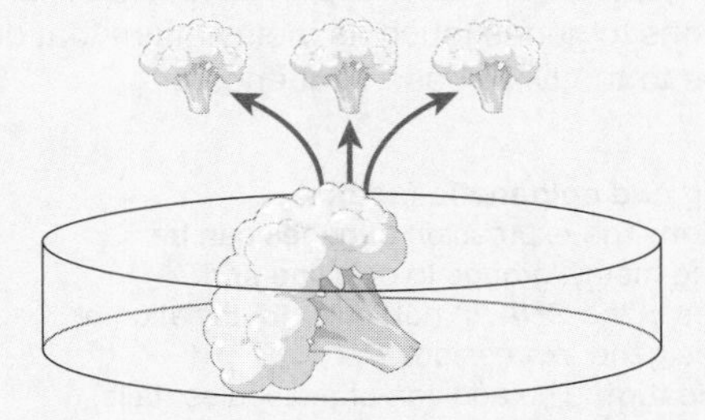
The floret is cut with a scalpel into three smaller pieces (**explants**) about 3-5mm.

A floret is removed from the cauliflower, and placed onto a clean petri dish.

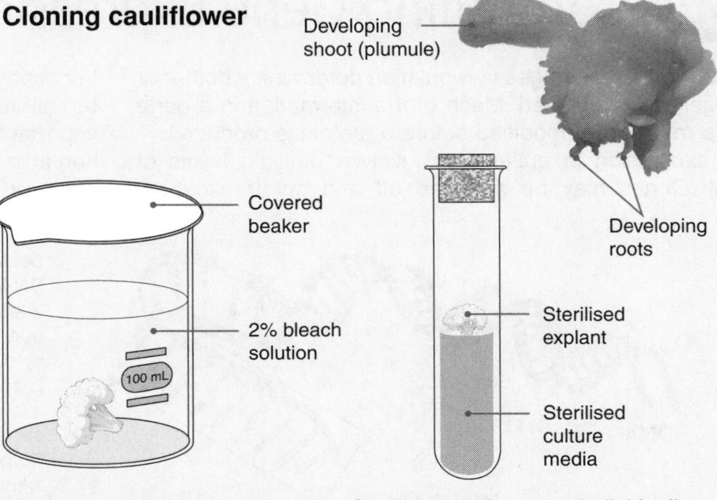

Cauliflower is often used in classrooms to demonstrate plant tissue culture because the stalked florets provide an inexpensive, readily available supply of floral meristems. The florets are also quite robust and can withstand the sterilisation process.

The explants are sterilised to remove any contaminants (e.g. fungi or bacteria). Care must be taken to use aseptic techniques and sterilised equipment from this point. Once sterilised the explants are rinsed several times in a covered container of sterilised water.

Sterilised explants are individually placed into covered test tubes containing sterilised culture medium. The stem of the explant is partially submerged in the growth media. Growth can be seen after about two weeks in a warm, well lit environment (photo, top).

1. What is the purpose of micropropagation (plant tissue culture)? ____________________

2. (a) What is a callus? ____________________

   (b) How can a callus be stimulated to initiate root and shoot formation? ____________________

3. Explain a potential problem with micropropagation in terms of long term ability to adapt to environmental change:

4. Describe two advantages of micropropagation compared with traditional methods of plant propagation such as grafting:

5. (a) Why is it possible to clone cauliflower florets? ____________________

   (b) Why were the cauliflower explants washed with bleach? ____________________

   (c) Why is the use of aseptic techniques important during cauliflower cloning? ____________________

   (d) If one of the explants became mouldy during the cloning process, what would you suspect? ____________________

ISBN: 978-1-927309-20-9

164 Controlling Gene Expression: An Overview

Key Idea: Different stages in regulation determine whether or not a gene is expressed. Much of the information in a gene may be removed or modified before a protein is produced.

Gene expression in eukaryotes involves multiple levels of control. Genes may be switched off and not transcribed, transcribed genes may be modified, and gene products may be altered after translation. Epigenetic modifications are important in the regulation of gene expression. These are heritable modifications to genes (such as methylation) that do not involve changes to the DNA base sequence.

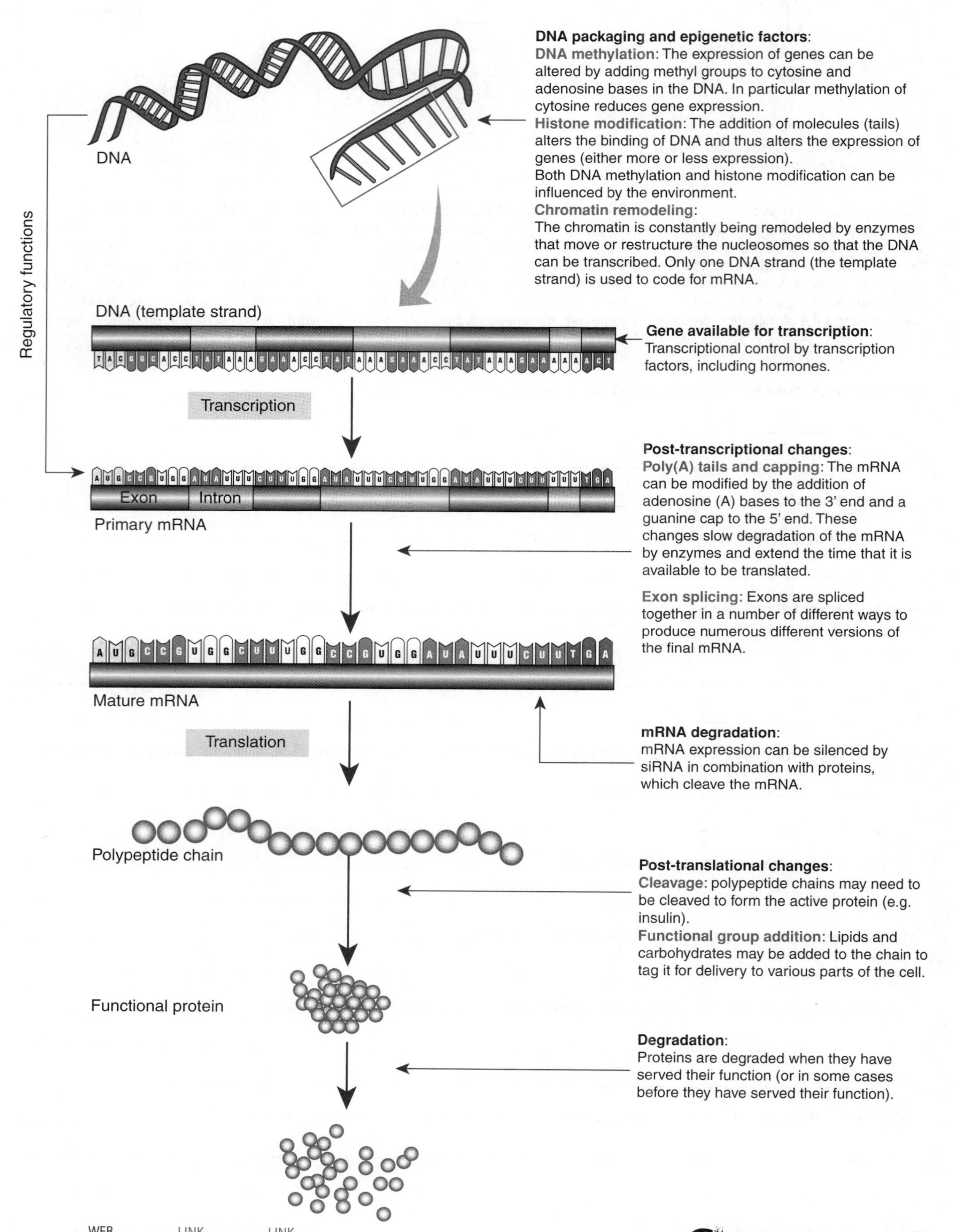

ISBN:978-1-927309-20-9

1. Identify three ways in which gene expression can be suppressed: ______

2. Identify a way in which gene expression can be increased: ______

3. Identical twins have identical DNA. However, as they grow and age, differences in personality and physical appearance can occur (people who are familiar with twins can easily tell them apart). How does DNA methylation and histone modification account for some of these differences?

4. What is the purpose of the poly(A) tails and guanine cap on mRNA? ______

5. How does exon splicing lead to a variety of possible polypeptide chains? ______

6. How do post-translational changes to the polypeptide chain affect its functioning? ______

7. Human DNA contains about 25 000 genes, but it is estimated that there are more than 250 000 proteins in the human body. In *Drosophila*, the Dscam gene can produce 38,016 different mRNA molecules. Explain how these vast numbers of proteins are produced from such a few genes.

8. What is the purpose of degradation of proteins? ______

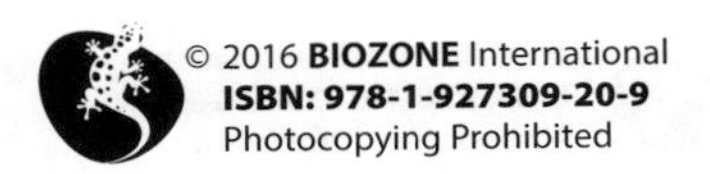

ISBN: 978-1-927309-20-9

165 DNA Packaging and Control of Gene Expression

Key Idea: A chromosome consists of DNA complexed with proteins to form a highly organised, tightly coiled structure.

The DNA in eukaryotes is complexed with histone proteins to form chromatin. The histones assist in packaging the DNA efficiently so that it can fit into the nucleus. Prior to cell division, the chromatin is at its most compact, forming the condensed metaphase chromosomes that can be seen with a light microscope.

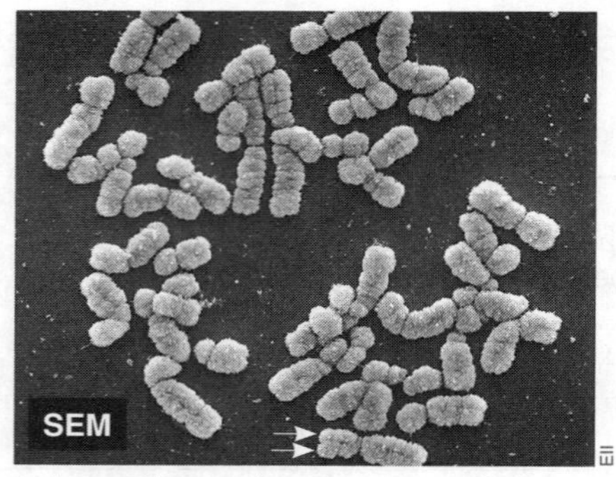

A cluster of human chromosomes seen during metaphase of cell division. Individual chromatids (arrowed) are difficult to discern on these double chromatid chromosomes.

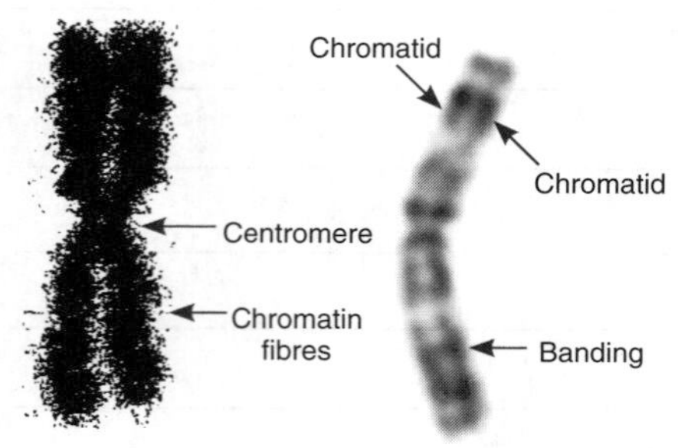

A human chromosome from a dividing white blood cell (above left). Note the compact organisation of the chromatin in the two chromatids. The LM photograph (above right) shows the banding visible on human chromosome 3.

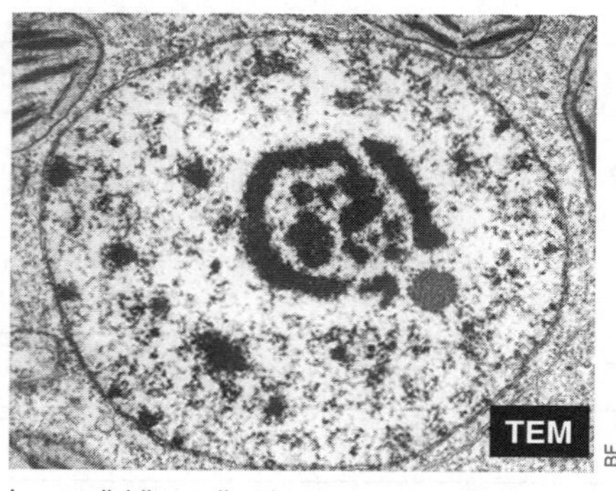

In non-dividing cells, chromosomes exist as single-armed structures. They are not visible as coiled structures, but are 'unwound' to make the genes accessible for transcription (above).

The evidence for the existence of looped domains comes from the study of giant lampbrush chromosomes in amphibian oocytes (above). Under electron microscopy, the lateral loops of the DNA-protein complex appear brushlike.

The packaging of chromatin

Chromatin structure is based on successive levels of DNA packing. **Histone proteins** are responsible for packing the DNA into a compact form. Without them, the DNA could not fit into the nucleus. Five types of histone proteins form a complex with DNA, in a way that resembles "beads on a string". These beads, or **nucleosomes**, form the basic unit of DNA packing.

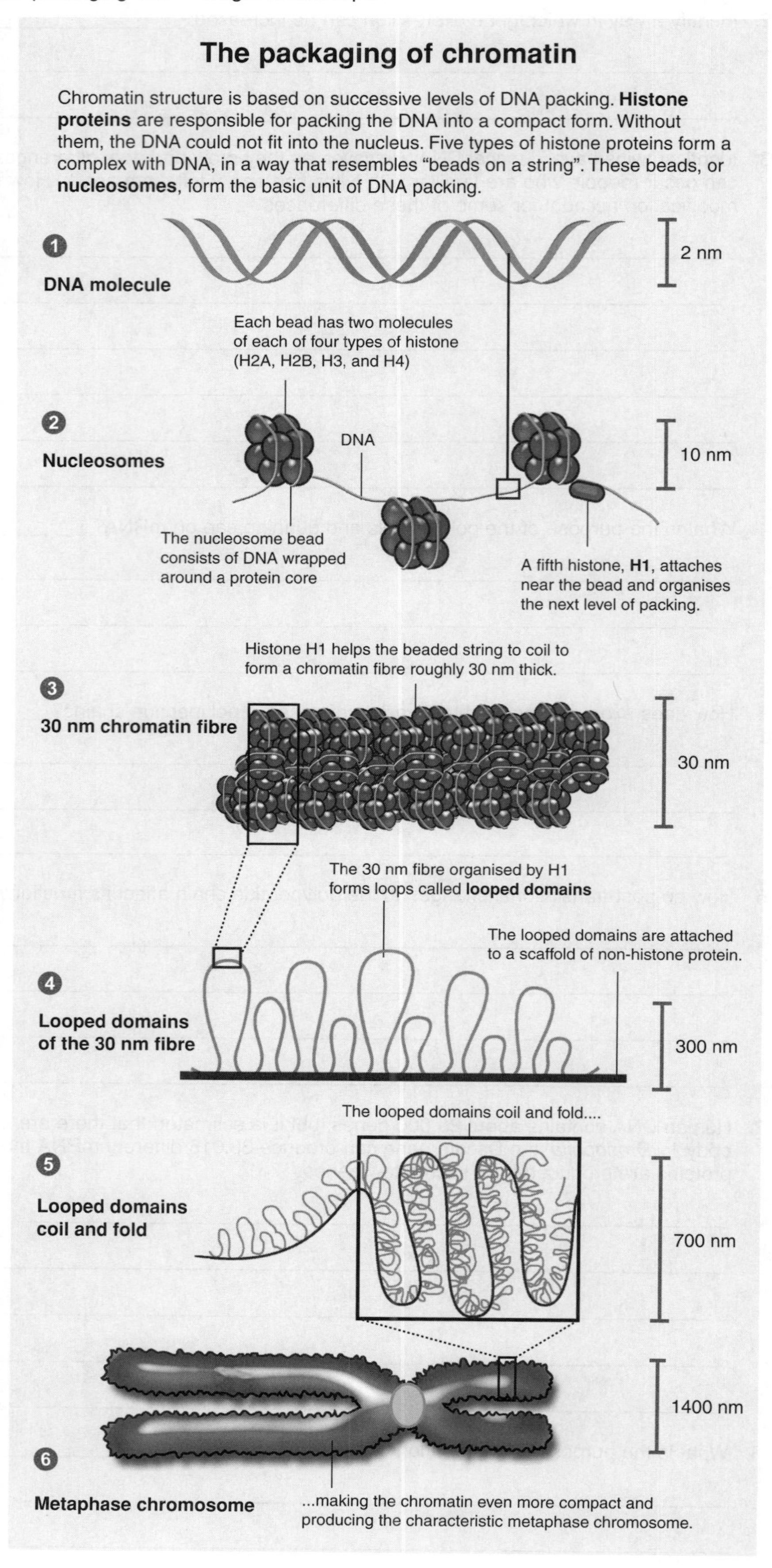

ISBN:978-1-927309-20-9

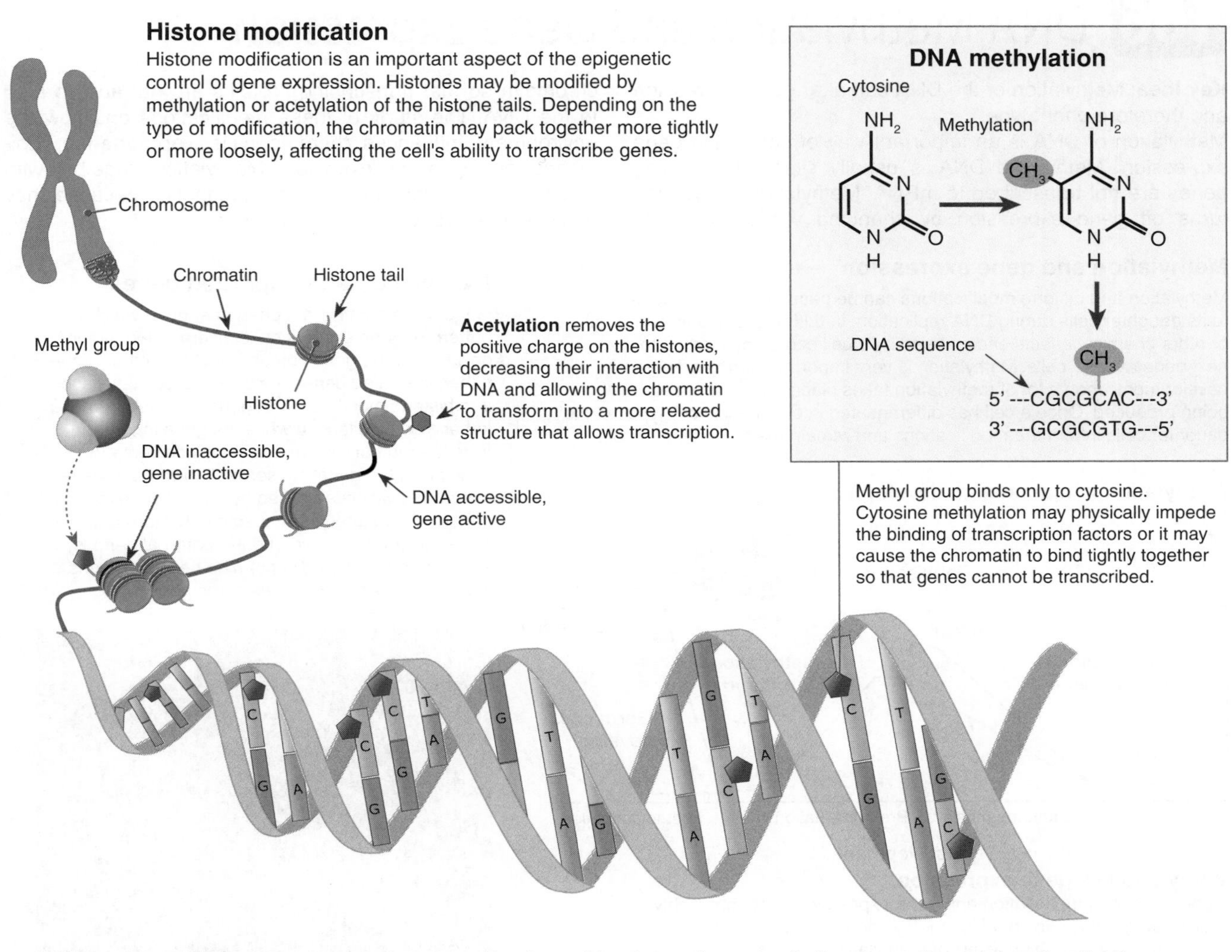

The packaging of DNA regulates gene expression either by making the nucleosomes in the chromatin pack together tightly (**heterochromatin**) or more loosely (**euchromatin**). This affects whether or not RNA polymerase can attach to the DNA and transcribe the DNA. Packaging of DNA is affected by histone modification (methylation and acetylation) and DNA methylation.

1. Explain the significance of the following terms used to describe the structure of chromosomes:

   (a) DNA: ______________________________

   (b) Chromatin: ______________________________

   (c) Histone: ______________________________

   (d) Nucleosome: ______________________________

2. (a) Describe the effect of histone modification and DNA methylation on DNA packaging: ______________________________

   (b) How do these processes affect transcription of the DNA? ______________________________

ISBN: 978-1-927309-20-9

166 DNA Methylation and Gene Expression

Key Idea: Methylation of the DNA can alter gene expression and therefore phenotype.

Methylation of DNA is an important way of controlling gene expression. Methylated DNA is usually silenced, meaning genes are not transcribed to mRNA. Methylation of cytosine turns off gene expression by changing the state of the chromatin so that transcribing proteins are not able to bind to the DNA. The study of these modifications and how the environment influences them is called **epigenetics**. Epi- means 'on top of' or 'extra to'. Methylation (together with histone modification) are part of epigenetics because they don't directly change the DNA sequence.

Methylation and gene expression

Methylation and histone modifications can be passed on from a cell's DNA to its daughter cells during DNA replication. In this way, any environmental or other chemical effects encountered by a cell can be passed on to the next generation of cells. Methylation is very important during embryonic development when a lot of methylation takes place and a lot of cells are being produced. Once a cell has differentiated into a specific cell type, all the daughter cells inherit the modifications and remain the same type of cell.

Methylation in mammalian development

Degree of methylation

Fertilisation

Zygote

Sperm

Somatic cells in adult

Methylation in germ cells

Demethylation in early embryo

Germ cells develop

Eggs

Germ cells develop

Gametogenesis | Pre-implantation | Post-implantation

Development

Methylation vs gene expression

A comparison of methylation and gene expression finds that highly expressed genes have very little methylation while genes that are not expressed have very high levels of methylation.

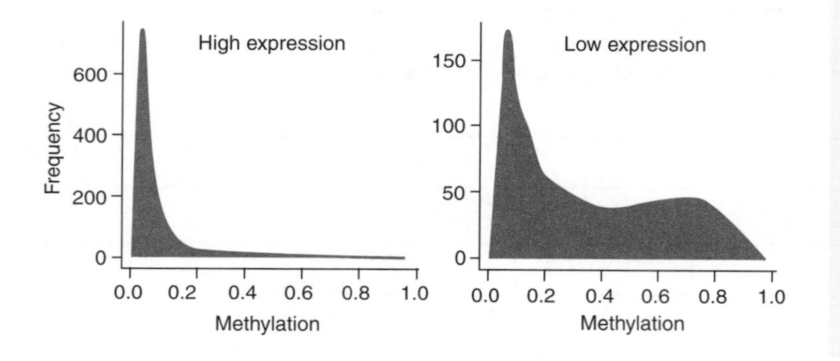

Methylation and imprinted genes

Genomic imprinting is a phenomenon in which the pattern of gene expression is different depending on whether the gene comes from the mother or the father. Imprinted genes are silenced by methylation and histone modification. A gene inherited from the father may be silenced while the gene inherited from the mother may be active or vice versa. An example is the genetic disorder known as Angelman syndrome, which is caused by a deletion in the maternally donated chromosome 15. Normally the maternal gene is active and the paternal gene is silenced (one working copy). The deletion means there are no working copies of the gene.

A liger

The effect of genomic imprinting can be seen in other mammals. Ligers (a cross between a male lion and a female tiger) although not naturally occurring, are the biggest of the big cats. However a tigon (a cross between a female lion and a male tiger) is no bigger than a normal lion. It is thought this difference in phenotype is due to the male lion carrying imprinted genes that result in larger offspring which are normally counteracted by genes from the female. Similarly the differences between a mule (male donkey + female horse) and a hinny (female donkey + male horse) may be to do with genomic imprinting.

1. (a) What is genomic imprinting? ______________________________

   (b) How is genomic imprinting achieved? ______________________________

   (c) What is its effect? ______________________________

2. Prader-Willi syndrome is caused when a mutated gene on chromosome 15 is inherited from the father. How does this tell us that the mother must therefore have donated the imprinted gene?

____

ISBN:978-1-927309-20-9

167 Epigenetic Factors and Phenotype

Key Idea: The environment or experiences of an individual can affect the development of following generations.

Studies of heredity have found that the environment, experiences, or lifestyle of an ancestor can have an effect on future generations. Certain environments or diets can affect the methylation and packaging of the DNA (rather than the DNA itself) determining which genes are switched on or off and so affecting the development of the individual. These effects can be passed on to offspring, and even on to future generations. It is thought that these inherited epigenetic effects may provide a rapid way to adapt to particular environmental situations, such as famine or chronic stress.

Epigenetics and inheritance

The destruction of New York's Twin Towers on September 11, 2001, traumatised thousands of people. In those thousands were 1700 pregnant women. Some of them suffered (often severe) post-traumatic stress disorder (PTSD), others did not. Studies on the mothers who developed PTSD found very low levels of the cortisol in their saliva. Cortisol helps the body adapt to stress. The children of these mothers also had much lower levels of cortisol than those whose mothers had not suffered PTSD. The environment of the mother had affected the offspring.

How can epigenetic effects be proved to be inherited? Anything the mother is exposed to will affect her and will also expose the fetus. In a female fetus, egg cells develop in the ovaries, so a third generation will be exposed also. If a fourth generation proves to be affected, then there is evidence that the epigenetic effect is inherited.

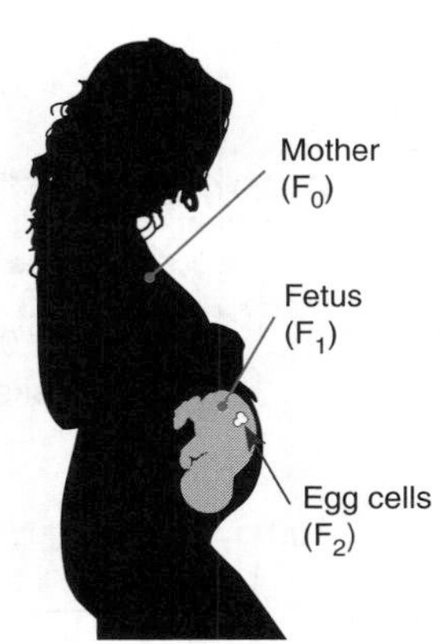

Rats and environmental effects

The effect of maternal environment and diet of on later generations exposed to a breast cancer trigger (a carcinogenic chemical) was investigated in rats fed a high fat diet or a diet high in oestrogen. The length of time taken for breast cancer to develop in later generations after the trigger for breast cancer was given was recorded and compared. The data are presented below.
F_1 = daughters, F_2 = granddaughters, F_3 = great granddaughters.

	Cumulative percent rats with breast cancer (high fat diet, HFD)					
	F_1%		F_2%		F_3%	
Weeks since trigger	HFD	Control	HFD	Control	HFD	Control
6	0	0	5	0	3	0
8	15	0	20	5	3	20
10	22	8	30	5	10	25
12	22	18	50	20	20	30
14	22	18	50	30	25	40
16	29	18	60	30	25	40
18	29	18	60	40	40	42
20	40	18	65	40	50	60
22	80	60	79	50	50	60

Data source: Science News April 6, 2013

	Cumulative percent rats with breast cancer (high oestrogen diet, HOD)					
	F_1%		F_2%		F_3%	
Weeks since trigger	HOD	Control	HOD	Control	HOD	Control
6	5	0	10	0	0	0
8	10	0	10	0	15	10
10	30	15	15	20	30	20
12	38	19	30	30	40	20
14	50	22	30	40	50	20
16	50	22	30	40	50	30
18	60	35	40	40	75	40
20	60	42	50	50	80	45
22	80	55	50	50	80	60

ISBN: 978-1-927309-20-9

LINK 166 KNOW

1. Use the data on the previous page to complete the graphs below. The first graph is done for you:

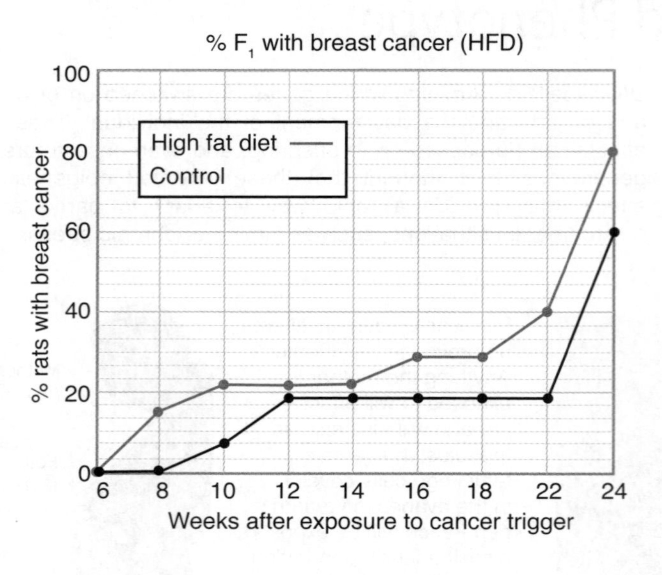

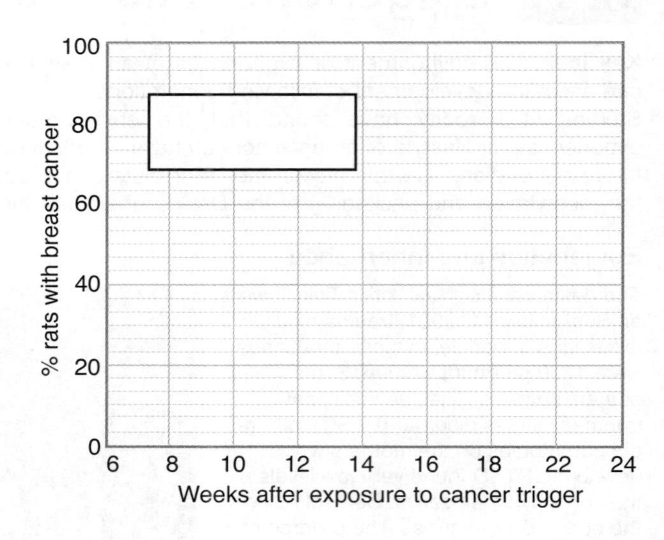

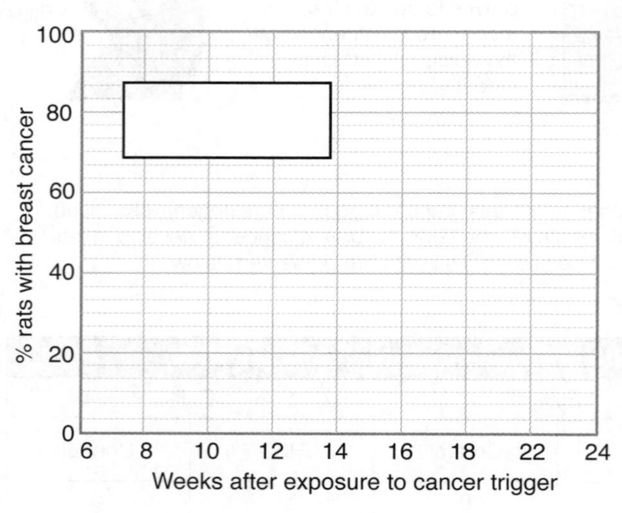

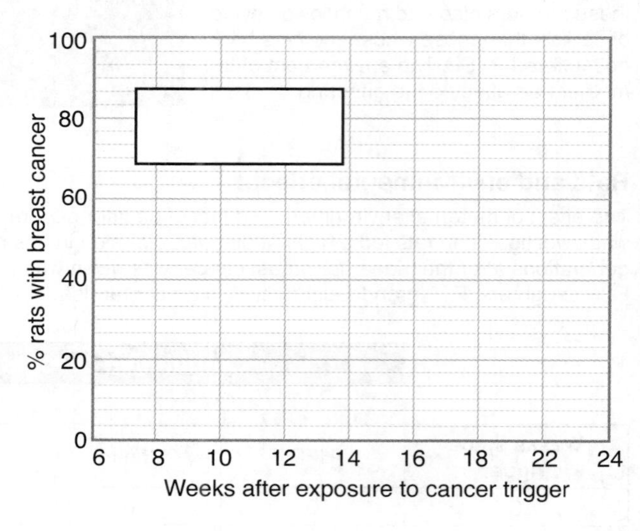

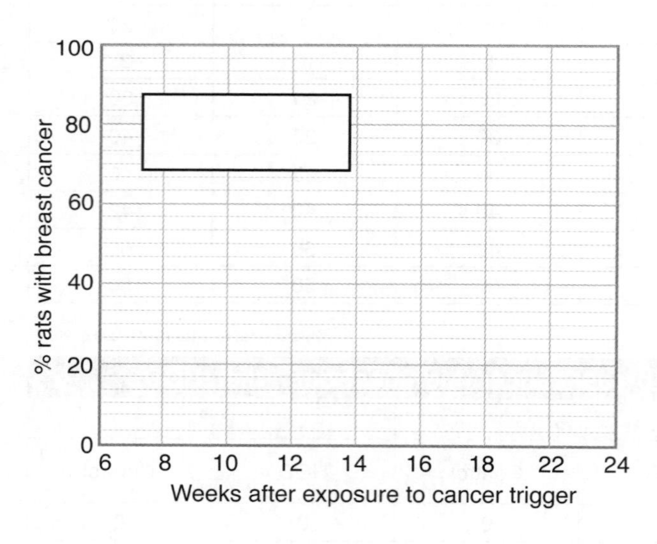

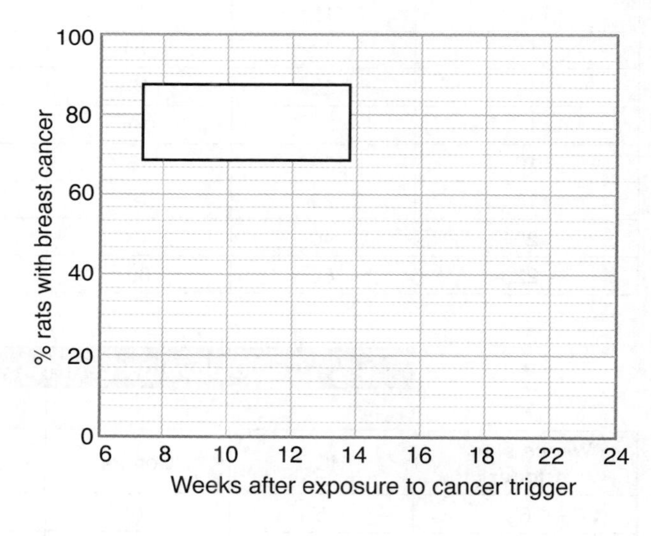

2. (a) Which generations are affected by the original mother eating a high fat diet? ______________________

(b) Which generations are affected by the mother eating a high oestrogen diet? ______________________

(c) Which diet had the longest lasting effect? ______________________

3. What do these experiments show with respect to epigenetic changes and inheritance? ______________________

____

____

ISBN:978-1-927309-20-9

168 The Role of Transcription Factors

Key Idea: Gene expression involves the interactions transcription factors with specific control regions of DNA.

All the cells in your body contain identical copies of your genetic instructions. Yet these cells appear very different (e.g. muscle, nerve, and epithelial cells have little in common). These morphological differences reflect profound differences in the expression of genes during the cell's development. For example, muscle cells express the genes for the proteins that make up the contractile elements of the muscle fibre. This diversity of cell structure and function reflects precise control over the time, location, and extent of expression of a huge variety of genes. The physical state of the DNA in or near a gene is important in helping to control whether the gene is even available for transcription. In the condensed **heterochromatin**, the transcription proteins cannot reach the DNA and the gene is not expressed. To be transcribed, a gene must first be unpacked from its condensed state. Once unpacked, control of gene expression involves the interaction of **transcription factors** with DNA sequences that control the specific gene. Initiation of transcription is the most important and universally used control point in gene expression. A simplified summary of this process is outlined below.

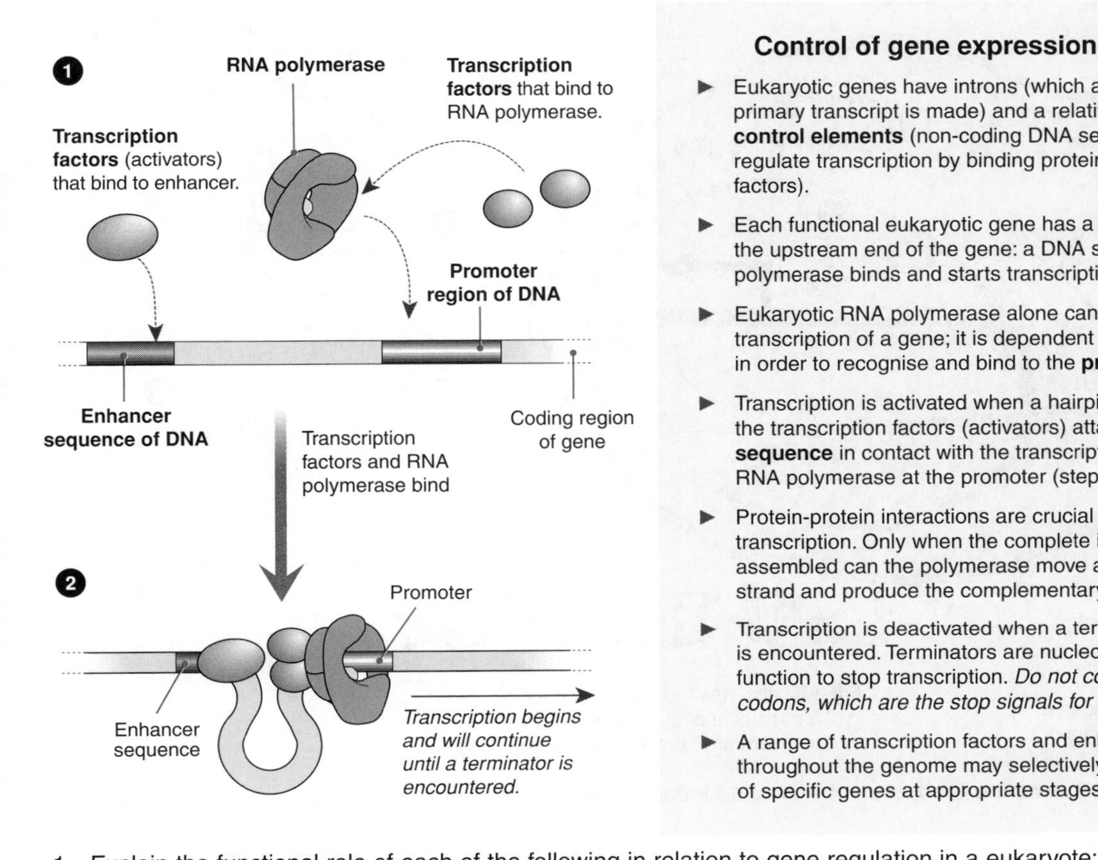

Control of gene expression in eukaryotes

- Eukaryotic genes have introns (which are removed after the primary transcript is made) and a relatively large number of **control elements** (non-coding DNA sequences that help regulate transcription by binding proteins called transcription factors).
- Each functional eukaryotic gene has a **promoter region** at the upstream end of the gene: a DNA sequence where RNA polymerase binds and starts transcription.
- Eukaryotic RNA polymerase alone cannot initiate the transcription of a gene; it is dependent on **transcription factors** in order to recognise and bind to the **promoter** (step 1).
- Transcription is activated when a hairpin loop in the DNA brings the transcription factors (activators) attached to the **enhancer sequence** in contact with the transcription factors bound to RNA polymerase at the promoter (step 2).
- Protein-protein interactions are crucial to eukaryotic transcription. Only when the complete initiation complex is assembled can the polymerase move along the DNA template strand and produce the complementary strand of RNA.
- Transcription is deactivated when a terminator sequence is encountered. Terminators are nucleotide sequences that function to stop transcription. *Do not confuse these with stop codons, which are the stop signals for translation.*
- A range of transcription factors and enhancer sequences throughout the genome may selectively activate the expression of specific genes at appropriate stages during cell development.

1. Explain the functional role of each of the following in relation to gene regulation in a eukaryote:

(a) Promoter: ______________________

(b) Transcription factors: ______________________

(c) Enhancer sequence: ______________________

(d) RNA polymerase: ______________________

(e) Terminator sequence: ______________________

2. What is the importance of chromatin in a condensed state? ______________________

ISBN: 978-1-927309-20-9

169 Oestrogen, Transcription, and Cancer

Key Idea: Oestrogen is important in initiating the immune response, but can also cause cancers of the immune system. Oestrogen, a steroid hormone found in high levels in women of reproductive age, has long been implicated in immune system function. Autoimmune diseases, in which the body attacks its own tissues, tend to be more common in women than in men and normal immune responses to infection are slightly faster in women than in men. Both of these responses have been linked to blood oestrogen levels. We now know that oestrogen acts as a switch to turn on the gene involved in antibody production. As it happens, activation of that gene is also linked to immune system cancer.

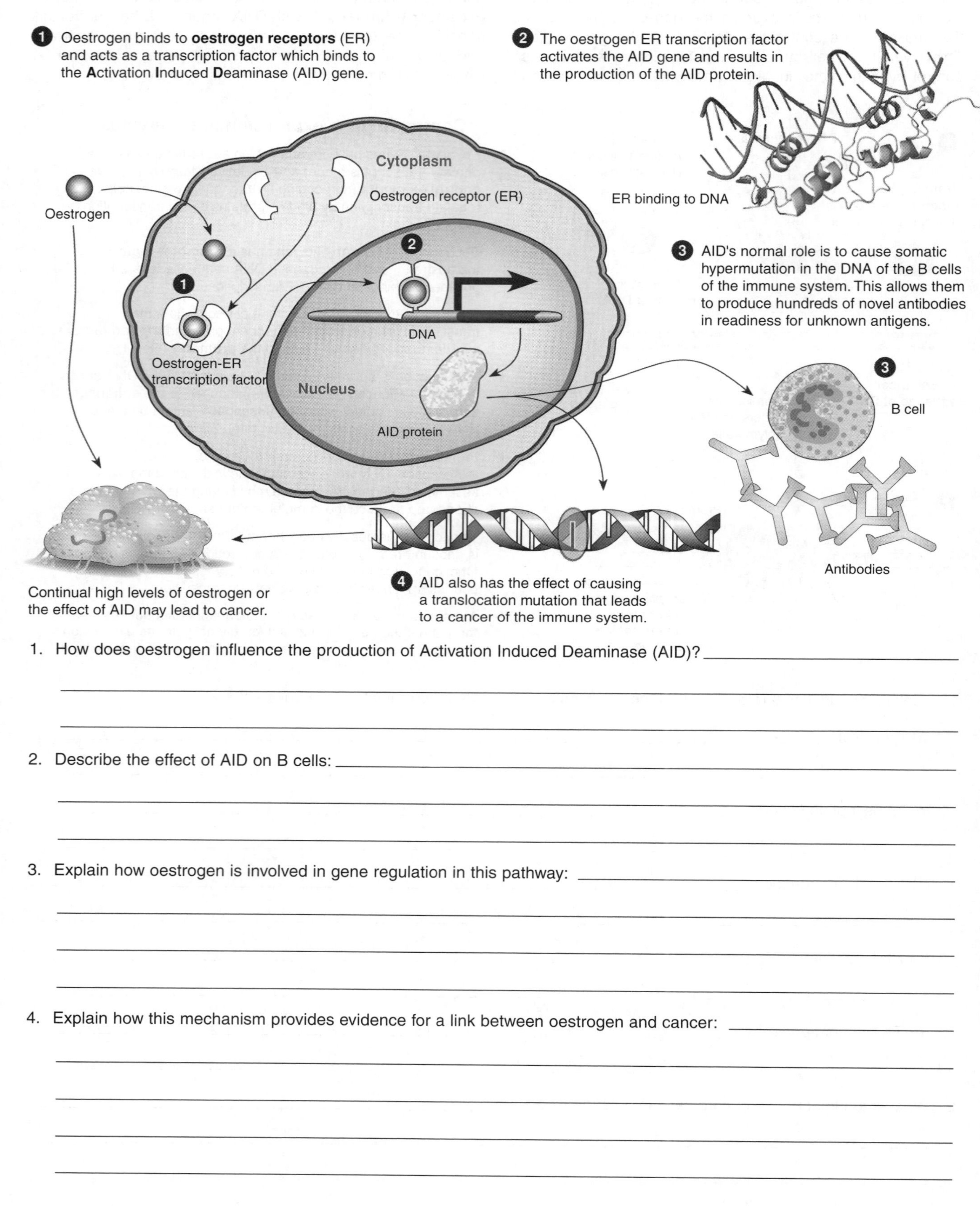

1. How does oestrogen influence the production of Activation Induced Deaminase (AID)? __________

2. Describe the effect of AID on B cells: __________

3. Explain how oestrogen is involved in gene regulation in this pathway: __________

4. Explain how this mechanism provides evidence for a link between oestrogen and cancer: __________

ISBN:978-1-927309-20-9

170 RNA Interference Models

Key Idea: RNA interference (RNAi) regulates gene expression through miRNAs and siRNAs, which act to silence genes.

RNA plays vital roles in transcribing and translating DNA, forming messenger RNA (mRNA), transfer RNA (tRNA), and ribosomal RNA (rRNA). RNA is also involved in modifying mRNA after transcription and regulating translation. RNA interference (RNAi) regulates gene expression through miRNAs and siRNAs, which bind to specific mRNA sequences, causing them to be cleaved. RNAi is important in regulating gene expression during development and in defense against viruses, which often use double-stranded RNA as an infectious vector. Regulation of translation is achieved by destroying specific mRNA targets using short RNA lengths, which may be exogenous (short interfering RNAs) or endogenous (microRNAs). Mechanisms of **RNA interference** (RNAi) are illustrated below.

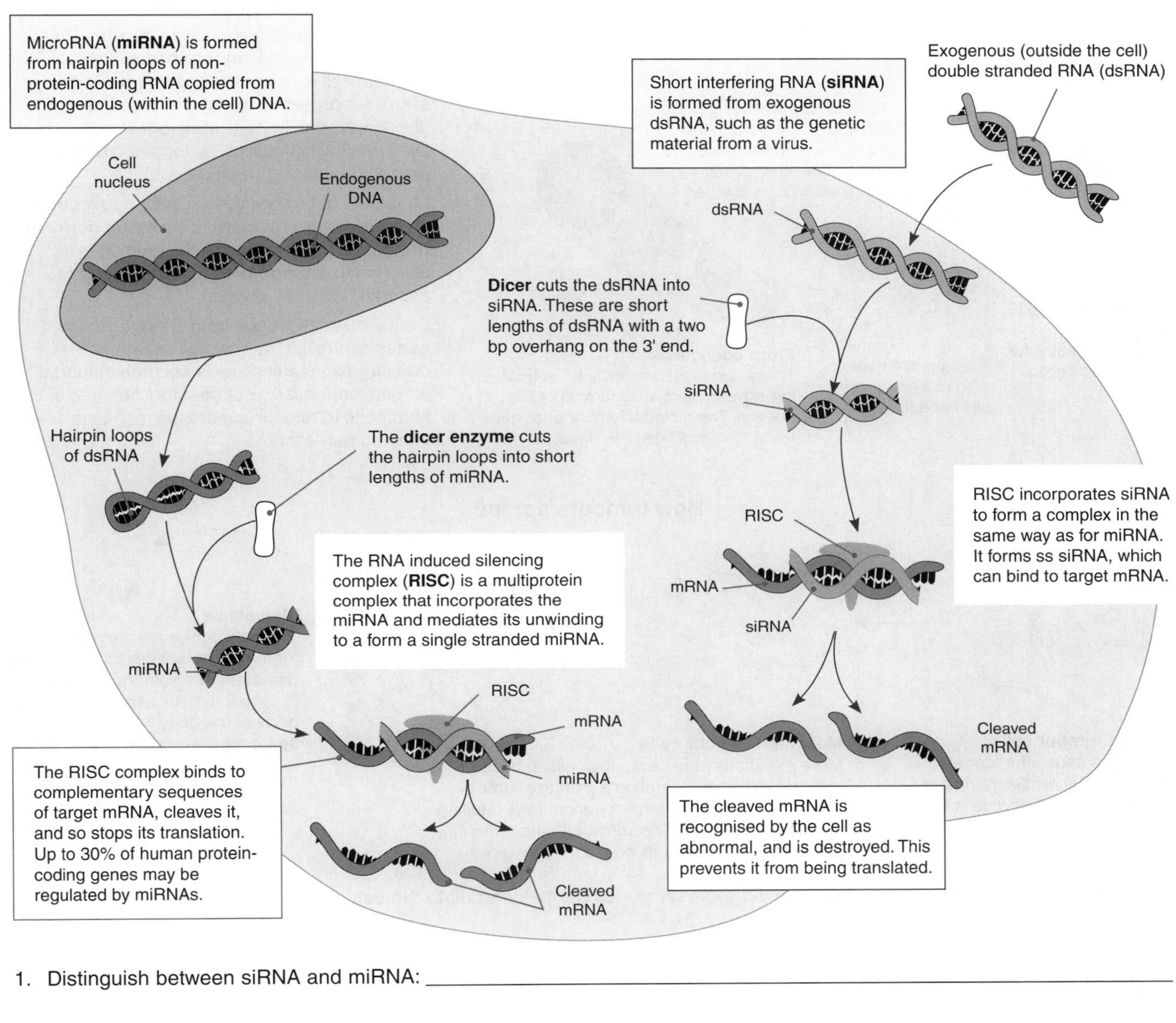

1. Distinguish between siRNA and miRNA: ______

2. How does RNAi regulate gene expression? ______

3. Under-expression of microRNAs is associated with tumour formation (uncontrolled cell growth). Based on your understanding of the role of microRNAs, account for this association:

ISBN: 978-1-927309-20-9

171 Oncogenes and Cancer

Key Idea: Cancer may be caused by carcinogens. Cancerous cells have lost their normal cellular control mechanisms.

Cells that become damaged beyond repair will normally undergo a programmed cell death (**apoptosis**), which is part of the cell's normal control system. Cancer cells evade this control and become immortal, continuing to divide without any regulation. **Carcinogens** are agents capable of causing cancer. Roughly 90% of carcinogens are also **mutagens**, i.e. they damage DNA. Long-term exposure to carcinogens accelerates the rate at which dividing cells make errors. Any one of a number of cancer-causing factors (including defective genes) may interact to induce cancer.

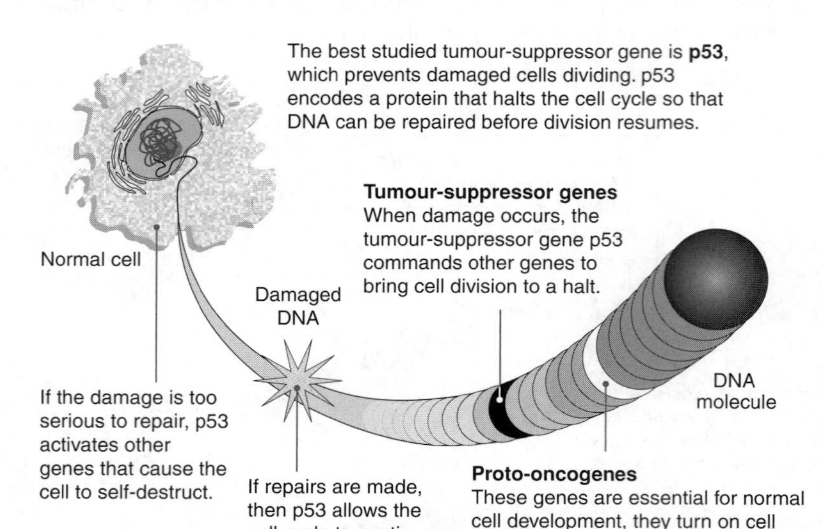

Cancer: cells out of control

Two types of gene are involved in controlling the cell cycle: **proto-oncogenes**, which start the cell division process and **tumour-suppressor genes**, which switch off cell division. Normally, both work together to perform vital tasks such as repairing defective cells and replacing dead ones.

Mutations (a change in the DNA sequence) in these genes can stop them operating normally. Proto-oncogenes, through mutation, can give rise to **oncogenes**; genes that lead to uncontrollable cell division.

Cancerous cells result from changes in the genes controlling normal cell growth and division. The resulting cells become immortal and no longer carry out their functional role. Mutations to tumour-suppressor genes initiate most human cancers.

How tumours spread

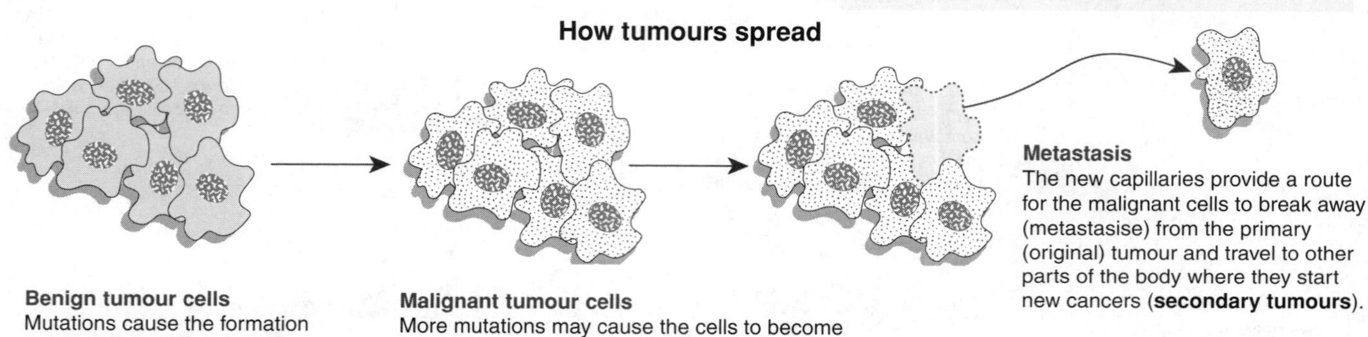

1. (a) How do proto-oncogenes and tumour suppresor genes normally regulate the cell cycle? ______________________

(b) How do oncogenes disrupt the normal cell cycle regulatory mechanisms? ______________________

2. What is a carcinogen? ______________________

3. How do cancers spread through the body? ______________________

ISBN:978-1-927309-20-9

172 Oestrogen and Cancer

Key Idea: Oestrogen is an important steroid hormone, but high levels over a lifetime increase the risk of breast cancer.

Oestrogen is an essential hormone. In women, it is responsible for normal primary and secondary sexual development. It is important in regulating the menstrual cycle and plays a role in the immune response. However oestrogen is implicated in the development of cancer, most commonly breast cancer, especially as women are often no longer pregnant or lactating for most of their reproductive lives, and oestrogen levels are higher for longer periods of time. Oestrogen is a term commonly used to encompass the four commonly occurring forms: oestriol, oestrone, oestradiol, and estetrol.

Risks of developing breast cancer

High levels of oestrogen are associated with an increased risk of cancer, particularly breast cancer. The levels of oestrogen in the body can vary throughout life, but the timing of certain events and the length of time exposed to high oestrogen levels play important roles in the risk of developing cancer. The table below shows the relative risk of breast cancer for various life events.

Event	Comparison	Risk group	Relative risk
Menarche (first menstruation)	< 12 years	12 years	0.9
		13 years	0.8
		14 years	0.9
		15 years	0.8
Age at birth of first child	Before 20 years	20-24 years	1.3
		25-29 years	1.6
		30+ years	1.9
		no children	1.9
Age at menopause	45-54 years	Before 45	0.7
		After 55	1.5
Obesity	BMI < 21	BMI 28+	1.6

Later ages of menarche expose the body to lower levels of oestrogen over a lifetime and so reduce the risk of cancer. Having child at a younger age reduces lifetime exposure to oestrogen because pregnancy and breast feeding suppress oestrogen levels. Not breast feeding increases the risk of breast cancer. The later occurrence of menopause also increases cancer risk again due to increased lifetime levels of oestrogen.

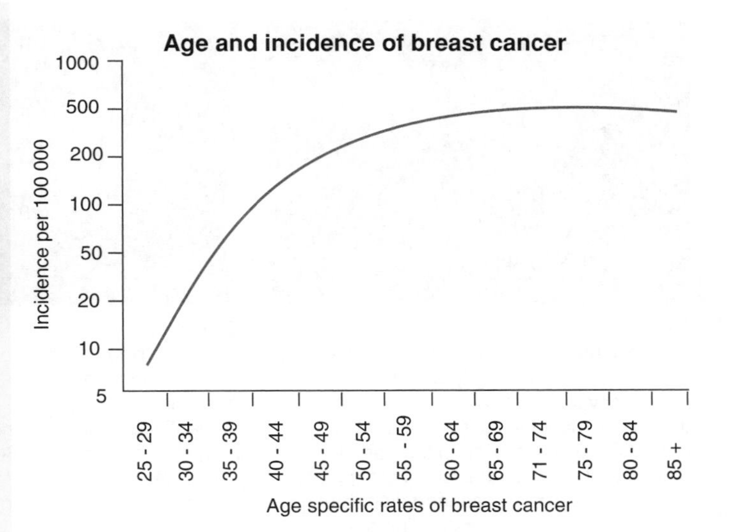

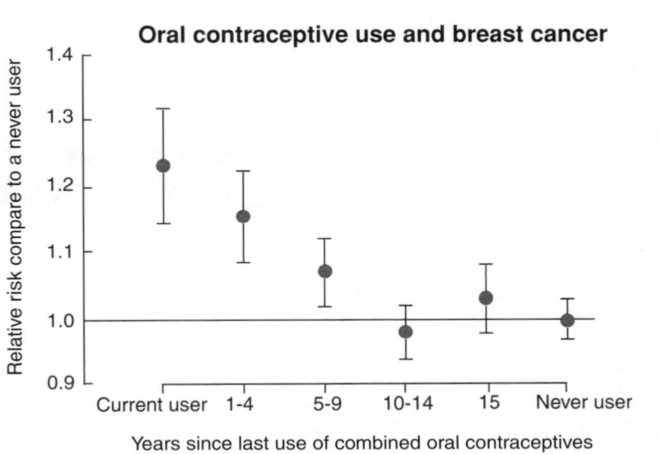

1. What is oestrogen? ______________________________

2. What is the relationship between oestrogen levels and the risk of breast cancer? ______________________________

3. What is the relationship between age and the incidence of breast cancer? ______________________________

4. (a) What is the effect of the age of first having children on the risk of developing breast cancer?

____

(b) Explain why: ______________________________

5. What is the effect of using of using the contraceptive pill on the risk of developing breast cancer?

____

ISBN: 978-1-927309-20-9

173 Genome Projects

Key Idea: Sequencing the genomes of numerous organisms will provide greater insight into gene function and evolution.

Probably the most famous genome sequencing project is the Human Genome Project, but there are many more genome sequencing projects. They include the Canine Genome Project, the Zebrafish Genome Project, and the 1000 Fungal Genomes Project. The goal of these projects is to obtain complete genome sequences and make comparisons both between individuals in the same taxonomic group (e.g. between different dog breeds) and between species. This helps establish evolutionary relationships and identify common genes and how they function in different species.

Cattle
The genome project for domestic cattle was completed in 2009. Analysis of its 22 000 genes may lead to milk or beef of higher quality and may tell us more about the impact of domestication on the genetics of a once wild animal. Analysis of the genetic differences between beef and dairy breeds may also help our understanding of the mechanisms of gene expression.

Zebrafish
Zebrafish have been studied for many years as a model organism for vertebrates. Analysis of their genome allows geneticists to identify genes with similar functions in humans and zebrafish. Sequencing has shown that about 70% of the zebrafish's genes have a related gene in humans.

Sea urchin
The sea urchin genome was sequenced in 2006. Seventy percent of its genes are similar to those in humans. Sea urchins can live as long as a human and have complex immune systems to protect them during their lifetime. Analysis of their genome could help in studying immune diseases and other aspects of the immune system.

The table below outlines the progress made since the beginning of the Human Genome Project (HGP). The push to sequence the human genome drove the development of faster, more powerful sequencing machines. At a cost of US$3 billion, it has produced an estimated economic gain of US$796 billion and created 310 000 jobs.

	Human Genome Project begins	Human Genome Project ends	10 years after HGP
Cost to generate human genome equivalent	US$1 billion	US$10-50 million	US$3-5 thousand
Time to generate human genome equivalent	6-8 years	3-4 months	1-2 days
Total DNA bases in GenBank	49 million bases	31 terabases (10^{12} bp)	150 terabases
Vertebrate genomes sequenced	0	3	112
Non-vertebrate genomes sequenced	0	14	455
Prokaryote genomes sequenced	1	167	8750
No. genes with known phenotype/ disease causing mutations	53	1474	2972
Drugs that can be "genetically tailored" to patients.	61	2264	4847

National Human Genome Research Institute

1. List a benefit for each of the three genome projects shown above:

(a) ______________________________

(b) ______________________________

(c) ______________________________

2. How has the cost and time to sequence a human genome equivalent changed since the beginning of the HGP?

____

____

____

3. How many genomes have been sequenced since the beginning of the HGP? ______________________________

____

REFER WEB 173 LINK 174 LINK 175

ISBN:978-1-927309-20-9

174 Genomes, Bioinformatics, and Medicine

Key Idea: Analysis of genomic information could lead to new and more efficient ways of treating or preventing disease. Bioinformatics has the potential to greatly increase the ability of medical technologies to identify, prevent, and treat disease.

Screening a pathogen's genome for antigenic genes and comparing them to already known genes may speed up the development of vaccines and antimicrobial drugs, and extend the tools available to fight disease.

Using bioinformatics in medicine

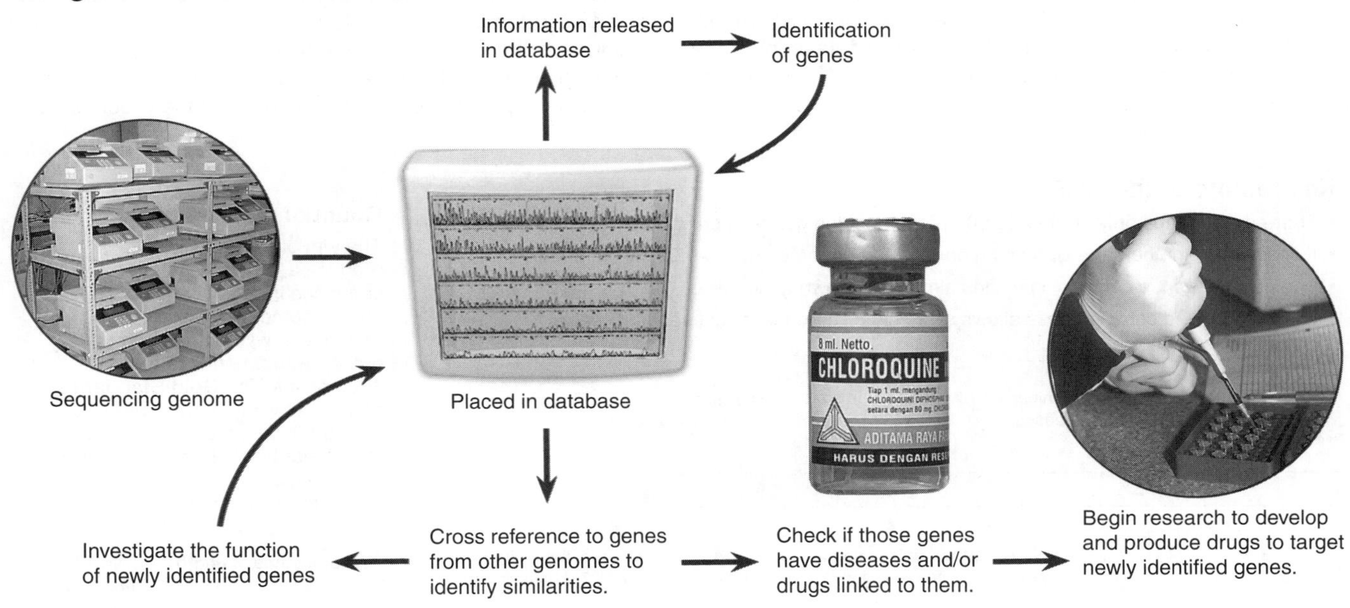

One of the new directions of research in medicine is the development of drugs targeting the gene function or gene products (proteins) of a pathogen. This begins with sequencing a genome (e.g. *Plasmodium*) and adding it to the database of already sequenced genomes. Genes are identified, cross referenced with other known similar genes, and their functions investigated. Any existing drugs targeting similar genes or gene products can then be identified and their effectiveness against the newly identified genes and their products tested. Novel genes can be identified and ways of exploiting them for medical purposes can be investigated.

Malaria and bioinformatics

Reverse vaccinology is a bioinformatics technique in which the entire genome of a pathogen is screened for genes that may produce antigenic properties, e.g. genes that code for extracellular products such as surface proteins. Once the gene is identified, the gene product (protein) is synthesised in the lab and tested in a model organism for an immune response. If successful, the product can then be used as the basis of a vaccine.

Plasmodium sporozoite

Malaria is a disease caused by the protozoan parasite *Plasmodium* of which *P. falciparum* is the most deadly. *Plasmodium* is becoming increasingly drug-resistant so a vaccine offers the best hope of controlling the disease.

The genome of *P. falciparum* was published in 2002. Fifteen loci have been identified as encoding antigens that may be useful in vaccines, including an antigen-rich region on chromosome 10. However only six of the loci appear to be similar to other *Plasmodium* species, reducing the likelihood of developing a single vaccine effective against all species.

National Human Genome Research Institute

1. Explain how bioinformatics and genome sequencing can help produce new vaccines or drugs for a disease:

2. Explain how the completion of *P. falciparum* genome has helped make the development of a malaria vaccine more likely:

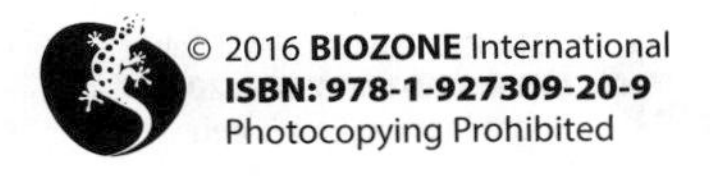

ISBN: 978-1-927309-20-9

175 The Human Genome Project

Key Idea: The Human Genome Project (HGP) was a publicly funded global venture to determine the sequence of bases in the human genome and identify and map the genes.

The HGP was completed in 2003, ahead of schedule, although analysis of it continues. Other large scale sequencing projects have arisen as a result of the initiative to sequence the human genome. In 2002, for example, the International HapMap Project was started with the aim of describing the common patterns of human genetic variation. The HGP has provided an immense amount of information, but it is not the whole story. The next task is to find out what the identified genes do. The identification and study of the protein products of genes (**proteomics**) is an important development of the HGP and will give a better understanding of the functioning of the genome. There is also increasing research in the area of the human **epigenome**, which is the record of chemical changes to the DNA and histone proteins that do not involve changes in the DNA base sequence itself. The epigenome is important in regulating genome function. Understanding how it works is important to understanding disease processes.

Key results of the HGP

- There are perhaps only 20 000-25 000 protein-coding genes in our human genome.
- It covers 99% of the gene containing parts of the genome and is 99.999% accurate.
- The new sequence correctly identifies almost all known genes (99.74%).
- Its accuracy and completeness allows systematic searches for causes of disease.

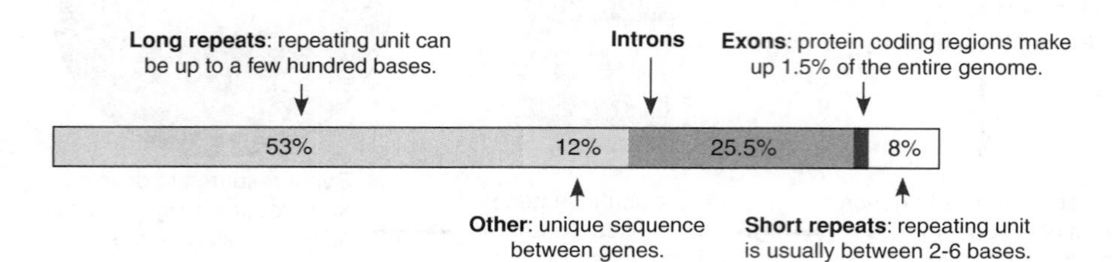

Examples of mapped genes

The positions of an increasing number of genes have been mapped onto human chromosomes (see below). Sequence variations can cause or contribute to identifiable disorders. Note that chromosome 21 (the smallest human chromosome) has a relatively low gene density, while others are gene rich. This is possibly why trisomy 21 (Down syndrome) is one of the few viable human autosomal trisomies.

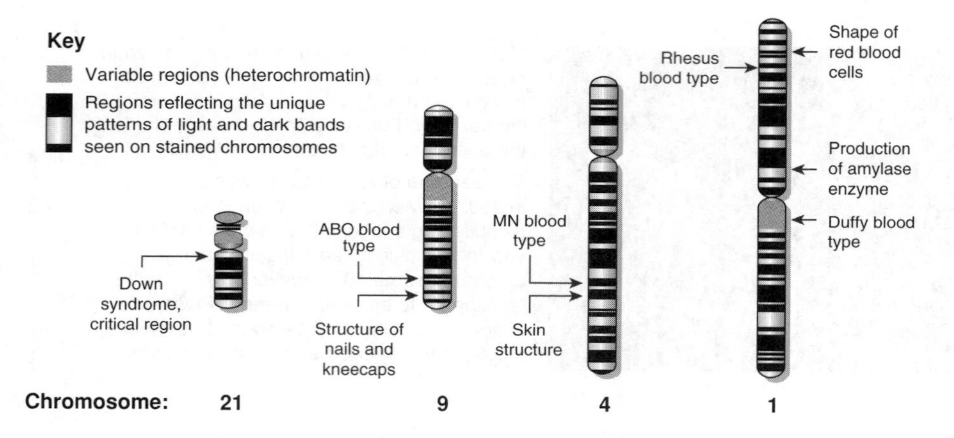

Count of mapped genes

The aim of the HGP was to produce a continuous block of sequence information for each chromosome. Initially the sequence information was obtained to draft quality, with an error rate of 1 in 1000 bases. The **Gold Standard** sequence, with an error rate of <1 per 100 000 bases, was completed in October 2004. This table shows the length and number of mapped genes for each chromosome.

Chromosome	Length (Mb)	No. of Mapped Genes
1	263	1873
2	255	1113
3	214	965
4	203	614
5	194	782
6	183	1217
7	171	995
8	155	591
9	145	804
10	144	872
11	144	1162
12	143	894
13	114	290
14	109	1013
15	106	510
16	98	658
17	92	1034
18	85	302
19	67	1129
20	72	599
21	50	386
22	56	501
X	164	1021
Y	59	122
	Total:	**19 447**

1. Briefly describe the objectives of the Human Genome Project (HGP): __________

2. (a) What percentage of the human genome is made up of long repeating units? __________

   (b) What percentage of the human genome is made up of short repeating units? __________

   (c) How would knowing the full human DNA sequence and position of genes help in identifying certain diseases?

   ____

3. On which chromosome is the gene associated with the ABO blood type found? __________

KNOW WEB 175 LINK 173 LINK 192

ISBN:978-1-927309-20-9

176 Making a Synthetic Gene

Key Idea: Synthetic genes are artificially produced DNA sequences made by piecing together nucleotides using conventional chemical processes.

The methods used to produce the DNA adapter sequences (oligonucleotides) for next generation DNA sequencing can also be used to produce synthetic genes. Today, if a gene has been sequenced, it can be produced artificially using a chemical process called solid-phase DNA synthesis. Unlike technologies such as PCR, no pre-existing DNA is needed as a template. Although there are currently some limitations, it is theoretically possible to produce any length of DNA in any base sequence. Already synthetic genes have been produced and transferred to yeasts and bacteria. The process is part of the wider field of **synthetic biology.**

Producing a synthetic gene

Short segments of DNA (oligonucleotides) can be produced using **solid-phase DNA synthesis** (below). Nucleotides are added in a stepwise fashion and are protected from incorrect reactions with blocking agents, which are removed at the end of the synthesis. Short fragments are joined together and *Taq* polymerase is used to complete the gene. The gene is then available for use, e.g. in recombinant DNA technology.

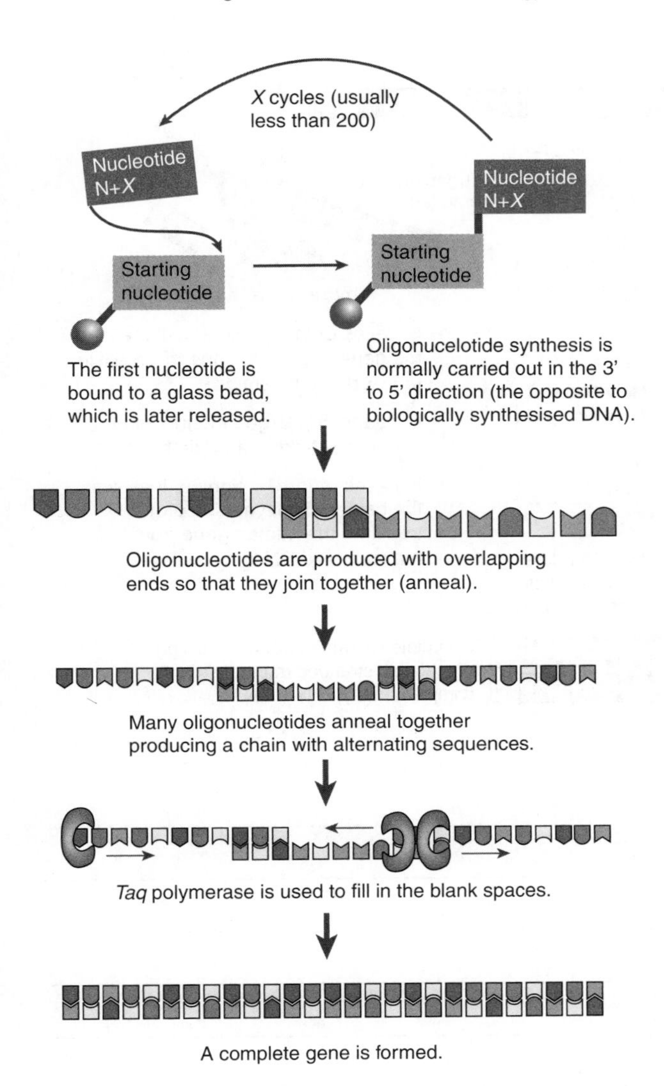

Uses of synthetic genes

Gene synthesis enables for production of novel genes that have not yet evolved in nature or it allows current genes to be refined and made more efficient. It can be used in the study and development of new drugs and vaccines, and in gene therapy.

In 2010, the J. Craig Venter Institute produced a synthetic bacterial chromosome consisting only of the genes shown to be essential for life. The gene was transplanted into a *Mycoplasma* bacterium from which the DNA had been removed. The new bacterium, called *Mycoplasma laboratorium*, replicated normally.

A new alphabet

The processes used to produce synthetic genes are not limited to the four bases normally found in DNA. Any type of nucleotide can be added providing it can be bound into the nucleotide chain in a stable way. Researchers have produced DNA with not four but six bases. The DNA was able to be replicated as a plasmid in *E. coli* until the new bases ran out, after which the bacterium replaced them with normal DNA bases.

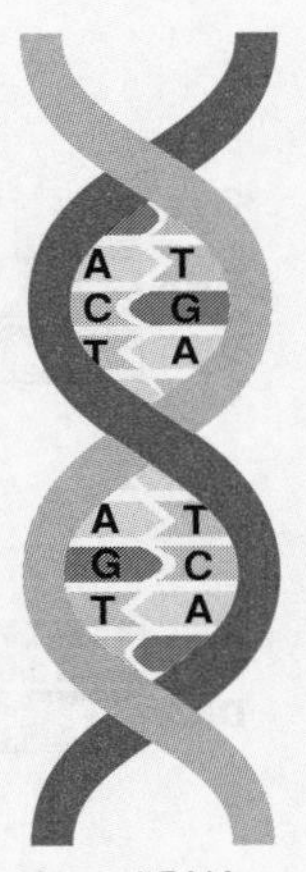

Natural DNA

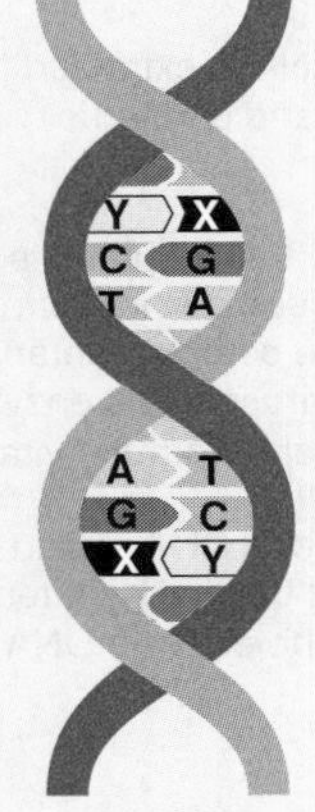

DNA with new base pairs

The DNA triplets based on the four bases of DNA directly encode the 20 amino acids commonly found in proteins. If every triplet coded for a different amino acid there could be a maximum 64 amino acids (4^3) but code degeneracy means this is not achieved. A code with six DNA bases could encode 172 amino acids (6^3 - 44 with existing code degeneracy). Potentially, this could be useful in the manufacture of new proteins and materials for use in medicine or industry. Currently, DNA sequences containing the new bases do not code for anything, as there are no RNAs that recognise the new bases.

1. (a) How is the production of a synthetic gene different from natural DNA synthesis? ______________________

   ____

   (b) Why does the synthetic gene need to be built up of oligonucleotides? ______________________

   ____

2. How might the production of synthetic genes be useful? ______________________

   ____

ISBN: 978-1-927309-20-9

177 Making a Gene from mRNA

Key Idea: Eukaryotic genes contain introns, which must be removed before a gene can be inserted into a prokaryotic cell for *in vivo* cloning.

The presence of introns (sequences of a gene's DNA that do not code for proteins) presents a problem when preparing a eukaryotic gene for insertion into a prokaryotic (bacterial) cell. Prokaryotes are the most commonly used organism in large scale culture of gene products but do not distinguish between introns and exons (coding sequences). The solution is to engineer a eukaryotic gene that can be transcribed and translated (expressed) by a prokaryote. This is achieved using the retroviral enzyme **reverse transcriptase**, which copies the mature mRNA (containing exons only) to produce a complementary strand of DNA. This task is important in both *in vivo* gene cloning and *ex vivo* gene cloning (by PCR) because it creates a gene ready for amplification.

Preparing a gene for cloning

1. Double stranded DNA of a gene from a eukaryotic organism (e.g. human) containing introns.
2. Transcription creates a **primary RNA** molecule as a part of the cell's normal gene expression.
3. The introns are removed by restriction enzymes to form a mature mRNA (now excluding the introns) that codes for the production of a single protein.
4. The mRNA is extracted from the cell and purified.
5. Reverse transcriptase is added which synthesises a single stranded DNA molecule complementary to the mRNA. In retroviruses, this enzyme makes a DNA strand from the viral ssRNA.
6. The second DNA strand is made by using the first as a template, and adding the enzyme DNA polymerase.

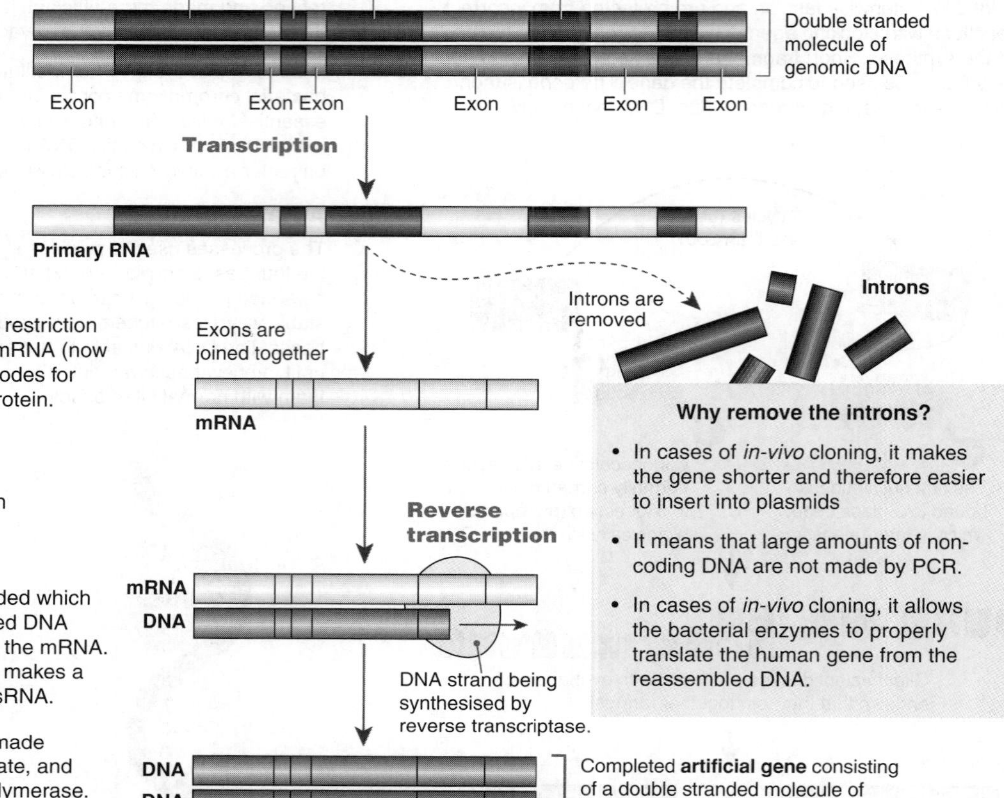

1. What is the role of restriction enzymes in preparing a clone? ______________________

____

____

2. (a) Why are introns removed before cloning a gene? ______________________

____

____

(b) What is the role of **reverse transcriptase** in this process? ______________________

____

____

3. What is the normal role of reverse transcriptase? ______________________

____

____

ISBN:978-1-927309-20-9

178 Making Recombinant DNA

Key Idea: Recombinant DNA (rDNA) is produced by first isolating a DNA sequence, then inserting it into the DNA of a different organism.

The production of rDNA is possible because the DNA of every organism is made of the same building blocks (**nucleotides**). rDNA allows a gene from one organism to be moved into, and expressed in, a different organism. Two important tools used to create rDNA are restriction digestion (chopping up the DNA) using **restriction enzymes** and DNA ligation (joining of sections of DNA) using the enzyme **DNA ligase**.

Information about restriction enzymes

1. A **restriction enzyme** is an enzyme that cuts a double-stranded DNA molecule at a specific **recognition site** (a specific DNA sequence). There are many different types of restriction enzymes, each has a unique recognition site.

2. Some restriction enzymes produce DNA fragments with two **sticky ends** (right). A sticky end has exposed nucleotide bases at each end. DNA cut in such a way is able to be joined to other DNA with matching sticky ends. Such joins are specific to their recognition sites.

3. Some restriction enzymes produce a DNA fragment with two **blunt ends** (ends with no exposed nucleotide bases). The piece it is removed from is also left with blunt ends. DNA cut in such a way can be joined to any other blunt end fragment. Unlike sticky ends, blunt end joins are non-specific because there are no sticky ends to act as specific recognition sites.

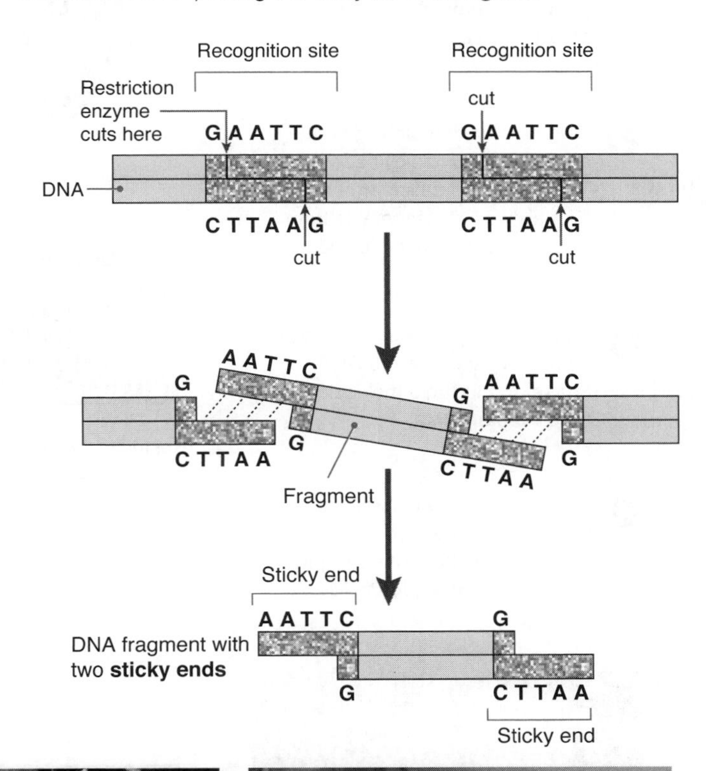

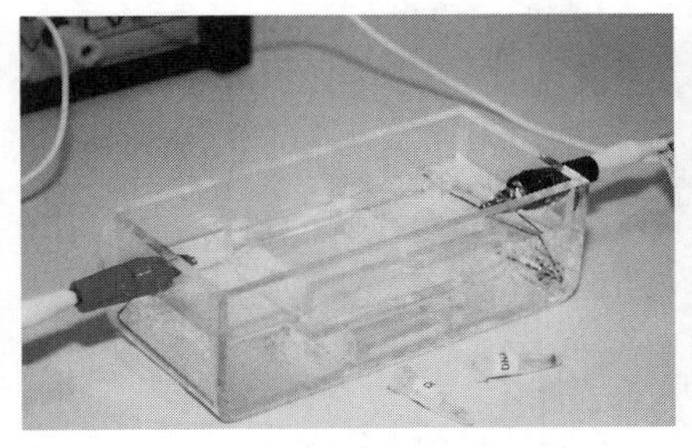

The fragments of DNA produced by the restriction enzymes are mixed with ethidium bromide, a molecule that fluoresces under UV light. The DNA fragments are then placed on an electrophoresis gel to separate the different lengths of DNA.

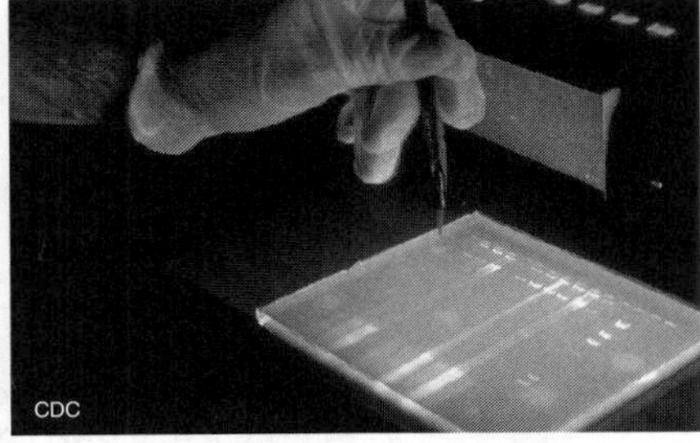

Once the DNA fragments are separated, the gel is placed on a UV viewing platform. The area of the gel containing the DNA fragments of the correct length is cut out and placed in a solution that dissolves the gel. This releases the DNA into the solution.

The solution containing the DNA is centrifuged at high speed to separate out the DNA. Centrifugation works by separating molecules of different densities. Once isolated, the DNA can be spliced into another DNA molecule.

1. What is the purpose of restriction enzymes in making recombinant DNA? ______

2. Distinguish between sticky end and blunt end fragments: ______

3. Why is it useful to have many different kinds of restriction enzymes? ______

ISBN: 978-1-927309-20-9

Creating a recombinant DNA plasmid

1. Two pieces of DNA are cut by the same restriction enzyme (they will produce fragments with matching **sticky ends**).

2. Fragments with matching sticky ends can be joined by base-pairing. This process is called **annealing.** This allows DNA fragments from different sources to be joined.

3. The fragments of DNA are joined together by the enzyme **DNA ligase**, producing a molecule of **recombinant DNA**.

4. The joined fragments will usually form either a linear or a circular molecule, as shown here (right) as recombinant **plasmid** DNA.

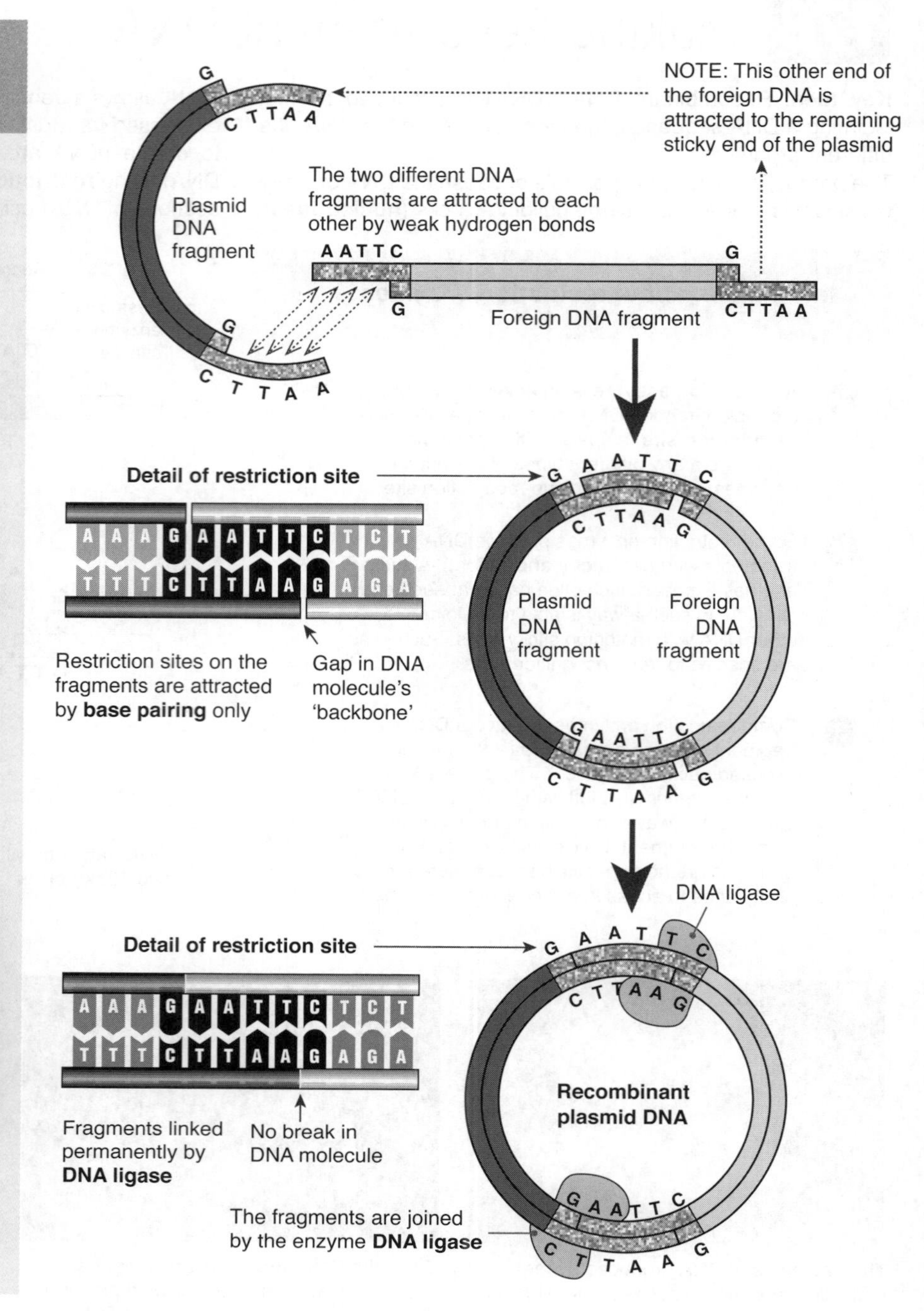

pGLO is a plasmid engineered to contain Green Fluorescent Protein (*gfp*). pGLO has been used to create fluorescent organisms, including the bacteria above (bright patches on agar plates).

4. Explain in your own words the two main steps in the process of joining two DNA fragments together:

   (a) Annealing: ______________________________

   ____

   (b) DNA ligase: ______________________________

   ____

5. Explain why **ligation** can be considered the reverse of the **restriction digestion** process:

   ____

6. Why can recombinant DNA be expressed in any kind of organism, even if it contains DNA from another species?

   ____

   ____

   ____

   ____

ISBN:978-1-927309-20-9

179 DNA Amplification Using PCR

Key Idea: PCR uses a polymerase enzyme to copy a DNA sample, producing billions of copies in a few hours.

Many procedures in DNA technology, e.g. DNA sequencing and profiling, require substantial amounts of DNA yet, very often, only small amounts are obtainable (e.g. DNA from a crime scene, from an extinct organism, or from an archaeological site). **PCR** (**polymerase chain reaction**) is a technique for reproducing large quantities of DNA in the laboratory from an original sample (**DNA amplification**). The technique is outlined below for a single cycle of replication. Subsequent cycles replicate DNA at an exponential rate, so PCR can produce billions of copies of DNA in a few hours.

A single cycle of PCR

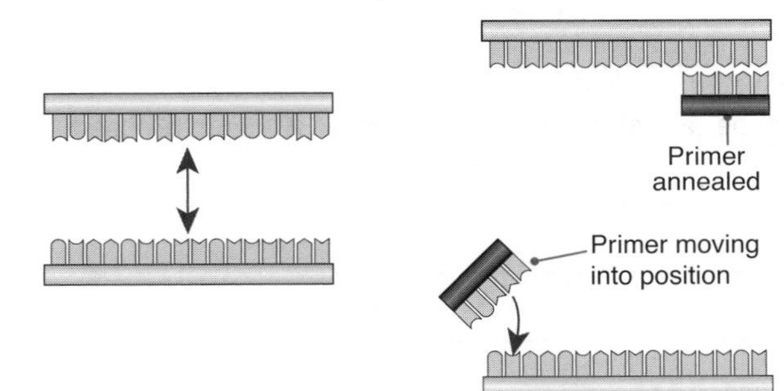

A DNA sample (called target DNA) is obtained. It is denatured (DNA strands are separated) by heating at 98°C for 5 minutes.

The sample is cooled to 60°C. Primers are annealed (bonded) to each DNA strand. In PCR, the primers are short strands of DNA; they provide the starting sequence for DNA extension.

Free nucleotides and the enzyme DNA polymerase are added. DNA polymerase binds to the primers andsynthesises complementary strands of DNA, using the free nucleotides.

After one cycle, there are now two copies of the original DNA.

Repeat for about 25 cycles

Repeat cycle of heating and cooling until enough copies of the target DNA have been produced

Loading tray
Prepared samples in PCR tubes are placed in the loading tray and the lid is closed.

Temperature control
Inside the machine are heating and refrigeration mechanisms to rapidly change the temperature.

Dispensing pipette
Pipettes with disposable tips are used to dispense DNA samples into the PCR tubes.

Thermal cycler

Amplification of DNA can be carried out with machines called thermal cyclers. Once a DNA sample has been prepared, the amount of DNA can be increased billions of times in just a few hours.

DNA quantitation
The amount of DNA in a sample can be determined by placing a known volume in this quantitation machine. For many genetic engineering processes, a minimum amount of DNA is required.

Controls
The control panel allows a number of different PCR programmes to be stored in the machine's memory. Carrying out a PCR run usually just involves starting one of the stored programmes.

RA

Reducing contamination: PCR will amplify all the DNA in the sample including unwanted DNA, so care must be taken not to contaminate the sample with DNA from the environment. Contamination is reduced by following strict protocols. The researcher must make sure that they are wearing appropriate clothing (hair net, gloves, coat) to stop their DNA contaminating the sample. In addition clean work surfaces, sterile solutions, and use of disposable equipment (e.g. pipettes and tubes) will help reduce contamination.

1. Explain the purpose of PCR: ______________________________

____

____

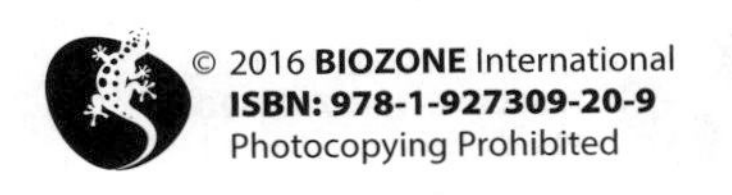

ISBN: 978-1-927309-20-9

LINK 194 LINK 177 WEB 179 KNOW

2. Describe how the **polymerase chain reaction** works:

3. Describe three situations where only very small DNA samples may be available for sampling and PCR could be used:

(a)

(b)

(c)

4. After only two cycles of replication, four copies of the double-stranded DNA exist. Calculate how much a DNA sample will have increased after:

(a) 10 cycles: (b) 25 cycles:

5. The risk of contamination in the preparation for PCR is considerable.

(a) Describe the effect of having a single molecule of unwanted DNA in the sample prior to PCR:

(b) Describe two possible sources of DNA contamination in preparing a PCR sample:

Source 1:

Source 2:

(c) Describe two precautions that could be taken to reduce the risk of DNA contamination:

Precaution 1:

Precaution 2:

6. Describe two other genetic engineering/genetic manipulation procedures that require PCR amplification of DNA:

(a)

(b)

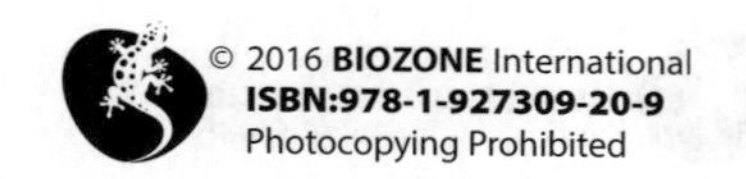

ISBN:978-1-927309-20-9

180 *In Vivo* Gene Cloning

Key Idea: *In vivo* cloning describes the insertion of a gene into an organism and using the replication machinery of that organism to multiply the gene or produce its protein product. Recombinant DNA techniques (restriction digestion and ligation) are used to insert a gene of interest into the DNA of a vector (e.g. plasmid or viral DNA). This produces a recombinant DNA molecule called a **molecular clone** that can transmit the gene of interest to another organism. To be useful, all vectors must be able to replicate inside their host organism, they must have one or more sites at which a restriction enzyme can cut, and they must have some kind of **genetic marker** that allows them to be identified. Bacterial plasmids are commonly used vectors because they are easy to manipulate, their restriction sites are well known, and they are readily taken up by cells in culture. Once the molecular clone has been taken up by bacterial cells, and those cells are identified, the gene can be replicated (cloned) many times as the bacteria grow and divide in culture.

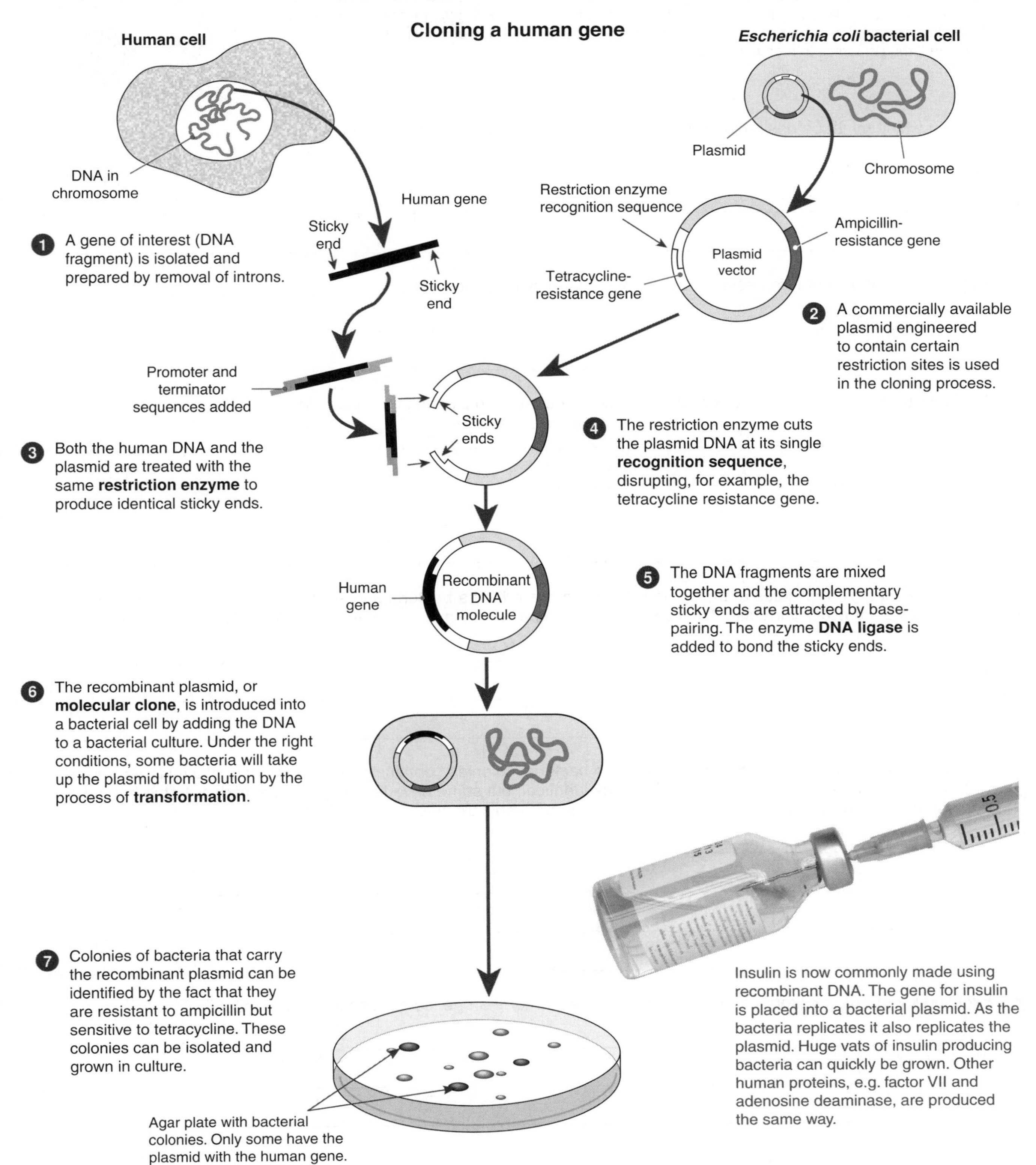

Insulin is now commonly made using recombinant DNA. The gene for insulin is placed into a bacterial plasmid. As the bacteria replicates it also replicates the plasmid. Huge vats of insulin producing bacteria can quickly be grown. Other human proteins, e.g. factor VII and adenosine deaminase, are produced the same way.

ISBN: 978-1-927309-20-9

Antibiotic resistance as a marker

Antibiotic resistant marker genes may be used to identify the bacteria that have taken up the foreign (e.g. human) DNA. The plasmid used often carries two genes that provide the bacteria with resistance to the antibiotics **ampicillin** and **tetracycline**. Without this plasmid, the bacteria have no antibiotic resistance genes. A single restriction enzyme recognition sequence lies within the tetracycline resistance gene. A foreign gene, spliced into this position, will disrupt the tetracycline resistance gene, leaving the bacteria vulnerable to this antibiotic. It is possible to identify the bacteria that successfully take up the recombinant plasmid by growing the bacteria on media containing ampicillin, and transferring colonies to media with both antibiotics.

gfp as a gene marker

Most often today, another gene acts as a marker instead of the tetracycline resistance gene. The gene for Green Fluorescent Protein (*gfp* above), isolated from the jellyfish *Aequorea victoria*, has become well established as a marker for gene expression in the recombinant organism. The *gfp* gene is recombined with the gene of interest and transformed cells can then be detected by the presence of the fluorescent product (cells with *gfp* present glow green under fluorescent light).

1. Explain why it might be desirable to use *in vivo* methods to clone genes rather than PCR:

2. Explain when it may not be desirable to use bacteria to clone genes:

3. Explain how a human gene is removed from a chromosome and placed into a plasmid:

4. A bacterial plasmid replicates at the same rate as the bacteria. If a bacteria containing a recombinant plasmid replicates and divides once every thirty minutes, calculate the number of plasmid copies there will be after twenty four hours:

5. When cloning a gene using **plasmid vectors**, the bacterial colonies containing the recombinant plasmids are mixed up with colonies that have none. All the colonies look identical, but some have taken up the plasmids with the human gene, and some have not. Explain how the colonies with the recombinant plasmids are identified:

6. Explain why the *gfp* marker is a more desirable gene marker than genes for antibiotic resistance:

ISBN:978-1-927309-20-9

181 Making Chymosin Using Recombinant Bacteria

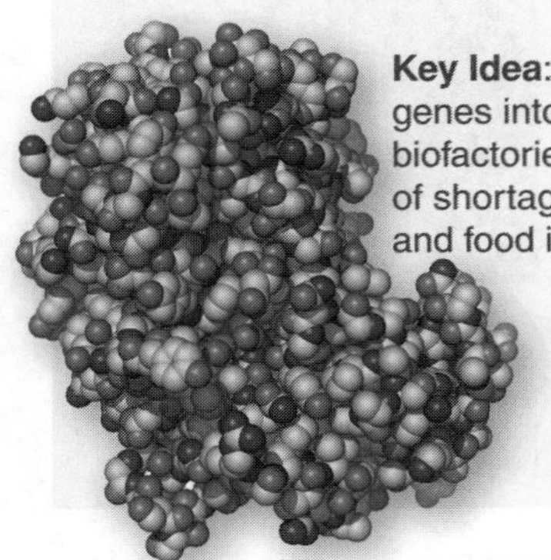

Key Idea: Inserting useful genes into bacteria to produce biofactories can solve the problem of shortages in the manufacturing and food industries.

The issue

- **Chymosin** (also known as **rennin**) is an enzyme that digests milk proteins. It is the active ingredient in rennet, a substance used by cheesemakers to clot milk into curds.
- Traditionally rennin is extracted from "chyme", i.e. the stomach secretions of suckling calves (hence its name of chymosin).
- By the 1960s, a shortage of chymosin was limiting the volume of cheese produced.
- Enzymes from fungi were used as an alternative but were unsuitable because they caused variations in the cheese flavour.

Concept 1

Enzymes are proteins made up of amino acids. The amino acid sequence of chymosin can be determined and the mRNA coding sequence for its translation identified.

Concept 2

Reverse transcriptase can be used to synthesise a DNA strand from the mRNA. This process produces DNA without the introns, which cannot be processed by bacteria.

Concept 3

DNA can be cut at specific sites using **restriction enzymes** and rejoined using **DNA ligase.** New genes can be inserted into self-replicating bacterial **plasmids**.

Concept 4

Under certain conditions, bacteria are able to lose or take up plasmids from their environment. Bacteria are readily grown in vat cultures at little expense.

Concept 5

The protein in made by the bacteria in large quantities.

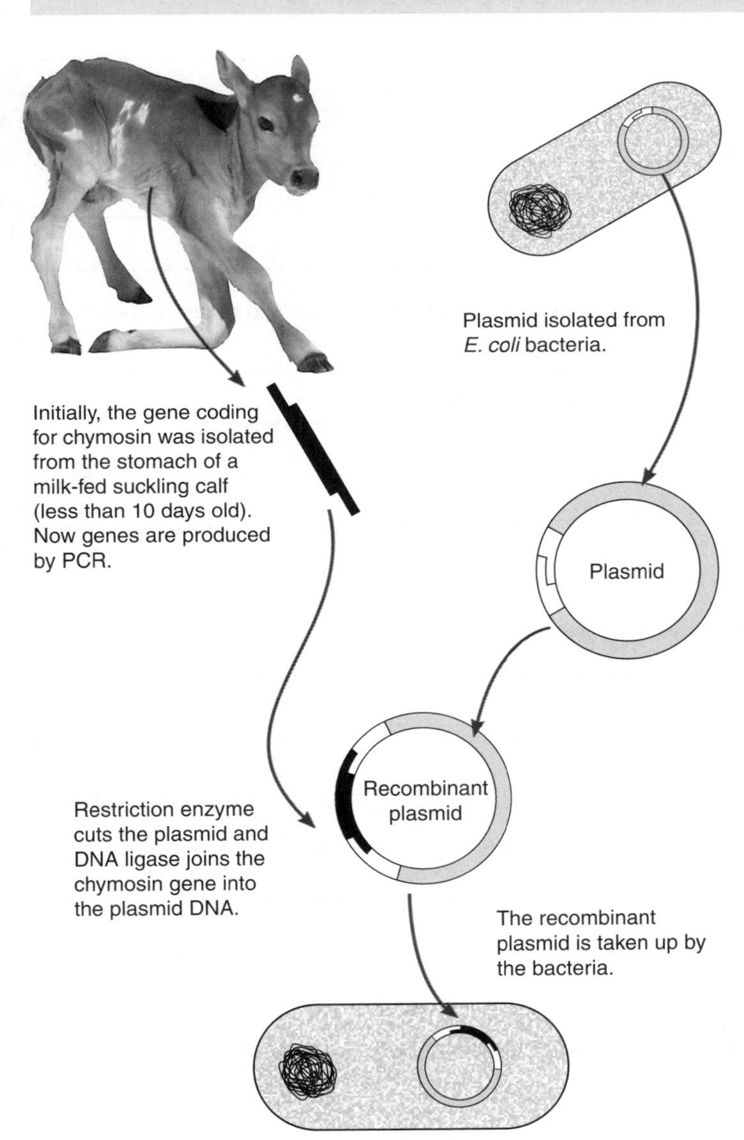

Transformed bacterial cells are grown in a vat culture

Techniques

The amino acid sequence of chymosin is first determined and the RNA codons for each amino acid identified.

mRNA matching the identified sequence is isolated from the stomach of young calves. **Reverse transcriptase** is used to transcribe mRNA into DNA. The DNA sequence can also be made synthetically once the sequence is determined.

The DNA is amplified using PCR.

Plasmids from *E. coli* bacteria are isolated and cut using **restriction enzymes.** The DNA sequence for chymosin is inserted using **DNA ligase**.

Plasmids are returned to *E. coli* by placing the bacteria under conditions that induce them to take up plasmids.

Outcomes

The transformed bacteria are grown in vat culture. Chymosin is produced by *E. coli* in packets within the cell that are separated during the processing and refining stage.

Recombinant chymosin entered the marketplace in 1990. It established a significant market share because cheesemakers found it to be cost effective, of high quality, and in consistent supply. Most cheese is now produced using recombinant chymosin such as CHY-MAX.

Further applications

A large amount of processing is required to extract chymosin from *E.coli.* There are now a number of alternative bacteria and fungi that have been engineered to produce the enzyme. Most chymosin is now produced using the fungi ***Aspergillus niger*** and ***Kluyveromyces lactis***. Both are produced in a similar way as that described for *E. coli.*

ISBN: 978-1-927309-20-9

LINK 182 LINK 178 KNOW

Enzymes from GMOs are widely used in the baking industry. Maltogenic alpha amylase from *Bacillus subtilis* bacteria is used as an anti-staling agent to prolong shelf life. Xylanase from the fungus *Aspergillus oryzae* and hemicellulases from *B. subtilis* are used for improvement of dough, crumb structure, and volume during baking.

Romain Behar

Lipase from *Aspergillus oryzae* is used in processing palm oil to produce low cost cocoa butter substitutes (above) with a similar 'mouth feel' to cocoa butter.

Dual Freq

Acetolactate decarboxylase from *B. subtilis* is an enzyme used in the brewing industry. It reduces maturation time of the beer by by-passing a rate-limiting step.

1. Describe the main use of chymosin: ____________________

____

2. What was the traditional source of chymosin? ____________________

____

3. Summarise the key concepts that led to the development of the technique for producing chymosin:

   (a) Concept 1: ____________________

   (b) Concept 2: ____________________

   (c) Concept 3: ____________________

   (d) Concept 4: ____________________

   (e) Concept 5: ____________________

4. Discuss how the gene for chymosin was isolated and how the technique could be applied to isolating other genes:

____

____

____

____

____

____

5. Describe three advantages of using chymosin produced by GE bacteria over chymosin from traditional sources:

   (a) ____________________

   ____

   (b) ____________________

   ____

   (c) ____________________

   ____

6. Explain why the fungus *Aspergillus niger* is now more commonly used to produce chymosin instead of *E. coli*:

____

____

____

ISBN:978-1-927309-20-9

182 Insulin Production

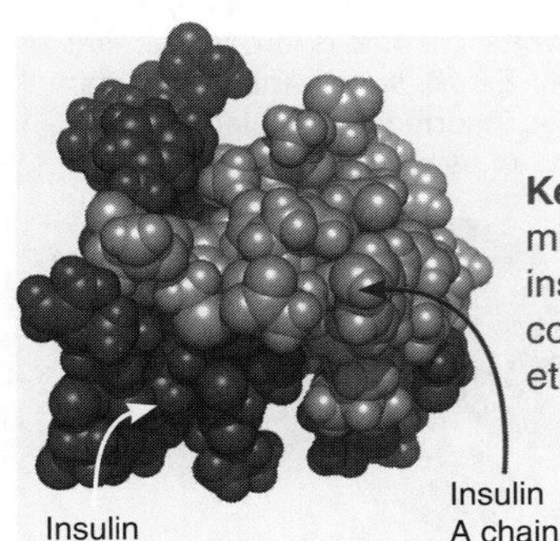

Key Idea: By using microorganisms to make human insulin, problematic issues of cost, allergic reactions, and ethics have been addressed.

The issue

- **Type I diabetes mellitus** is a metabolic disease caused by a lack of **insulin**. Around 25 people in every 100 000 suffer from type I diabetes.
- It is treatable only with injections of insulin.
- In the past, insulin was taken from the pancreases of cows and pigs and purified for human use. The method was expensive and some patients had severe allergic reactions to the foreign insulin or its contaminants.

Concept 1

DNA can be cut at specific sites using **restriction enzymes** and joined together using **DNA ligase**. Genes can be inserted into self-replicating bacterial **plasmids** at the point where the cuts are made.

Concept 2

Plasmids are small, circular pieces of DNA found in some bacteria. They usually carry genes useful to the bacterium. *E. coli* plasmids can carry promoters required for the transcription of genes.

Concept 3

Under certain conditions, Bacteria are able to lose or pick up plasmids from their environment. Bacteria can be readily grown in vat cultures at little expense.

Concept 4

The DNA sequences coding for the production of the two polypeptide chains (A and B) that form human insulin can be isolated from the human genome.

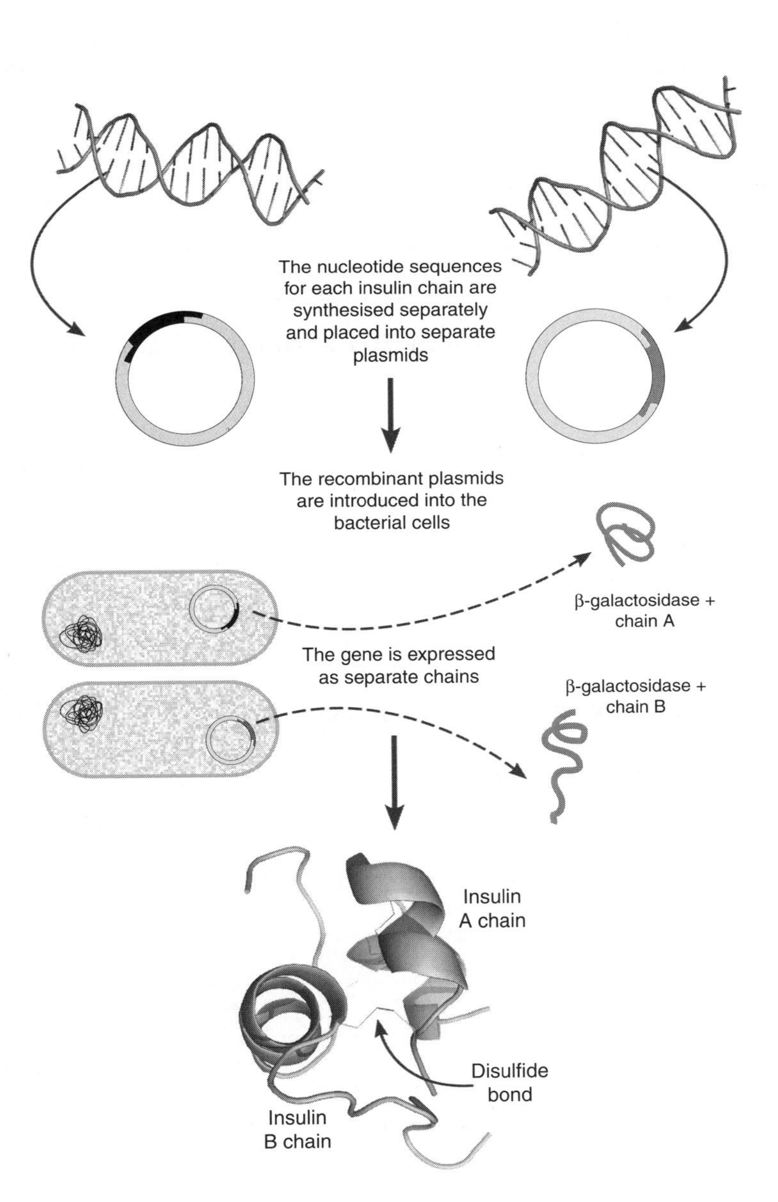

Techniques

The **gene** is **chemically synthesised** as two nucleotide sequences, one for the **insulin A chain** and one for the **insulin B chain**. The two sequences are small enough to be inserted into a plasmid.

Plasmids are extracted from *Escherichia coli*. The gene for the bacterial enzyme **β-galactosidase** is located on the plasmid. To make the bacteria produce insulin, the insulin gene must be linked to the **β-galactosidase** gene, which carries a promoter for transcription.

Restriction enzymes are used to cut plasmids at the appropriate site and the A and B insulin sequences are inserted. The sequences are joined with the plasmid DNA using **DNA ligase**.

The **recombinant plasmids** are inserted back into the bacteria by placing them together in a culture that favours plasmid uptake by bacteria.

The bacteria are then grown and multiplied in vats under carefully controlled growth conditions.

Outcomes

The product consists partly of β-galactosidase, joined with either the A or B chain of insulin. The chains are extracted, purified, and mixed together. The A and B insulin chains connect via **disulfide cross linkages** to form the functional insulin protein. The insulin can then be made ready for injection in various formulations.

Further applications

The techniques used to produce human insulin from genetically modified bacteria can be applied to a range of human proteins and hormones. Proteins currently being produced include human growth hormone, interferon, and factor VIII.

ISBN: 978-1-927309-20-9

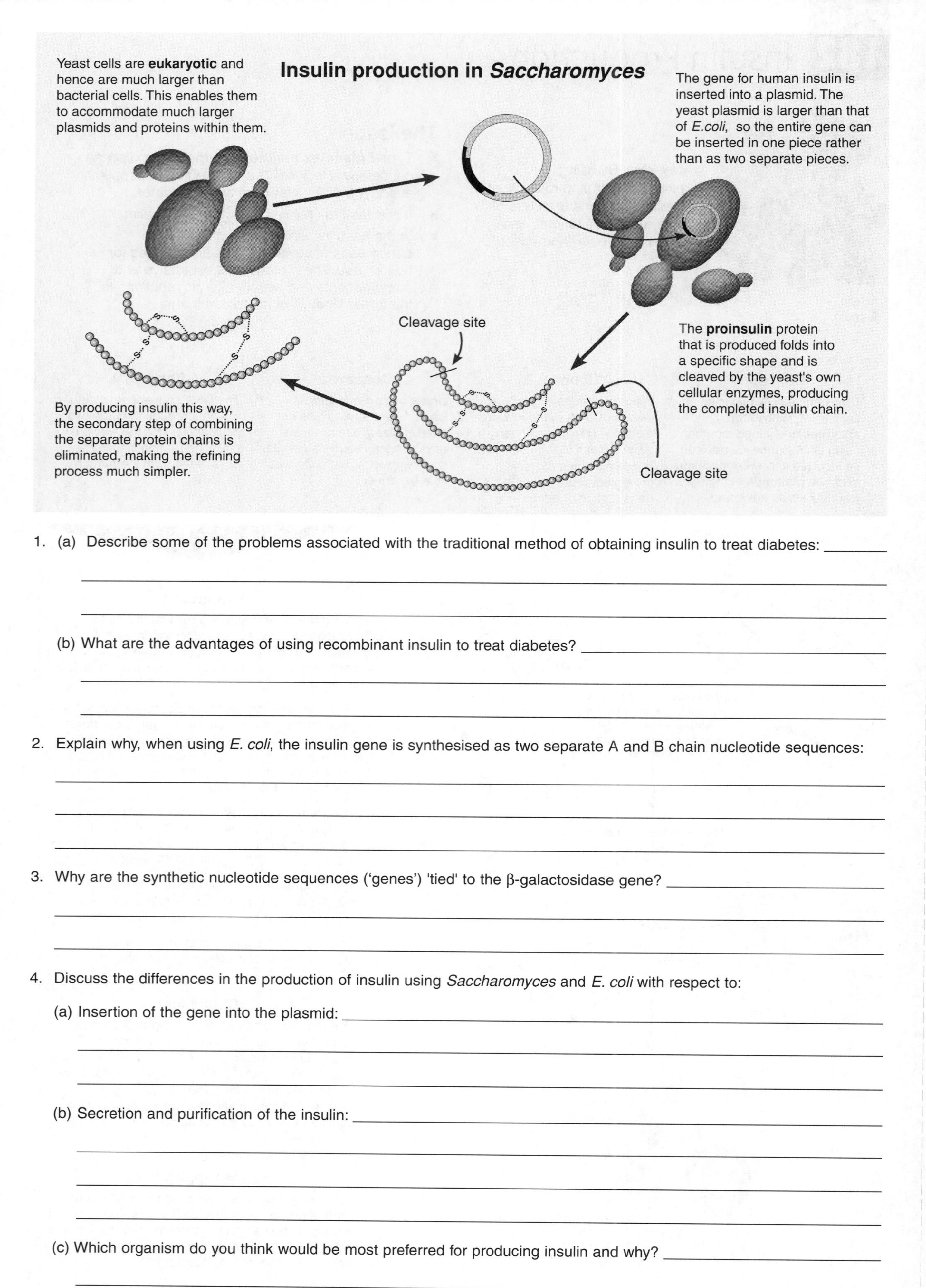

1. (a) Describe some of the problems associated with the traditional method of obtaining insulin to treat diabetes:

   (b) What are the advantages of using recombinant insulin to treat diabetes?

2. Explain why, when using *E. coli*, the insulin gene is synthesised as two separate A and B chain nucleotide sequences:

3. Why are the synthetic nucleotide sequences ('genes') 'tied' to the β-galactosidase gene?

4. Discuss the differences in the production of insulin using *Saccharomyces* and *E. coli* with respect to:

   (a) Insertion of the gene into the plasmid:

   (b) Secretion and purification of the insulin:

   (c) Which organism do you think would be most preferred for producing insulin and why?

ISBN:978-1-927309-20-9

183 Genetically Modified Food Plants

Key Idea: The use of recombinant DNA to build a new metabolic pathway has greatly increased the nutritional value of a variety of rice.

The issue

- **Beta-carotene** (β-carotene) is a precursor to **vitamin A** which is involved in many functions including vision, immunity, fetal development, and skin health.
- Vitamin A deficiency is common in developing countries where up to 500 000 children suffer from night blindness, and death rates due to infections are high due to a lowered immune response.
- Providing enough food containing useful quantities of β-carotene is difficult and expensive in many countries.

Concept 1

Rice is a staple food in many developing countries. It is grown in large quantities and is available to most of the population, but it lacks many of the essential nutrients required by the human body for healthy development. It is low in β-carotene.

Concept 2

Rice plants produce β-carotene but not in the edible rice endosperm. Engineering a new biosynthetic pathway would allow β-carotene to be produced in the endosperm. Genes expressing enzymes for carotene synthesis can be inserted into the rice genome.

Concept 3

The enzyme **carotene desaturase** (CRT1) in the soil bacterium *Erwinia uredovora*, catalyses multiple steps in carotenoid biosynthesis. **Phytoene synthase** (PSY) overexpresses a colourless carotene in the daffodil plant *Narcissus pseudonarcissus*.

Concept 4

DNA can be inserted into an organism's genome using a suitable vector. *Agrobacterium tumefaciens* is a tumour-forming bacterial plant pathogen that is commonly used to insert novel DNA into plants.

The development of golden rice

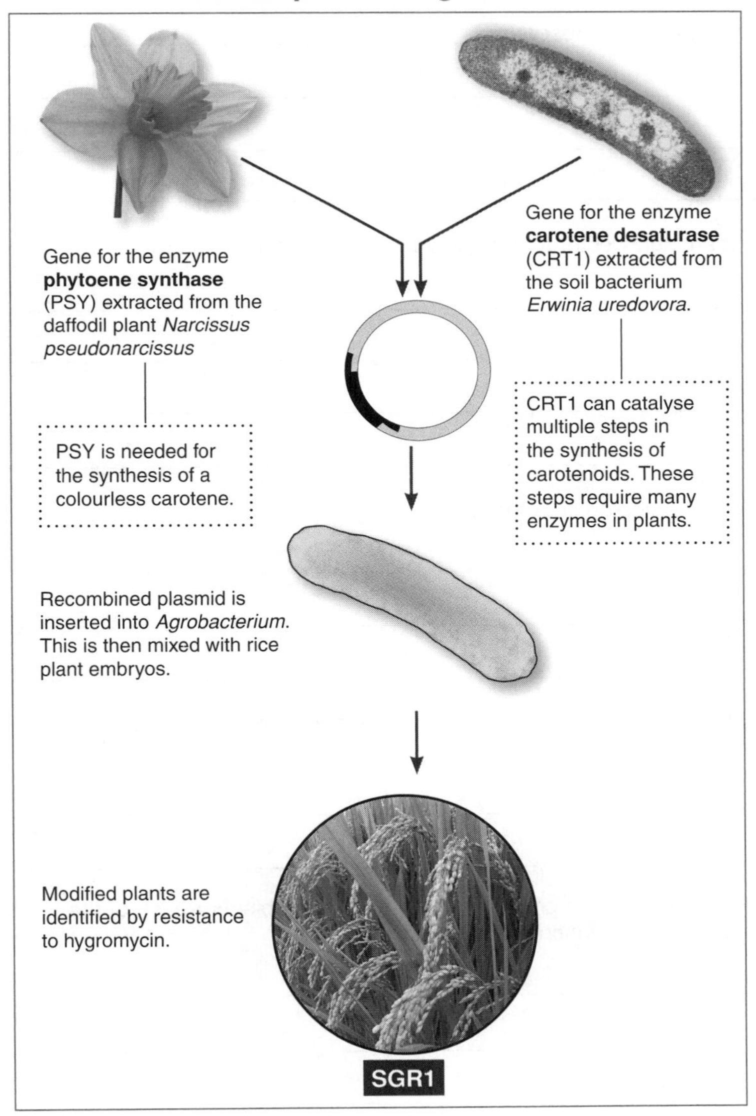

Techniques

The **PSY** gene from daffodils and the **CRT1** gene from *Erwinia uredovora* are sequenced.

DNA sequences are synthesised into packages containing the CRT1 or PSY gene, terminator sequences, and **endosperm specific promoters** (these ensure expression of the gene only in the edible portion of the rice).

The ***Ti* plasmid** from *Agrobacterium* is modified using restriction enzymes and DNA ligase to delete the tumour-forming gene and insert the synthesised DNA packages. A gene for resistance to the antibiotic **hygromycin** is also inserted so that transformed plants can be identified later. The parts of the *Ti* plasmid required for plant transformation are retained.

Modified *Ti* plasmid is inserted into the bacterium.

Agrobacterium is incubated with rice plant embryo. Transformed embryos are identified by their resistance to hygromycin.

Outcomes

The rice produced had endosperm with a distinctive yellow colour. Under greenhouse conditions golden rice (**SGR1**) contained 1.6 μg per g of carotenoids. Levels up to five times higher were produced in the field, probably due to improved growing conditions.

Further applications

Further research on the action of the PSY gene identified more efficient methods for the production of β-carotene. The second generation of golden rice now contains up to 37 μg per g of carotenoids. Golden rice was the first instance where a complete biosynthetic pathway was engineered. The procedures could be applied to other food plants to increase their nutrient levels.

ISBN: 978-1-927309-20-9

The ability of *Agrobacterium* to transfer genes to plants is exploited for crop improvement. The tumour-inducing *Ti* plasmid is modified to delete the tumour-forming gene and insert a gene coding for a desirable trait. The parts of the *Ti* plasmid required for plant transformation are retained.

Soybeans are one of the many food crops that have been genetically modified for broad spectrum herbicide resistance. The first GM soybeans were planted in the US in 1996. By 2007, nearly 60% of the global soybean crop was genetically modified; the highest of any other crop plant.

GM cotton was produced by inserting the gene for the BT toxin into its genome. The bacterium *Bacillus thuringiensis* naturally produces BT toxin, which is harmful to a range of insects, including the larvae that eat cotton. The BT gene causes cotton to produce this insecticide in its tissues.

1. Describe the basic methodology used to create golden rice: ____________________

2. Explain how scientists ensured β-carotene was produced in the endosperm: ____________________

3. What property of *Agrobacterium tumefaciens* makes it an ideal vector for introducing new genes into plants?

4. (a) How could this new variety of rice reduce disease in developing countries? ____________________

   (b) Absorption of vitamin A requires sufficient dietary fat. Explain how this could be problematic for the targeted use of golden rice in developing countries:

5. As well as increasing nutrient content as in golden rice, other traits of crop plants are also desirable. For each of the following traits, suggest features that could be desirable in terms of increasing yield:

   (a) Grain size or number: ____________________

   (b) Maturation rate: ____________________

   (c) Pest resistance: ____________________

ISBN:978-1-927309-20-9

184 Gene Therapy

Key Idea: Gene therapy aims to replace faulty genes by using a vector to transfer the correctly functioning gene into a patient's DNA.

Gene therapy uses gene technology to treat disease by correcting or replacing faulty genes. Although the details vary, all gene therapies are based around the same technique. The correct non-faulty gene is inserted into a vector, a carrier which transfers the DNA into the patient's cells (**transfection**). The vector is introduced into a sample of the patient's cells, and these are cultured to **amplify** the correct gene. The cultured cells are then transferred back to the patient. The use of altered stem cells instead of mature somatic cells has achieved longer lasting results in many patients. The treatment of somatic cells or stem cells is therapeutic (provides a benefit) but the changes are not inherited. **Germline therapy** (modification of the gametes) would enable genetic changes to be passed on. Gene therapy has had limited success because transfection of targeted cells is inefficient, and the side effects can be severe or even fatal. However, gene therapy to treat SCID, a genetic disease affecting the immune system, has had some success.

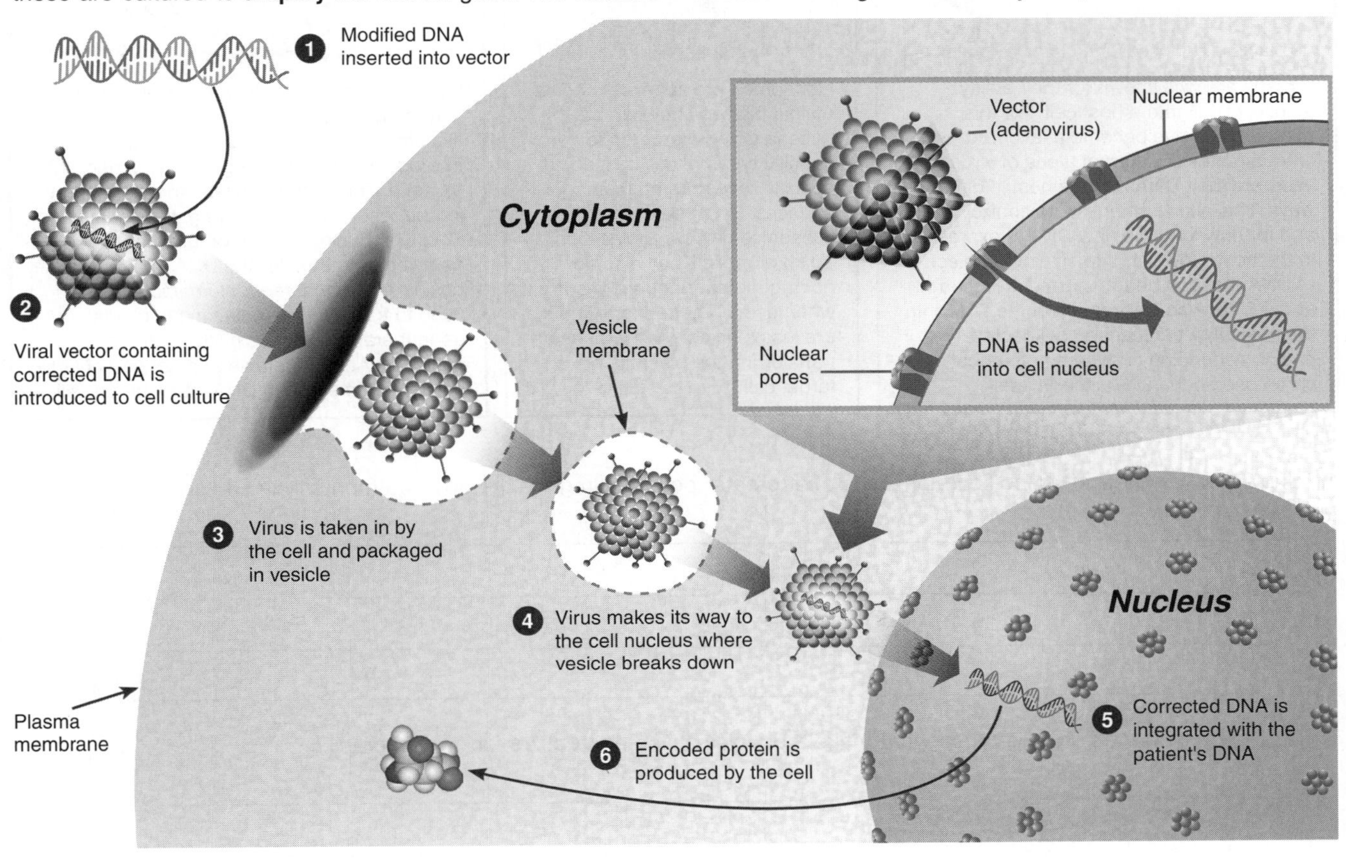

1. (a) Describe the general principle of gene therapy: ______

   (b) Describe the medical areas where gene therapy might be used: ______

2. Explain the significance of transfecting **germline cells** rather than **somatic cells**: ______

3. Explain the purpose of **gene amplification** in gene therapy: ______

ISBN: 978-1-927309-20-9

Vectors that can be used for gene therapy

Viruses

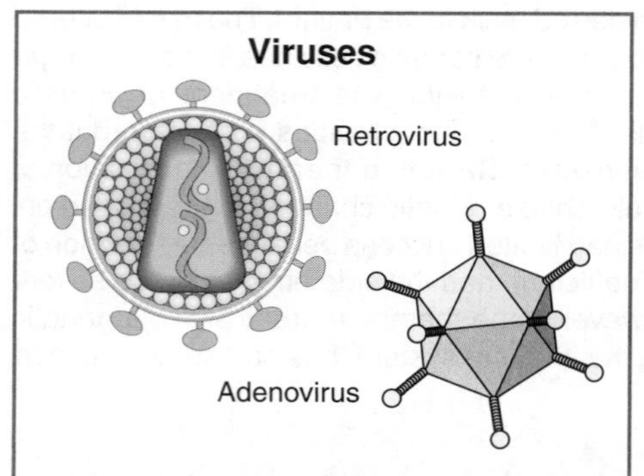

Viruses are well known for their ability to insert DNA into a host cell. For this reason they have become a favoured tool in transgenesis. Different types of viruses integrate their DNA into the host in different ways. This allows scientists to control where and for how long the new DNA is expressed in the host. However, the size of the piece of DNA that can be transferred is limited to about 8 kb. Also, integration of the DNA into the host DNA can cause unexpected side effects depending on where in the host's chromosome the DNA inserts itself.

Liposomes

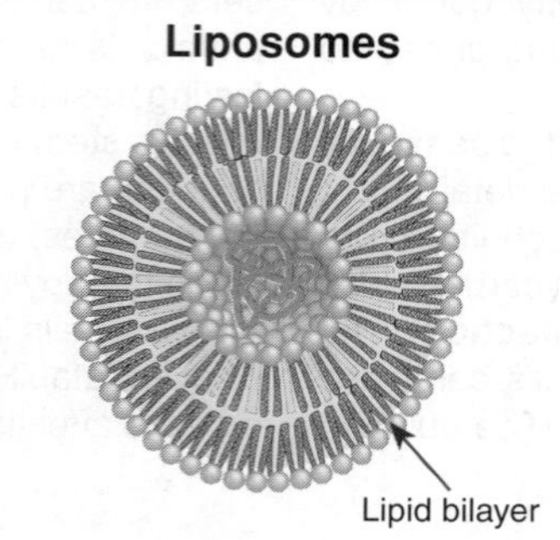

Liposomes are spherical bodies of lipid bilayer. They can be quite large and targeted to specific types of cell by placing specific receptors on their surfaces. Because of their size, liposomes can carry plasmids 20 kb or more. They also do not trigger immune responses when used in gene therapy, but are less efficient than viruses at transferring the plasmid into a target cell.

Plasmids

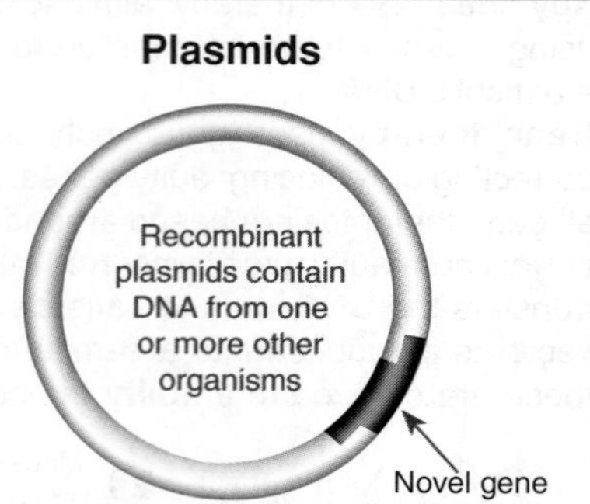

Plasmids are circular lengths of DNA that can be up to 1000 kb long (1 kb = 1000 bp). Recombinant plasmids are frequently used to produce transgenic organisms, especially bacteria. The bacteria maybe the final target for the recombinant DNA (e.g. transgenic *E. coli* producing insulin) or it can be used as a vector to transfer the DNA to a different host (e.g. *Agrobacterium tumefaciens* is used to transfer the *Ti* plasmid to plants). In gene therapy, plasmids by themselves, as naked DNA, are unstable and not particularly efficient at integrating DNA into a target cell.

4. Explain why genetically modified stem cells offer greater potential in gene therapy treatments than GM somatic cells:

5. (a) Describe the features of viruses that make them well suited as **vectors** for gene therapy:

   (b) Describe two problems with using viral vectors for gene therapy:

6. (a) Explain why it may be beneficial for a (therapeutic) gene to integrate into the patient's chromosome:

   (b) Explain why this has the potential to cause problems for the patient:

7. (a) Explain why naked DNA is likely to be unstable within a patient's tissues:

   (b) Explain why enclosing the DNA within liposomes might provide greater stability:

ISBN:978-1-927309-20-9

185 Gene Delivery Systems

Key Idea: The delivery of genes into target cells and then into patients has proved technically difficult, limiting the use of gene therapy.

It remains technically difficult to deliver genes successfully to a patient, limiting the success rate of gene therapy treatments. Any improvements have been mostly short-lived, or counteracted by adverse side effects. The inserted genes may reach only about 1% of target cells. Those that reach their target may work inefficiently and produce too little protein, too slowly to be of benefit. Many patients also have immune reactions to the vectors used in gene transfer. One of the first gene therapy trials was for cystic fibrosis (CF). CF was an obvious candidate for gene therapy because, in most cases, the disease is caused by a single, known gene mutation. However, despite its early promise, gene therapy for this disease has been disappointing (below right). Severe Combined Immune Deficiency (SCID) is another candidate for gene therapy, again because the disease is caused by single, known mutation (below left). Gene therapies for this disease have so far proved promising.

Treating SCID using gene therapy

The most common form of **SCID** (Severe Combined Immune Deficiency) is **X-linked SCID**, which results from mutations to a gene on the X chromosome encoding the **common gamma chain**, a protein forming part of a receptor complex for numerous types of leucocytes. A less common form of the disease, (**ADA-SCID**) is caused by a defective gene that codes for the enzyme adenosine deaminase (ADA).

Both of these types of SCID lead to immune system failure. A common treatment for SCID is bone marrow transplant, but this is not always successful and runs the risks of infection from unscreened viruses. **Gene therapy** appears to hold the best chances of producing a cure for SCID because the mutation affects only one gene whose location is known. DNA containing the corrected gene is placed into a **gutted retrovirus** and introduced to a sample of the patient's **bone marrow.** The treated cells are then returned to the patient.

In some patients with ADA-SCID, treatment was so successful that supplementation with purified ADA (produced using genetically modified bacteria) was no longer required. The treatment carries risks though. In early trials, two of ten treated patients developed leukaemia when the corrected gene was inserted next to a gene regulating cell growth.

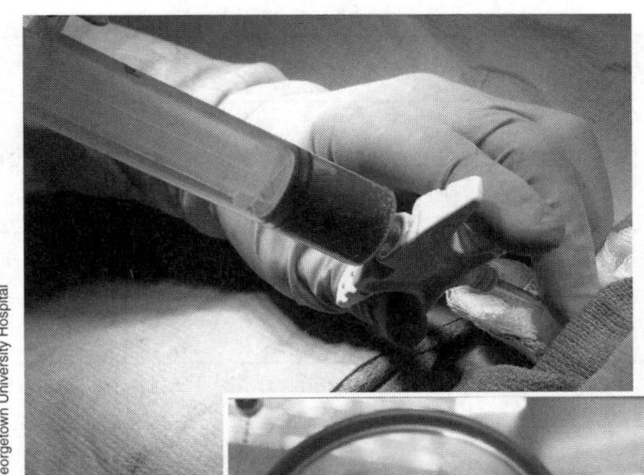

Samples of bone marrow being extracted prior to treatment with gene therapy.

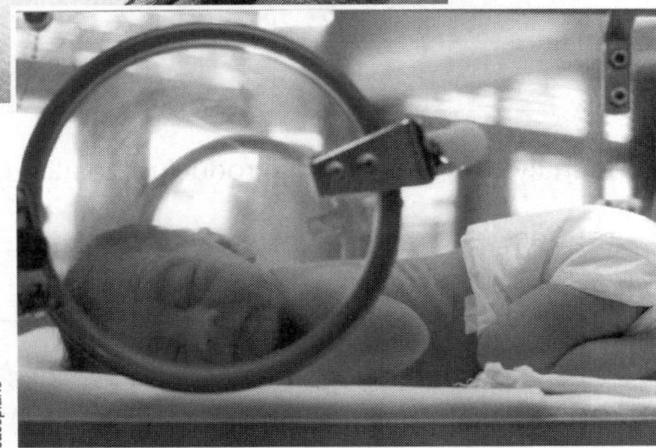

Detection of SCID is difficult for the first months of an infant's life due to the mother's antibodies being present in the blood. Suspected SCID patients must be kept in sterile conditions at all times to avoid infection.

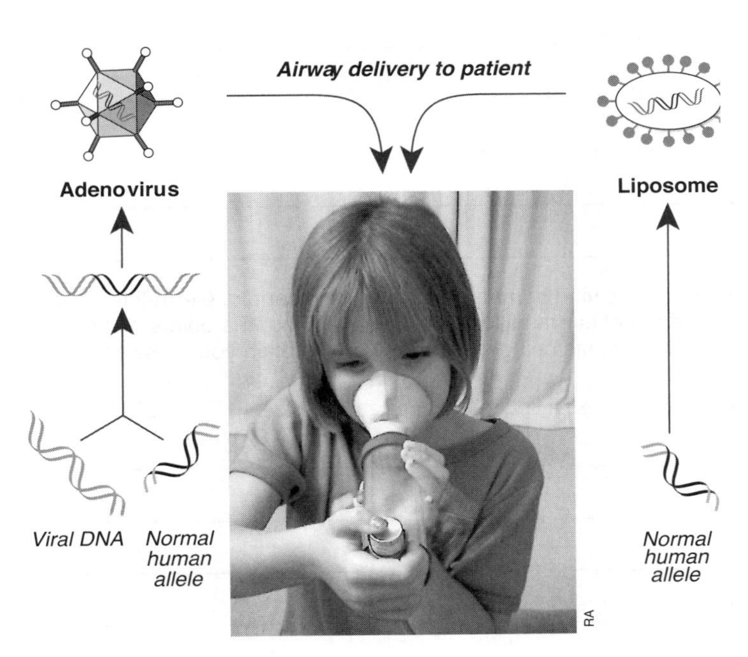

An **adenovirus** that normally causes colds is genetically modified to make it safe and to carry the normal (unmutated) CFTR ('cystic fibrosis') gene.

Liposomes are tiny fat globules. Normal CF genes are enclosed in liposomes, which fuse with plasma membranes and deliver the genes into the cells.

Gene therapy - potential treatment for cystic fibrosis?

Cystic fibrosis (CF) is caused by a mutation to the gene coding for a chloride ion channel important in creating sweat, digestive juices, and mucus. The dysfunction results in abnormally thick, sticky mucus that accumulates in the lungs and intestines. The identification and isolation of the CF gene in 1989 meant that scientists could look for ways in which to correct the genetic defect rather than just treating the symptoms using traditional therapies.

The main target of CF gene therapy is the lung, because the progressive lung damage associated with the disease is eventually lethal.

In trials, normal genes were isolated and inserted into patients using vectors such as **adenoviruses** and **liposomes**, delivered via the airways (left). The results of trials were disappointing: on average, there was only a 25% correction, the effects were short lived, and the benefits were quickly reversed. Alarmingly, the adenovirus used in one of the trials led to the death of one patient.

Source: Cystic Fibrosis Trust, UK.

ISBN: 978-1-927309-20-9

1. A great deal of current research is being devoted to discovering a gene therapy solution to treat **cystic fibrosis** (CF):

(a) Describe the symptoms of CF:

(b) Why has this particular genetic disease been so eagerly targeted by geneticists?

(c) Outline some of the problems so far encountered with gene therapy for CF:

2. Identify two vectors for introducing healthy CFTR genes into CF patients.

(a) Vector 1:

(b) Vector 2:

3. (a) Describe the difference between X-linked SCID and ADA-SCID:

(b) Identify the vector used in the treatment of SCID:

4. Briefly outline the differences in the gene therapy treatment of CF and SCID:

5. Changes made to chromosomes as a result of gene therapy involving somatic cells are not inherited. Germ-line gene therapy has the potential to cure disease, but the risks and benefits are still not clear. For each of the points outlined below, evaluate the risk of germ-line gene therapy relative to somatic cell gene therapy and explain your answer:

(a) Chance of interfering with an essential gene function:

(b) Misuse of the therapy to selectively alter phenotype:

ISBN:978-1-927309-20-9

186 The Ethics of GMO Technology

Key Idea: There are many potential benefits, risks, and ethical questions in using genetically modified organisms.

Genetically modified organisms (GMOs) have many potential benefits, but their use raises a number of biological and ethical concerns. Some of these include risk to human health, animal welfare issues, and environmental safety. Currently a matter of concern to consumers is the adequacy of government regulations for labelling food products with GMO content. In some countries GM products must be clearly labelled, while other countries have no requirements for GM labelling at all. This can take away consumer choice about the types of products they buy. The use of GM may also have trade implications for countries exporting and importing GMO produce.

Potential benefits of GMOs

1. Increase in crop yields, including crops with more nutritional value and that store for longer.
2. Decrease in use of pesticides, herbicides and animal remedies.
3. Production of crops that are drought tolerant or salt tolerant.
4. Improvement in the health of the human population and the medicines used to achieve it.
5. Development of animal factories for the production of proteins used in manufacturing, the food industry, and health.

Potential risks of GMOs

1. Possible (uncontrollable) spread of transgenes into other species of plants, or animals.
2. Concerns that the release of GMOs into the environment may be irreversible.
3. Animal welfare and ethical issues: GM animals may suffer poor health and reduced life span.
4. GMOs may cause the emergence of pest, insect, or microbial resistance to traditional control methods.
5. May create a monopoly and dependence of developing countries on companies who are seeking to control the world's commercial seed supply.

Issues and solutions

Issue: The accidental release of GMOs into the environment.

Problem: Recombinant DNA may be taken up by non-target organisms. These then may have the potential to become pests or cause disease.

Solution: Rigorous controls on the production and release of GMOs. GMOs could have specific genes deleted so that their growth requirements are met only in controlled environments.

Issue: A new gene or genes may disrupt normal gene function.

Problem: Gene disruption may trigger cancer. Successful expression of the desired gene is frequently very low.

Solution: A combination of genetic engineering, cloning, and genetic screening so that only those cells that have been successfully transformed are used to produce organisms.

Issue: Targeted use of transgenic organisms in the environment.

Problem: Once their desired function, e.g. environmental clean-up, is completed, they may be undesirable invaders in the ecosystem.

Solution: GMOs can be engineered to contain "suicide genes" or metabolic deficiencies so that they do not survive for long in the new environment after completion of their task.

© Greenpeace

GMO protestors are arrested

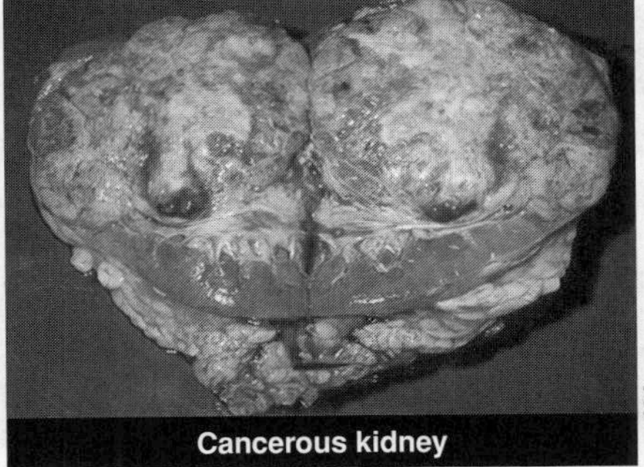

CDC

Cancerous kidney

© Greenpeace

Protest against GMOs in the environment

1. Describe an advantage and a problem with the use of plants genetically engineered to be resistant to crop pests:

 (a) Advantage: ______________________

 ____

 ____

 (b) Problem: ______________________

 ____

 ____

2. Some years ago, Britain banned the import of a GM, pest resistant corn variety containing marker genes for ampicillin antibiotic resistance. Suggest why the use of antibiotic-resistance genes as markers is no longer common practice:

 ____

 ____

 ____

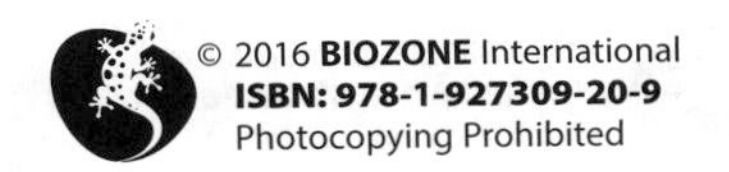

ISBN: 978-1-927309-20-9

187 Food for the Masses

Key Idea: Genetic engineering has the potential to solve many of the world's food shortage problems by producing crops with greater yields than those currently grown.

Currently 1/6 of the world's population are **undernourished**. If trends continue, 1.5 billion people will be at risk of starvation by 2050 and, by 2100 (if global warming is taken into account), nearly half the world's population could be threatened with food shortages. The solution to the problem of food production is complicated. Most of the Earth's arable land has already been developed and currently uses 37% of the Earth's land area, leaving little room to grow more crops or farm more animals. Development of new fast growing and high yield crops appears to be part of the solution, but many crops can only be grown under a narrow range of conditions or are susceptible to disease. Moreover, the farming and irrigation of some areas is difficult, costly, and can be environmentally damaging. **Genetic modification** of plants may help to solve some of these problems by producing plants that will require less intensive culture or that will grow in areas previously considered not arable.

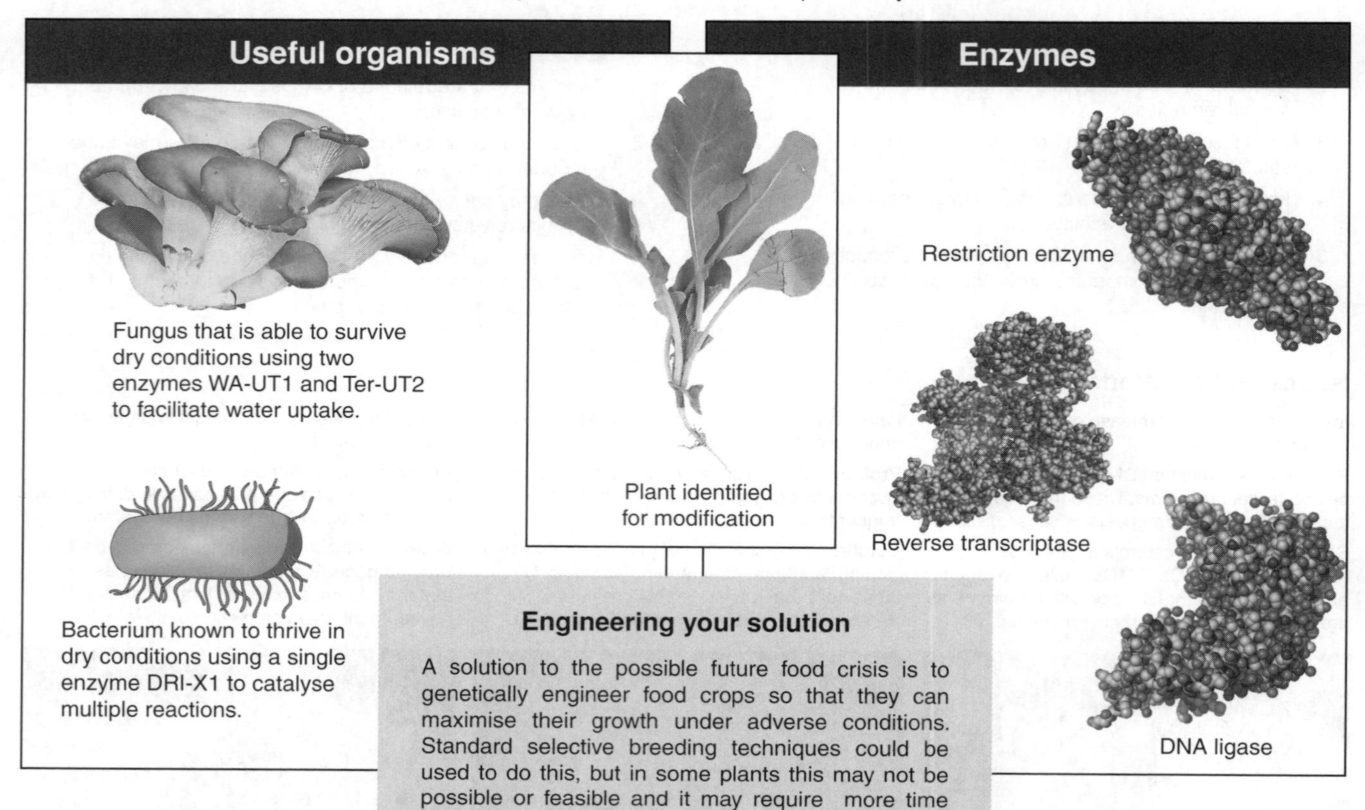

Engineering your solution

A solution to the possible future food crisis is to genetically engineer food crops so that they can maximise their growth under adverse conditions. Standard selective breeding techniques could be used to do this, but in some plants this may not be possible or feasible and it may require more time than is available. A selection of genetic tools and organisms with useful characteristics are described. **Your task** is to use the items shown to devise a technique to successfully create a plant that could be successfully farmed in semi-desert environments such as sub-Saharan Africa. The following page will take you through the procedure. Not all the items will need to be used.

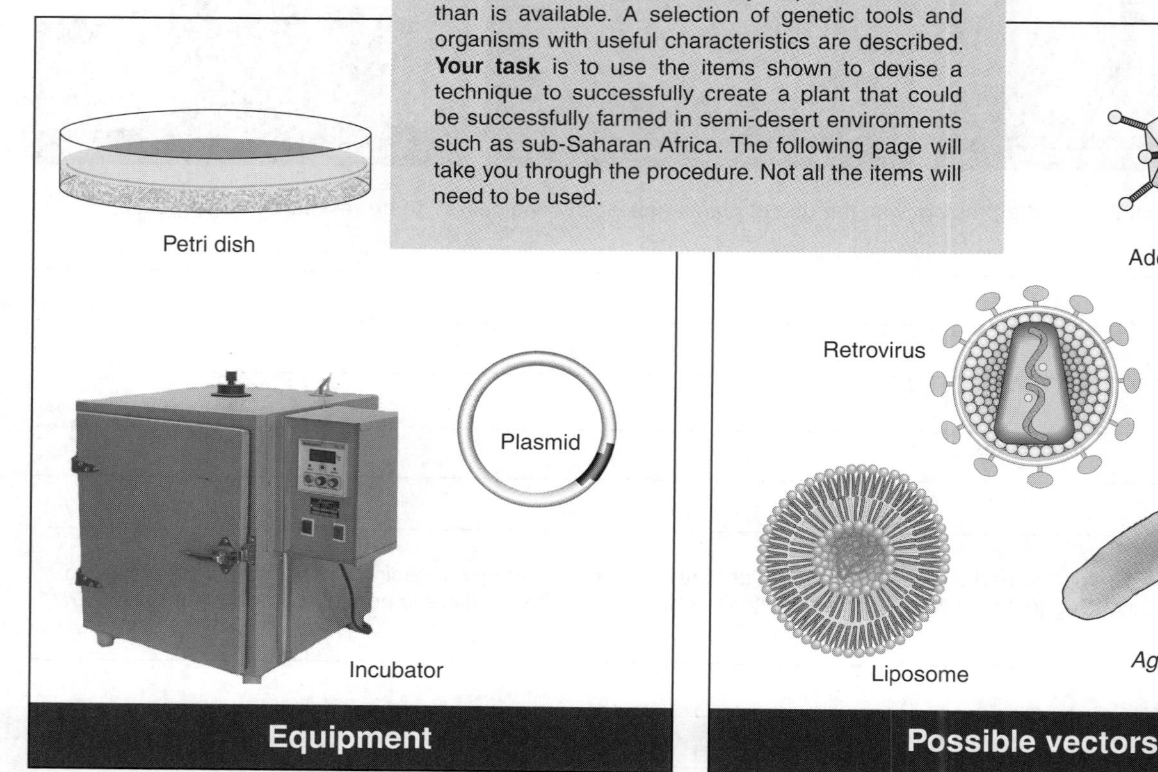

ISBN:978-1-927309-20-9

1. Identify the organism you would chose as a 'donor' of drought survival genes and explain your choice:

2. Describe a process to identify and isolate the required gene(s) and identify the tools to be used:

3. Identify a vector for the transfer of the isolated gene(s) into the crop plant and explain your decision:

4. Explain how the isolated gene(s) would be integrated into the vector's genome:

5. (a) Explain how the vector will transform the identified plant:

   (b) Identify the stage of development at which the plant would most easily be transformed. Explain your choice:

6. Explain how the transformed plants could be identified:

7. Explain how a large number of plants can be grown from the few samples that have taken up the new DNA:

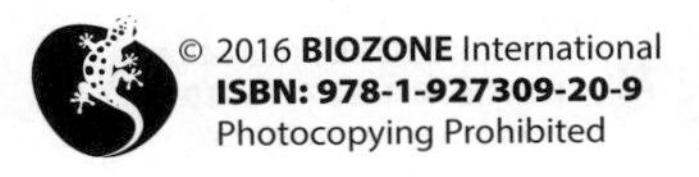

ISBN: 978-1-927309-20-9

188 Gel Electrophoresis

Key Idea: Gel electrophoresis is used to separate DNA fragments on the basis of size.

DNA can be loaded onto an electrophoresis gel and separated by size. DNA has an overall negative charge, so when an electrical current is run through a gel, the DNA moves towards the positive electrode. The rate at which the DNA molecules move through the gel depends primarily on their size and the strength of the electric field. The gel they move through is full of pores (holes). Smaller DNA molecules move through the pores more quickly than larger ones. At the end of the process, the DNA molecules can be stained and visualised as a series of bands. Each band contains DNA molecules of a particular size. The bands furthest from the start of the gel contain the smallest DNA fragments.

Analysing DNA using gel electrophoresis

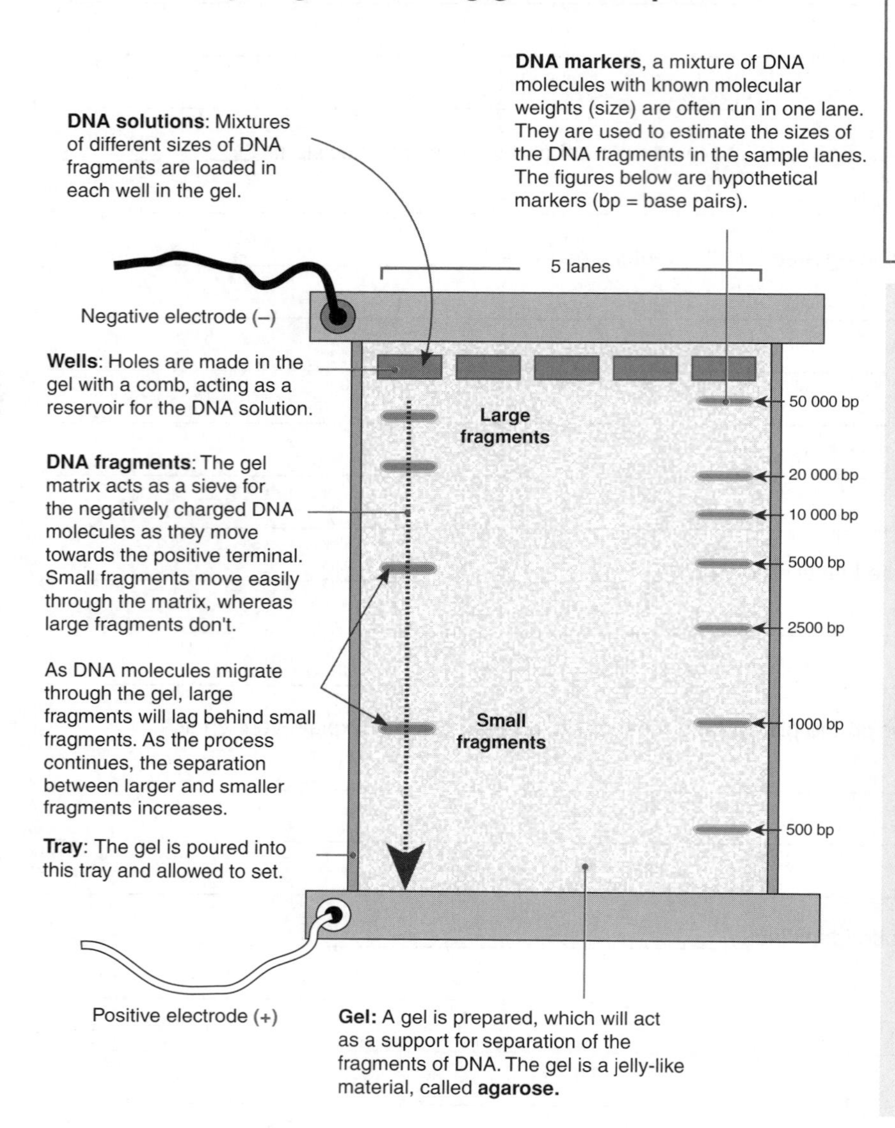

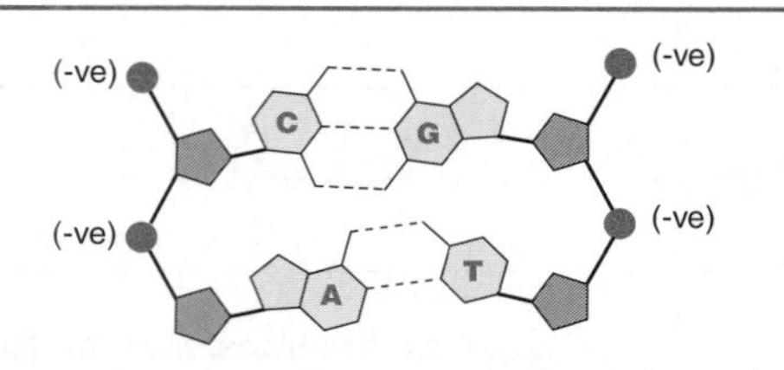

DNA is **negatively charged** because the phosphates (blue) that form part of the backbone of a DNA molecule have a negative charge.

Steps in the process of gel electrophoresis of DNA

1. A tray is prepared to hold the gel matrix.
2. A gel comb is used to create holes in the gel. The gel comb is placed in the tray.
3. Agarose gel powder is mixed with a buffer solution (this stabilises the DNA). The solution is heated until dissolved and poured into the tray and allowed to cool.
4. The gel tray is placed in an electrophoresis chamber and the chamber is filled with buffer, covering the gel. This allows the electric current from electrodes at either end of the gel to flow through the gel.
5. DNA samples are mixed with a "loading dye" to make the DNA sample visible. The dye also contains glycerol or sucrose to make the DNA sample heavy so that it will sink to the bottom of the well.
6. The gel is covered, electrodes are attached to a power supply and turned on.
7. When the dye marker has moved through the gel, the current is turned off and the gel is removed from the tray.
8. DNA molecules are made visible by staining the gel with **methylene blue** or ethidium bromide which binds to DNA and will fluoresce in UV light.

1. What is the purpose of gel electrophoresis? ______________________

____

____

2. Describe the two forces that control the speed at which fragments pass through the gel:

(a) ______________________

(b) ______________________

3. Why do the smallest fragments travel through the gel the fastest? ______________________

____

____

ISBN:978-1-927309-20-9

189 Interpreting Electrophoresis Gels

Key Idea: The banding pattern on an electrophoresis gel can give information about genetic variation and relationships.

Once made, an electrophoresis gel must be interpreted. If a specific DNA base sequence was being investigated, then the band pattern can be used to determine the DNA sequence and the protein that it encoded. Alternatively, depending on how the original DNA was treated, the banding pattern may be used as a profile for a species or individual. Commonly, the gene for cytochrome oxidase I (COXI), a mitochondrial protein, is used to distinguish animal species. The genetic information from this gene is both large enough to measure differences between species and small enough to have the differences make sense (i.e. the differences occur in small regions and aren't hugely varied).

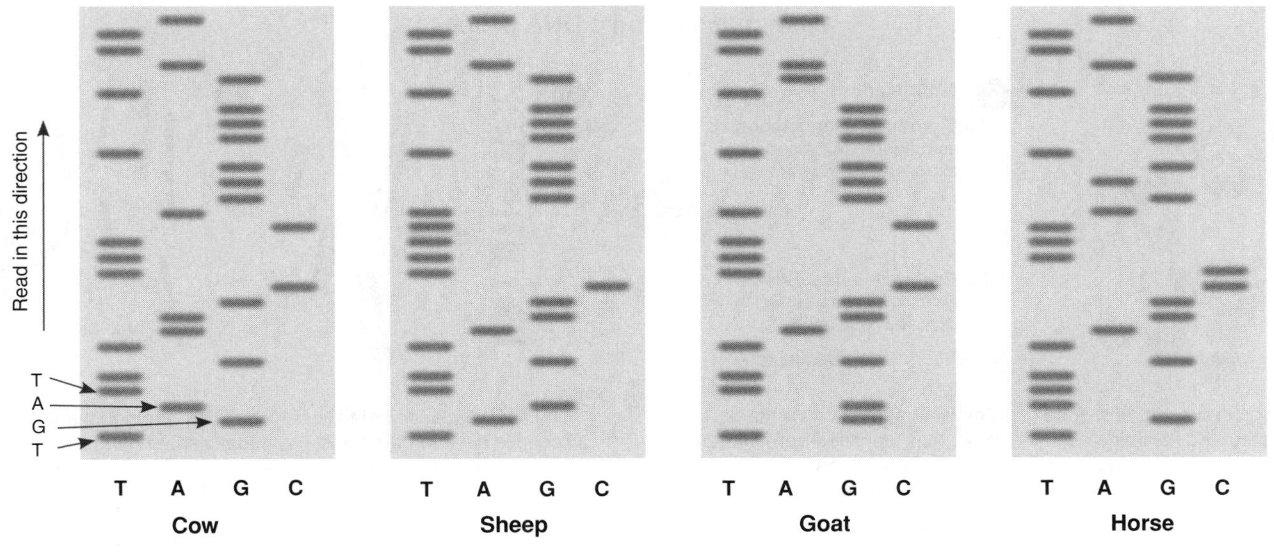

1. For each of the species above:

   (a) Determine the sequence of **synthesised DNA** in the gel in the photographs above. The synthesised DNA is what is visible on the gel. It is complementary to the sample DNA.
   (b) Convert it to the complementary sequence of the sample DNA. This is the DNA that is being investigated.

   Cow: **synthesised DNA**: ____________________

   sample DNA: ____________________

   Sheep: **synthesised DNA**: ____________________

   sample DNA: ____________________

   Goat: **synthesised DNA**: ____________________

   sample DNA: ____________________

   Horse: **synthesised DNA**: ____________________

   sample DNA: ____________________

   Based on the number of differences in the DNA sequences:

   (c) Identify the two species that are most closely related: ____________________

   (d) Identify the two species that are the least closely related: ____________________

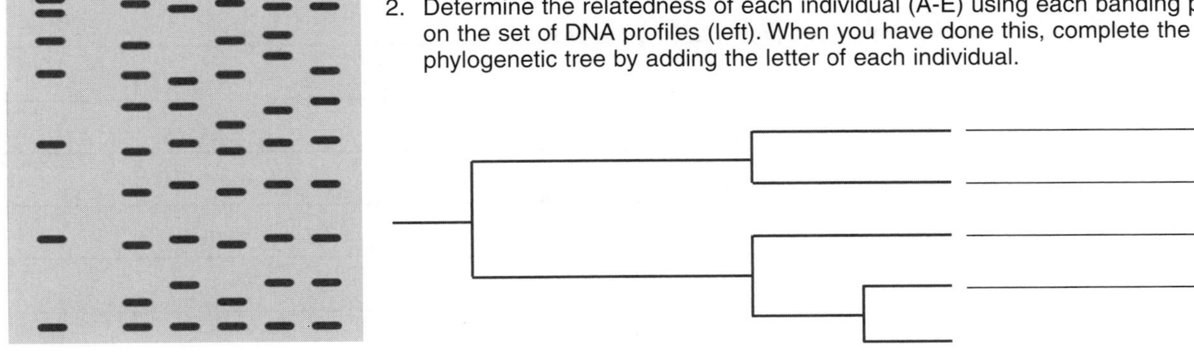

2. Determine the relatedness of each individual (A-E) using each banding pattern on the set of DNA profiles (left). When you have done this, complete the phylogenetic tree by adding the letter of each individual.

ISBN: 978-1-927309-20-9

190 Screening for Genes

Key Idea: DNA probes use attached markers (tags) to identify the presence and location of individual genes.

A DNA probe is a single stranded DNA sequence, with a base sequence that is complementary to a gene of interest. DNA probes target specific DNA sequences so they can be used to determine whether a person has a gene for a specific genetic disease, or to construct a gene map of a chromosome. DNA probes have either a radioactive label (e.g. ^{32}P) or a fluorescent dye so that they can be visualised on an electrophoresis gel or X-ray film.

Making and using a DNA probe

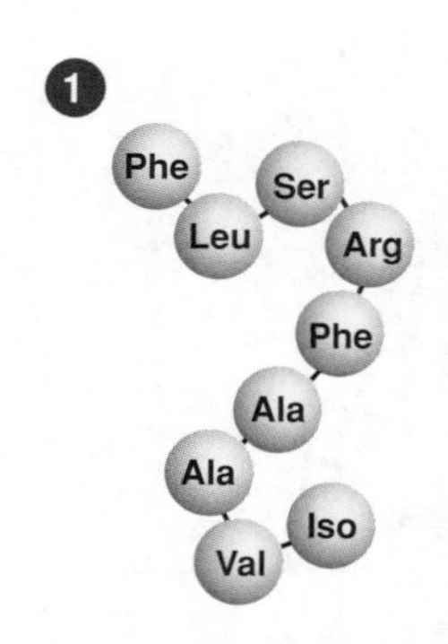

The protein product of a gene is isolated and its amino acid sequence is determined.

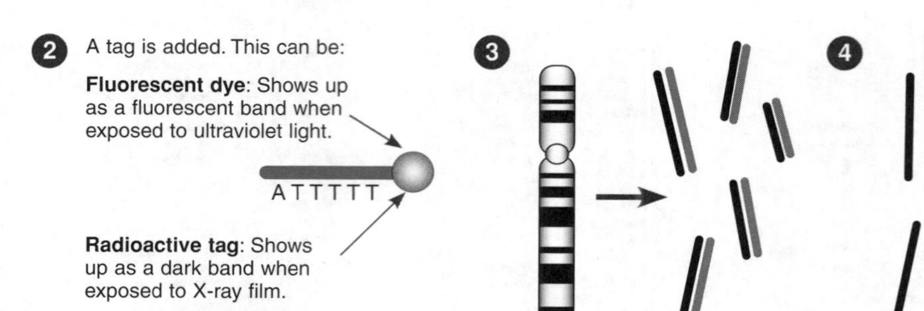

The DNA sequence for the protein product is identified from the amino acid sequence. The DNA sequence is artificially manufactured.

The DNA sequence being probed is cut into fragments using restriction enzymes.

The DNA fragments are denatured, forming single stranded DNA. The probe is added to the DNA fragments.

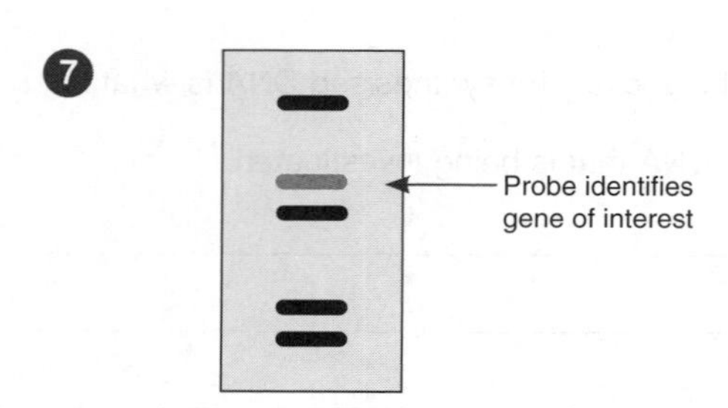

The gel is viewed by fluorescent light or on X-ray film (depending on the type of probe used). If the probe has bound to a gene, the tag makes it visible.

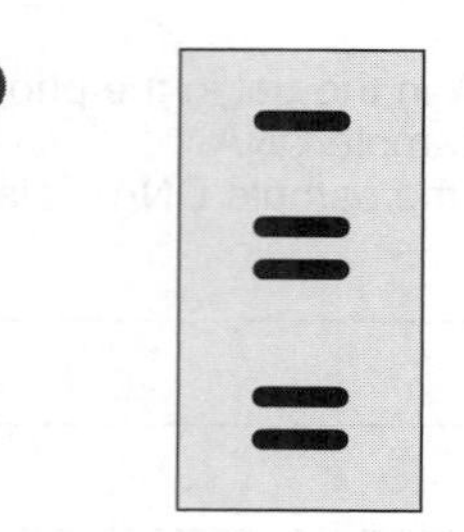

The DNA fragments are run on an electrophoresis gel. The fragments are separated by size.

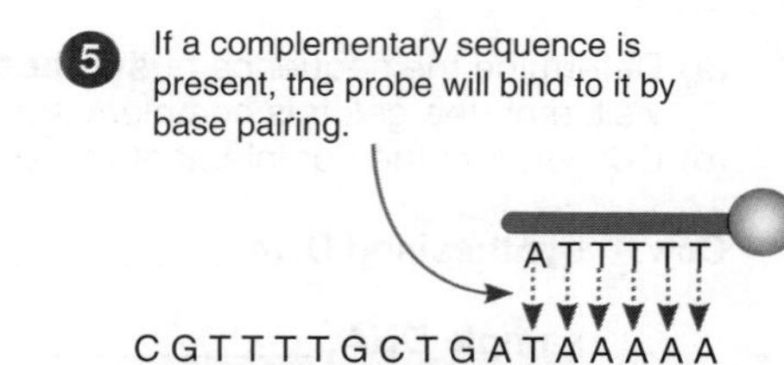

Target DNA strand (contains the complementary sequence to that of the probe).

1. What is the purpose of a DNA probe? ___________

2. Explain why a DNA probe can be used to identify a gene or DNA sequence: ___________

3. Why does the DNA have to be denatured before adding the probe? ___________

4. How is the presence of a specific DNA sequence or gene visualised? ___________

ISBN:978-1-927309-20-9

191 Hunting for a Gene

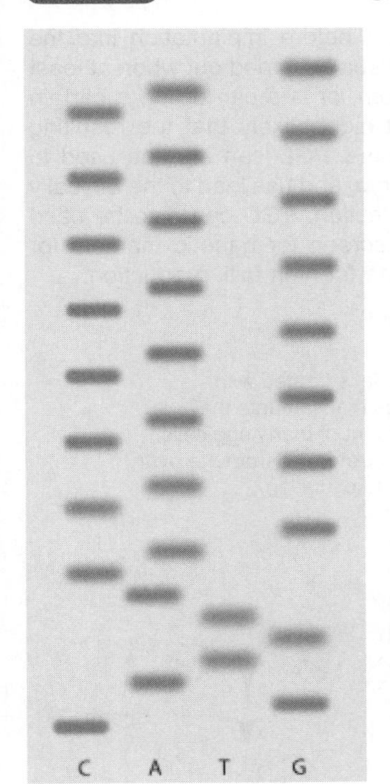

Key Idea: Huntington's disease is caused by a repeating section of DNA. The longer the repeating pattern, the earlier the disease tends to appear and the worse its symptoms.

Huntington's disease (HD) is a genetic neuro-degenerative disease that normally does not affect people until about the age of 40. Its symptoms usually appear first as a shaking of the hands and an awkward gait. Later manifestations of the disease include serious loss of muscle control and mental function, often ending in dementia and ultimately death.

All humans have the huntingtin (**HTT**) gene, which in its normal state produces a protein with roles in gene transcription, synaptic transmission, and brain cell survival. The mutant gene (**mHTT**) causes changes to and death of the cells of the cerebrum, the hippocampus, and cerebellum, resulting in the atrophy (reduction) of brain matter. The gene was discovered by Nancy Wexler in 1983 after ten years of research working with cell samples and family histories of more than 10,000 people from the town of San Luis in Venezuela, where around 1% of the population have the disease (compared to about 0.01% in the rest of the world). Ten years later the exact location of the gene on the chromosome 4 was discovered.

The identification of the HD gene began by looking for a gene probe that would bind to the DNA of people who had HD, and not to those who didn't. Eventually a marker for HD, called **G8**, was found. The next step was to find which chromosome carried the marker and where on the chromosome it was. The researchers hybridised human cells with those of mice so that each cell contained only one human chromosome, a different chromosome in each cell. The hybrid cell with chromosome 4 was the one with the G8 marker. They then found a marker that overlapped G8 and then another marker that overlapped that marker. By repeating this many times, they produced a map of the genes on chromosome 4. The researchers then sequenced the genes and found people who had HD had one gene that was considerably longer than people who did not have HD, and the increase in length was caused by the repetition of the base sequence CAG.

The HD mutation (mHTT) is called a trinucleotide repeat expansion. In the case of mHTT, the base sequence CAG is repeated multiple times on the short arm of chromosome 4. The normal number of CAG repeats is between 6 and 30. The mHTT gene causes the repeat number to be 35 or more and the size of the repeat often increases from generation to generation, with the severity of the disease increasing with the number of repeats. Individuals who have 27 to 35 CAG repeats in the HTT gene do not develop Huntington disease, but they are at risk of having children who will develop the disorder. The mutant allele, mHTT, is also dominant, so those who are homozygous or heterozygous for the allele are both at risk of developing HD.

New research has shown that the mHTT gene activates an enzyme called JNK3, which is expressed only in the neurones and causes a drop in nerve cell activity. While a person is young and still growing, the neurones can compensate for the accumulation of JNK3. However, when people get older and neurone growth stops, the effects of JNK3 become greater and the physical signs of HD become apparent. Because of mHTT's dominance, an affected person has a 50% chance of having offspring who are also affected. Genetic testing for the disease is relatively easy now that the genetic cause of the disease is known. While locating and counting the CAG repeats does not give a date for the occurrence of HD, it does provide some understanding of the chances of passing on the disease.

American singer-songwriter and folk musician Woody Guthrie died from complications of HD

1. Describe the physical effects of Huntington's disease: ______

2. Describe how the mHTT gene was discovered: ______

3. Discuss the cause of Huntington's disease and its pattern of increasing severity with each generation: ______

ISBN: 978-1-927309-20-9

192 Genetic Screening and Embryo Selection

Key Idea: Genetic screening is used to identify specific (often harmful) genes in embryos. Embryos for implantation into the mother can be selected based on the presence, or absence, of the gene of interest.

Numerous diseases, such as Huntington's disease, are known to be caused by specific genes. The **genetic screening** of gametes, embryos, or individuals for some diseases is now possible. **Pre-implantation genetic diagnosis** (or pre-implantation genetic screening) tests gametes and embryos for a specific genetic disorder before implantation into the uterus. Genetic screening is usually carried out when at least one of the parents suffers from, or is a carrier for, a certain disease. Screening makes it highly likely that the resulting baby will be free of the disease. PGD can also be used to determine the sex of the embryo, and has lead to the ethically debatable practice of sex selection. PGD can also be used in livestock improvement to screen for genetic markers for preferred genes, such as those for high milk production.

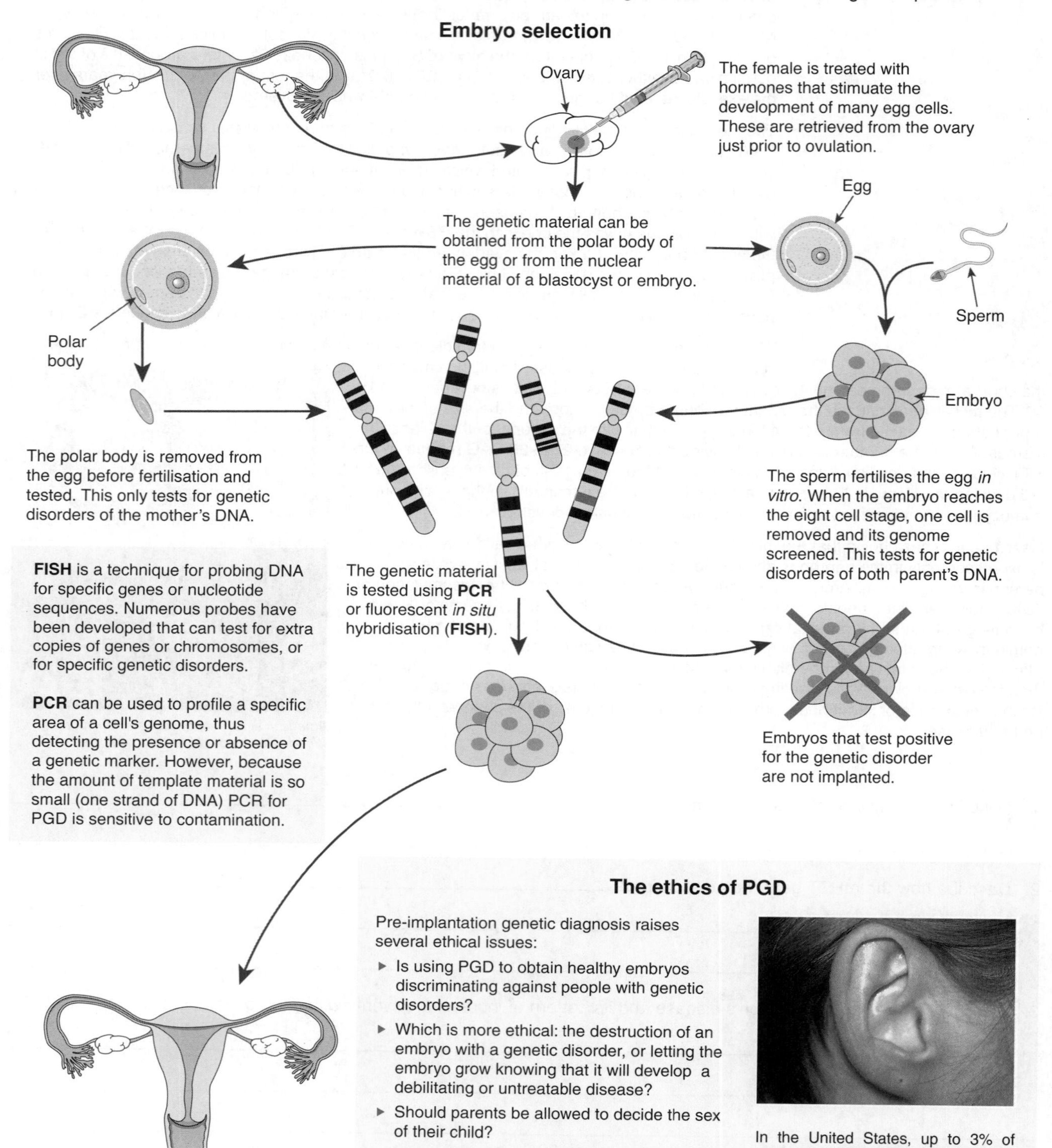

The ethics of PGD

Pre-implantation genetic diagnosis raises several ethical issues:

- Is using PGD to obtain healthy embryos discriminating against people with genetic disorders?
- Which is more ethical: the destruction of an embryo with a genetic disorder, or letting the embryo grow knowing that it will develop a debilitating or untreatable disease?
- Should parents be allowed to decide the sex of their child?
- PGD allows affected gametes and embryos to be discarded prior to implantation. Is this a better option than terminating a pregnancy once a genetic disorder has been discovered?

In the United States, up to 3% of PGD is used to select embryos *with* a genetic disorder. This most often occurs when parents have a particular disorder and want their child to have it too. For example, genetic deafness.

ISBN:978-1-927309-20-9

1. What is the main purpose of pre-implantation genetic diagnosis (PGD)?

2. Write a short paragraph that would explain the procedure for PGD to prospective parents:

3. Explain why using the polar body to test for genetic disorders does not give a full diagnosis of the resulting embryo:

4. Explain why using PCR in PGD could give a false response:

5. How can PGD be used for livestock improvement?

6. Discuss some of the ethical issues involved in PGD. Discuss what limitations (if any) should be placed on its use:

ISBN: 978-1-927309-20-9

193 DNA Profiling Using PCR

Key Idea: Short units of DNA that repeat a different number of times in different people can be used to produce individual genetic profiles.

In chromosomes, some of the DNA contains simple, repetitive sequences. These non-coding nucleotide sequences repeat over and over again and are found scattered throughout the genome. Some repeating sequences, called **microsatellites** or **short tandem repeats** (STRs), are very short (2-6 base pairs) and can repeat up to 100 times. The human genome has many different microsatellites. Equivalent sequences in different people vary considerably in the numbers of the repeating unit. This phenomenon has been used to develop **DNA profiling**, which identifies the natural variations found in every person's DNA. Identifying these DNA differences is a useful tool for forensic investigations. DNA testing in the UK is carried out by the Forensic Science Service (FSS). The FSS targets 10 STR sites; enough to guarantee that the odds of someone else sharing the same result are extremely unlikely; about one in a thousand million (a billion). DNA profiling has been used to help solve previously unsolved crimes and to assist in current or future investigations. DNA profiling can also be used to establish genetic relatedness (e.g. in paternity disputes or pedigree disputes), or when searching for a specific gene (e.g. screening for disease).

Microsatellites (short tandem repeats)

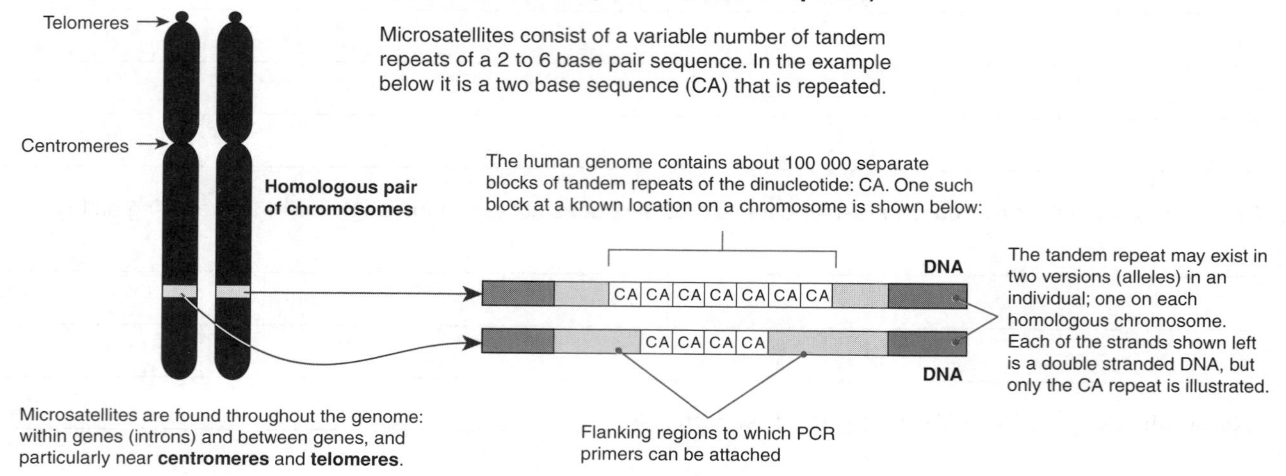

How short tandem repeats are used in DNA profiling

This diagram shows how three people can have quite different microsatellite arrangements at the same point (locus) in their DNA. Each will produce a different DNA profile using gel electrophoresis:

1 Extract DNA from sample

A sample collected from the tissue of a living or dead organism is treated with chemicals and enzymes to extract the DNA, which is separated and purified.

2 Amplify microsatellite using PCR

Specific primers (arrowed) that attach to the flanking regions (light grey) either side of the microsatellite are used to make large quantities of the micro-satellite and flanking regions sequence only (no other part of the DNA is amplified/replicated).

3 Visualise fragments on a gel

The fragments are separated by length, using **gel electrophoresis**. DNA, which is negatively charged, moves toward the positive terminal. The smaller fragments travel faster than larger ones.

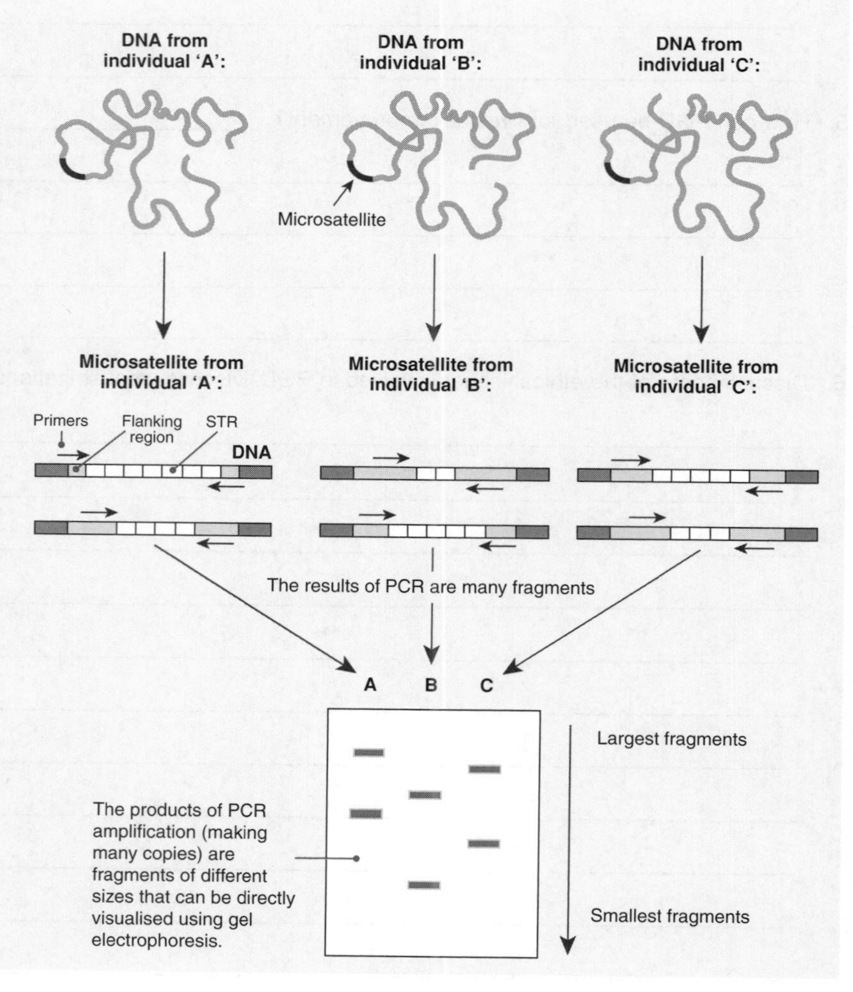

ISBN:978-1-927309-20-9

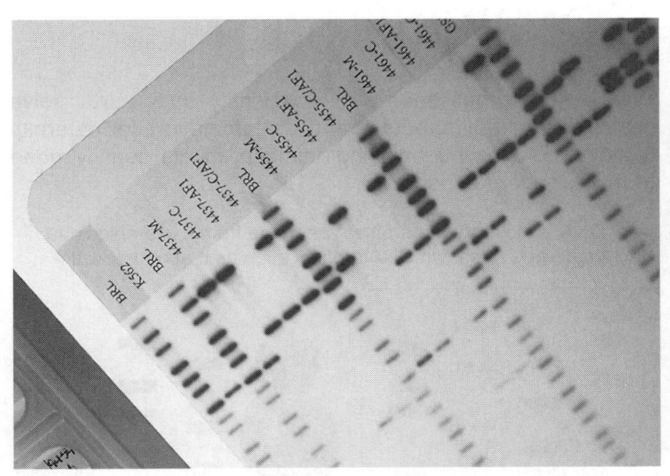

The photo above shows a film output from a DNA profiling procedure. Those lanes with many regular bands are used for calibration; they contain DNA fragment sizes of known length. These calibration lanes can be used to determine the length of fragments in the unknown samples.

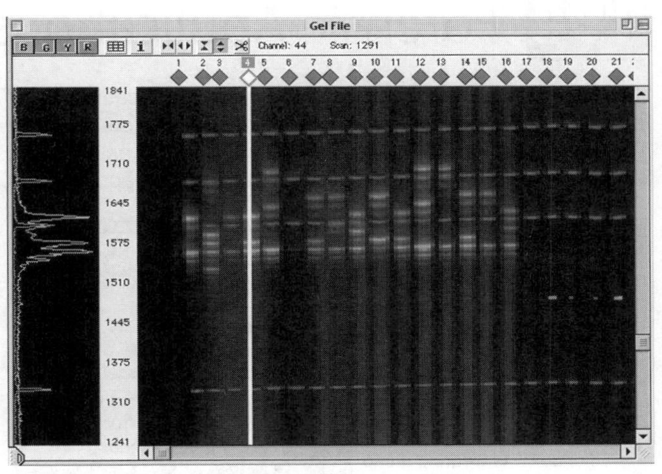

DNA profiling can be automated in the same way as DNA sequencing. Computer software is able to display the results of many samples run at the same time. In the photo above, the sample in lane 4 has been selected and displays fragments of different length on the left of the screen.

1. Describe the properties of **short tandem repeats** that are important to the application of **DNA profiling** technology:

2. Explain the role of each of the following techniques in the process of DNA profiling:

   (a) Gel electrophoresis:

   (b) PCR:

3. Describe the three main steps in DNA profiling using PCR:

   (a)

   (b)

   (c)

4. Explain why as many as 10 STR sites are used to gain a DNA profile for forensic evidence:

ISBN: 978-1-927309-20-9

194 Forensic Applications of DNA Profiling

Key Idea: DNA profiling has many forensic applications, from identifying criminal offenders to saving endangered species. The use of DNA as a tool for solving crimes such as homicide is well known, but it can also has several other applications. DNA evidence has been used to identify body parts, solve cases of industrial sabotage and contamination, for paternity testing, and even in identifying animal products illegally made from endangered species.

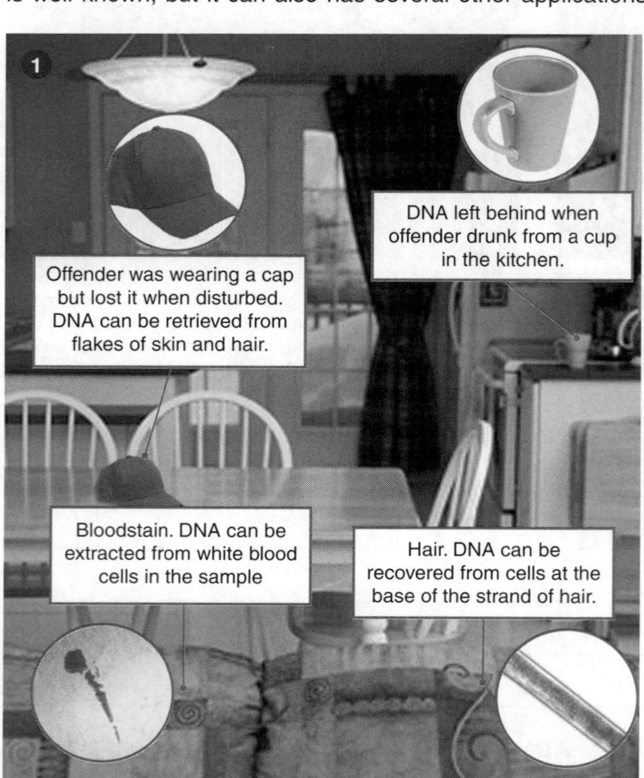

During the initial investigation, samples of material that may contain DNA are taken for analysis. At a crime scene, this may include blood and body fluids as well as samples of clothing or objects that the offender might have touched. Samples from the victim are also taken to eliminate them as a possible source of contamination.

2 DNA is isolated and profiles are made from all samples and compared to known DNA profiles such as that of the victim.

A B C D

Calibration

Profiles of collected DNA

Investigator (C)

Victim (D)

3 Unknown DNA samples are compared to DNA databases of convicted offenders and to the DNA of the alleged offender.

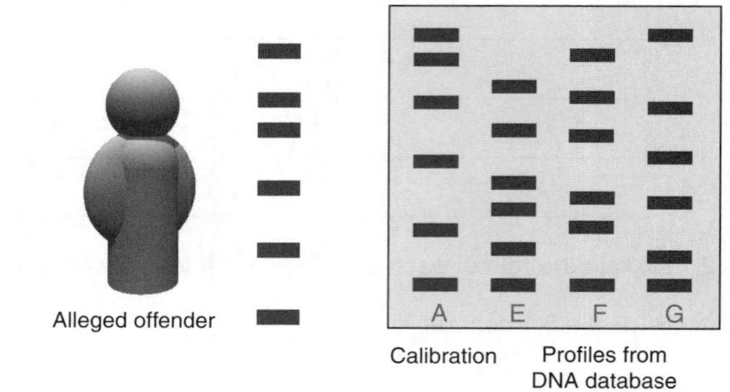

4 Although it does not make a complete case, DNA profiling, in conjunction with other evidence, is one of the most powerful tools in identifying offenders or unknown tissues.

The role of frequency and probability

Every person has two copies of each chromosome and therefore two copies (alleles) of every testable DNA marker. For example, the short tandem repeat (STR) known as CSF1PO contains between 7 and 15 repeats of GATA and has 9 possible alleles. Some alleles (and therefore genotypes) are more common in the population that others. For the CSF1PO STR, the frequency of the genotype 10,11 (allele 10 and allele 11) is 0.1270, i.e. it appears in 12.7% of the population. When DNA is tested, a number of STRs are sampled (the exact number varies between countries). When the data from all STRs is considered, levels of probability that the DNA came from a certain person can be calculated to 1 in 500 trillion.

Allele frequencies of the CSF1PO STR

Allele (number of repeats)	Frequency	Allele (number of repeats)	Frequency
7	0.0232	12	0.3446
8	0.0212	13	0.0656
9	0.0294	14	0.0092
10	0.2321	15	0.0010
11	0.2736		

1. Why are DNA profiles obtained for both the victim and investigator? ______________________

2. Use the evidence to decide if the alleged offender is innocent or guilty and explain your decision:

3. What is the frequency of the following CSF1PO alleles:

(a) 9: ______________________ (b) 12: ______________________

(c) The 9, 12 genotype (*hint, use the Hardy-Weinberg equation*): ______________________

ISBN:978-1-927309-20-9

Paternity testing

DNA profiling can be used to determine paternity (and maternity) by looking for matches in alleles between parents and children. This can be used in cases such as child support or inheritance. DNA profiling can establish the certainty of paternity (and maternity) to a 99.99% probability of parentage.

Every STR allele is given the number of its repeats as its name, e.g. 8 or 9. In a paternity case, the mother may be 11, 12 and the father may be 8, 13 for a particular STR. The child will have a combination of these. The table below illustrates this:

DNA marker	Mother's alleles	Child's alleles	Father's alleles
CSF1PO	7, 8	8, 9	9, 12
D10S1248	14, 15	11, 14	10, 11
D12S391	16, 17	17, 17	17, 18
D13S317	10, 11	9, 10	8, 9

The frequency of the each allele occurring in the population is important when determining paternity (or maternity). For example, DNA marker CSF1PO allele 9 has a frequency of 0.0294 making the match between father and child very significant (whereas allele 12 has a frequency of 0.3446, making a match less significant). For each allele, a paternity index (PI) is calculated. These indicate the significance of the match. The PIs are combined to produce a probability of parentage. 10-13 different STRs are used to identify paternity. Mismatches of two STRs between the male and child is enough to exclude the male as the biological father.

Whale DNA: tracking illegal slaughter

Under International Whaling Commission regulations, some species of whales can be captured for scientific research and their meat sold legally. Most, including humpback and blue whales, are fully protected and to capture or kill them for any purpose is illegal. Between 1999 and 2003 Scott Baker and associates from Oregon State University's Marine Mammal Institute investigated whale meat sold in markets in Japan and South Korea. Using DNA profiling techniques, they found around 10% of the samples tested were from fully protected whales including western grey whales and humpbacks. They also found that many more whales were being killed than were being officially reported.

4. For the STR D10S1248 in the example above, what possible allele combinations could the child have?

5. A paternity test was carried out and the abbreviated results are shown below:

DNA marker	Mother's alleles	Child's alleles	Man's alleles
CSF1PO	7, 8	8, 9	9, 12
D10S1248	14, 15	11, 14	10, 11
D19S433	9, 10	10,15	14, 16
D13S317	10, 11	9, 10	8, 9
D2S441	7, 15	7, 9	14, 17

(a) Could the man be the biological father?

(b) Explain your answer:

6. (a) How could DNA profiling be used to refute official claims of the **type** of whales captured and sold in fish markets?

(b) How could DNA profiling be used to refute official claims of the **number** of whales captured and sold in fish markets?

ISBN: 978-1-927309-20-9

195 Profiling for Effect of Drug Response

Key Idea: DNA profiling can be used to assess disease risk or to determine how well someone will respond to a treatment. DNA profiling relies on variations in the length of DNA fragments produced by PCR. This variation can be used to identify areas of DNA related to genetic diseases and match a person to a drug treatment. The key is to identify variation in the DNA that is related to specific disease types or drug susceptibility.

Identifying a relationship

To study the genetic component of a disease, researchers require two main study groups: people with the disease and people without the disease (known as phenotype first).

A genome wide association study (GWAS) is carried out on each individual in each group. The study looks for single nucleotide polymorphisms or **SNPs** (changes to single base pairs) to see there are differences between the control group and the afflicted group. There are over 100 million SNPs spread throughout the human genome and there are numerous ways to identify them, including DNA sequencing and using restriction enzymes.

Once the SNPs are identified, their frequencies are analysed to see if any are significantly different between the groups. SNPs found to be associated with a disease do not always just appear in people with the disease. Often they can be found in people without the disease. This lack of a definitive link makes linking a disease to a specific DNA profile a matter of probability and risks.

Using profiling to analyse disease risk

Using DNA profiling as a tool to analyse disease risk is a relatively new field. Although it promises a new way of diagnosing disease, to date it has produced inconsistent results because of the complexity of diseases involving large numbers of genes. One promising use of DNA profiling in disease analysis is the ability to determine response to albuterol in the treatment of asthma.

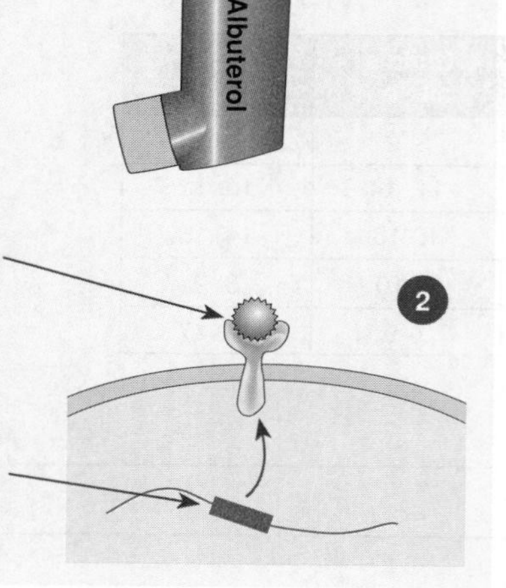

Albuterol is used to relieve the symptoms of asthma. It works well on some people but not on others.

Albuterol binds to the beta2 adrenergic receptor in a cell's plasma membrane and causes the relaxation of smooth muscle in the airways.

The beta2 AR protein is encoded by the ADRB2 gene. Scientists wondered if genetic differences in or near this gene affected how well albuterol worked.

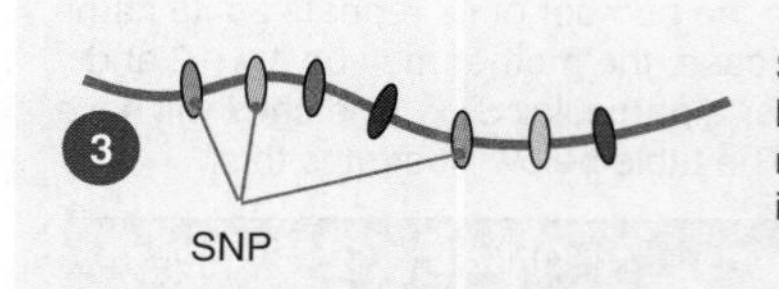

Scientists analysed a 3000 base pair piece of DNA near the ADRB2 gene and identified 13 SNPs.

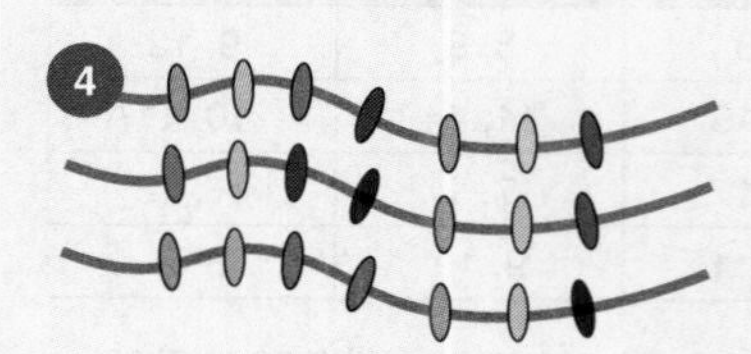

The SNPs are found arranged in combinations called haplotypes. 12 different haplotypes have been identified.

% haplotypes in population

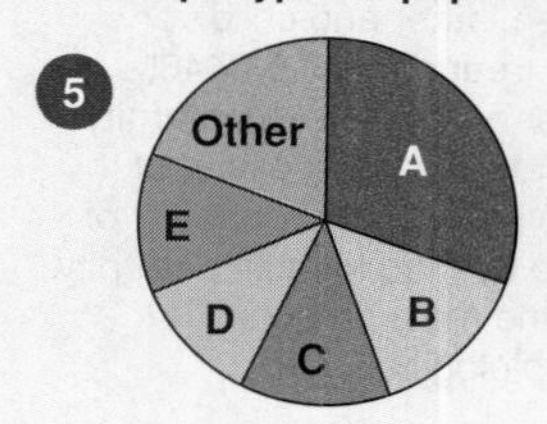

Every person has a haplotype profile relating the ADRB2 gene. Five of the haplotypes are relatively common.

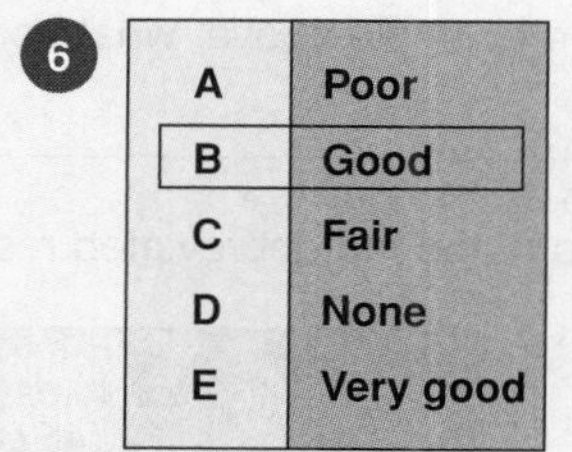

A	Poor
B	Good
C	Fair
D	None
E	Very good

Some of the haplotypes seem to be related to the effect of albuterol on people. Some haplotypes are found mainly in people who have a poor response to albuterol (D), while others are found in people who respond well to the drug (B, E).

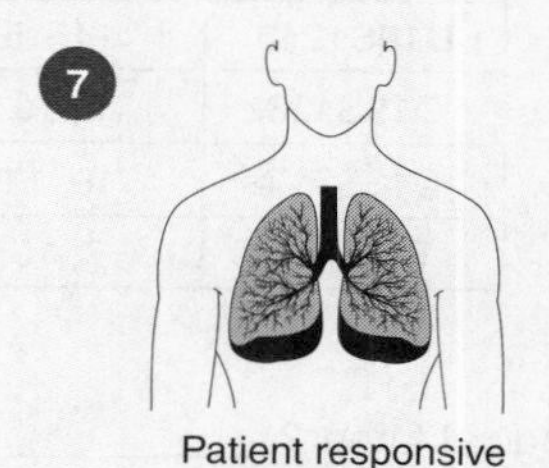

Patient responsive to albuterol.

This data may be able to be used by doctors when prescribing albuterol. In the future, other profiles could be used to determine the effect of other drugs on people.

1. (a) Why is a set of SNPs (a haplotype) used when identifying a genetic disease, rather than just a single SNP?

   (b) Why is the use of SNPs not a definitive way to assess the risk of a genetic disease?

2. How can personal genetic profiles be used to improve the treatment outcomes of patients?

ISBN:978-1-927309-20-9

196 Finding the Connection

Key Idea: Using specific DNA sequences to identify breeds and the relationships between them can be useful in developing future livestock.

In the UK there are a number of different sheep breeds, each of which has been bred for a specific purpose (e.g. meat of wool production). While some of these breeds are very old, others are recent developments. As new sheep breeds are developed by selective breeding, older breeds become less profitable and are eventually replaced. These older breeds are important because they carry traits that could be valuable but are now absent in more recent breeds. The relationships between the older and newer breeds are important as they show the development of breeds and help farmers and breeders to plan for the future development of their livestock.

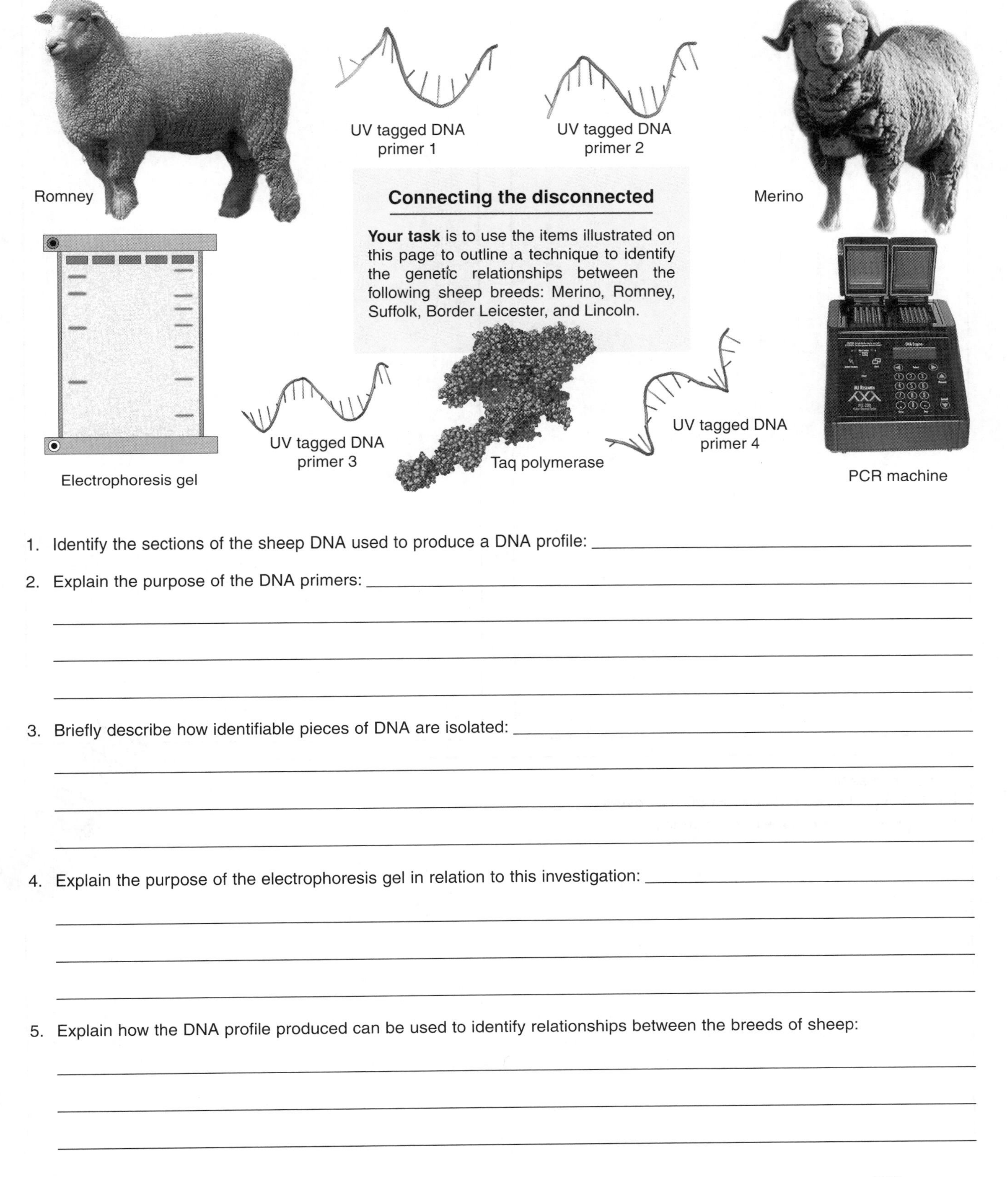

1. Identify the sections of the sheep DNA used to produce a DNA profile: ___

2. Explain the purpose of the DNA primers: ___

3. Briefly describe how identifiable pieces of DNA are isolated: ___

4. Explain the purpose of the electrophoresis gel in relation to this investigation: ___

5. Explain how the DNA profile produced can be used to identify relationships between the breeds of sheep: ___

ISBN: 978-1-927309-20-9

197 Chapter Review

Summarise what you know about this topic under the headings and sub-headings provided. You can draw diagrams or mind maps, or write short notes to organise your thoughts. Use the images and hints to help you and refer back to the introduction to check the points covered:

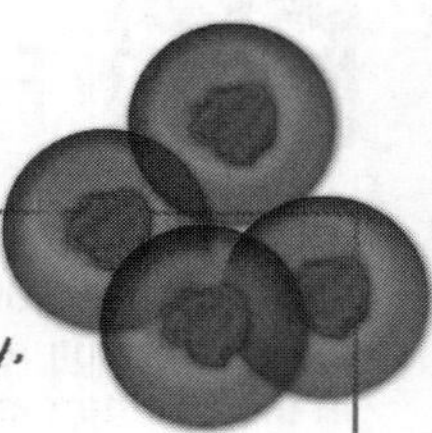

Mutations
HINT: Types of mutation

Stem cells
HINT: Types of stems cells, their potency, and medical applications

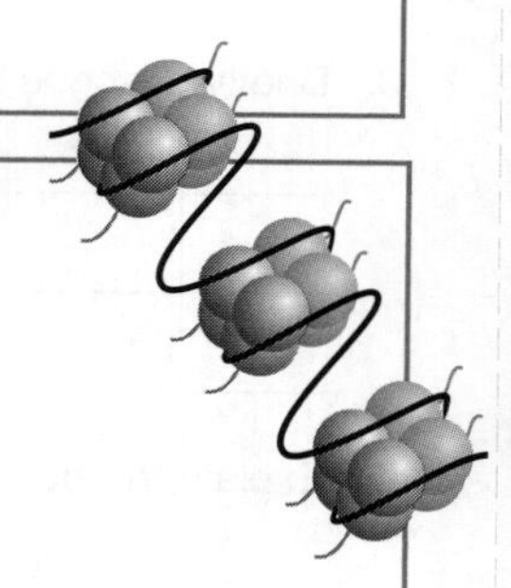

Gene expression
HINT: DNA packaging and control of gene expression. Consequences of errors in gene expression.

ISBN:978-1-927309-20-9

Gene technology

HINT: Genome sequencing and recombinant DNA technology.

Genetic screening and DNA profiling

HINT: Gel electrophoresis, DNA probes and gene screening, DNA profiling and its applications.

ISBN: 978-1-927309-20-9

198 KEY TERMS AND IDEAS: Did You Get It?

1. Test your vocabulary by matching each term to its definition, as identified by its preceding letter code.

adult stem cells ....

annealing ....

DNA amplification ....

DNA ligation ....

DNA polymerase ....

embryonic stem cells ....

gel electrophoresis ....

GMO ....

marker gene ....

microsatellite ....

mutation ....

PCR ....

primer ....

recognition site ....

recombinant DNA ....

restriction enzyme ....

sticky end ....

vector ....

A An organism or artificial vehicle that is capable of transferring a DNA sequence to another organism.

B The pairing (by hydrogen bonding) of complementary single-stranded nucleic acids to form a double-stranded polynucleotide. The term is applied to making recombinant DNA, to the binding of a DNA probe, or to the binding of a primer to a DNA strand during PCR.

C A cut in a length of DNA by a restriction enzyme that results in two strands of DNA being different lengths with one strand overhanging the other.

D A short length of DNA used to identify the starting sequence for PCR so that polymerase enzymes can begin amplification.

E An enzyme that is able to cut a length of DNA at a specific sequence or site.

F The site or sequence of DNA at which a restriction enzyme attaches and cuts.

G A gene, with an identifiable effect, used to determine if a piece of DNA has been successfully inserted into the host organism.

H A reaction that is used to amplify fragments of DNA using cycles of heating and cooling (abbreviation).

I A process that is used to separate different lengths of DNA by placing them in a gel matrix placed in a buffered solution through which an electric current is passed.

J The process of producing more copies of a length of DNA, normally using PCR.

K DNA that has had a new sequence added so that the original sequence has been changed.

L The repairing or attaching of fragmented DNA by ligase enzymes.

M A short (normally two base pairs) piece of DNA that repeats a variable number of times between people and so can be used to distinguish between individuals.

N An organism that has had part of its DNA sequence altered either by the removal or insertion of a piece of DNA.

O An enzyme that is able to replicate DNA and commonly used in PCR to amplify a length of DNA.

P Undifferentiated, multipotent cells found in tissues or organs that may give rise to several cell types, but are primarily involved in the maintenance and repair of the tissue in which they are found.

Q Pluripotent stem cells derived from the inner cell mass of the blastocyst (early-stage embryo).

R A change to the DNA sequence of an organism. This may be a deletion, insertion, duplication, inversion or translocation of DNA in a gene or chromosome.

2. An original DNA sequence is shown right: **GCG TGA TTT GTA GGC GCT CTG**

For each of the following DNA mutations, state the type of mutation that has occurred:

(a) **GCG TGT TTG TAG GCG CTC TG** ____________________

(b) **GCG TGA TTT GTA AGG CGC TCT G** ____________________

(c) **GCG TGA TTT GGA GGC GCT CTG** ____________________

(d) **GCG TGA TTT GTA TCG CGG CTG** ____________________

(e) **GCG TGA GTA GGC GCT CTG** ____________________

ISBN:978-1-927309-20-9

3. Three partial DNA sequences are shown below for the turkey, emu, and ostrich.

(a) Identify the sequence differences between the three species (the first one has been done for you):

Ostrich	A T G G C **[C]** C C C A A C A T T C G A A A A T C G C A C C C C C T G C T C A A A A T T A T C A A C
Emu	A T G G C **[C]** C C T A A C A T C C G A A A A T C C C A C C C T C T A C T C A A A A T C A T C A A C
Turkey	A T G G C **[A]** C C C A A T A T C C G A A A A T C A C A C C C C C T A T T A A A A A C A A T C A A C

(b) How many differences between the ostrich and emu? ______________________

(c) How many differences between the ostrich and turkey? ______________________

(d) Which two species are most closely related? ______________________

(e) Explain your answer: ______________________

____

(f) Looking at the photographs, does the sequence information support similarities based on appearance? Explain:

____

____

4. Give an example of each of the following stem cell types:

(a) Unipotent: ______________________

(b) Multipotent: ______________________

(c) Pluripotent: ______________________

(d) Totipotent: ______________________

5. Use the following word list to label the diagram below: *chromatid, chromatin, chromosome, DNA bases, gene, histone, nucleosome:*

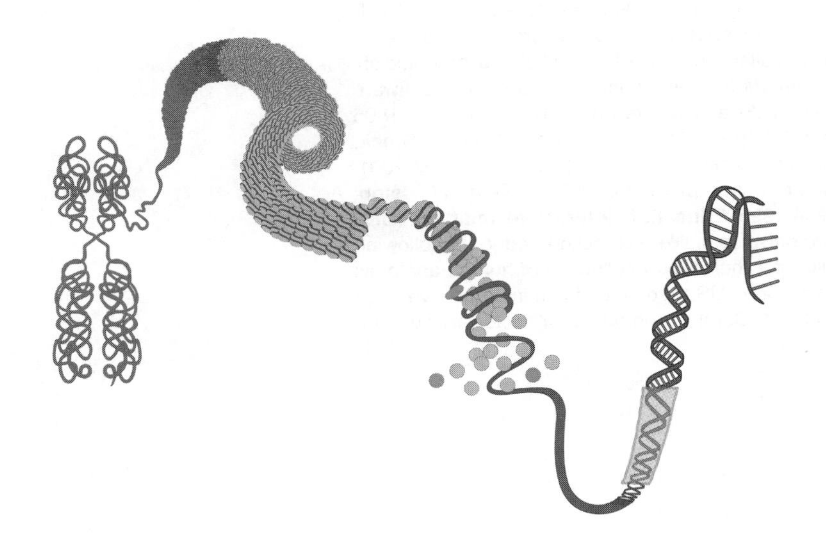

ISBN: 978-1-927309-20-9

Image credits

The writing team would like to thank the following people and organisations who have kindly provided photographs or illustrations for this edition:

• D. Fankhauser, University of Cincinnati, Clermont College for the image of the Pacinian corpuscle • WMRCVM (Virginia-Maryland college of veterinary medicine for the image of the pancreas • UC Regents David campus • Wellington Harrier Athletic Club for the photo of the sprinters • Helen Hall for the photo of the marathon runner • Dartmouth college for the electron micrograph images of the chloroplast • Louisa Howard Dartmouth college for the electron micrograph of the mitochondria • Wintec for the sports testing image • Ed Uthman for the image of the human fetus • Marc King for the photos of the chicken combs • Aptychus for the photo of the Tamil girl • Rita Willaert for the photo of the Nuba woman • Dept. of Natural Resources, Illinois for the photo of the Illinois prairie chicken • Allan and Elaine Wilson for the photo of Harris' antelope squirrel • Dr David Wells, AgResearch • C Gemmil for the transect sampling photo • Kent Pryor for the photo of the rocky shore • Conrad Pilditch for images of rocky shore organisms • Rhys Barrier for the photo and data of the mudfish • C Gemmil for the transect sampling photo • Greenpeace for photos of GMO protesting

We also acknowledge the photographers who have made images available through **Wikimedia Commons** under Creative Commons Licences 2.0, 2.5, 3.0, or 4.0: • Ragesoss • RM Hunt • Ildar Sagdejev • Pöllö • Jfoldmei • Jpogi • Solimena Lab • Roadnottaken • Tangopaso • Ryan Somma • Kristian Peters • Woutergroen • Dan Ferber • Johnmaxmena • Piotr Kuczynski • Stem cell scientist • Matthias Zepper • it:Utente:Cits • NYWTS • Dr Graham Beards • Jpbarrass • Jim Conrad • Kamal Ratna Tuladhar • KTBN • Velela • Aviceda • UtahCamera • Bruce Marlin • Lorax • Onno Zweers • AKA • Dirk Beyer • Masur • Kahuroa • Citron • Janus Sandsgaard • Jan Ainali • Crulina 98 • Madprime • Peter Halasz • Sagt • Gina Mikel • Shirley Owens MSU • Wojsyl • Janke • Rasbak • Daderot • Mikrolit • Graham Bould • Rudolph89 •Andreas Trepte • Stemonitis • Olaf Leillinger • Tuxyso • Arterm Topchiy • Dual Freq • Romain Behar • Jacoplane • Mikael Häggström • Onderwijsgek • Luc Viatour www.Lucnix.be • Marshal Hedin • Sharon Loxton • IRRI • Ute Frevert • 25kartika • Georgetown University Hospital

Contributors identified by coded credits:

BF: Brian Finerran (Uni. of Canterbury) **BH:** Brendan Hicks (Uni. of Waikato), **CDC**: Centers for Disease Control and Prevention, Atlanta, USA, **EII**: Education Interactive Imaging, **FRI**: Forest and Research Industry, **NASA**: National Aeronautics and Space Administration, **NIH**: National Institute of Health, **NYSDEC**: New York State Dept of Environmental Conservation, **RA**: Richard Allan, **RCN**: Ralph Cocklin, **TG**: Tracey Greenwood, **WBS**: Warwick Silvester (Uni. of Waikato),**WMU**: Waikato Microscope Unit, **USDA**: United States Department of Agriculture, **USGS**: United States Geological Survey

Image libraries:

We also acknowledge our use of royalty-free images, purchased by BIOZONE International Ltd from the following sources: **Corel** Corporation from various titles in their Professional Photos CD-ROM collection; Dollar Photo Club, dollarphotoclub.com; istock photos, istockphoto.com; **IMSI** (International Microcomputer Software Inc.) images from IMSI's MasterClips® and MasterPhotosTM Collection, 1895 Francisco Blvd. East, San Rafael, CA 94901-5506, USA; ©1996 **Digital Stock**, Medicine and Health Care collection; **©Hemera** Technologies Inc, 1997-2001; © 2005 JupiterImages Corporation www.clipart.com; ©1994., **©Digital Vision**; Gazelle Technologies Inc.; ©1994-1996 **Education Interactive Imaging** (UK), **PhotoDisc®**, Inc. USA, www.photodisc.com. We also acknowledge the following clipart providers: TechPool Studios, for their clipart collection of human anatomy: Copyright ©1994, TechPool Studios Corp. USA (some of these images have been modified); Totem Graphics, for clipart; Corel Corporation, for vector art from the Corel MEGAGALLERY collection.

Index